Biometry for Forestry and Environmental Data

Chapman & Hall/CRC
Applied Environmental Series

Recently Published Titles

Environmental Statistics with S-PLUS
Steven P. Millard, Nagaraj K. Neerchal

Statistical Tools for Environmental Quality Measurement
Douglas E. Splitstone, Michael E. Ginevan

Sampling Strategies for Natural Resources and the Environment
Timothy G. Gregoire, Harry T. Valentine

Sampling Techniques for Forest Inventories
Daniel Mandallaz

Statistics for Environmental Science and Management
Second Edition
Bryan F.J. Manly

Statistical Geoinformatics for Human Environment Interface
Wayne L. Myers, Ganapati P. Patil

Introduction to Hierarchical Bayesian Modeling for Ecological Data
Eric Parent, Etienne Rivot

Handbook of Spatial Point-Pattern Analysis in Ecology
Thorsten Wiegand, Kirk A. Moloney

Introduction to Ecological Sampling
Bryan F.J. Manly, Jorge A. Navarro Alberto

Future Sustainable Ecosystems: Complexity, Risk, and Uncertainty
Nathaniel K Newlands

Environmental and Ecological Statistics with R
Second Edition
Song S. Qian

Statistical Methods for Field and Laboratory Studies in Behavioral Ecology
Scott Pardo, Michael Pardo

Biometry for Forestry and Environmental Data with Examples in R
Lauri Mehtätalo, Juha Lappi

For more information about this series, please visit: https://www.crcpress.com/Chapman--HallCRC-Applied-Environmental-Statistics/book-series/ CRCAPPENVSTA

Biometry for Forestry and Environmental Data

with Examples in R

Lauri Mehtätalo
Juha Lappi

CRC Press is an imprint of the
Taylor & Francis Group, an **informa** business
A CHAPMAN & HALL BOOK

First edition published 2020
by CRC Press
6000 Broken Sound Parkway NW, Suite 300, Boca Raton, FL 33487-2742

and by CRC Press
2 Park Square, Milton Park, Abingdon, Oxon, OX14 4RN

First issued in paperback 2022

Visit the Taylor & Francis Web site at
http://www.taylorandfrancis.com

and the CRC Press Web site at
http://www.crcpress.com

Library of Congress Cataloging-in-Publication Data

Names: Mehtätalo, Lauri, author.
Title: Biometry for forestry and environmental data with examples in R /
Lauri Mehtätalo and Juha Lappi.
Description: Boca Raton, FL : CRC Press, 2020. | Series: Chapman & Hall/CRC
applied environmental statistics | Includes bibliographical references
and index.
Identifiers: LCCN 2020006643 (print) | LCCN 2020006644 (ebook) | ISBN
9781498711487 (hardback) | ISBN 9780429173462 (ebook)
Subjects: LCSH: Forests and forestry--Statistical methods. | Biometry. | R
(Computer program language)
Classification: LCC SD387.B53 M44 2020 (print) | LCC SD387.B53 (ebook) |
DDC 577.3072--dc23
LC record available at https://lccn.loc.gov/2020006643
LC ebook record available at https://lccn.loc.gov/2020006644

ISBN: 978-0-367-50845-6 (pbk)
ISBN: 978-1-498-71148-7 (hbk)
ISBN: 978-0-429-17346-2 (ebk)

DOI: 10.1201/9780429173462

Typeset in CMR
by Nova Techset Private Limited, Bengaluru & Chennai, India

Contents

Preface

This book is based on our original lecture materials. In 1993, Juha wrote a 182-page booklet entitled "Methods for Forest Biometry" that described intermediate and advanced statistics for Finnish foresters. About 15 years later, Lauri took over responsibility for the biometrics course at the Universities of Helsinki and Joensuu, and wrote his own lectures, which covered a smaller number of detailed topics, but where everything was demonstrated using the statistical program R. In 2013, motivated by a suggestion from Timo Tokola, we decided to combine the essential contents of these two material sources for an international audience. Writing a book was much more time consuming than we expected. Now after many years, the book is finally published.

The contents of the book are quite diverse. However, we have tried to keep it as self-sufficient as possible. Therefore, the first chapters cover the basics of probability calculus and estimation theory, while in the later chapters there is also some new material. Originally, we had planned to cover many more topics, but we had to drop some of them due to limitations in time and space. For example, the widely used $k-$NN method is only briefly noted in Section 13.2.2 and Chapter 11. We are not including references to those results of basic statistical theory that are well described in publicly available high-quality sites in the Internet.

There are many references to our own publications throughout the book, even though many equally relevant papers by other authors might have been available. A simple reason for this approach is that we have used the references we know best. This book is not a review of the applications of the presented methods in forestry and environmental sciences; it is a book about statistical methods, where the individual applied papers are just examples of case studies where the presented ideas have been utilized.

This book could not have been completed without the support and help of many of our colleagues. Over the years, many colleagues and students have commented on the lecture material. These colleagues include Timothy G. Gregoire, Jeffrey H. Gove, Andrew Robinson, Leonardo Grilli, Juha Heikkinen, Mika Hujo, Ville Hallikainen, Mikko Sillanpää, Lucas Ferreira, Juho Kopra and many students of our courses. A script written by Jari Miina was used as template in Web Example 9.6. In recent years, drafts of the chapters have been used as teaching material at the University of Eastern Finland, and also in several intensive courses in Finland, Sweden, Germany and Austria. Feedback based on those courses has also been very useful in revising the material. Many students and colleagues also used our material in our collaborative research, thus acting as guinea-pigs for the material and helping us further develop it. These colleagues include Sergio de Miguel, Jari Vauhkonen, Edgar de Souza Vismara, Ruben Valbuena, Karol Bronistz, Tomi Karjalainen, Paula Thitz, Lauri Kortelainen and Annika Kangas. The editors, David Grubbs and Lara Spieker, have patiently trusted us, even though we were forced on many occasions to request more time to finish the book. Shashi Kumar from Taylor&Francis has helped with technical LaTeX issues. David Wilson has reviewed the English language of the manuscript. Funding from the Metsämiesten Säätiö Foundation allowed us to hire Mikko Härkönen to edit, synchronize and index the material. The Society of Finnish Foresters provided a grant for language revision of the book. University of Eastern Finland has supported writing of this book through covering the costs of the book website, copyright fees and Lauri's salary. The contributions of all the above-mentioned persons and institutions are highly appreciated. We are also very grateful to our families for the support and understanding that they have constantly shown during this project.

Joensuu and Suonenjoki, February 2020

Lauri Mehtätalo and *Juha Lappi*

1

Introduction

Forestry and environmental research is commonly based on quantitative data, which can be used for various purposes. These include, for example, assessments of natural resources, finding and quantifying associations between some variables of interest, estimating the effects of certain factors on the variable of interest, and prediction of the variable of interest for units that have not been observed, such as new spatial locations or points in time. Quantitative data are commonly analyzed using statistical methods.

In this book, we focus on statistical methods that are widely applicable with forestry and environmental data. Those methods, such as the regression analysis, are very general in the sense that the same methods are applied in many fields. However, when comparing quantitative forestry and environmental data to data sets from other fields, such as health sciences, social sciences, or economics, there are some differences. For example, forest and environmental data sets are often collected from spatially variable locations, possibly over time. They are strongly affected by the spatial locations themselves, and by the weather and climatic conditions of the site. Very often the data sets have a grouped structure, for example, they may consist of trees measured on sample plots that are measured at different locations at different points in time. Sometimes, the grouped structure may exist only in the collected data set, not in the population from which the data were sampled. For example, it may be caused by smooth changes in the studied phenomena over space, which has a very wide range compared to the size of the sample plots. The types of variables of interest also vary, including binary indicators, counts, percentages, and other continuous quantitative measures. When analyzing forest tree data, the applied statistical models should take into account what we already know about the shapes and dimensions of the trees or, more generally, our knowledge of the natural processes behind the variable of interest. Sometimes the data are based on controlled experiments in a laboratory. However, more often they are based on field-measurements. In such data sets, there are lot of factors that are not under the control of the experimenter and have a large effect on the data analysis, even when the field experiment has been well planned.

The methods, concepts and results presented in this book are illustrated throughout with examples, which are mostly implemented using the open-source statistical software R. The examples in R are used for two reasons. First, in many cases ideas are better communicated with the use of examples, and by showing the script associated with these examples, communication should be easier and more transparent. Second, the hands-on examples of implementations using R shown in this book should make it easy even for non-statisticians to carry out similar analysis. However, the aim of this book is not to describe in detail the use of R in modeling environmental data, but rather to describe the theory and applications of certain statistical methods and illustrate their use and basic concepts through examples in R. The examples are separated from the main part of text so that the theory and methods can be also understood without the examples.

We emphasize that even though one might be able to make a statistical analysis by just editing the R-scripts from our examples, or by clicking the correct buttons in some other software, such analysis is very error-prone and should be avoided in serious scientific work. In addition, such analysis often underutilizes the capabilities of the statistical methods, does not show all the valuable information contained in the data, and may also be badly misleading. Therefore, a good understanding of the fundamentals of statistics is an invaluable asset to any researcher who wants to carry out applied research in forestry and environmental sciences. To further underline this point, we list below examples of a number of issues that are commonly misunderstood or badly recognized by researchers in applied fields:

- A linear model does not assume normality of the residual errors. The role of normality is often overly emphasized in applied sciences (see Section 4.6.3). This can lead, for example, to the use of non-parametric tests (which we do not address here) in many such data sets where parametric tests would be much better justified. For example, a parametric test (linear mixed-effects model) is more justified for grouped non-normal data than a non-parametric analysis that does not require normality even in small samples but ignores the dependency caused by the grouping. In general, the model assumptions have an order of importance: a good model for the expected value is more important than the model for the variance-covariance structure of the data, which in turn is more important than the assumptions about the shape of the distribution.

- A generalized linear model (GLM) does not utilize the information about the shape of the distribution in estimation and inference; it only utilizes the implicit variance-mean ratio (see Chapter 8). The excellence of generalized linear (mixed) models (GL(M)Ms) is, therefore, often overly emphasized. Any approach that properly models the error variance, and takes into account the range of the mean, may be equally well justified.

- Root mean squared error (RMSE) and the coefficient of determination are not acceptable criteria for comparing models fitted by different methods (see Box 4.3, p. 95) or evaluating whether random effects should be used or not. The model should be based on the structure of the data and previous knowledge of the process being modeled. In general, the use of coefficient of determination in model comparison has several pitfalls (see Section 4.5.3).

- The distribution of the y-variable in the context of regression models means that the distribution is conditional on the predictors. The marginal distribution of the y-variable is not useful in evaluating whether the stated assumption about the distribution has been met.

- If one wants to learn only one statistical concept well, we recommend the theory of linear prediction. It provides a large number of statistical concepts as a special case, as we will discuss in Section 3.5.2 and demonstrate throughout the book.

- The high p-value from a test indicates that the test failed to reject the null hypothesis, but it does not indicate that the null hypothesis is true (see Section 3.6). This is an issue that everyone learns in a basic course in statistics, but seems to be easily forgotten during a research career, probably because so many research papers misinterpret the high p-values (see Amrhein *et al.* (2019) and comments thereof).

- Many researchers regard that the purpose of the statistical analysis is to obtain low, (i.e. significant) p-values. We emphasize that the modeling should be based on valid assumptions, not on assumptions that produce low p-values. Also, an honestly significant p-value does not imply that the effect is significant in practice.

This book is aimed at students and researchers in forestry and environmental studies. We assume that the readers have a basic understanding of statistics. We also assume that they are familiar with basic matrix algebra; readers who are not familiar with matrices should read, e.g. Appendix A in Fahrmeir *et al.* (2013) or any other similar appendices on the basics of matrices. Moreover, we use a lot of `R`-examples in this book; readers who are not familiar with the `R` software package should consult, for example, "An introduction to R" (`https://www.r-project.org/`).

There are approximately 170 examples in this textbook. A proportion of them are web examples, which are just briefly mentioned in the book and are available in their entirety from the book website at

`http://www.biombook.org.`

In addition, the full scripts for all `R`-examples are available from the book website. The data sets and some specific functions that are used in the examples are available in the `R`-package `lmfor`, which is available at the comprehensive `R` archive network (CRAN).

The subsequent chapters of the book are organized as follows: Chapters 2 and 3 summarize the necessary preliminaries for a sufficiently deep understanding of the main ideas, capabilities and constraints of the methods described in the subsequent chapters. Chapter 2 presents the basic mathematical tools that are used to formulate a (theoretical) model for a process of interest and Chapter 3 presents the general principles about how the process parameters can be estimated using observed data. In particular, Section 3.5 describes the linear predictor, the generality of which is emphasized and demonstrated in almost all subsequent chapters of the book.

Chapters 4–10, are devoted to regression models in different contexts. Chapter 4 covers the linear model. A difference between our text and many other textbooks on the topic is that we present the model directly for a general variance-covariance structure. The use of that model in prediction is also illustrated, which links the linear model to the prediction of time series and geostatistics. Chapters 5 and 6 cover the linear mixed-effect models for a data set with a single level of grouping. In contrast to many other textbooks on the subject, we devote considerable space to illustrate how the model is formulated in the matrix form. One reason for this is the common application in forest sciences, where a previously fitted model is used to predict the random effects in a new group of interest. Sections 6.3 – 6.5 include topics that we have not seen previously discussed in textbooks. Chapters 7, 8 and 9 generalize the ideas of Chapters 4 – 6 to non-normal, nonlinear and multivariate models. The similarity between the nonlinear and generalized linear models is emphasized. Chapter 9 discusses the multivariate models and shows through examples how it is formulated as a univariate system, and how the cross-model correlations can be utilized in prediction. Chapter 10 extends the discussion of regression models by addressing some topics that are common to all the models described in the previous chapters.

Chapters 11–14 discuss some specific topics related to our own experiences in modeling forest data sets. These include the modeling of tree size, stem taper, measurement errors, and a short discussion on analysis and planning of forest experiments.

Our notations differ slightly between chapters. The most important difference is probably the meaning of capital and lowercase bolded symbols. In Chapters 1 and 2, uppercase symbols are used for a random variable and lowercase symbols for its realization. For random vectors, these are further printed in bold font, i.e. $\boldsymbol{X}$ is used for a random vector and $\boldsymbol{x}$ for its realized value. This notation follows the standard principles used in elementary textbooks of statistical inference. Starting from Chapter 3, we show the distinction between a random variable and its realization only occasionally. Lowercase bolded symbols are used for vectors and uppercase bolded symbols for matrices. Another notation to point out is the use of hats and tildes. The hat (e.g. $\widehat{\boldsymbol{\beta}}$) denotes an estimate of a parameter $\boldsymbol{\beta}$ that is thought as fixed. The tilde (e.g. $\widetilde{\boldsymbol{b}}$) is used on a random variable $\boldsymbol{b}$ that has been predicted using the procedures described in Section 3.5.2.

2

Random Variables

2.1 Introduction to random variables

Random variables are variables that can exhibit different values depending on the outcome of a *random process*. The value assigned to a random variable due to a specific outcome of an experiment is called *realization*. The random variable itself cannot be observed, but the realized value can be. However, realization of a random variable does not include all the information of the properties of the random variable. Furthermore, several realizations include more information than just a single one. A variable that is not random is *fixed*.

Terms such as *measurement* and *observation* are often used synonymously with realization. However, 'observation' can also mean *the unit or the subject* from which we make measurements, and we can measure several variables for each unit. Measurements of a given property of a unit are called *variables*. For example, trees of a certain area may be the units selected for measurements of diameter, height, species, volume and biomass. Some quantities cannot be measured for practical or theoretical reasons, and they are called *latent variables*.

Whenever the distinction between the random variable and the realized/observed value is explicitly shown, a common practice is to denote the random variable by an uppercase letter and the realized value by a lowercase letter. For example, $X = x$ and $Y = 2$ means that values x and 2 were assigned to random variables X and Y, respectively, by some random processes.

2.1.1 Sources of randomness

An intuitive reason for the randomness may be sampling: the unit was selected from a population of units randomly, and therefore any characteristic that one observes from that particular unit is affected by the unit selected. Another way of thinking is to assume that there is a certain, usually unknown random process, which generated the value for the unit of interest.

Example 2.1. Consider the random variable X, "the measured diameter at breast height of a tree stem in a certain forest stand". Two different ways of thinking for the randomness can be taken.

a **Randomness through sampling.** For each tree in the stand, the tree diameter is a fixed value. A tree is selected randomly, for example by using simple random sampling. The realized value of X is affected only by which tree was selected, and the sampling is the only source of randomness.

b **Randomness through an underlying random process.** The diameter of a tree is a random variable, and there is an underlying random process that generated the realized diameter. In this context, this process may be affected by the site, origin, and past growth conditions. The effects of all these processes cannot be quantified explicitly, and therefore we just accept that they jointly cause such variation in tree diameter, which we observe as random variation.

If the same process of stand growth is run again for the same tree (which is naturally impossible), the resulting tree diameter would be different. A random variable would express the set of possible values the tree diameter can get and the associated probabilities, if the process was run infinitely many times.

If randomness is thought as a result of sampling only, then measuring the entire population (e.g. callipering all trees in the previous example) would give us complete information about the random variable in question. Approaches where randomness is taken purely as a result of sampling from a fixed population are called *design-based*. They are discussed in the sampling literature (see e.g. Gregoire and Valentine (2008) and Mandallaz (2008) for discussion in forestry and environmental context, Thompson (2012) for an easily accessible and Särndal *et al.* (1992) for a more detailed general text).

In this book, we regard randomness to be a result of a random process. Therefore, the methods covered can be generally called *model-based approaches*. We assume that there is a model that generates the data we observe. The model generates a random vector where the elements may be either independent or dependent. The realized value of the vector is used for statistical inference about the assumed model. From the design-based point of view, a realization of a random variable can be seen as a simple random sample from the infinite population described by the model. The underlying model is, however, just a useful concept to mathematically handle the uncertainty involved in the observed data. Because information about the model is based on data that includes uncertainty, the true model can never be known for sure, although the more data that we have, the less uncertain we are about the true model. In many forest-related data sets, the randomness results both from sampling and a random process. The model-based approach is suitable for such approach as well.

The model-based approach may be confusing at first. For example, what we call the true model is, at its best, only a realistic description of the process behind the data. It is the responsibility of a researcher to formulate a realistic model, and all inference in the model-based approach is conditional on the assumptions made in the model. Therefore, models with mild assumptions should, in general, be preferred over models with strong assumptions. For more discussion, see Section 3.2.

2.1.2 Univariate and multivariate random variables

The random variables may be univariate or multivariate. A *univariate random variable* describes only a certain property of the unit of interest, such as the diameter of a tree stem. A *multivariate random variable* is needed when several univariate random variables are observed, and they are related to each other. For example, the univariate components may describe several properties from the same unit, such as diameter and height of a tree. They may also be otherwise related, such as measurements of soil depth at different locations, where the relationship may get stronger with decreasing distance between the locations. Also repeated measurements of tree growth may be related, due to the similarity of observations from a certain tree, or due to the similarity of growth of different trees in a certain calendar year due to common weather conditions. The multivariate random variable is handled as a *random vector*. For example the random vector of tree diameter and height can be defined as

$$\boldsymbol{X} = \begin{pmatrix} X_1 \\ X_2 \end{pmatrix}$$

where X_1 is the scalar random variable "tree diameter", and X_2 is "tree height".

Univariate random variables are discussed in more detail in Section 2.2. The discussion is generalized to random vectors in Section 2.3.

2.2 Univariate random variables

2.2.1 Sample space and support set

For a formal definition of a random variable, consider an *experiment* producing a value specified by a random process. The set of possible outcomes from such an experiment is called the *sample space*. For example, experiment "Tree status" has the sample space *"dead"*, *"alive"*, the experiment "Number of trees on a sample plot" has the sample space $0, 1, 2, \ldots$, and experiment "Tree diameter" has the sample space of positive real numbers $(0, \infty)$.

A random variable is a mapping of the sample space of an experiment to a numeric scale, which is called the *range* or *support set* (or just support) of the random variable. The probabilities of the original experiment define the probabilities related to the random variable.

For example, one may define the random variable Y ("Tree status") as $Y = 0$, if a tree is dead and $Y = 1$, if the tree is alive, so that the support set of Y includes values $\{0, 1\}$. For random variable Z, "Tree species" in a three-species mixed forest stand, one may define

$$Z = \begin{cases} 1 & \text{for Scots Pine} \\ 2 & \text{for Norway Spruce} \\ 3 & \text{for Silver Birch} \end{cases}$$

and the support set is $\{1, 2, 3\}$. If the sample space of an experiment is already numeric, then it can be directly used as the support set of a random variable.

There are two types of random variables: discrete and continuous. For a *discrete random variable*, the range is a list of numbers. The list of discrete numbers can be either finite (e.g. $\{0, 1, 2, \ldots, 10\}$) or infinite (e.g. $\{0, 1, 2, \ldots\}$). For a *continuous random variable*, the range is an interval of the real axis. Examples of discrete random variables with a finite range are variables Y and Z above. An example of a discrete random variable with an infinite range is the number of trees on a sample plot, which can be assigned any non-negative integer values. Examples of a continuous random variable are tree diameter, which has the range $[0, \infty)$, and the measurement error in tree diameter, which has the range $[-d, \infty)$, where d is the true diameter of the tree, assuming that negative measurements are not possible.

A random variable can also be a combination of discrete and continuous variables. For example, let Y describe the proportion of a sample plot covered by a certain plant, such as lingonberry (*Vaccinum vitis idaea*). The existence of lingonberry may be first described by a discrete variable "Does lingonberry exist on the plot", with sample space $\{YES, NO\}$ and a corresponding support set $\{0, 1\}$. If the berry exists on the plot, then the coverage is defined as a continuous random variable with the support $(0, 1]$.

2.2.2 Distribution function and density

The support set defines the values that a random variable can be assigned, but we still need to define how frequently that the different values occur. For example, a die has the support $\{1, 2, 3, 4, 5, 6\}$, with equal probabilities for these values if the die is fair. In the case of tree diameters, diameters close to the average diameter of the plot are more common than extremely big trees. The probabilities associated with specific values of a univariate random variable are described by the *distribution* (also called *cumulative distribution function, cdf*). It is a function that expresses the probability that the random variable X takes on a value x below a specified value:

$$F(x) = P(X \leq x) \quad \text{for all } x\,.$$

The notation $F(x)$ for the *cdf* does not explicitly show the random variable X, because it is evident from the context. However, if there is some risk of confusion, the random variable in question can be shown as a subscript to F, e.g. $F_X(x)$ is probability $P(X \leq x)$, and $F_X(u) = P(X \leq u)$.

The definition does not make a distinction between the continuous and discrete random variables. However, the outlook of the distribution functions differs, as illustrated in the left panel of Figure 2.1. The distribution function of a discrete variable has "jumps" at the values that belong to the support of the random variable, and it is flat for values that are outside the support. If a discrete variable does not have a natural ordering of classes, the distribution function will look very different, depending on how the random variable is mapped from the sample space of the experiment. In contrast, the distribution function of a continuous random variable is a continuous function within the support set.

For a function $F(x)$ to be a *cdf*, the following three conditions need to hold (Casella and Berger 2001).

1. $\lim_{x \to -\infty} F(x) = 0$ and $\lim_{x \to \infty} F(x) = 1$.
2. $F(x)$ is a non-decreasing function of x.
3. $F(x)$ is right continuous, i.e. for any x, $\lim_{x_0 \to x+} F(x_0) = F(x)$, where $x_0 \to x+$ means that x_0 approaches x from the right.

Any function that fulfills these conditions is a proper distribution function. Thus, the function may have jumps (see the distribution functions of Y and Z in Figure 2.1). The function may also be a mixture of continuous "pieces" and "jumps", as in the example of lingonberry cover discussed above.

The *cdf* can be used to find the probability for a random variable to take on a value between specified limits $(x_0, x]$ as follows:

$$P(x_0 < X \leq x) = P(X \leq x) - P(X \leq x_0) = F(x) - F(x_0) \,. \qquad (2.1)$$

An alternative but equivalent way to express the probability distribution for the discrete random variable is through the *probability mass function (pmf)*. It gives the probability that random variable X will take on value x:

$$f(x) = P(X = x) \quad \text{for all } x \,.$$

Thus, the *pmf* gives the *point probabilities* for x, with non-zero values only for those x that belong to the support set of X. These point probabilities are the heights of the jumps in the cdf at $X = x$.

For a continuous random variable, the probability of each event $X = x$ is 0. Therefore, the *pmf* cannot be defined, and non-zero probabilities can be defined only for intervals with non-zero length. Those probabilities are computed using Equation (2.1). However, we can define a function that intuitively corresponds to the *pmf*, the *probability density function (pdf))*. It is defined as

$$f(x) = F'(x) \,,$$

where $F'(x)$ denotes the first derivative of function $F(x)$ with respect to x. An essential difference to *pmf* is that the value of the *pdf*, $f(x)$ cannot be interpreted as a probability for the event $X = x$. For example, if tree diameter is a continuous random variable, the probability for finding a tree with a diameter of exactly 20 cm is always zero: all tree diameters differ from this value, at least by an infinitesimal amount. However, we have a non-zero probability of finding a tree with diameter between 19.95 and 20.05 cm, which would be classified to 20.0 cm when the 1-mm measurement accuracy is used when callipering.

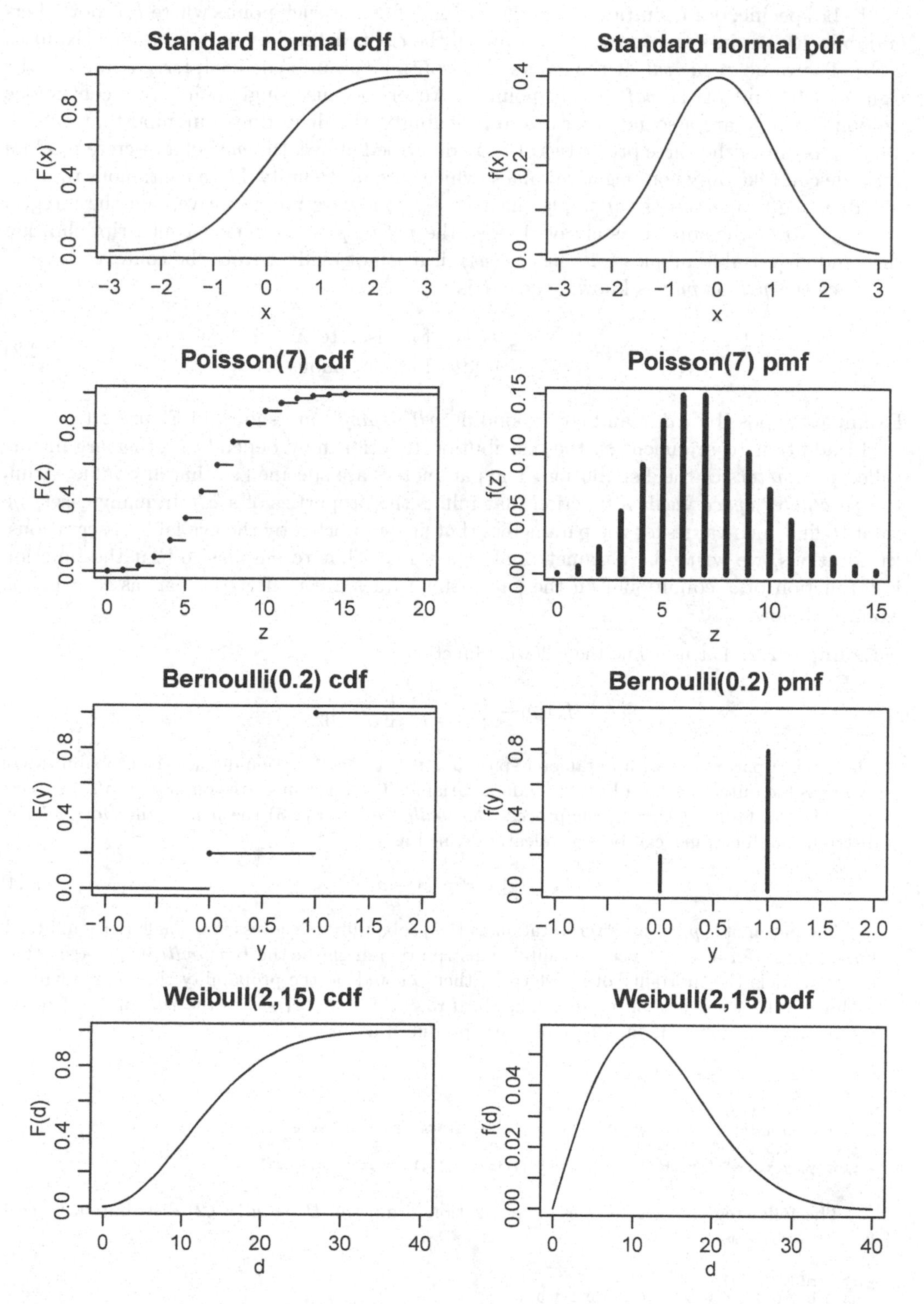

FIGURE 2.1 Examples of four different cumulative distribution functions (*cdf*) (on the left) and the corresponding probability mass function (*pmf*) or probability density functions (*pdf*) (on the right).

To be specific, our definition leaves the *pdf* undefined at such points where F is not differentiable. Such points may exist, for example, if the *cdf* is defined by parts (see Web Example 2.23). The value of the *pdf* at the points of discontinuity can be defined, for example, as the right or left limit of the *pdf* at that point. However, because single points of a continuous random variable are associated with zero probability, the definition is unimportant.

As a result of the three properties of $F(x)$ described above, the *pmf* of a discrete random variable can take only non negative values, which sum up to unity. For a continuous random variable, the properties of *cdf* imply that the *cdf* is always non-negative, and the integral of the *pdf* over the support is always 1. Also the *pdf* or *pmf* may have a subscript showing the random variable in question (e.g. $f_X(u)$), if it is not evident from the context.

Once the *pdf* or *pmf* is known, the *cdf* is

$$F(x) = \begin{cases} \sum_{t \le x} f(t) & \text{for discrete } X \\ \int_{-\infty}^{x} f(t)dt & \text{for continuous } X\,. \end{cases} \tag{2.2}$$

Examples of possible *cdf*'s and corresponding *pdf*'s/*pmf*'s are shown in Figure 2.1.

In addition to argument x, the distribution function may depend on other arguments called *parameters* of the distribution. The parameters are specified so that any value within the *parameter space* yields a function that fulfills the properties of a *cdf*. In many cases, we want to find such values for the parameters that are supported by the available observations. In notations, we write the parameters after a vertical bar to emphasize that the function is a function of x conditional on the parameter values given after the bar, as seen in the example below.

Example 2.2. Let us define the following function

$$f(x|p) = \begin{cases} p & \text{if } x = 1 \\ 1-p & \text{if } x = 0\,. \end{cases} \tag{2.3}$$

where p is a parameter with parameter space $0 < p < 1$. The function defines the probabilities of two possible outcomes for a binary random variable. The outcomes are commonly called success ($X = 1$) and failure ($X = 0$), the process *Bernoulli trial* and (2.3) the *pmf* of the *Bernoulli*(p) distribution. The *pmf* can be equivalently defined as

$$f(x|p) = p^x(1-p)^{1-x} \tag{2.4}$$

The parameter p has an interpretation as the probability of success in a single Bernoulli trial. For example, for a toss of a coin, a suitable distribution might be the *Bernoulli*(0.5) distribution. If one models the mortality of forest trees, then p could be the probability that a tree will die within the time interval of interest. The third row of Figure 2.1 shows the *pdf* and *pmf* of the *Bernoulli*(0.2) distribution. The plots were produced using

```
> # Bernoulli(0.2) cdf and pmf
> y<-c(0,1)
> Fy<-c(0,0.2,1.0)
> plot(stepfun(y,Fy),xlab="y",ylab="F(y)",verticals=FALSE,pch=16,main=NA)
> fy<-c(0.2,0.8)
> plot(y,fy,xlab="y",ylab="f(y)",type="h",ylim=c(0,1),lwd=2,xlim=c(-1,2))
```

The following script generates 20 realizations from the *Bernoulli*(0.2) distribution using **R** function **rbinom**.

```
> rbinom(20,1,0.2)
 [1] 0 0 0 0 0 0 0 0 0 0 0 0 0 1 0 0 1 1 0 0 0
```

The situation becomes much more realistic and interesting if one defines the parameter p as a function of tree properties, e.g. the age and species of the tree. For example, one may define that the probability depends on tree age according to $p = \beta_1 + \beta_2 Age$. Here β_0 and β_1 are now

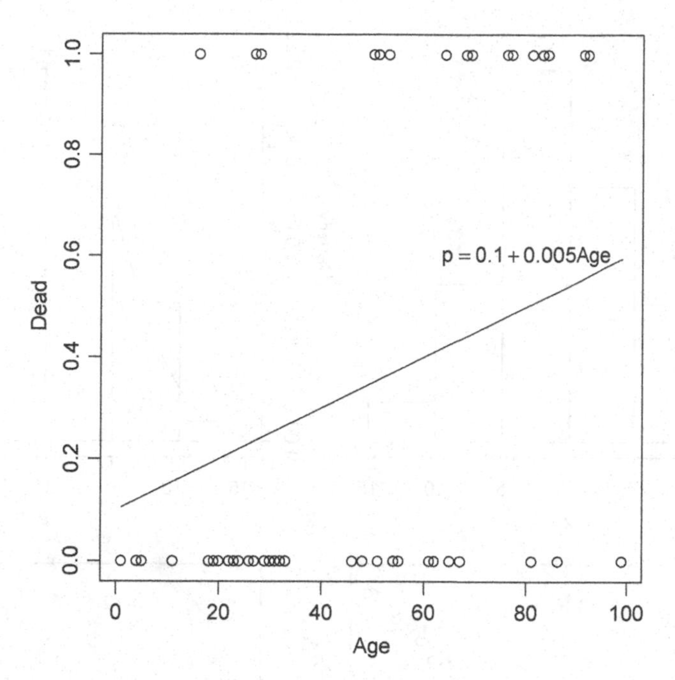

FIGURE 2.2 Observed values of tree mortality (0=alive, 1=dead) as a function of tree age in a hypothetical example. The underlying relationship between tree age and death probability p is shown by the straight line.

the parameters of a model that may be of interest when one wants to understand the process behind tree mortality, as illustrated by the hypothetical example in Figure 2.2, where $\beta_1 = 0.1$ and $\beta_2 = 0.005$. This thinking leads to the binary models, such as the widely used binary logistic regression, which is covered in Chapter 8.

For a better formulation of this model, the expression for p should be specified in such a way that the probability falls within $(0, 1)$ for all possible values of *Age*. In generalized linear models, this restriction is formulated through a link function.

Figure 2.2 was produced by:

```
> Age<-c( 1, 4, 5,11,16,18,19,20,22,23,23,24,26,27,27,28,29,29,30,30,31,32,33,46,48,
+        50,51,51,53,54,55,61,62,62,62,64,65,67,68,69,76,77,81,81,83,84,86,91,92,99)
> p<-0.1+0.005*Age
> Dead<-rbinom(50,1,p)
> plot(Age,Dead)
> lines(Age,p)
> text(80,0.6,expression(p == 0.1 + 0.005 * Age))
```

Example 2.3. Another important distribution family is the *normal distribution*. The distribution has two parameters: μ and σ. A common shorthand notation for the distribution is $N(\mu, \sigma)$. The support of a normally distributed random variable is the entire real axis, $(-\infty, \infty)$, with the *pdf* defined as

$$f(x|\mu, \sigma) = \frac{1}{\sqrt{2\pi}\sigma} e^{-\frac{(x-\mu)^2}{2\sigma^2}} . \tag{2.5}$$

The function is a proper *pdf* if $-\infty < \mu < \infty$ and $\sigma > 0$. The first row of Figure 2.1 shows the normal *cdf* and *pdf* with $\mu = 0$ and $\sigma = 1$ which is commonly called the *standard normal distribution*. The parameter μ is related to the location of the distribution and σ to the spread of the distribution.

The corresponding *cdf* cannot be written in a closed form, but it can be computed numerically as the value of integral in Equation (2.2), in `R` one can use the function `pnorm`. For example, for a random variable with distribution $N(10, 5)$ the probability of a value below 11 is 0.58, the probability of a value below 8 is 0.345, and the probability of a value between 8 and 11 is 0.235.

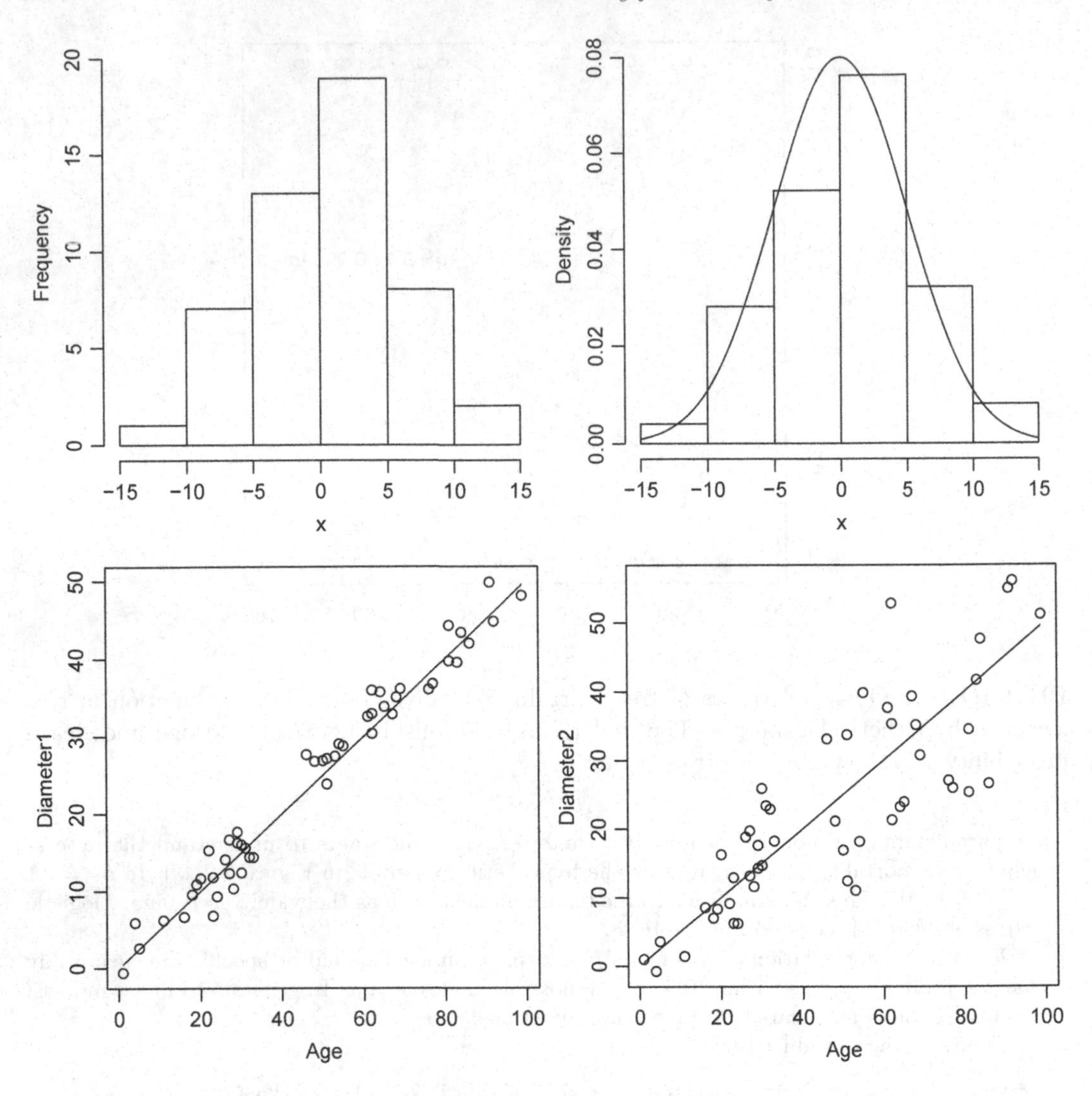

FIGURE 2.3 The upper graph shows 50 realizations from $N(0, 5)$ distribution, plotted using frequencies (left) and relative densities (right) together with the true underlying probability density function. The lower graphs show examples where the parameters of the normal distribution are functions of tree age (see Example 2.3).

```
> print(a<-pnorm(11,10,5))
[1] 0.5792597
> print(b<-pnorm(8,10,5))
[1] 0.3445783
> a-b
[1] 0.2346815
```

The upper graphs in Figure 2.3 illustrate the realizations of a normally distributed random variable; they were generated using function `rnorm`. The height of the bars in the left histogram show the total number of realizations within that class. The right-hand histogram is scaled so that the total area of the bars equals to 1, so that t is comparable to the true model illustrated by the normal density function.

```
> x<-rnorm(50,0,5)
> hist(x,xlim=c(-15,15))
```

```
> hist(x,freq=FALSE,xlim=c(-15,15),ylim=c(0,0.1))
> norm05<-function(x) dnorm(x,0,5) # A one-argument function for function curve
> curve(norm05,-15,15,add=TRUE)    # You could also use function lines here
```

Corresponding to the Bernoulli distribution in Example 2.2, the parameters of a normal distribution may also be functions of some predictor variables. For example, consider a hypothetical Diameter-Age relationship of forest trees. Assume that the diameter follows the normal distribution, where the expected value depends on tree age according to $\mu = 0.5Age$ so that $D \sim N(0.5Age, 2)$. Such data, together with the true relationship of age and μ, is illustrated by the lower left-hand graph of Figure 2.3.

The parameter σ may also be a function. We may have, for example, $D \sim Normal(0.5Age, 2+0.1Age)$. A simulated data set from such model, together with the true relationship of age and μ, is illustrated in the lower right-hand graph of Figure 2.3. These ideas lead to the linear model for normally distributed data (see Chapter 4).

Figure 2.3 was produced using:

```
> Age<-c( 1, 4, 5,11,16,18,19,20,22,23,23,24,26,27,27,28,29,29,30,30,31,32,33,46,48,
+          50,51,51,53,54,55,61,62,62,62,64,65,67,68,69,76,77,81,81,83,84,86,91,92,99)
> mu=0.5*Age
> sigma=2+0.1*Age
> Diameter1<-rnorm(Age,mu,2)
> plot(Age,Diameter1)
> abline(0,0.5)
> Diameter2<-rnorm(Age,mu,sigma)
> plot(Age,Diameter2)
> abline(0,0.5)
```

Example 2.4. The Poisson distribution is a discrete distribution of random counts without an upper limit. For example, it is the distribution of the number of points within a line segment or within a 2- or 3-dimensional area under the *complete spatial randomness (CSR)*, which means that the points are uniformly distributed over the space with intensity λ (points per length/area or volumetric unit) and the locations of "new" points are not affected by the locations of "existing" points (i.e. the locations are independent from each other). Therefore, the number of trees on a fixed-area sample plot of area $|A|$ is distributed according to $Poisson(\mu)$ where $\mu = \lambda|A|$, if the trees are uniformly distributed and the locations do not interact. The Poisson *pmf* is

$$f(x|\mu) = \frac{e^{-\mu}\mu^x}{x!} \qquad x = 0, 1, 2, \ldots; \mu > 0\,, \tag{2.6}$$

where parameter μ specifies the mean number of trees per plot. The second row of Figure 2.1 on page 9 shows the *cdf* and *pmf* of the $Poisson(7)$ distribution.

Examples of computations using that distribution are $P(X = 3)$

```
> dpois(3,7)
[1] 0.05212925
```

and $P(X = 3 \text{ or } X = 4)$

```
> dpois(3,7)+dpois(4,7)
[1] 0.1433554
> ppois(4,7)-ppois(2,7)
[1] 0.1433554
```

Web Example 2.5. The lower graphs of Figure 2.1 on page 9 show the *cdf* and *pdf* of a Weibull(5, 15) distribution.

Web Example 2.6. Deriving the distribution of airborne laser hits on a tree crown with a given crown shape.

2.2.3 Transformations of random variables

Transformation of a random variable means the application of a certain function $g(X)$ to the random variable X to get a new random variable Y. The distribution of the new random variable is determined by the distribution of the original random variable. The commonly used transformations are monotonic one-to-one transformations, which means that a single value of the original variable X always corresponds to only one value of Y. Let X have *cdf* $F_X(x)$ and let $Y = g(X)$. The *cdf* of Y is

$$F_Y(y) = \begin{cases} F_X(g^{-1}(y)) & \text{if } g(x) \text{ is increasing} \\ 1 - F_X(g^{-1}(y)) & \text{if } g(x) \text{ is decreasing.} \end{cases} \tag{2.7}$$

For a continuous random variable, the *pdf* results by differentiating the *cdf* with respect to y. Let $f_X(x)$ be the *pdf* of X and $g(X)$ a monotone transformation, whose inverse $g^{-1}(y)$ has a continuous derivative. Then the *pdf* of Y within the support of Y is

$$f_Y(y) = f_X(g^{-1}(y)) \left\| \frac{d}{dy} g^{-1}(y) \right\| , \tag{2.8}$$

and 0 elsewhere.

Example 2.7. A forestry student has made measurements of tree root biomass. The data are skewed to the right and all observations are naturally positive. Based on these observations, she decides to use the lognormal distribution as a starting point for the analysis. The lognormal *pdf* is defined as

$$f_X(x) = \frac{1}{\sqrt{2\pi\sigma^2}} \frac{e^{-(\log(x)-\mu)^2/(2\sigma^2)}}{x} .$$

For modeling purposes, she applies the logarithm transformation $Y = g(X) = \log(X)$ to the observations. The inverse function of the applied transformation $g^{-1}(y) = e^y$ is continuous and increasing, and has derivative $\frac{d}{dy} g^{-1}(y) = e^y$. Therefore the distribution of logarithmic observations is

$$\begin{aligned} f_Y(y) &= \frac{1}{\sqrt{2\pi\sigma^2}} \frac{e^{-(y-\mu)^2/(2\sigma^2)}}{e^y} e^y \\ &= \frac{1}{\sqrt{2\pi\sigma^2}} e^{-(y-\mu)^2/(2\sigma^2)} , \end{aligned}$$

which is the *pdf* of the normal distribution.

Web Example 2.8. Deriving the distribution of height when tree diameter follows the Weibull distribution and the height-diameter relationship is described by the power function.

2.2.4 Expected value

In many cases, the distribution of a random variable is unknown, and we do not want to make strong assumptions about it. However, we may want to make inference about the behavior of the random variable. In such a case, we can summarize the most interesting and important characteristics to summary figures, which describe the most important properties of the random variable in question, but do not specify the entire distribution.

The most commonly used characteristic for describing the properties of a random variable is the *expected value*, which is also called the *mean* of the random variable. It describes the center of gravity for the distribution, but does not reveal how much the individual realized values may vary around it. If an infinitely large number of independent realizations of X are available, their average is the expected value. Notice the difference between the mean of a random variable and *sample mean* (see Section 3.4.2), which is often used as an estimator of the mean.

The expected value of random variable X is the weighted sum/integral of the random variable, the weight being defined by the *pmf*/*pdf*:

$$\mathrm{E}\,X = \begin{cases} \sum_j p_j x_j & \text{for discrete } X \\ \int_{-\infty}^{\infty} x f_X(x) dx & \text{for continuous } X\,. \end{cases} \tag{2.9}$$

It is possible that the expected value is infinite, if the tails of the distribution are sufficiently thick (e.g. the t- distribution with $df = 1$). The expected value of transformation $g(X)$ is defined analogously as:

$$\mathrm{E}[g(X)] = \begin{cases} \sum_j p_j g(x_j) & \text{for discrete } X \\ \int_{-\infty}^{\infty} g(x) f_X(x) dx & \text{for continuous } X\,. \end{cases} \tag{2.10}$$

The parentheses can be dropped from expected value if there is no risk of confusion; therefore notations $\mathrm{E}(X)$ and $\mathrm{E}\,X$ are equivalent. In some cases, a subscript is used to show which distribution is used in computation. For example, $\mathrm{E}_X\,[g(x)]$ uses the *pdf*/*pmf* of X.

An important consequence from (2.10) is that, generally, $\mathrm{E}\,[g(X)] \neq g\,(\mathrm{E}\,X)$: the expected value of a transformed random variable is not found by making the applied transformation to the expected value. However, if $g(X)$ is a *linear* function of X, e.g. $g(X) = a + bX$, then $\mathrm{E}\,[g(X)] = g\,[\mathrm{E}\,X]$.

Let c be a fixed constant and X and Y random variables. The expected value has the following properties

$$\mathrm{E}(c) = c \tag{2.11}$$
$$\mathrm{E}(cX) = c\,\mathrm{E}\,X \tag{2.12}$$
$$\mathrm{E}(X + Y) = \mathrm{E}\,X + \mathrm{E}\,Y\,. \tag{2.13}$$

Example 2.9. Let $Y \sim$ *Bernoulli*(p). The expected value is

$$\mathrm{E}(Y) = p \times 1 + (1 - p) \times 0 = p\,.$$

Example 2.10. The expected value of the normal distribution is the parameter μ. We could compute it analytically from the following equation

$$\int_{-\infty}^{\infty} x \frac{1}{\sqrt{2\pi\sigma^2}} e^{-\frac{(x-\mu)^2}{2\sigma^2}} dx$$

to see that the result of the integral is μ. This is left as an exercise. (Let Z be a standard normal random variable, which has $\mu = 0$ and $\sigma^2 = 1$. Find $\mathrm{E}\,Z$ initially and thereafter use $X = \mu + \sigma Z$ to find $\mathrm{E}\,X$.)

Quite often, a quick solution using numerical methods is sufficient. The `R`- function `integrate` evaluates the integral of a specified function over a given interval. The code below evaluates numerically the expected value of the $N(10, 2)$ distribution.

```
> fun<-function(x){
+       x*dnorm(x,10,2)
+       }
> integrate(fun,-Inf,Inf)
10 with absolute error < 0.0011
```

The code below evaluates numerically the expected value of the *Weibull*$(2.98, 15.26)$ distribution

```
> funw<-function(x){
+        x*dweibull(x,2.98,15.26)
+        }
> integrate(funw,0,Inf)
13.62288 with absolute error < 0.00085
```

Example 2.11. Let random variable X, the tree species in a forest stand, take on values 1 (Scots Pine), 2 (Norway Spruce), 3 (Silver Birch) and 4 (Pubescent Birch) with probabilities 0.5, 0.3, 0.15, 0.05, respectively. The mean height by species is expressed as

$$g(x) = \begin{cases} 18 & \text{if } x = 1 \\ 15 & \text{if } x = 2 \\ 12 & \text{if } x = 3 \\ 22 & \text{if } x = 4 . \end{cases}$$

The mean height of all trees in the stand is

$$\mathrm{E}\, H = \sum_j p_j g(X_j) = 0.5 \times 18 + 0.3 \times 15 + 0.15 \times 12 + 0.05 \times 22 = 16.4 .$$

```
> h<-c(18,15,12,22)
> p<-c(0.5,0.3,0.15,0.05)
> sum(p*h)
[1] 16.4
```

2.2.5 Variance and standard deviation

In addition to the mean of a random variable, we may be interested in how much the values of a random variable vary around the mean. This variation is described by the *variance*, which is defined as the expected value of the squared difference between the random variable and its mean:

$$\mathrm{var}\,(X) = \mathrm{E}\,(X - \mathrm{E}\,X)^2 \;;$$

notation $\mathrm{E}\,(X - \mathrm{E}\,X)^2$ means $\mathrm{E}\left[(X - \mathrm{E}X)^2\right]$, but the outermost parentheses are usually not written explicitly. To simplify the notations, denote constant $\mathrm{E}\,X$ by μ. A computationally better form results through simple algebra:

$$\begin{aligned} \mathrm{var}\,(X) &= \mathrm{E}\,(X - \mu)^2 \\ &= \mathrm{E}\left(X^2 - 2\mu X + \mu^2\right) \\ &= \mathrm{E}\,X^2 - 2\mu\,\mathrm{E}\,X + \mu^2 \\ &= \mathrm{E}\,X^2 - \mu^2 . \end{aligned} \qquad (2.14)$$

Variance is not easy to interpret because the units of measure for variance are the squared units of X. An easier-to-interpret measure for the variation around the mean is the *standard deviation*, which is the square root of the variance

$$\mathrm{sd}(X) = \sqrt{\mathrm{var}(X)} . \qquad (2.15)$$

It has the same unit as X. The standard deviation measures the spread of the observations around the expected value. However, it is different from the expected value of absolute differences; standard deviation fulfills $\mathrm{sd}(X) \geq \mathrm{E}(|X - \mathrm{E}(X)|)$. The difference between these two depends on the shape of the distribution.

In practice, variance is used in computations, and the results are transformed to standard deviations for reporting and interpretation. To avoid confusion and to aid in interpretation, we report the numeric values of variances in the form sd^2. For example, $\mathrm{var}(X) = 4$ is reported in the form $\mathrm{var}(X) = 2^2$, which shows directly that the standard deviation is 2, and there is no doubt whether the reported value is variance or standard deviation. The *coefficient of variation* is the ratio of standard deviation and expected value:

$$\mathrm{vc}(X) = \frac{\mathrm{sd}(X)}{\mathrm{E}\,X} .$$

Example 2.12. To find the standard deviation of the *Weibull*(2.98, 15.26) distribution, we first compute $E(X^2)$ numerically. Thereafter, we use the previously computed EX (see Example 2.10) in (2.14) to find the variance and standard deviation.

```
> funw2<-function(x){
+           x^2*dweibull(x,2.98,15.26)
+           }
> integrate(funw2,0,Inf)$value->EX2
> EX2
[1] 210.3929
> integrate(funw,0,Inf)$value->EX
> EX
[1] 13.62288
> EX2-EX^2
[1] 24.81001
> sqrt(EX2-EX^2)
[1] 4.980965
```

These results mean that if we randomly select trees from a stand with the assumed Weibull distribution, the mean diameter would be 13.6 cm in the long run. The second power of the deviation from 13.6 cm would be, on average, 5^2 cm^2 in the long run, which means that the sampled tree diameters would differ by no more than 5 cm on average from 13.6 cm in the long run.

Example 2.13. If $X \sim$ *Bernoulli*(p), then $E(X) = p$ and $\text{var}(X) = p(1-p)$. If $X \sim$ *Poisson*(μ), then $E(X) = \mu$ and $\text{var}(X) = \mu$. If $X \sim N(\mu, \sigma)$, then $E(X) = \mu$ and $\text{var}(X) = \sigma^2$. Proofs and illustrations are left as an exercise.

2.2.6 Moments

Expected value and variance are special cases of quantities that are called *moments* of the distribution. Let n be an integer. The nth moment of random variable X is

$$\mu'_n = E(X^n)\,.$$

The first moment μ'_1 is therefore the expected value. The first and second moments together define the variance through $\text{var}(X) = \mu'_2 - (\mu'_1)^2$, which is called *the second central moment.* In general, the nth central moment is

$$\mu_n = E(X - \mu)^n\,,$$

where $\mu = EX = \mu'_1$. The third and fourth central moments are related to skewness and kurtosis of the distribution.

The properties of a random variable related to the first moments only are called the first-order properties. Properties that are related to the first and second moments are correspondingly called second-order properties. They include the variance and also include covariance, which we will define shortly. As many statistical modeling approaches are based on the first- and second-order properties only, the shape of the distribution is often unimportant.

2.3 Multivariate random variables

2.3.1 Joint, conditional and marginal distributions

Consider a *multivariate random variable*

$$\boldsymbol{X} = \begin{pmatrix} X_1 \\ X_2 \\ \vdots \\ X_p \end{pmatrix},$$

where components $X_1, X_2, \ldots, X_p$ are univariate random variables. $\boldsymbol{X}$ can also be called a *random vector*. The distribution of a random vector is expressed by a *multivariate distribution*, also called *the joint distribution* of $\boldsymbol{X}$. The joint distribution defines the probabilities of all possible value combinations of the component variables. In this section, $\boldsymbol{X}$ refers to a random vector and $\boldsymbol{x}$ its realization.

The definition of the joint distribution function is analogous to the univariate case:

$$\begin{aligned} F(\boldsymbol{x}) &= P(\boldsymbol{X} \leq \boldsymbol{x}) \\ &= P(X_1 \leq x_1, X_2 \leq x_2, \ldots, X_p \leq x_p). \end{aligned}$$

The limiting value of F as all x_i's approach $-\infty$ is 0, and correspondingly it is 1 as all x_i's approach ∞. Furthermore, F is non-decreasing with respect to all x_i.

If all components of $\boldsymbol{X}$ are discrete, then the joint *pmf* is defined as

$$f(\boldsymbol{x}) = P(\boldsymbol{X} = \boldsymbol{x}) = P(X_1 = x_1, X_2 = x_2, \ldots, X_p = x_p).$$

The function is a proper joint *pmf* if all probabilities are non-negative and sum to one.

If all components of the random vector are continuous, then the joint *pdf* of $\boldsymbol{X}$ is defined as a derivative of the joint distribution function with respect to all components of $\boldsymbol{x}$:

$$f(\boldsymbol{x}) = f(x_1, x_2, \ldots, x_p) = \frac{\partial F(x_1, x_2, \ldots, x_p)}{\partial x_1 \partial x_2 \ldots, \partial x_p}.$$

Similar to the case of univariate densities, non-zero probabilities are defined only for non-zero length intervals of the random variable by integrating over the interval, which is defined for each component variable. Denote the lower and upper bounds of the interval for component X_i as l_i and u_i. The probability of the random vector $\boldsymbol{X}$ to fall between $\boldsymbol{l} = \begin{pmatrix} l_1 & \ldots & l_p \end{pmatrix}'$ and $\boldsymbol{u} = \begin{pmatrix} u_1 & \ldots & u_p \end{pmatrix}'$ is

$$P(l_1 < X_1 \leq u_1, \ldots, l_p < X_p \leq u_p) = \int_{l_1}^{u_1} \ldots \int_{l_p}^{u_p} f(x_1, \ldots, x_p) \mathrm{d}x_p \ldots \mathrm{d}x_1$$

or simply

$$P(\boldsymbol{l} < \boldsymbol{X} \leq \boldsymbol{u}) = \int_{\boldsymbol{l}}^{\boldsymbol{u}} f(\boldsymbol{x}) \mathrm{d}\boldsymbol{x} = F(\boldsymbol{u}) - F(\boldsymbol{l}).$$

The expected value of a random vector is a vector of expected values of the component variables:

$$\mathrm{E}(\boldsymbol{X}) = \begin{pmatrix} \mathrm{E}(X_1) \\ \vdots \\ \mathrm{E}(X_p) \end{pmatrix}.$$

Let us partition the p-dimensional random variable $\boldsymbol{X}$ into two components: $\boldsymbol{X}_1$ with dimension q and $\boldsymbol{X}_2$ with dimension $p-q$. The joint *pdf* of the partitioned random vector is $f(\boldsymbol{x}) = f(\boldsymbol{x}_1, \boldsymbol{x}_2)$. An extreme special case occurs when $\boldsymbol{X}_1$ is a scalar random variable, including only one component of $\boldsymbol{X}$, and $\boldsymbol{X}_2$ includes the remaining components. For $\boldsymbol{X}_1$, two different q-variate random variables can be defined: the *conditional* and *marginal* versions of $\boldsymbol{X}_1$.

The conditional version of $\boldsymbol{X}_1$, denoted by $\boldsymbol{X}_1|\boldsymbol{X}_2 = \boldsymbol{x}_2$ is a random variable $\boldsymbol{X}_1$ in a special situation where $\boldsymbol{X}_2$ takes the value $\boldsymbol{x}_2$. Its distribution, *the conditional distribution* of $\boldsymbol{X}_1$ given that $\boldsymbol{X}_2 = \boldsymbol{x}_2$, is denoted by $F(\boldsymbol{X}_1|\boldsymbol{X}_2 = \boldsymbol{x}_2)$ and the *pdf/pmf* correspondingly by $f(\boldsymbol{X}_1|\boldsymbol{X}_2 = \boldsymbol{x}_2)$. All possible values of $\boldsymbol{X}_2$ may provide a different conditional distribution.

For a discrete $\boldsymbol{X}$, the conditional *pmf* of $\boldsymbol{X}_1$ is defined for all such values of $\boldsymbol{X}$ for which $\boldsymbol{X}_2 = \boldsymbol{x}_2$ as:

$$f(\boldsymbol{x}_1|\boldsymbol{X}_2 = \boldsymbol{x}_2) = \frac{f(\boldsymbol{x}_1, \boldsymbol{x}_2)}{\sum_{\boldsymbol{x}_1 \in \boldsymbol{S}_1} f(\boldsymbol{x}_1, \boldsymbol{x}_2)}$$

where $\boldsymbol{S}_1$ is the support of $\boldsymbol{X}_1$. The denominator on the right-hand side rescales those probabilities from the joint density that are used in the numerator so that they sum up to unity. If all components of $\boldsymbol{X}$ are continuous, then the conditional *pdf* is defined analogously

$$f(\boldsymbol{x}_1|\boldsymbol{x}_2) = \frac{f(\boldsymbol{x}_1, \boldsymbol{x}_2)}{\int_{\boldsymbol{S}_1} f(\boldsymbol{x}_1, \boldsymbol{x}_2) d\boldsymbol{x}_1} = \frac{f_{\boldsymbol{X}}(\boldsymbol{x}_1, \boldsymbol{x}_2)}{f_{\boldsymbol{X}_2}(\boldsymbol{x}_2)} \tag{2.16}$$

where $f_{\boldsymbol{X}_2}(\boldsymbol{x}_2)$ is the marginal pdf of $\boldsymbol{X}_2$.

The *marginal distribution* of a component variable $\boldsymbol{X}_1$ describes the distribution of $\boldsymbol{X}_1$ without reference to the other component variables. In that case, the joint *pmf/pdf* of $\boldsymbol{X}_1$ is summed/integrated over the distribution of $\boldsymbol{X}_2$. For a discrete $\boldsymbol{X}$, the marginal *pmf* of $\boldsymbol{X}_1$ is

$$f(\boldsymbol{x}_1) = \sum_{\boldsymbol{x}_2 \in \boldsymbol{S}_2} f(\boldsymbol{x}_1, \boldsymbol{x}_2).$$

Because the sum over the joint probability mass function is one and each point probability of the joint *pmf* will be included exactly once in the sum of the corresponding value of $\boldsymbol{x}_1$, the marginal probabilities sum up to one. The marginal *pdf* for a continuous $\boldsymbol{X}$ is defined in an analogous way, by integrating the joint *pdf* $f(\boldsymbol{x})$ over the range of $\boldsymbol{x}_2$:

$$f(\boldsymbol{x}_1) = \int_{\boldsymbol{x}_2 \in \boldsymbol{S}_2} f(\boldsymbol{x}_1, \boldsymbol{x}_2) d\boldsymbol{x}_2. \tag{2.17}$$

The integral of the marginal density over $\boldsymbol{x}_1$ directly yields the integral over the entire joint *pdf* which ensures a unit area under the marginal density.

The marginal distribution of a random variable does not show the random variables that are marginalized out from the distribution. This is because they are unknown: integrating or summing them out from the joint *pdf/pmf* yields a low-dimensional joint *pdf/pmf* from which the effect of the marginalized component is removed. The variance of the marginal distribution is always higher or equal to the variance of the conditional distribution.

Example 2.14. Trees were measured for species and health status. The species were coded in the classes described in Example 2.11 (p. 16), and health in classes 1, 2 and 3, indicating dead, weakened and healthy, respectively. The joint *pmf* of the two discrete variables is shown in Table 2.1. The conditional *pmf* of $X|Y = 1$, i.e. the distribution of Scots pine trees in three classes, can be obtained by dividing the values of the first column by the marginal probability of Scots pine. The conditional probabilities are $0.1/0.5 = 0.2$, $0.1/0.5 = 0.2$ and $0.3/0.5 = 0.6$ for dead, weakened and healthy, respectively.

The conditional *pmf* $Y|X = 1$ is the distribution of dead trees by species. The conditional probabilities, obtained by dividing the values of the first row with the row sum of 0.21, are

		Y				
		1	2	3	4	$\sum$
	1	0.1	0	0.1	0.01	0.21
X	2	0.1	0	0.05	0.02	0.17
	3	0.3	0.3	0	0.02	0.62
	$\sum$	0.5	0.3	0.15	0.05	

TABLE 2.1 The joint distribution of tree species and health status in Example 2.14.

$0.1/0.21 = 0.476$, $0/0.21 = 0.000$, $0.1/0.21 = 0.476$, and $0.01/0.21 = 0.048$ for tree species 1, 2, 3 and 4, respectively.

The marginal distributions of tree species are given in the last column and row of the table.

```
> jointd<-cbind(c(0.1,0.1,0.3),c(0,0,0.3),c(0.1,0.05,0),c(0.01,0.02,0.02))
> jointd
     [,1] [,2] [,3] [,4]
[1,]  0.1  0.0 0.10 0.01
[2,]  0.1  0.0 0.05 0.02
[3,]  0.3  0.3 0.00 0.02
> sum(jointd)
[1] 1
> condr1<-jointd[1,]/sum(jointd[1,])
> condr1
[1] 0.47619048 0.00000000 0.47619048 0.04761905
# The marginal distribution of tree species
> margr<-jointd[1,]+jointd[2,]+jointd[3,]
> margr
[1] 0.50 0.30 0.15 0.05
# A more general alternative for the same purpose
> apply(jointd,2,sum)
[1] 0.50 0.30 0.15 0.05
# The marginal distribution of health
> margc<-apply(jointd,1,sum)
> margc
[1] 0.21 0.17 0.62
```

It is also possible that the joint distribution includes both discrete and continuous components. In that case, the definitions of marginal and conditional densities become technically more complicated, but definitions are straightforward generalizations of the results presented above. In particular, the total probability for a continuous random variable is computed through integration and over a discrete random variable through summation (see Section 6.4 for an example).

If all conditional distributions of $\boldsymbol{X}_1$ and the marginal distribution of $\boldsymbol{X}_2$ are known, the joint distribution can be constructed. Especially, solving (2.16) for $f(\boldsymbol{x})$ gives

$$f(\boldsymbol{x}_1, \boldsymbol{x}_2) = f(\boldsymbol{x}_1|\boldsymbol{x}_2) f(\boldsymbol{x}_2) \,. \tag{2.18}$$

The *Bayes rule* follows from using this result again in (2.16):

$$f(\boldsymbol{x}_2|\boldsymbol{x}_1) = \frac{f(\boldsymbol{x}_1, \boldsymbol{x}_2)}{f(\boldsymbol{x}_1)} = \frac{f(\boldsymbol{x}_1|\boldsymbol{x}_2) f(\boldsymbol{x}_2)}{f(\boldsymbol{x}_1)} \,, \tag{2.19}$$

where the denominator is often written in the integral form $f(\boldsymbol{x}_1) = \int_{\mathcal{X}_2} f(\boldsymbol{x}_1|\boldsymbol{x}_2) f(\boldsymbol{x}_2) d\boldsymbol{x}_2$, which does not necessarily have a closed-form solution. Eq. 2.19 specifies the conditional distribution of $\boldsymbol{X}_2|\boldsymbol{X}_1$ once the conditional distribution of $\boldsymbol{X}_1|\boldsymbol{X}_2$ and the marginal distribution of $\boldsymbol{X}_1$ are known. Equations (2.18) and (2.19) are also valid if the *pdf*/*pmf* is replaced by the corresponding *cdf*. Consequently, they are also valid if the functions are replaced by any probabilities based on the *cdf*'s.

The joint distribution of $\boldsymbol{X}$ cannot be derived from the marginal distributions of component random variables alone, unless additional assumptions on the relationship between the component variables $\boldsymbol{X}_1$ and $\boldsymbol{X}_2$ are made. This leads to the terms *dependence* and *independence* of random variables. However, before we can define those terms, we need an expression for the expected value of a many-to-one transformation.

2.3.2 Bivariate transformations

A bivariate transformation is a scalar-valued transformation of a bivariate random vector. The distribution for such a transformation will not be covered in general terms here, but we will define the expected value for such a transformation. Furthermore, the distribution of the simplest case, i.e. sum of random variables, is discussed Section 2.10.

The expectation of a bivariate transformation $g(X, Y)$ of a bivariate random vector (X, Y) is analogous to (2.10):

$$\mathrm{E}[g(X,Y)] = \begin{cases} \sum_i \sum_j p_{ij} g(x_i, y_j) & \text{for discrete } (X,Y) \\ \int_{-\infty}^{\infty} \int_{-\infty}^{\infty} g(x,y) f(x,y) dyx & \text{for continuous } (X,Y) \end{cases} \tag{2.20}$$

Example 2.15. Let the joint distribution of tree diameter X and height Y on a forest sample plot be expressed by a bivariate density $f(x, y)$. The marginal density of diameter is

$$f_X(x) = \int_{-\infty}^{\infty} f(x,y) dy \,,$$

and the expected value of the diameter is the expected value over that distribution

$$\mathrm{E}(X) = \int_{-\infty}^{\infty} x f_X(x) dx \,.$$

Correspondingly, the marginal density and the mean of height become $f_Y(y) = \int_{-\infty}^{\infty} f(x,y)dx$ and $\mathrm{E}(Y) = \int_{-\infty}^{\infty} y f_Y(y) dy$.

Assume that the volume as a function of diameter and height is given by $v(x, y)$. The mean volume is the expected value of the transformation $v(x, y)$

$$\mathrm{E}(v(X,Y)) = \int_{-\infty}^{\infty} \int_{-\infty}^{\infty} v(x,y) f(x,y) dx dy \,.$$

2.3.3 Independence and dependence

Two scalar random variables X_1 and X_2 are *independent* if a realized value of X_2 does not provide any information on X_1. The formal definition uses the marginal and joint *pdf*s or *pmf*s. The random variables X_1 and X_2 are independent if

$$f(x_1, x_2) = f(x_1) f(x_2) \qquad \text{for all } x_1, x_2 \,. \tag{2.21}$$

That is, if the value of the joint *pmf*/*pdf* for all values of x_1 and x_2 results from the multiplication of the corresponding components of the marginal *pmf*/*pdf.* The definition could be equivalently stated in terms of *cdf.* The same definition holds for both discrete and continuous cases. If random variables are not independent, then they are *dependent.* The definition above implies that under independence of X_1 and X_2, the conditional distribution of X_1 is not affected by the realized values of X_2. This is seen directly when comparing (2.18) and (2.21). Therefore, all conditional distributions of X_1 are similar to each other and to the marginal distribution of X_1.

Random vectors $\boldsymbol{X}_1$ and $\boldsymbol{X}_2$ are independent if the joint *pdf*/*pmf* of $\boldsymbol{X} = (\boldsymbol{X}_1', \boldsymbol{X}_2')'$ can be written as follows:

$$f(\boldsymbol{x}_1, \boldsymbol{x}_2) = f(\boldsymbol{x}_1) f(\boldsymbol{x}_2) \qquad \text{for all } \boldsymbol{x}_1, \boldsymbol{x}_2 \,. \tag{2.22}$$

This means means that independence holds for all such pairs of scalar random variables, where one random variable is a component of $\boldsymbol{X}_1$ and the other a component of $\boldsymbol{X}_2$. A random vector $\boldsymbol{X}$ is *mutually* or *jointly independent* if all partitions $\boldsymbol{X} = (\boldsymbol{X}_1', \boldsymbol{X}_2')'$ are independent.

Example 2.16. The products of the marginal probabilities of Example 2.14 are easily computed as the outer product $\boldsymbol{c}\boldsymbol{r}'$ of the vectors that include the marginal probabilities for the columns and rows, respectively:

```
> margc%o%margr
      [,1]  [,2]   [,3]   [,4]
[1,] 0.105 0.063 0.0315 0.0105
[2,] 0.085 0.051 0.0255 0.0085
[3,] 0.310 0.186 0.0930 0.0310
```

These are all different from the probabilities of the joint distribution. Therefore, health status and tree species are not independent in this example. Note that a difference in even one of the probabilities of the joint distribution and product of marginals would show the dependence.

2.3.4 Covariance and correlation

A specific type of dependence between random variables is *linear association*, which is called *correlation*. Random variables X_1 and X_2 are said to be correlated if the expected values of the two random variables have a linear relationship of form $\mathrm{E}(X_1) = a + b\,\mathrm{E}(X_2)$, where a and $b \neq 0$ are constants. The degree of linear association is measured through *covariance*, which is defined as:

$$\mathrm{cov}(X_1, X_2) = \mathrm{E}\left[(X_1 - \mathrm{E}\,X_1)(X_2 - \mathrm{E}\,X_2)\right] . \tag{2.23}$$

Due to the commutativity of the terms in the product on the right-hand side, we have $\mathrm{cov}(X_1, X_2) = \mathrm{cov}(X_2, X_1)$. Computing the product by components and simplifying gives an alternative formulation as:

$$\mathrm{cov}(X_1, X_2) = \mathrm{E}(X_1 X_2) - \mathrm{E}\,X_1\,\mathrm{E}\,X_2 \,.$$

Note that if either of the two variables has the expected value zero, the formula simplifies to $\mathrm{cov}(X_1, X_2) = \mathrm{E}(X_1 X_2)$.

Covariance is difficult to interpret, because it depends on the variances of X_1 and X_2. Scaling using the standard deviations gives *correlation*, which is easier to interpret:

$$\mathrm{cor}(X_1, X_2) = \frac{\mathrm{cov}(X_1, X_2)}{\mathrm{sd}(X_1)\,\mathrm{sd}(X_2)} \,. \tag{2.24}$$

The correlation is always between -1 and 1. The greater the absolute value of the correlation, the stronger is the linear association between X_1 and X_2. If the absolute value of the correlation is 1, then there is a perfect, deterministic linear relationship between X_1 and X_2. A positive value of correlation/covariance implies that an increase in X_1 is associated with an increase in X_2, and negative value means that an increase in X_1 is associated with a decrease in X_2. If correlation (and therefore covariance as well) is 0, then the random variables are said to be *uncorrelated*.

For independent random variables the correlation is zero. However, zero correlation does not necessarily imply independence, because correlation measures the degree of linear association only. It is possible to establish such a nonlinear relationship that leads to a zero slope for a linear relationship; examples can be easily found on the Internet.

The covariance of scalar random variable X with itself gives the variance:

$$\begin{aligned} \text{cov}(X,X) &= \text{E}(XX) - \text{E}(X)\,\text{E}(X) \\ &= \text{E}(X^2) - [\text{E}(X)]^2 \\ &= \text{var}(X)\,. \end{aligned} \tag{2.25}$$

Therefore, it is justified to define the variance of a random vector through a *variance-covariance matrix*, which pools all pairwise covariances of vector $\boldsymbol{X}$ into a single matrix. In the literature, the variance-covariance matrix is often called a variance matrix or a covariance matrix; here, we use the term variance-covariance matrix. It is defined as:

$$\begin{aligned} \text{var}(\boldsymbol{X}) &= \text{E}\{[\boldsymbol{X} - \text{E}(\boldsymbol{X})][\boldsymbol{X} - \text{E}(\boldsymbol{X})]'\} \\ &= \begin{pmatrix} \text{var}(X_1) & \text{cov}(X_1,X_2) & \cdots & \text{cov}(X_1,X_p) \\ \text{cov}(X_2,X_1) & \text{var}(X_2) & \cdots & \text{cov}(X_2,X_p) \\ \vdots & \vdots & \ddots & \vdots \\ \text{cov}(X_p,X_1) & \text{cov}(X_p,X_2) & \cdots & \text{var}(X_p) \end{pmatrix}. \end{aligned}$$

Because $\text{cov}(X_1, X_2) = \text{cov}(X_2, X_1)$, the variance-covariance matrix is symmetric. In addition, the values in the matrix need to be mutually logical so that a random vector with the defined dependencies can exist. Mathematically, this means that the matrix needs to be *positive semidefinite*. This requirement can be seen as a generalization of the requirement for a scalar random variable to have non-negative variance. When the variance-covariance matrix of $\boldsymbol{X}$ is positive semidefinite, then any choice of a fixed vector $\boldsymbol{a}$ leads to a non-negative variance for $\boldsymbol{a}'\boldsymbol{X}$.

A *correlation matrix* can be used for interpretation of the variance-covariance matrix. The matrix is of the same size and structure as the variance-covariance matrix, but includes the pairwise correlations in the place of covariances. The diagonal elements of a correlation matrix are always 1. In `R`, the correlation matrix corresponding to a given covariance matrix can be computed using function `cov2cor`.

Example 2.17. Recall example 2.15. To find the covariance between X_1 and X_2. one needs to compute the expected value of the product of diameter and height

$$\text{E}(XY) = \int_{-\infty}^{\infty}\int_{-\infty}^{\infty} xyf(x,y)dydx\,.$$

The covariance can be computed as $\text{cov}(X,Y) = \text{E}(XY) - \text{E}(X)\,\text{E}(Y)$.

2.4 Calculation rules for expected value, variance and covariance

Let a, b, c, d be scalar constants and X, Y, Z and W scalar random variables. The variances and covariances for linear transformations can be found using the following calculation rules. The results follow directly from the algebraic rules for real numbers, which are shown on the left. The corresponding rules for the expected value were given in Equations (2.11), (2.12) and (2.13) on p. 15.

$$
\begin{aligned}
& \operatorname{var}(X+a)=\operatorname{var}(X) && (2.26)\\
(ax)^2=a^2x^2 \Rightarrow\ & \operatorname{var}(aX)=a^2\operatorname{var}(X) && (2.27)\\
(x+y)^2=x^2+y^2+2xy \Rightarrow\ & \operatorname{var}(X+Y)=\operatorname{var}(X)+\operatorname{var}(Y)+2\operatorname{cov}(X,Y) && (2.28)\\
(x-y)^2=x^2+y^2-2xy \Rightarrow\ & \operatorname{var}(X-Y)=\operatorname{var}(X)+\operatorname{var}(Y)-2\operatorname{cov}(X,Y) && (2.29)\\
& \operatorname{cov}(X+a,Y+b)=\operatorname{cov}(X,Y) && (2.30)\\
xy=yx \Rightarrow\ & \operatorname{cov}(X,Y)=\operatorname{cov}(Y,X) && (2.31)\\
x(y+z)=xy+xz \Rightarrow\ & \operatorname{cov}(X,Y+Z)=\operatorname{cov}(X,Y)+\operatorname{cov}(X,Z) && (2.32)\\
(ax)(by)=abxy \Rightarrow\ & \operatorname{cov}(aX,bY)=ab\operatorname{cov}(X,Y)\,. && (2.33)
\end{aligned}
$$

Combining (2.28) and (2.29) with (2.27) gives

$$\operatorname{var}(aX+bY)=a^2\operatorname{var}(X)+b^2\operatorname{var}(Y)+2ab\operatorname{cov}(X,Y) \tag{2.34}$$

$$\operatorname{var}(aX-bY)=a^2\operatorname{var}(X)+b^2\operatorname{var}(Y)-2ab\operatorname{cov}(X,Y)\,, \tag{2.35}$$

and (2.32) with (2.33) gives

$$\operatorname{cov}(aX+bY,cZ+dW)=ac\operatorname{cov}(X,Z)+ad\operatorname{cov}(X,W)+bc\operatorname{cov}(Y,Z)+bd\operatorname{cov}(Y,W)\,. \tag{2.36}$$

Let $\boldsymbol{a}_p$ and $\boldsymbol{b}_q$ be fixed vectors, and $\boldsymbol{X}_p$ and $\boldsymbol{Y}_q$ be random vectors of lengths p and q. Equations (2.12), (2.27) and (2.33) generalize to random vectors as follows:

$$\mathrm{E}(\boldsymbol{a}'\boldsymbol{X})=\boldsymbol{a}'\,\mathrm{E}(\boldsymbol{X}) \tag{2.37}$$

$$\operatorname{var}(\boldsymbol{a}'\boldsymbol{X})=\boldsymbol{a}'\operatorname{var}(\boldsymbol{X})\boldsymbol{a} \tag{2.38}$$

$$\operatorname{cov}(\boldsymbol{a}'\boldsymbol{X},\boldsymbol{b}'\boldsymbol{Y})=\boldsymbol{a}'\operatorname{cov}(\boldsymbol{X},\boldsymbol{Y}')\boldsymbol{b}\,. \tag{2.39}$$

The rules further generalize to products of random vectors and fixed matrices. Let $\boldsymbol{A}$ and $\boldsymbol{B}$ be fixed matrices with p and q columns, respectively. Now

$$\mathrm{E}(\boldsymbol{AX})=\boldsymbol{A}\,\mathrm{E}(\boldsymbol{X}) \tag{2.40}$$

$$\operatorname{var}(\boldsymbol{AX})=\boldsymbol{A}\operatorname{var}(\boldsymbol{X})\boldsymbol{A}' \tag{2.41}$$

$$\operatorname{cov}(\boldsymbol{AX},(\boldsymbol{BY})')=\boldsymbol{A}\operatorname{cov}(\boldsymbol{X},\boldsymbol{Y}')\boldsymbol{B}'\,. \tag{2.42}$$

Consider random vector $\boldsymbol{X}=(X_1,X_2,\ldots,X_p)'$. Equations (2.13) and (2.28) generalize to

$$\mathrm{E}\left(\sum_{i=1}^{p}X_i\right)=\sum_{i=1}^{p}\mathrm{E}\,X_i \tag{2.43}$$

$$\operatorname{var}\left(\sum_{i=1}^{p}X_i\right)=\sum_{i=1}^{p}\sum_{j=1}^{p}\operatorname{cov}(X_i,X_j)\,; \tag{2.44}$$

recall that $\operatorname{cov}(X_i,X_i)=\operatorname{var}(X_i)$.

2.5 Conditional and marginal expected value and variance

The *conditional expected value* $\mathrm{E}(\boldsymbol{X}_1|\boldsymbol{X}_2=\boldsymbol{x}_2)$ and *conditional variance* $\operatorname{var}(\boldsymbol{X}_1|\boldsymbol{X}_2=\boldsymbol{x}_2)$ are the expected value and variance of $\boldsymbol{X}_1$ over the conditional distribution $f(\boldsymbol{x}_1|\boldsymbol{x}_2)$. They

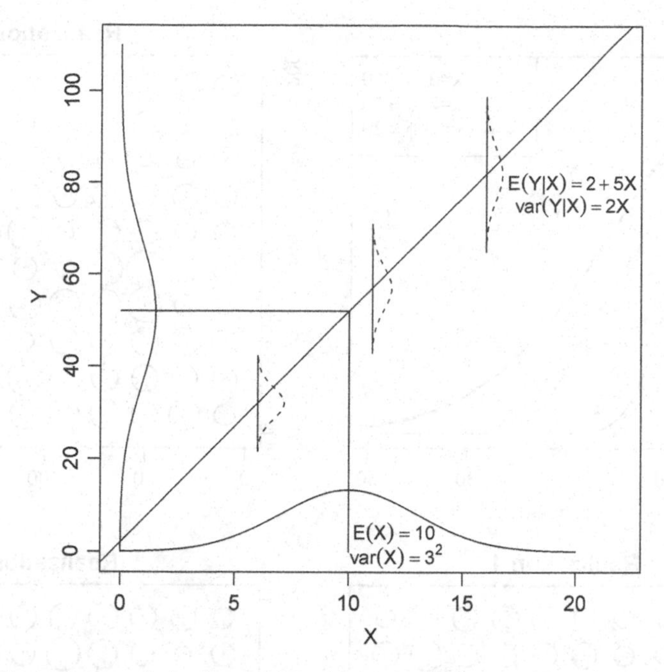

FIGURE 2.4 Illustration of the laws of total expectation and total variance. For the purpose of illustration, normality is unnecessarily assumed for X and $Y|X$. Rescaled marginal densities of X and Y are illustrated by solid lines, and rescaled conditional densities of $Y|X = 6$, $Y|X = 11$ and $Y|X = 16$ by dashed lines.

are functions of $\boldsymbol{x}_2$. The *marginal expected value* and *marginal variance* are the expected values over the marginal distribution, $\mathrm{E}(\boldsymbol{X})$ and $\mathrm{var}(\boldsymbol{X})$.

The expected value of a marginal distribution can always be expressed by conditioning on some random variable and using the following *law of total expectation*:

$$\mathrm{E}(Y) = \mathrm{E}_X\left[\mathrm{E}(Y|X)\right] \tag{2.45}$$

i.e. take the initial expected value of the conditional distribution of $Y|X$ which is a function of X. Thereafter, take the (marginal) expected value of it over the distribution of X. The corresponding *law for total variance* comprises the mean of conditional variances and variance of conditional means:

$$\mathrm{var}(Y) = \mathrm{E}_X\left[\mathrm{var}(Y|X)\right] + \mathrm{var}_X\left[\mathrm{E}(Y|X)\right]\,. \tag{2.46}$$

Example 2.18. Consider random variables X and Y with the following properties $\mathrm{E}(X) = 10$, $\mathrm{var}(X) = 3^2$, $\mathrm{E}(Y|X) = 2 + 5X$ and $\mathrm{var}(Y|X) = 2X$. The (marginal) expected value of Y is

$$\mathrm{E}(Y) = \mathrm{E}(\mathrm{E}(Y|X)) = \mathrm{E}(2 + 5X) = 2 + 5(\mathrm{E}(X)) = 52$$

and the (marginal) variance of Y is

$$\begin{aligned}
\mathrm{var}(Y) &= \mathrm{E}(\mathrm{var}(Y|X)) + \mathrm{var}(\mathrm{E}(Y|X)) \\
&= \mathrm{E}(2X) + \mathrm{var}(2 + 5X) \\
&= 2\,\mathrm{E}(X) + 5^2\,\mathrm{var}(X) \\
&= 20 + 5^2 \times 3^2 = 245 = 15.65^2\,.
\end{aligned}$$

Figure 2.4 illustrates these random variables, by using the unnecessary assumption of normality for the marginal and conditional distributions.

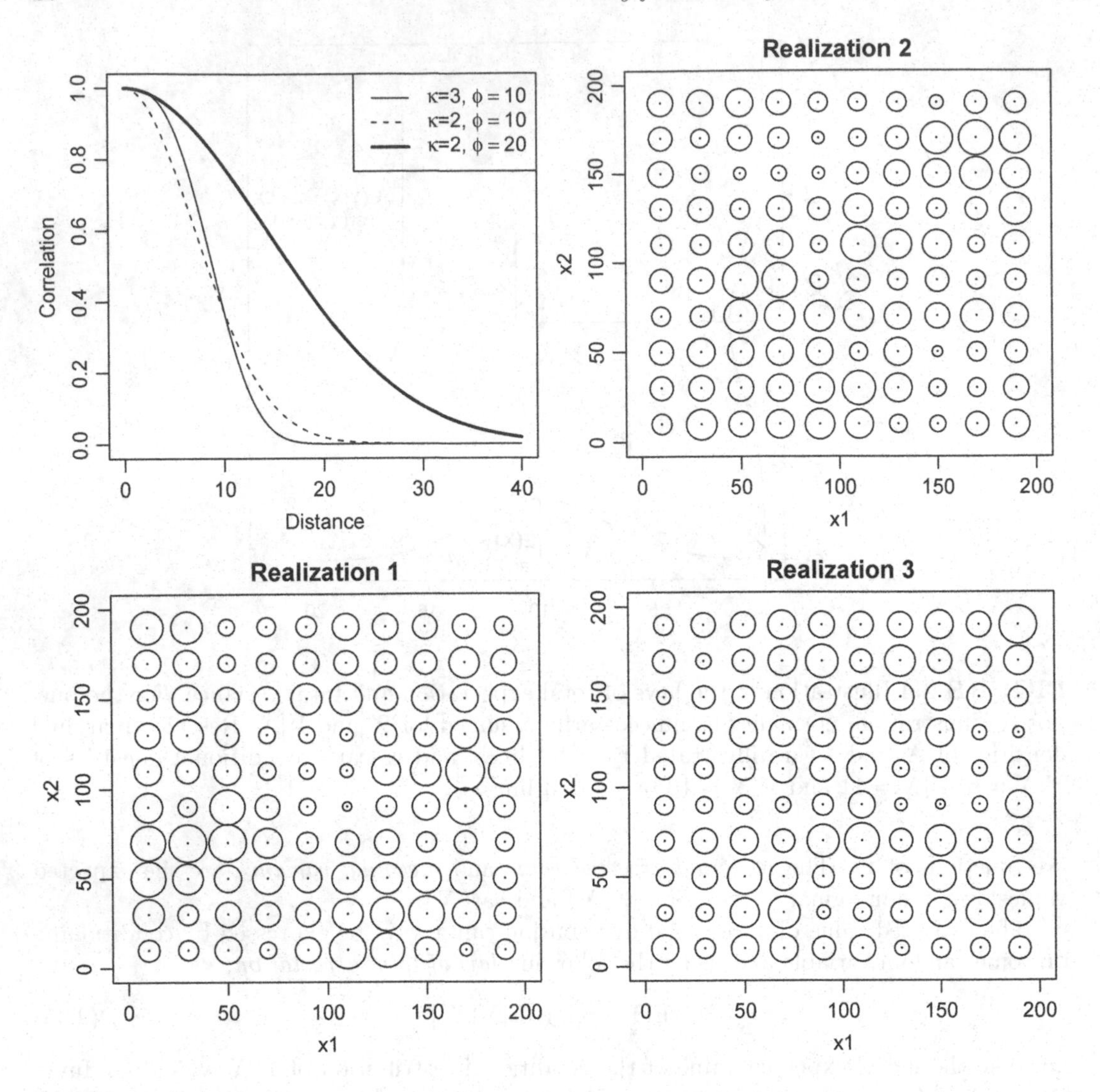

FIGURE 2.5 The powered exponential correlation function using three combinations of κ and ϕ (top left) and three realizations of a two-dimensional Gaussian random field when $\kappa = 2$ and $\phi = 20$.

2.6 Multivariate normal distribution

A widely applicable continuous multivariate distribution is the *multivariate normal distribution*, also known as *multivariate Gaussian distribution*. Let $\boldsymbol{X}$ be a p-length random vector with the expected value $\mathrm{E}(\boldsymbol{X}) = \boldsymbol{\mu}$ and the variance-covariance matrix $\mathrm{var}(\boldsymbol{X}) = \boldsymbol{\Sigma}$. $\boldsymbol{X}$ is said to follow the multivariate normal distribution if the joint density is

$$f(\boldsymbol{x}|\boldsymbol{\mu}, \boldsymbol{\Sigma}) = \frac{1}{\sqrt{(2\pi)^n}} |\boldsymbol{\Sigma}|^{-1/2} e^{-\frac{1}{2}(\boldsymbol{x}-\boldsymbol{\mu})'\boldsymbol{\Sigma}^{-1}(\boldsymbol{x}-\boldsymbol{\mu})} . \tag{2.47}$$

Notation $|\boldsymbol{\Sigma}|$ denotes a scalar called *the determinant* of matrix $\boldsymbol{\Sigma}$ (see the Internet for a definition). The two parameters of the distribution are the mean vector $\boldsymbol{\mu}$ and the positive definite variance-covariance matrix $\boldsymbol{\Sigma}$. The formula yields the univariate normal density (2.5), when $p = 1$. The (multivariate) normality of $\boldsymbol{X}$ is denoted by $\boldsymbol{X} \sim N(\boldsymbol{\mu}, \boldsymbol{\Sigma})$.

In the unrestricted case, a unique expected value and variance is defined for each component variable and a unique covariance for each pair of components. This gives a total of $2p + \frac{1}{2}(p)(p-1)$ unique scalar parameters, which is too much for many applications. Implementing meaningful restrictions to $\boldsymbol{\mu}$ and $\boldsymbol{\Sigma}$ leads to a more parsimonious model: a model where $\boldsymbol{\mu}$ and $\boldsymbol{\Sigma}$ are parameterized by few well-justified parameters. For example, all components of $\boldsymbol{\mu}$ and all variances in the diagonal of $\boldsymbol{\Sigma}$ may be assumed to be equal for all i, or they may be assumed to have some systematic dependence on some known properties of the unit of observation. In addition, the covariances within $\boldsymbol{\Sigma}$ can be parameterized in terms of the time lag between repeated measurements of a longitudinal data set, in terms of the spatial distances between the locations of measurements in a spatial data set, or according to group membership in a grouped data set, to give a few common examples. These parameterizations lead to a wide class of models that includes, e.g. the linear and nonlinear regression models, linear and nonlinear mixed-effects models, and Gaussian random fields. An example of a two-dimensional Gaussian random field is given below.

Example 2.19. Consider a multivariate normal random variable $\boldsymbol{Y} = (Y_1, Y_2, \ldots, Y_p)'$, defined at locations $\boldsymbol{x}_1, \boldsymbol{x}_2, \ldots, \boldsymbol{x}_p$ in a two-dimensional space, where $\boldsymbol{x}_i$ includes the two coordinates of point i, $i = 1, \ldots, p$. $\boldsymbol{Y}$ might describe, for example, the topsoil depth or site index at points $\boldsymbol{x}_i$ within a region of interest. Consider locations $\boldsymbol{x}_i$ at the nodes of a square grid over a square (200 m $\times$ 200 m region); the distance between rows and columns is 20 m. The resulting 100 random variables Y_i can be arranged in a vector $\boldsymbol{Y}$. They can be arranged in any order; we have selected the starting point as the top-left corner of the area and include the points by rows.

We assume that all Y_i have the marginal normal distribution, and they are identically distributed with a common expected value μ and variance σ^2 for $i = 1, \ldots, 100$. However, the realized values y_i at adjacent locations tend to be similar to each other and the correlation depends on the euclidian distance $|\boldsymbol{x}_i - \boldsymbol{x}_j|$ between locations of Y_i and Y_j according to the powered exponential function:

$$\mathrm{cor}(Y_i, Y_j) = -\exp(-(|\boldsymbol{x}_i - \boldsymbol{x}_j|/\phi)^{\kappa})$$

where ϕ and κ are two parameters that describe the behavior of the correlation as a function of the distance (see Figure 2.5). The powered exponential is one such function that leads to a positive definite variance-covariance matrix for any set of locations $\boldsymbol{x}_i$. We have now been able to reparameterize the 5150 scalar parameters of a 100-variate normal distribution by using 4 scalar parameters: μ, σ^2, ϕ, and κ.

Assuming that $\mu = 6$, $\sigma^2 = 1.5^2$, $\kappa = 2$ and $\phi = 20$, the 100-variate random vector $\boldsymbol{Y}$ has the multivariate normal distribution with parameters

$$\boldsymbol{\mu} = \begin{pmatrix} 6 \\ 6 \\ \vdots \\ 6 \end{pmatrix}_{100\times 1} \quad \text{and} \quad \boldsymbol{\Sigma} = \begin{pmatrix} 2.250 & 0.828 & 0.041 & \ldots & 0.000 \\ 0.828 & 2.250 & 0.828 & \ldots & 0.000 \\ 0.041 & 0.828 & 2.250 & \ldots & 0.000 \\ \vdots & \vdots & \vdots & \ddots & \vdots \\ 0.000 & 0.000 & 0.000 & \ldots & 2.250 \end{pmatrix}_{100\times 100} .$$

Figure 2.5 illustrates three realizations of $\boldsymbol{Y}$ so that the radii of the circles at each x_i illustrate y_i. The y_i for adjacent locations are usually similar, but different realizations place the small and large values in different parts of the area. The smoothness of these changes as a function of distance is controlled by parameters κ and ϕ. Figure 2.5 was produced using the following R-script:

```
> # Generate a 10 by 10 lattice
> x1<-rep(seq(10, 190, 20),10)
> x2<-rep(seq(10, 190, 20),each=10)
```

```
> mu<-rep(6,100)
> var<-1.5^2

> # Illustrate the correlation functions
> d<-seq(0,40,0.1)
> corA<-exp(-(d/10)^3)
> corB<-exp(-(d/10)^2)
> corC<-exp(-(d/20)^2)

> plot(d,corA,type="l",xlab="Distance",ylab="Correlation")
> lines(d,corB,lty="dashed")
> lines(d,corC,lwd=2)
> legend("topright",legend=c(expression(kappa*"=3, " * phi==10),expression(kappa*"=2, " * phi==10),
+                         expression(kappa*"=2, " * phi==20)),
+        lty=c("solid","dashed","solid"),lwd=c(1,1,2))

> # Compute the distances between all pairs of points into a 100 by 100 matrix
> # and compute the variance-covariance matrix Sigma.
> dist<-matrix(NA,ncol=100,nrow=100)
> for (i in 1:100) {
+       dist[i,]<-sqrt((x1[i]-x1)^2+(x2[i]-x2)^2)
+       }
> Sigma<-var*exp(-(dist/20)^2)
>
> # Generate the first realization
> y<-mvrnorm(1,mu,Sigma)
> plot(x1,x2,cex=0.1,xlim=c(0,200),ylim=c(0,200),main="Realization 1")
> symbols(x1,x2,circles=y,add=T,inches=F)
```

Consider any partition $\boldsymbol{X} = \begin{pmatrix} \boldsymbol{X}_1 \\ \boldsymbol{X}_2 \end{pmatrix}$ of the multivariate normal $\boldsymbol{X}$, where the length of $\boldsymbol{X}_1$ and $\boldsymbol{X}_2$ are q and $p - q$. We also partition $\boldsymbol{\mu}$ and $\boldsymbol{\Sigma}$ accordingly:

$$\boldsymbol{\mu} = \begin{pmatrix} \boldsymbol{\mu}_1 \\ \boldsymbol{\mu}_2 \end{pmatrix} \quad \text{and} \quad \boldsymbol{\Sigma} = \begin{pmatrix} \boldsymbol{\Sigma}_1 & \boldsymbol{\Sigma}_{12} \\ \boldsymbol{\Sigma}_{21} & \boldsymbol{\Sigma}_2 \end{pmatrix}$$

so that $\mathrm{E}(\boldsymbol{X}_1) = \boldsymbol{\mu}_1$, $\mathrm{E}(\boldsymbol{X}_2) = \boldsymbol{\mu}_2$, $\mathrm{var}(\boldsymbol{X}_1) = \boldsymbol{\Sigma}_1$, $\mathrm{var}(\boldsymbol{X}_2) = \boldsymbol{\Sigma}_2$ and $\mathrm{cov}(\boldsymbol{X}_1, \boldsymbol{X}_2') = \boldsymbol{\Sigma}_{12} = \boldsymbol{\Sigma}_{21}'$.

For example, we might want to partition the 100-unit vector in Example 2.19 so that $\boldsymbol{X}_1$ and $\boldsymbol{\mu}_1$ include the first 10 elements of $\boldsymbol{X}$ and $\boldsymbol{\mu}$, and $\boldsymbol{X}_2$ and $\boldsymbol{\mu}_2$ the remaining elements $11, \ldots, 100$. Now $\boldsymbol{\Sigma}_1$ would include the 10×10 matrix from the first 10 rows and columns of $\boldsymbol{\Sigma}$. $\boldsymbol{\Sigma}_2$ would include the 90×90 matrix from rows and columns $11 - 100$ of $\boldsymbol{\Sigma}$. $\boldsymbol{\Sigma}_{12}$ would include the 10×90 matrix from the first 10 rows and the last 90 columns of $\boldsymbol{\Sigma}$. $\boldsymbol{\Sigma}_{21}$ would be the transpose of $\boldsymbol{\Sigma}_{12}$.

An important property of the multivariate normal distribution is that all conditional distributions also have a normal distribution. For the conditional random variable $\boldsymbol{X}_1 | \boldsymbol{X}_2 = \boldsymbol{x}_2$, the mean and variance of the q-variate normal distribution are

$$\mathrm{E}\left(\boldsymbol{X}_1 | \boldsymbol{X}_2 = \boldsymbol{x}_2\right) = \boldsymbol{\mu}_1 + \boldsymbol{\Sigma}_{12}\boldsymbol{\Sigma}_2^{-1}\left(\boldsymbol{x}_2 - \boldsymbol{\mu}_2\right) \tag{2.48}$$

$$\mathrm{var}\left(\boldsymbol{X}_1 | \boldsymbol{X}_2 = \boldsymbol{x}_2\right) = \boldsymbol{\Sigma}_1 - \boldsymbol{\Sigma}_{12}\boldsymbol{\Sigma}_2^{-1}\boldsymbol{\Sigma}_{21}\,. \tag{2.49}$$

The formula for the conditional expected value (2.48) is linear with respect to $\boldsymbol{x}_2$. An important result from the linearity is that the zero correlation between two components of a multivariate normal random variable implies independence of these components. Another result worth noting is that the conditional variance (2.49) does not depend on $\boldsymbol{x}_2$; therefore, the variance is equal for all realized values of $\boldsymbol{x}_2$. We will return to these equations in almost every chapter of this book.

All marginal distributions of a multivariate normal distribution are also normal. However, the multivariate distribution whose marginal distributions are normal is not necessarily normal.

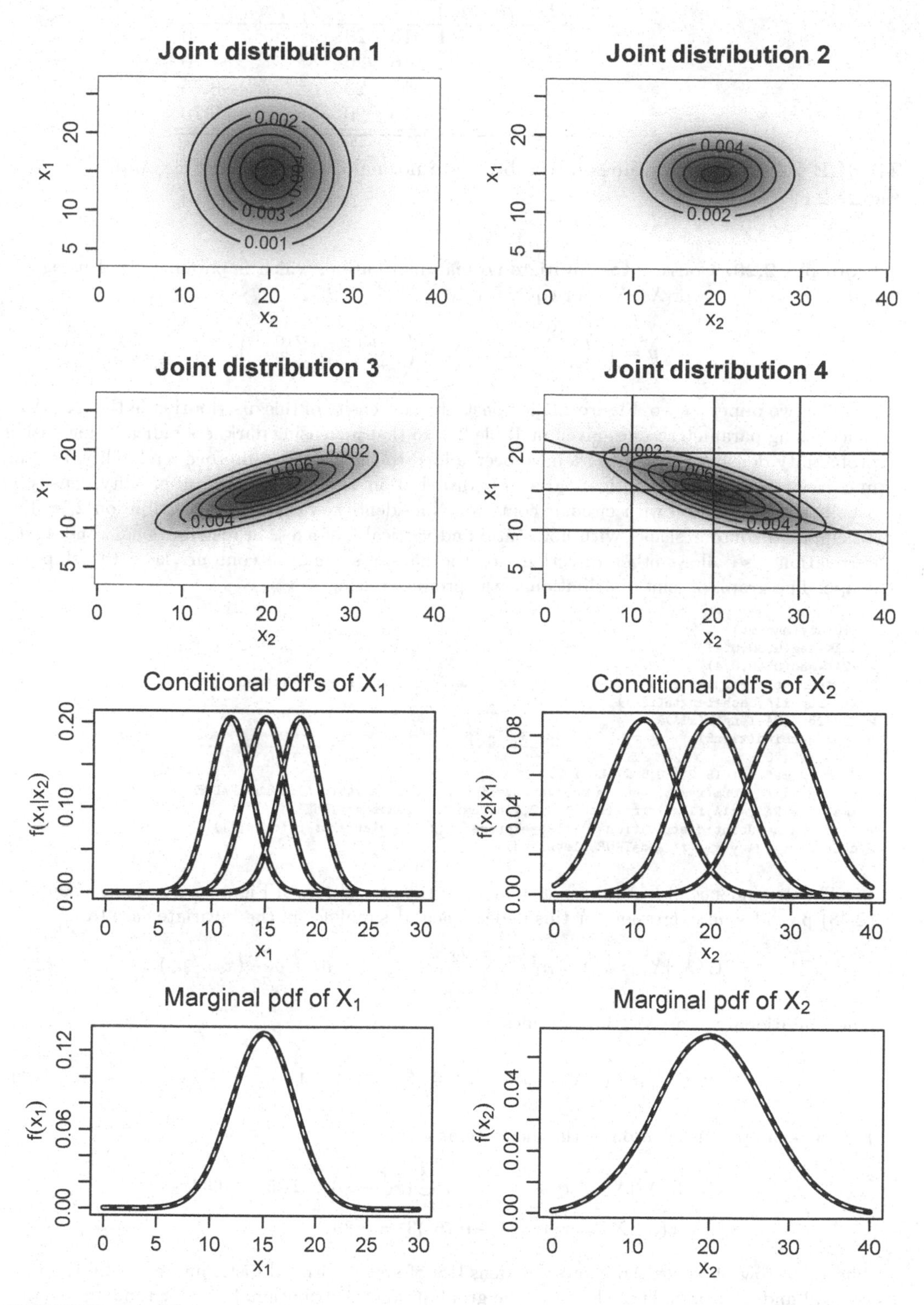

FIGURE 2.6 Illustration of four bivariate normal densities and the conditional and marginal densities for one of them in Example 2.20.

Distribution	μ_1	μ_2	σ_1^2	σ_2^2	ρ
1	15	20	5^2	5^2	0
2	15	20	3^2	7^2	0
3	15	20	3^2	7^2	0.76
4	15	20	3^2	7^2	-0.76

TABLE 2.2 Parameter values of the bivariate normal distributions of Example 2.20 and Figure 2.6.

Example 2.20. Consider the bivariate normal distribution, which is parameterized using μ_1, μ_2, σ_1, σ_2 and $\rho = \text{cor}(X_1, X_2)$, or equivalently

$$\boldsymbol{\mu} = \begin{pmatrix} \mu_1 \\ \mu_2 \end{pmatrix} \quad \text{and} \quad \boldsymbol{\Sigma} = \begin{pmatrix} \sigma_1^2 & \rho\sigma_1\sigma_2 \\ \rho\sigma_1\sigma_2 & \sigma_2^2 \end{pmatrix}.$$

The two upper rows of Figure 2.6 demonstrate the density of this distribution in the (X_1, X_2)-space, using parameter values given in Table 2.2, so that increasing darkness indicates increasing probability density. Five contours have been added to the graph to improve readability. For the uncorrelated pair with identical variances (distribution 1), the bivariate probability density is a bell-shaped function with circular contours. Non-identical variances in distribution 2 lead to an elliptical contour shape, with horizontal and vertical half-axes. In distributions 3 and 4, the correlation also allows other directions for the half-axes, but the contours keep the elliptical shape. The graph of joint distribution 1 was produced using:

```
> library(mnormt)
> x1A<-seq(0,30,0.4)
> x2A<-seq(0,40,0.4)
> # Generate a regular grid of points at which the density will be evaluated
> x1<-rep(x1A,each=length(x2A))
> x2<-rep(x2A,length(x1A))
> xymat<-cbind(x1,x2)
> mu<-c(15,20)
> Sigma1<-matrix(c(5^2,0,0,5^2),ncol=2)
> f1<-matrix(dmnorm(xymat, mean = mu, varcov=Sigma1),ncol=length(x1A),byrow=FALSE)
> image(x=x2A,y=x1A,f1,col=gray(seq(1,0.2,length=100),useRaster=TRUE),
+       main="Joint distribution 1",xlab=expression(x[2]),ylab=expression(x[1]))
> contour(x=x2A,y=x1A,f1,add=TRUE,nlevels=5)
```

The dependence of $\text{E}(X_1|X_2 = x_2)$ on x_2 in the second row of Figure 2.6 is linear. Equation (2.48) provides an expression for this dependence. It simplifies in the bivariate case to:

$$\text{E}\left(X_1|X_2 = x_2\right) = \mu_1 + \frac{\rho\sigma_1\sigma_2}{\sigma_2^2}\left(x_2 - \mu_2\right) = \mu_1 + \rho\frac{\sigma_1}{\sigma_2}\left(x_2 - \mu_2\right). \tag{2.50}$$

The conditional variance (2.49) becomes:

$$\text{var}\left(X_1|X_2 = x_2\right) = \sigma_1^2 - \frac{(\rho\sigma_1\sigma_2)^2}{\sigma_2^2} = (1 - \rho^2)\sigma_1^2. \tag{2.51}$$

For the example distribution 4, these equations give

$$\text{E}\left(X_1|X_2 = x_2\right) = 15 - 0.76\frac{3}{7}\left(x_2 - 20\right) = 21.51 - 0.33x_2$$
$$\text{var}\left(X_1|X_2 = x_2\right) = (1 - 0.76^2)3^2 = 3.80.$$

The third row of Figure 2.6 shows the densities of six conditional distributions along the three vertical and horizontal lines shown in the graph of joint distribution 4, i.e. for random variables $X_1|X_2 = 10$, $X_1|X_2 = 20$, $X_1|X_2 = 30$ (left), and for $X_2|X_1 = 10$, $X_2|X_1 = 15$, and $X_2|X_1 = 20$ (right). The thick black lines show the *pdf*s computed using the general formula (2.16) and the following R-code.

```
> x1<-seq(0,30,0.1)
> mu<-c(15,20)
> Sigma<-matrix(c(3^2,-4^2,-4^2,7^2),ncol=2)

> # Using the general formula for conditional density
> plot(x1,dmnorm(cbind(x1,20),mu,Sigma)/dnorm(20,mean=mu[2],sd=sqrt(Sigma[2,2])),
+       xlab=expression(x[1]),ylab=expression(f*(x[1]*"|"*x[2])),type="l",
+       main=expression("Conditional pdf's of "*X[1]),lwd=3)
> lines(x1,dmnorm(cbind(x1,10),mu,Sigma)/dnorm(10,mean=mu[2],sd=sqrt(Sigma[2,2])),lwd=3)
> lines(x1,dmnorm(cbind(x1,30),mu,Sigma)/dnorm(30,mean=mu[2],sd=sqrt(Sigma[2,2])),lwd=3)
```

The white dashed lines on the thicker line show the univariate normal density with mean and variance based on (2.50) and (2.51). One of them was produced using

```
> lines(x1,dnorm(x1,mean=21.51-0.33*30,sd=sqrt(3.80)),lty="dashed",lwd=1,col="white")
```

Finally, the lower row in Figure 2.6 shows the marginal distributions of X_1 (left) and X_2 (right). The thick black lines were based on (2.17) and the white dashed lines are the univariate densities of X_1 and X_2. The left graph was produced using

```
> f3fun<-function(x1,x2){
+       x<-cbind(x1,x2)
+       dmnorm(x,mean=mu,varcov=Sigma)
+       }

> # Integrate x2 numerically out from the joint density
> margx1<-function(x1) integrate(function(x2) f3fun(x1,x2),0,100)$value
> marg1<-sapply(x1, margx1)
> plot(x1,marg1,type="l",main=expression("Marginal pdf of "*X[1]),
+       xlab=expression(x[1]),ylab=expression(f(x[1])),lwd=3)

> # Use the knowledge of marginal normality
> lines(x1,dnorm(x1,mean=mu[1],sd=sqrt(Sigma[1,1])),lty="dashed",lwd=1, col="white")
```

2.7 Quantile function

The pth *quantile* of a random variable X, denoted by ξ_p, is the value of X such that $P(X \leq \xi_p) = p$. The condition is equivalent to $F(\xi_p) = p$, and solving it for p yields the *quantile function*, which is inverse of the *cdf*:

$$Q(p) = F^{-1}(p) .$$

Some important special cases are the median, $Q(0.5)$ (i.e. $\xi_{0.5}$), and the lower and upper quartiles of the distribution; $Q(0.25)$ and $Q(0.75)$. Quantiles are commonly needed in statistical inference to determine prediction intervals, credible intervals, confidence intervals and p-values; see Section 3.6.

Example 2.21. A $100(1-p)\%$ *prediction interval* of random variable X is an interval that includes the value of X with the probability $1-p$. One possible prediction interval is the interval $[Q(p/2), Q(1-p/2)]$, where we have the probability $p/2$ that X is below the lower bound of the interval, and the probability $p/2$ that it is above the upper bound. Another $100(1-p)\%$ prediction interval is the interval $[Q(0), Q(1-p)]$, where the probability of a value below the lower bound is 0 and the probability of a value above the upper bound is p. Infinitely many other alternatives are naturally possible. Usually the prediction intervals are constructed so that both tails have equal probability.

For example, the 80% prediction interval with equal tail probabilities for X with $X \sim N(0,1)$ is $[-1.28, 1.28]$. This interval is symmetric around μ due to the symmetry of the normal *pdf*. An 80% one-sided prediction interval is $[-\infty, 0.84]$. Corresponding intervals with 95% and 90% coverage probabilities are $[-1.96, 1.96]$ and $[-\infty, 1.64]$.

```
> qnorm(c(0.1,0.9),0,1)
[1] -1.281552  1.281552

> qnorm(c(0,0.8),0,1)
[1]      -Inf 0.8416212

> qnorm(c(0.025,0.975),0,1)
[1] -1.959964  1.959964

> qnorm(c(0,0.95),0,1)
[1]      -Inf 1.644854
```

Example 2.22. The $F(df1, df2)$-distribution is widely used in testing in the contexts of regression and analysis of variance. The parameters of the *F-distribution* are denoted by $df1$ and $df2$ because they can be interpreted as the degrees of freedom in models being compared (see Section 2.11.3). Consider a situation where the parameters are $df1 = 2$ and $df2 = 100$, so that $Y \sim F(2, 100)$. We want to find y such that $P(Y \leq y) = 0.95$.

```
> qf(0.95,2,100)
[1] 3.087296
```

The F-distribution is commonly used in statistical hypothesis testing, see Section 3.6. In that context, we have defined a test statistic F_{obs}, a random variable that is known to have the F-distribution when our null hypothesis is true. The test statistic is a random variable, and the collected data set provides one realized value of it. If that observed statistic appears to be very far in the right tail of the F-distribution, we take this as evidence against the null hypothesis.

One important use of the quantile function is in generating a sample from a specified distribution. Let us define the *pdf* for the continuous $Uniform(a, b)$ distribution, which has the range $[a, b]$, and all values within that range have equal probability. The *pdf* is

$$f(u)) = \begin{cases} \frac{1}{b-a} & \text{if} \quad a \leq u \leq b \\ 0 & \text{otherwise}. \end{cases}$$

If one has a tool to generate random realizations of *the standard uniform random variable* $U \sim Uniform(0, 1)$, then a random sample from a distribution with quantile function $Q(p)$ can be generated by using the realized values of U in $Q(u)$, as illustrated in Example 2.23.

Quantiles can be used to define a distribution of a random variable, or an approximation of it. The approximation improves as the number of percentiles increases. In the context of tree size modeling, such distribution is called *percentile-based distribution* (see Chapter 11). The percentiles are another name for quantiles: the pth quantile is the $100p^{th}$ percentile. We may specify some fixed values for certain quantiles of the distribution and interpolate the *cdf* between these points using a continuous, increasing continuous function, such as a straight line.

Example 2.23. Let us specify that the 0th, 25th, 50th, 75th and 100th tree diameter percentiles in a forest stand are 5, 10, 13, 17 and 24 cm, respectively. Therefore, the *cdf* satisfies $F(5) = 0$, $F(10) = 0.25$, $F(13) = 0.5$, $F(17) = 0.75$, $F(24) = 1$. Furthermore assume that the *cdf* for intermediate values is linear between these points. We get

$$F(x) = \begin{cases} 0 & \text{if } x < 5 \\ -0.25 + 0.050x & \text{if } 5 \leq x < 10 \\ -0.58 + 0.083x & \text{if } 10 \leq x < 13 \\ -0.31 + 0.063x & \text{if } 13 \leq x < 17 \\ 0.14 + 0.036x & \text{if } 17 \leq x < 24 \\ 1 & \text{if } x \geq 24\,. \end{cases}$$

The upper left-hand graph in Figure 2.7 illustrates that the function fulfills the three conditions of a proper distribution function. The corresponding *pdf* (see Equation (11.3), p. 345

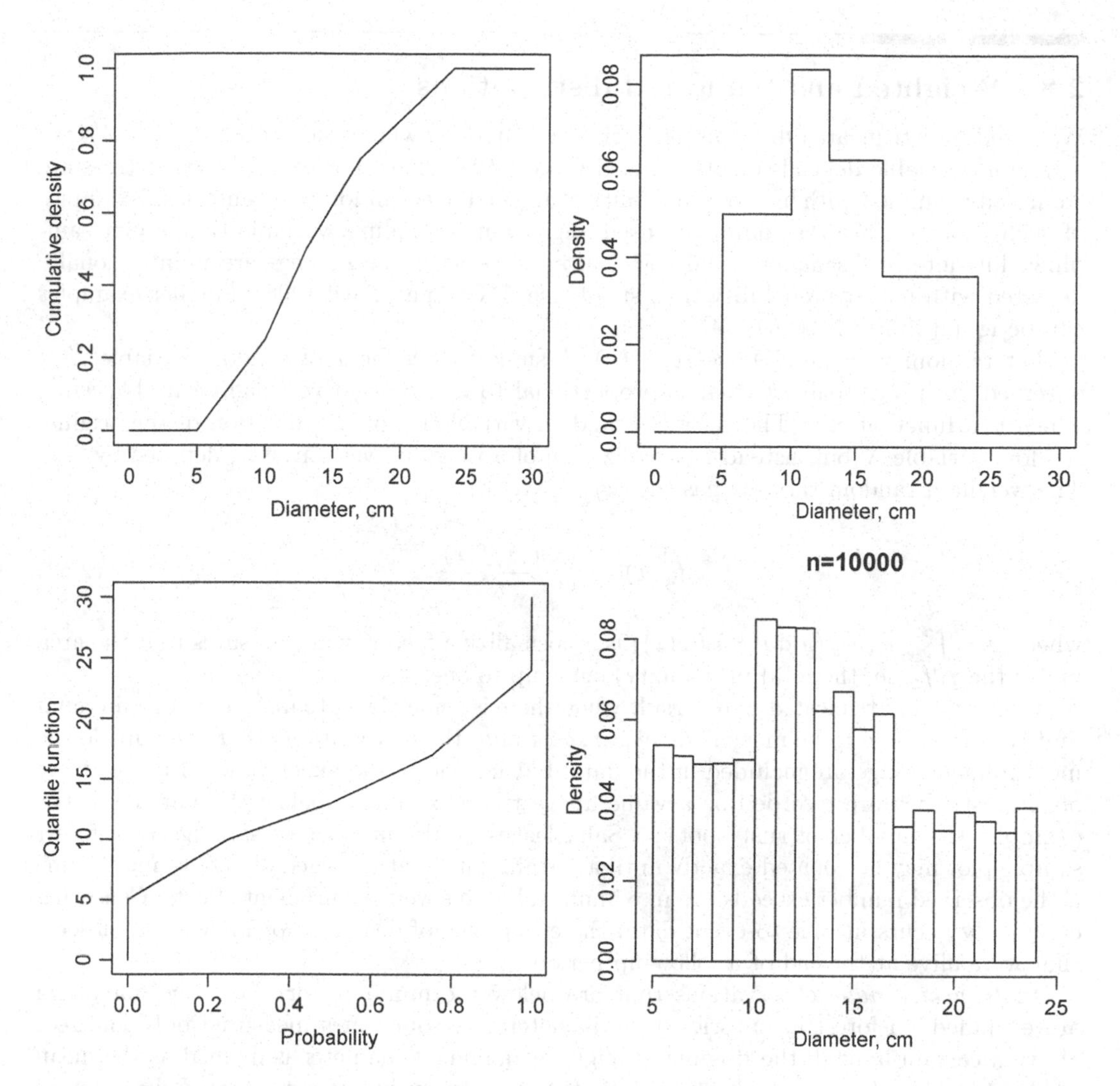

FIGURE 2.7 Illustration of the *cdf* (upper left), *pdf* (upper right), quantile function (bottom left) and a generated sample from the percentile-based distribution in Example 2.23.

for mathematical presentation) is shown in the upper right-hand graph. The quantile function becomes

$$Q(u) = \begin{cases} 5 + 20u & \text{if } 5 \leq u < 10 \\ 7 + 12u & \text{if } 10 \leq u < 13 \\ 5 + 16u & \text{if } 13 \leq u < 17 \\ -4 + 28u & \text{if } 17 \leq u < 24\,. \end{cases}$$

The graph of this function is shown in the lower left-hand corner of Figure 2.7. The lower right-hand graph illustrates a sample of 10000 random numbers from the percentile-based distribution. It was generated by first generating 10000 realizations from $Uniform(0, 1)$ variables and then transforming them using the percentile-based quantile function.

```
> d<-c(5,10,13,17,24)
> p<-c(0,0.25,0.5,0.75,1)
> yvec<-approx(x=p,y=d,xout=runif(10000,0,1))$y
```

2.8 Weighted and truncated distributions

Weighted and truncated distributions arise in situations where the probability to observe a random variable depends on its realized value (Patil 2006). For example, large trees are commonly sampled with higher probability than small trees in forest inventories. Examples of such protocols are the commonly used angle-count sampling, variable radius plot sampling, line-intersect sampling, and aerial inventories where large trees are unintentionally detected with higher probability than small trees (Gove and Patil 1998). Further examples can be found in Patil (2006).

Let random variable X have *pdf* $f(x)$. Assume that value x of random variable X is observed with a probability that is proportional to a non-negative weight variable $w(x)$, which is a function of x. The recorded random variable is not a realization of the original random variable X but instead is a realization of a weighted version of it, denoted by X^w. The weighted random variable has the *pdf*

$$f_w(x) = \frac{w(x)f(x)}{\omega}, \tag{2.52}$$

where $\omega = \int_{-\infty}^{\infty} w(u)f(u)du = \mathrm{E}(w(x))$ is a normalizing factor, which ensures that the area under the *pdf* (i.e. the total probability) sums up to one.

Censored and truncated data result when there is some absolute maximum or minimum limit for a variable to be measured. In *censored data* the observations that are outside the measurement range are included in the data, but instead of the exact value of the variable of interest, they are recorded as a value above the maximum / below the minimum. For example, the number of rust spots in Salix leaves or the number of saplings in a forest sample plot may be counted exactly until a certain limit, but recorded as "above maximum" if the observed number exceeds the maximum value to save measurement efforts. Censoring commonly occurs in time-to-event data: the exact time of death is not known for subjects that were alive at the end of a follow-up period.

In *truncated data*, observations that are below the minimum or above the maximum are excluded. In forest inventories, tree diameters are sometimes measured only for trees above a certain limit. If the dominant height or dominant diameter is defined as the mean of the 100 largest trees per hectare, it is the expected value of a truncated diameter or height distribution (see Example 11.14, p. 351). A truncated random variable results as a special case of the weighted random variable, when zero weight is used for some values and a constant positive weight for the others. Examples of weighted, truncated and censored distributions in the context of tree size modeling are given in Sections 11.4.2 and 11.4.3.

If the weighting is done using weight $v(z)$ based on another variable Z, which is related to X, then the weighted random variable X has the pdf (see Patil and Rao (1978) or Lappi and Bailey (1987)):

$$f_w(x) = \frac{\int_{-\infty}^{\infty} v(z)\, p(z,x)\, dz}{\omega} \tag{2.53}$$

where p is the joint density of Z and X and $\omega = \mathrm{E}(v(z))$ is the normalizing factor.

2.9 Compound distributions

A useful type of multivariate distribution results from a model where the data are assumed to be independent samples from a univariate distribution, but the parameters of the distribution are also random variables. Therefore, for each unit of the population, there is a random vector $(\Theta, X)'$ where Θ is the random parameter and X the variable of primary interest. However, Θ is a latent variable and only X can be observed. Therefore, the observations are generated by the marginal distribution of X. The support of Θ should naturally include only values that are within the range of the parameters of the distribution of X.

Nice mathematical properties for the marginal distribution of X are obtained by using the *conjugate* distributions for the parameters. The compound distributions are useful in modeling overdispersion in generalized linear models, and in the specification of an appropriate random effect-distribution for generalized linear mixed-effects models. The same distributions are used also in Bayesian statistics, where the interest lies in the prior and posterior distribution of the parameters. If conjugate priors are used, the prior and posterior distributions are of the same mathematical form and differ only in the values of the parameters.

The following example presents an example of a Beta-Binomial distribution, where the parameter p of the Binomial distribution has the Beta distribution, which is the conjugate distribution for the success probability (Skellam 1948). Another example of a compound distribution is the negative binomial distribution (see Example 8.7, p. 267).

Example 2.24. Consider a Bernoulli trial, where the probability of success P, is a random variable with support $0 \leq P \leq 1$. Let P have the two-parameter $Beta(\alpha, \beta)$ distribution, with shape parameters α and β and density $f_P(p) = \frac{1}{B(\alpha,\beta)} p^{\alpha-1}(1-p)^{\beta-1}$. The normalizing constant $B(\alpha, \beta) = \int_0^1 x^{\alpha-1}(1-x)^{\beta-1} dx$ is called *the beta function*. The beta function cannot be expressed in closed form and needs to be evaluated numerically, e.g. by using R-function `beta`.

Conditional on $P = p$, the number of successes, X in n trials has the $Binomial(n, p)$ distribution with *pmf* $P(X = x|P = p) = \binom{n}{x} p^x (1-p)^{n-x}$. The marginal probability for $X = x$ is the expected value of $P(X = x|P = p)$ over the distribution of P.

$$\begin{aligned}
P(X = x) &= E_P(P(X = x|P = p)) = \int_0^1 f_P(u) P(X = x|P = u) \\
&= \binom{n}{x} \frac{1}{B(\alpha, \beta)} \int_0^1 u^{\alpha-1}(1-u)^{\beta-1} u^x (1-u)^{n-x} du \\
&= \binom{n}{x} \frac{1}{B(\alpha, \beta)} \int_0^1 u^{\alpha+x-1}(1-u)^{\beta+n-x-1} du \\
&= \binom{n}{x} \frac{B(\alpha + x, \beta + n - x)}{B(\alpha, \beta)}.
\end{aligned}$$

This marginal *pmf* is called the *Beta-Binomial*(α, β) *pmf* with parameters α and β.

```
> n<-30
> x<-seq(0,30,1)
> plot(x+0.1,db<-dbinom(x,n,mup),xlab="x",ylab="density",type="h",ylim=c(0,0.20))
> points(x-0.1,dbb<-choose(n,x)*beta(alpha+x,beta+n-x)/beta(alpha,beta),
> col=gray(0.5),type="h")
```

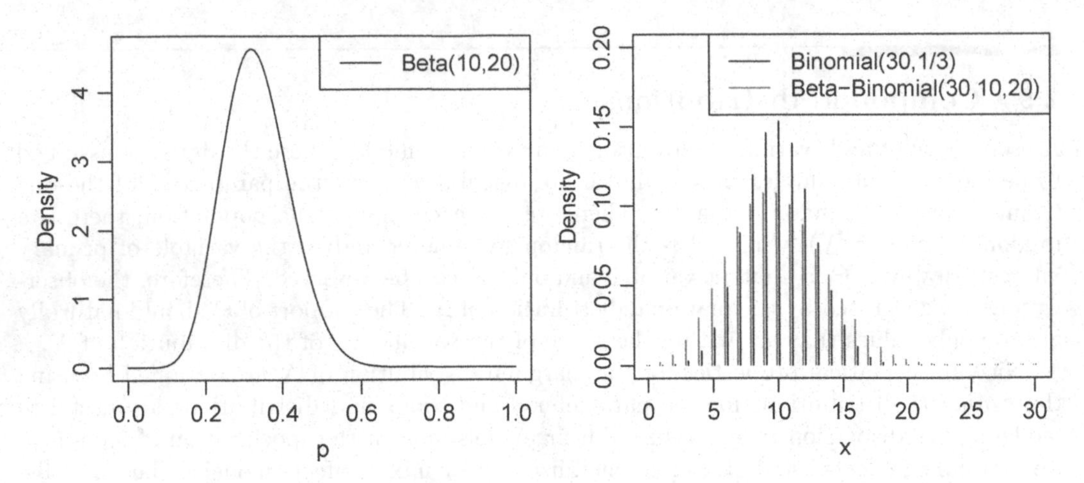

FIGURE 2.8 Illustration of Beta, Binomial and Beta-Binomial densities with parameters $\alpha = 10$, $\beta = 20$, $n = 30$, and $p = 1/3$.

2.10 Sums of independent random variables

Distribution of many-to-one transformations of a multivariate random variable can easily become complicated, but such distributions are commonly needed in reality. Let us consider one of the simpler cases: the sum of two independent random variables. Such a situation arises, e.g. when an additive measurement error is associated with a measurement of a random variable.

Let X_i, $i = 1, \ldots, n$ be random variables with expected values μ_i and variances σ_i^2. Consider the random variable $Y = \sum_{i=1}^{n} X_i$. The expected value and variance of a sum are not affected by the shape of the distributions, and Eq. (2.43) gives $\mathrm{E}(Y) = \sum_i \mu_i$. The variance is $\mathrm{var}\,(Y) = \sum_{i=1}^{p} \sum_{j=1}^{p} \mathrm{cov}\,(X_i, X_j)$ (Eq. (2.44), p. 24). If the random variables X_i are uncorrelated, this gives $\mathrm{var}(Y) = \sum_i \sigma_i^2$.

Sometimes the expected values and variance are not sufficient, and the shape of the distribution is of interest. Let X and Y be two random variables and consider the random variable $Z = X + Y$. Let us first fix Y to value y. The *conditional cdf* of Z is

$$P(Z \leq z | Y = y) = P(X + y \leq z) = P(X \leq z - y)\,.$$

The *marginal cdf* is the expected value of $P(Z \leq z|Y = y)$ over the distribution of Y, which gives $F(z) = P(Z \leq z) = \mathrm{E}_Y(P(Z \leq z|Y = y))$. If Y is continuous, we get

$$F(z) = \int_{y \in \mathcal{Y}} f_Y(y) P(X \leq z - y) dy\,,$$

where f_Y is the *pdf* of Y, $\mathcal{Y}$ is the range of Y, and $P(X \leq z - y) = F_X(z - y)$ is the *cdf* of X evaluated at $z - y$. Differentiating with respect to z yields the *marginal pdf* of Z as

$$\begin{aligned} f(z) &= F'(z) \\ &= \int_{y \in \mathcal{Y}} f_Y(y) \frac{\partial}{\partial z} P(X \leq z - y) dy \\ &= \int_{y \in \mathcal{Y}} f_Y(y) f_X(z - y) dy\,, \end{aligned}$$

where the second-row results from that the order of integration and differentiation can be changed in this case. This formula is commonly called the *convolution* of two random variables.

If both X and Y are discrete, we can do similar conditioning to the *pmf* directly to get the *conditional pmf* of Z as

$$P(Z = z|Y = y) = P(X = z - y).$$

The *marginal pmf* results as the expected value over the distribution of Y as

$$P(Z = z) = \sum_{y \in \mathcal{Y}} P(Y = y)P(X = z - y).$$

Example 2.25. Consider the penetration of narrow light rays in a homogeneous tree canopy, which is assumed to consist of independently and uniformly placed leaves in a random orientation. The penetration has the *exponential distribution* with *pdf*

$$f(x) = \lambda e^{-\lambda x}.$$

The expected value of the distribution is $1/\lambda$.

Assume then that the canopy is not homogeneous, but comprised of branches, and needles within branches. The penetration of a light ray in such a canopy is comprised of two components: the penetration between branches, and the penetration into branches. Assume that these components X_1 and X_2 are independent and that both have exponential distributions with parameters λ_1 and λ_2, so that X_1 is the component with the larger mean ($\lambda_1 \geq \lambda_2$). The density of total penetration $Z = X_1 + X_2$ is

$$fz(z) = \int_{-\infty}^{\infty} f_1(z - y) f_2(y) dy.$$

Because $X_1, X_2 \geq 0$, the maximum of y is z and the integral is evaluated over $[0, z]$. Writing the exponential densities and simplifying we get

$$\begin{aligned} fz(z) &= \int_0^z \lambda_1 e^{-\lambda_1(z-y)} \lambda_2 e^{-\lambda_2 y} dy \\ &= \lambda_1 \lambda_2 e^{-\lambda_1 z} \int_0^z e^{(\lambda_1 - \lambda_2) y} dy \\ &= \frac{\lambda_1 \lambda_2}{\lambda_1 - \lambda_2} \left(e^{-\lambda_2 z} - e^{-\lambda_1 z} \right). \end{aligned}$$

The special case $\lambda_1 = \lambda_2$ is undefined. Conducting the above derivation for this special case gives

$$fz(z) = \lambda_1^2 z e^{-\lambda_1 z}.$$

Therefore,

$$fz(z) = \begin{cases} \frac{\lambda_1 \lambda_2}{\lambda_1 - \lambda_2} \left(e^{-\lambda_2 z} - e^{-\lambda_1 z} \right) & \text{if } \lambda_1 > \lambda_2 \\ \lambda_1^2 z e^{-\lambda_1 z} & \text{if } \lambda_1 = \lambda_2 \end{cases}$$

where $z, \lambda_1, \lambda_2 > 0$ Figure 2.9 illustrates the total penetration when the two components have means of 1.5 m and 0.5 m. Essential parts of the R-code are shown below.

```
> lambda1<-1/1.5 # The first component penetrates 1.5 meters on average
> lambda2<-1/0.5 # The second component penetrates 0.5 meters on average
> x1<-rexp(1000,lambda1)
> x2<-rexp(1000,lambda2)

> # component 1
> hist(x1,freq=FALSE,ylim=c(0,lambda1),xlab=expression(x[1]),main="",breaks=seq(0,max(x1)+0.5,0.5))
> x0<-seq(0,10,0.01)
> lines(x0,dexp(x0,lambda1))

> # the sum
> hist(x1+x2,freq=FALSE,ylim=c(0,0.40),xlab=expression(x[1]+x[2]),main="",
+      breaks=seq(0,max(x1+x2)+0.5,0.5))
> lines(x0,lambda1*lambda2/(lambda1-lambda2)*(exp(-lambda2*x0)-exp(-lambda1*x0)))
```

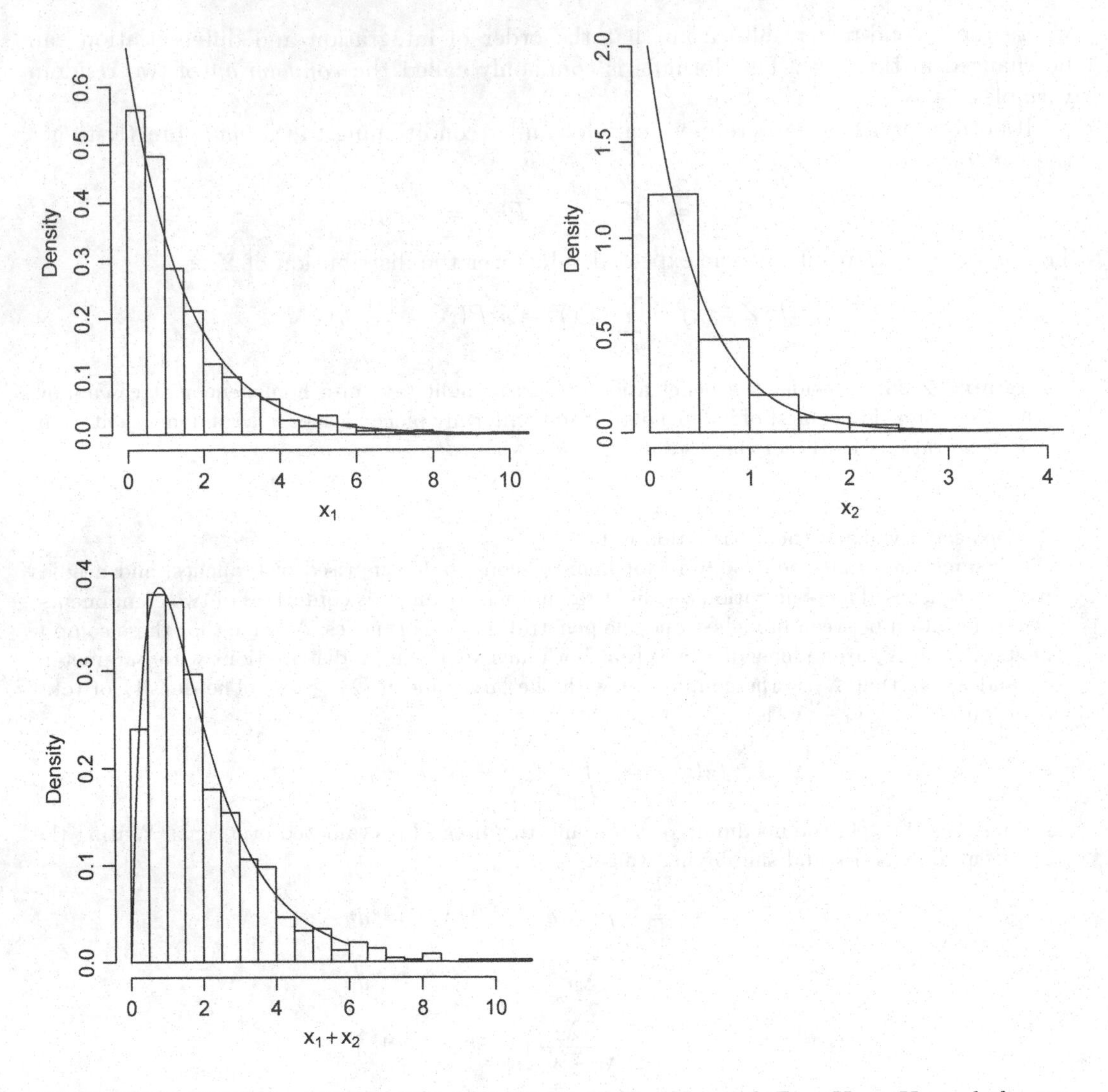

FIGURE 2.9 Histogram of 1000 realizations of X_1, X_2, and $Z = X_1 + X_2$ and the true underlying *pdf* when $\lambda_1 = 1/1.5$ and $\lambda_2 = 1/0.5$ in Example 2.25.

The formulas and example above only considered the case where Z is a sum of two components. Distributions for sums of more than two components are found by repeating the same steps for the sums. For example, the distribution for sum $X + Y + W$ results from the application of the above-specified formulas to the sum $Z + W$. An important special case is the normal distribution, for which the sum is also normally distributed.

2.11 Transformations of a standard normal variable

This section presents distributions for some important transformations of a standard normal variate, which are widely used in constructing confidence and prediction intervals, and in testing hypotheses about population parameters. Examples of such uses were already given in Examples 2.21 and 2.22. Usually, the collected data set is summarized to one single summary statistic, which is a random variable with one of these distributions under the null hypothesis.

2.11.1 χ^2-distribution

Let Z_i be independent random variables from the standard normal distribution $N(0,1)$. The sum of squares $Y = \sum_{i=1}^{p} Z_i^2$ follows the $\chi^2(p)$-distribution with p degrees of freedom. It has density

$$f(y|p) = \frac{1}{\Gamma(p/2)2^{p/2}} x^{p/2-1} e^{-y/2}, \quad 0 \leq y < \infty, \quad p = 1, 2, \ldots,$$

where $\Gamma(\alpha) = \int_0^\infty t^{\alpha-1} e^{-t} dt$ is the gamma function that needs to be evaluated numerically. The expected value and variance of $\chi^2(p)$-distribution are p and $2p$, respectively. The expectation results directly from

$$\begin{aligned} \mathrm{E}(Y) &= \mathrm{E}\left(\sum_{i=1}^{p} Z_i^2\right) = \sum_{i=1}^{p} \mathrm{E}\left(Z_i^2\right) \\ &= p(\mathrm{var}(Z) + \mathrm{E}(Z_i)^2) = p(1+0) = p. \end{aligned}$$

χ^2 distribution is used, for example, in the likelihood ratio, Wald, and Rao's test procedures for models fitted using the method of maximum likelihood. `R` has functions `dchisc`, `pchisc`, `rchisc` and `qchisc` for different computations with the χ^2-distribution. The function `gamma` can be used for evaluating the gamma function, which appears in the denominator of the *pdf*. If $X_i \sim \chi^2(p_i)$, the sum $\sum_{i=1}^{k} X_i \sim \chi^2(\sum_{i=1}^{k} p_i)$.

2.11.2 Student's t-distribution

If $Z \sim N(0,1)$ and $Y \sim \chi^2(p)$, independent of Z, then the ratio $X = \frac{Z}{\sqrt{Y/p}}$ follows the Student's t-distribution, $t(p)$, with p degrees of freedom. The *pdf* is

$$f(x|p) = \frac{\Gamma(\frac{p+1}{2})}{\Gamma(\frac{p}{2})} \frac{1}{\sqrt{p\pi}} \frac{1}{(1+\frac{x^2}{p})^{\frac{1}{2}(p+1)}}, \quad -\infty < x < \infty, \quad p = 1, 2, \ldots.$$

The mean and variance are $\mathrm{E}(X) = 0$ (for $p > 1$) and $\mathrm{var}(X) = \frac{p}{p-2}$ (for $p > 2$). The *pdf* is bell-shaped, and symmetric as normal distribution, but it is wider than the standard normal distribution.

The t-distribution was originally developed to test whether the expected value μ of an independent, identically distributed sample $W_1, W_2, \ldots, W_n$ of size n from a normally distributed population significantly differs from a fixed pre-specified value. If the variance of the population, $\mathrm{var}(W_i) = \sigma^2$, were known, such test would be based on the normal distribution. However, the variance is unknown in practice and needs to be estimated from a sample as $\widehat{\sigma}^2 = \frac{1}{n-1}\sum_{i=1}^{n}\left(W_i - \bar{W}\right)^2$. This estimate is a random variable and $Y = \frac{n-1}{\sigma^2}\widehat{\sigma}^2 \sim \chi^2(n-1)$. Let us now define $Z = \frac{\bar{W}-\mu}{\sigma/\sqrt{n}}$. Now $\sim N(0,1)$ exactly for a normally distributed population and asymptotically otherwise (see Section 3.4.2). The ratio $X = \frac{Z}{\sqrt{Y/(n-1)}}$ gets the form $X = \frac{\bar{W}-\mu}{\sqrt{\widehat{\sigma}^2/n}}$, which has the $t(n-1)$ distribution. The idea generalizes to testing the differences in the means of two groups and generally to testing regression

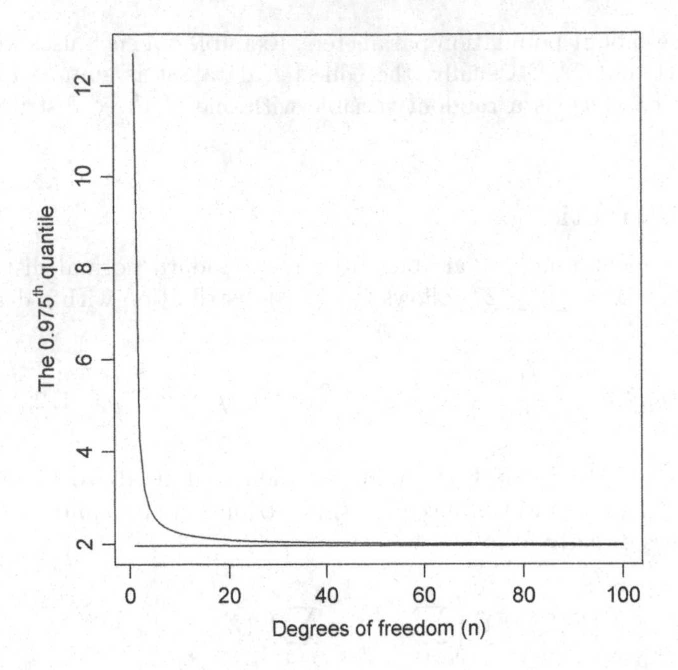

FIGURE 2.10 The 0.975th quantile of $t-$ distributions as a function of degrees pf freedom. The horizontal line at 1.96 level shows the 0.975th quantile of the standard normal distribution.

coefficients and linear functions of them. As n gets large, the t-distribution approaches the standard normal distribution. Thus, normal distribution leads to a very similar inference as t-distribution with large sample sizes, and tests on a large sample can often be based on normal distribution (Figure 2.10). R has functions `dt`, `pt`, `qt` and `rt` for computations with Student's t-distribution.

2.11.3 *F*-distribution

Let X_1 and X_2 be independent random variables that follow the χ^2-distribution: $X_1 \sim \chi^2(p)$ and $X_2 \sim \chi^2(q)$ The distribution of ratio $R = \frac{X_1/p}{X_2/q}$ follows the $F(p,q)$- distribution with p and q degrees of freedom. The density is

$$f(r|p,q) = \frac{\Gamma(\frac{p+q}{2})}{\Gamma(\frac{p}{2})\Gamma(\frac{q}{2})}\left(\frac{p}{q}\right)^{p/2}\frac{r^{\frac{1}{2}(p-2)}}{(1+\frac{p}{q}r)^{\frac{1}{2}(p+q)}}$$

where $0 \le z < \infty$ and $m,n = 1,2,\ldots$. The mean and variance are $\mathrm{E}(Z) = \frac{n}{n-2}$ for $(n > 2)$ and $\mathrm{var}(Z) = 2\left(\frac{n}{n-2}\right)^2 \frac{m+n-2}{m(n-4)}$ for $(n > 4)$.

The F-distribution is commonly used in testing two nested linear regression models against each other. The numerator of the $F-$statistic is based on the difference in the explained sum of squares between the two models, and the denominator on the unexplained residual sum of squares in the larger model (see Section 4.6.2). R has functions `pf`, `df`, `qf` and `rf` for computations with the F-distribution.

2.12 Random functions: stochastic and spatial processes

Random functions deal with collections of random variables. The random variables are indexed by a set D, which belongs to n-dimensional space R^n. The index set can be discrete or continuous, leading to *discrete parameter random functions*, or *continuous parameter random functions.* An example of a one-dimensional discrete index set is the index $i = 1, 2, 3, \ldots$ of a sequence ordered with respect to time. An example of a two-dimensional continuous set is a set of spatial coordinates (x_1, x_2). The theory of random functions deals with the multivariate distributions of these indexed random variables. Thus, they are a generalization of 'ordinary' multivariate distributions. What makes the theory of random functions special is that it analyzes how the joint properties of the multivariate distributions are related to the indexes of the random variables. If the index set D is one-dimensional, then the random functions are called *stochastic processes.* Usually the one-dimensional index refers to time, and the process is denoted as $X(t)$, or by X_t. Random functions with a higher dimensional index set are called *random fields*. Usually the index set refers to locations in two-dimensional or three-dimensional space, and the corresponding random fields are called *spatial processes* (a spatial process on a line is a stochastic process). The distinction between stochastic and spatial processes may appear strange because spatial processes are also, of course, stochastic, but so the common terminology goes. In *spatio-temporal processes,* one index refers to time and the others to location. The machine-specific time series in Example 4.5 (p. 86) provides an example of a stochastic process. Examples 2.19 and 4.14 (p. 27 and 108) are examples of a two-dimensional random field.

In this section, we present the basic concepts of the stochastic and spatial processes. In forestry, spatial *point processes* are also important, but we do not deal with them in this book (except for Example 8.6 on p. 266) because they require different types of methods to those presented in this book. For interested readers, the methods are well described in Illian *et al.* (2008).

A stochastic process is *strictly stationary* if the multivariate distribution for any set of index values is the same as the multivariate distribution where the same constant is added to each index value. Intuitively, this condition says that both the distributions of the random variables and their dependencies remain the same over time. A stochastic process is *weakly stationary* or *covariance stationary* if it has finite second moments and $\text{cov}(X_t, X_{t+h})$ depends only on the distance h between the indexes of the variables. The covariances and correlations can thus be expressed with the *autocovariance* or *autocorrelation* function, which depends only on h. It is a common mis-belief that weak stationarity requires that the mean function $\text{E}(X_t)$ is constant. Often a process is not stationary, although the increments $X_{t+h} - X_t$ may be so.

In principle, there are two different ways to infer the properties of random processes: either we can observe one or a few processes for a long time, or we can observe several processes for short time (which leads to *panel data*). One can infer the properties of the process from one series, if the stochastic process is *ergodic.* A necessary condition for ergodicity is that the correlations die out sufficiently fast so that distant observations from the process are uncorrelated.

In forestry, stochastic process concepts (e.g. autocorrelation) are usually applied for residual errors of regression models. In typical applications, the assumption of stationary is very strong, and it may not be valid as often as the assumption is made (either implicitly or explicitly). In general, tree growth has a sigmoidal shape, and it is hard to believe that during different phases of the growth the covariances or correlations have a similar structure. The stand development changes the competitive status of trees so that the properties of growth processes of different trees will change. In the analysis of stem curves, the assumption of a standard autocorrelation function is quite implausible. Standard assumptions are more often plausible in forestry application of spatial processes.

Let us then briefly describe some models for stochastic processes. *Autoregressive process* of order p, denoted as AR(p), is defined as

$$X_t = c + \sum_{i=1}^{p} \phi_i X_{t-i} + \epsilon_i \tag{2.54}$$

where $\phi_1, \ldots, \phi_p$ are fixed parameters and ϵ_i is *white noise*, i.e. uncorrelated random variables with a mean of 0 and variance σ^2. Often c is 0. The most important process is the AR(1) process, which is stationary if (dropping the subindex of ϕ) $|\phi| < 1$. If $|\phi| \geq 1$ the process diverges, and the variance grows to infinity. A stationary process has the mean:

$$\mathrm{E}(X_t) = \frac{c}{1-\phi}$$

and variance

$$\mathrm{var}(X_t) = \frac{\sigma^2}{1-\phi^2}.$$

It is left as an exercise to derive the mean and variance from the stationarity of the process, i.e. using the property that $\mathrm{E}(X_t) = \mathrm{E}(X_{t-1})$ and $\mathrm{var}(X_t) = \mathrm{var}(X_{t-1})$. The process has the autocorrelation function:

$$\mathrm{cor}(X_t, X_{t-k}) = \phi^k \tag{2.55}$$

Note that the autocorrelation function is equivalent to $\exp(-\alpha k)$ where $\alpha = -\log(\phi)$ is a positive constant. Both forms of the same function appear in the literature. There are many methods for estimating the parameters of the AR(p) process, the ordinary least squares (OLS) (to be discussed later) fit to the linear model of form 2.54 being the easiest to apply. The OLS is biased because the error terms and regressor variables are correlated. However, the OLS method is consistent, because the error term of observation t is uncorrelated with the regressors of the same observation (see Section 10.1.2).

The *moving average* process of order q, denoted as MA(q) is defined as:

$$X_t = \mu + \sum_{i=1}^{q} \theta_i \epsilon_{t-i} + \epsilon_t \tag{2.56}$$

The variance of the MA(1) process is seen to be $(1+\theta^2)\sigma^2$ and the correlation $\mathrm{cor}(X_t, X_{t-k}) = \frac{\theta}{1+\theta}$ if $k = 1$ and 0 otherwise. Comparing the correlations of the AR(1) and MA(1) processes, we note that the random noise added at some time point has an effect up to infinity in the AR(1) process. In the MA(1) process, the influence has already died out by distance 2. In MA(q), the autocorrelation dies out at distance $q+1$. The parameters of a MA process cannot be estimated using OLS regression, because in Equation (2.56), θ parameters multiply latent random variables. In an ARMA(p,q)-process, there are p autoregressive terms and q moving average terms. To learn more about the properties of time series and more advanced models, see Box *et al.* (2015). For possible definitions in the spatial context, see Cressie (1993), Diggle and Ribeiro (2007) and Bivand *et al.* (2008), for example.

3

Statistical Modeling, Estimation and Prediction

3.1 Concepts

Chapter 2 presented some basic aspects related to probability theory and random variables. These properties are useful in formulating the model behind the observed data. We also presented examples of realizations from some simple models. This chapter presents the basic principles for using observed data to estimate the parameters of an assumed model.

In empirical sciences we try to learn some regularities of the world by putting empirical measurements and probability theory together. This is called *statistical modeling.* The method we use to get values for unknown fixed parameters in an assumed model is called an *estimator*, and the obtained value is called an *estimate*. The distinction between an estimator and an estimate is similar to the distinction between a random variable and its realization.

Often, we may also want to guess the value of an irregular measurement, i.e. a random variable that we have not measured. This is called *statistical prediction.* When parameters are considered to be random variables, as happens in Bayesian analysis or in mixed models, they are also predicted.

In statistical modeling we describe some measurements using a function that can have constant arguments that are called *parameters.* Mathematically, parameters are just variables; the term parameter is used for variables that describe the more general properties of our model of the world than ordinary variables. Parameters are usually also *latent variables*, i.e. they cannot be measured directly. The simplest case is when we have a number of independent observations from a given distribution, and we want to estimate the parameters of the distribution. In *regression modeling,* the main emphasis is to estimate the conditional mean of a given random variable, called a *dependent variable*, given the parameters and *independent variables.* The distributional properties of assumed random errors also need to be considered. In *generalized linear models* the conditional mean of the dependent variable is not modeled directly using a model that includes independent variables but the connection is defined using a separate *link function.*

3.2 The role of assumptions in modeling

It is important to understand the role of assumptions in modeling. Assumptions are something that are taken as true when the parameters are estimated and tests on a model are conducted. Therefore, it is important to understand what properties of our estimates are based on the assumptions that we made in model formulation, and what are based on the data. However, we do not know much for sure: we do not know the true model behind the observed data. Neither do we know the expected values, the variances nor the covariances of the random variables that we have observed. However, by making some assumptions about the model, we may be able to estimate the parameters that describe the properties of interest. For example, we may assume a certain distribution family for the data, probably with assumed dependencies on some predictor variables, as illustrated in Examples 2.2 and 2.3 (p. 10 and 11). The validity of the assumptions can usually be evaluated only after parameter estimation. This makes the modeling task an iterative procedure, which iterates between the three steps: model formulation, parameter estimation, and evaluation of the underlying assumptions.

Some methods can be used with fewer assumptions, whereas others require stronger assumptions. In particular, it is quite often possible to perform an estimation, assuming only the first- and second-order properties (mean, variance and covariance) of the distribution that generated the data. A linear model, for example, relies mostly on these. The assumptions are the choices made by the modeler. Stronger assumptions can help to obtain results that would not be possible with weaker assumptions, and may lead, for example, to more powerful tests. However, there is a risk of wrong conclusions if the strong assumptions are not met. For example, parametric tests require stronger assumptions about the underlying population than non-parametric tests. It is important to recognize that the assumptions are made about the population. If the data are small, it is not necessarily possible to properly evaluate whether the stated assumptions have been met. On the other hand, powerful analysis tools are beneficial especially with small sample sizes.

Independence is a commonly assumed key property of the data in many modeling contexts, such as regression modeling. In particular, it is often assumed that the observations are independent and are identically distributed realizations of the same random variable. These are two very strong assumptions.

The observations are identically distributed if the distribution for every observation is the same. However, the data includes only one realization for each observation, thus this assumption cannot be exactly evaluated. One possibility to evaluate this assumption is to group the data according to a predictor variable and check if the (empirical) marginal distribution within each group is similar. If the data are identically distributed, any grouping that is not based on the variable of interest itself should lead to similar distributions for all groups.

A weaker assumption than independence is the assumption that the observations are uncorrelated. However, recall that if observations are uncorrelated and that they have a multivariate normal distribution, then they are also independent. It is often not possible to observe the dependence from empirical data if the process of data collection is not known. Common reasons for dependence are

- Data have a *grouped structure*, such as many trees on a sample plot, many branches from a tree, many locations from the same tracked animal, several newt captures around the same breeding pond, several growth measurements on the same calendar year and so on. With such data sets, observations from the same group are often similar to each other, which implies dependence.

- The data of repeated measurements of the same unit, such as growth rings of a tree, may have *temporal dependence*. The observations from successive years are dependent, but the dependence gets weaker and probably vanishes as the time lag gets longer.

- The observations in an xy plane (e.g. on a map) may have *spatial dependence*. For example, they may tend to become more similar as the distance between observations gets smaller.

If the observations are not independent or uncorrelated, it may be possible to use a modeling approach that takes the dependence explicitly into account, e.g. by assuming dependence according to some justified model. Alternatively, the data can be reduced or aggregated so that the resulting subsets or aggregates are independent: however, this approach wastes information and, therefore, is not generally suggested. In some cases, it may be justified to assume that the possible dependence is so weak that it does not affect the conclusions, or the research question may be such that the faulty assumption of independence may not be too harmful. In that case, it is advisable to communicate the possible problems and to justify the reasons for ignoring the dependence.

If we have an estimator for a parameter vector, then the variance and bias of the estimator are the most important measures of the obtained estimates. The variance of the estimator tells us what the variance of obtained estimates would be, if we collect independently new data sets with the same design from the same population, or from new independent realizations of the same random process. The bias is the difference between the expected value of the estimator and the true value. It keeps us humble that the assumed model is never true. "All models are wrong, but some are useful", as noted by George Box. We can also never resample from the same biological or human population. "No man ever steps in the same river twice" (Heraclitus).

3.3 Scales of measurement

The kind of values that a random, and also a fixed variable can get is related to the *scale of measurement*. The scale of measurement can be defined by determining the kind of transformations that maintain the information contained in the measurements. The lowest measurement scale is the *nominal scale*. A nominal scale measurement merely determines the class that the subject will be assigned. Any one-to-one transformation of the measurement maintains the original information of the measurements. Note that random variables always receive numeric values. Thus, colors, for example, cannot be measured by classifying subjects as black or white, so numeric values need to be assigned.

Subjects are measured on an *ordinal scale* if they can be ordered according to the measured values. Any monotonously increasing transformation maintains the information of the measurement. The *interval scale* means that differences between subjects can be compared using the measurements. An example is the measurement of temperature using Celsius degrees: the difference between 10°C and 20°C is as large as the difference between 30°C and 40°C. Linear transformations $(a \times x + b)$ are legitimate ways to change the measurement units. A *ratio scale* has a meaningful zero. This means that sizes of subjects can be compared using ratios of the measurements. Examples of ratio scales are the scales for length, area, age, etc. Temperatures can also be measured with the ratio scale using Kelvin degrees. Multiplication with a positive number is the only legitimate way to change measurement units in the ratio scale. In the *absolute scale*, there are no legitimate transformations. Counting is the only example of the absolute scale.

The kind of statistical methods that can be legitimately applied depends on the scales of measurement. For instance, if a variable is measured on the ordinal scale, then the variable should not be used in principle as the dependent or explanatory variable in the regression analysis. Often, however, e.g. when measuring preferences, we can think that there is a background interval scale that is just rounded to integers. So we do not want to take an orthodox view in this matter.

3.4 Estimation using observed data

3.4.1 Comparing alternative estimators

Let us then turn to some more formal definitions. For an estimator producing estimates $\widehat{\alpha}$ for a fixed (i.e. nonrandom) parameter α using the available data, the $\text{var}(\widehat{\alpha})$ describes its precision. Note that $\text{var}(\widehat{\alpha}) = \text{var}(\widehat{\alpha} - \alpha)$. That is, the variance of the estimator refers to the variance of the estimation error. When we discuss random parameters in later chapters, we will see that the variance of the predictor is not equal to the variance of the prediction error. If α is a scalar parameter, then the variance or the standard deviation can be used to compare precision. The standard deviation is on the same scale as the parameter itself, so it should be used in reports, even if the statistical derivations are based on the variance.

If $\boldsymbol{a}$ is a vector and $\widehat{\boldsymbol{a}}_1$ and $\widehat{\boldsymbol{a}}_2$ are two estimators, then it is defined that

$$\text{var}(\widehat{\boldsymbol{a}}_1) \leq \text{var}(\widehat{\boldsymbol{a}}_2) \tag{3.1}$$

if for any fixed $\boldsymbol{w}$

$$\text{var}(\boldsymbol{w}'\widehat{\boldsymbol{a}}_1) \leq \text{var}(\boldsymbol{w}'\widehat{\boldsymbol{a}}_2) \tag{3.2}$$

The condition says that matrix $\text{var}(\widehat{\boldsymbol{a}}_2) - \text{var}(\widehat{\boldsymbol{a}}_1)$ is positive semidefinite. It may happen that the matrix $\text{var}(\widehat{\boldsymbol{a}}_2) - \text{var}(\widehat{\boldsymbol{a}}_1)$ is indefinite in which case $\text{var}(\widehat{\boldsymbol{a}}_2)$ and $\text{var}(\widehat{\boldsymbol{a}}_1)$ cannot be compared.

The *bias* of an estimator is the difference between the expected value (in repeated estimations) of the estimate and the true value:

$$\text{bias}(\widehat{\alpha}) = \text{E}(\widehat{\alpha}) - \alpha \tag{3.3}$$

An estimator is said to be *unbiased* if the bias is zero. A weaker property is *consistency*: an estimator is consistent if the bias approaches zero as the sample size increases towards ∞.

Mean Square Error (MSE) combines both variance and bias by measuring the expected value of squared estimation error:

$$\begin{aligned} MSE(\widehat{\alpha}) &= E\left((\widehat{\alpha} - \alpha)^2\right) \\ &= \text{E}\left((\widehat{\alpha} - \text{E}(\widehat{\alpha}) + \text{E}(\widehat{\alpha}) - \alpha)^2\right) \\ &= \text{var}(\widehat{\alpha}) + \text{bias}(\widehat{\alpha})^2 \end{aligned} \tag{3.4}$$

The MSE of a parameter vector is just the vector of the component MSEs. One could easily think that unbiased estimators would inevitably have smaller MSEs than biased estimators. But this is not the case. There are cases where unbiasedness must be paid with increased variance, and with small bias one can obtain smaller MSE. Examples are the James-Stein estimator and ridge regression, described on Boxes 4.1 and 4.2 (p. 90 and 91). If an estimator is unbiased, then MSE is equal to the variance.

Root Mean Square Error is defined as

$$RMSE(\widehat{\alpha}) = \sqrt{MSE(\widehat{\alpha})} \tag{3.5}$$

RMSE is on the same scale as the parameter itself, so it can be compared to the obtained value of the estimate.

If we have measured a random vector $\boldsymbol{y}$, then an estimator of parameter α is called a *linear estimator* if:

$$\widehat{\alpha} = a_0 + \boldsymbol{w}'\boldsymbol{y} \tag{3.6}$$

where a_0 or $\boldsymbol{w}$ are not functions of $\boldsymbol{y}$. Linear estimators are nice because the expected value of the estimator can be obtained using the equations in Section 2.2.4:

$$\mathrm{E}(\widehat{\alpha}) = \mathrm{E}(a_0 + \boldsymbol{w}'\boldsymbol{y}) = a_0 + \boldsymbol{w}'\,\mathrm{E}(\boldsymbol{y}) \tag{3.7}$$

where $E(\boldsymbol{y})$ can be obtained by substituting the assumed model for $\boldsymbol{y}$, and using probability calculus to derive $\mathrm{E}(\boldsymbol{y})$.

The variance is obtained using the formulas in Section 2.4:

$$\mathrm{var}(\widehat{\alpha}) = \mathrm{var}(a_0 + \boldsymbol{w}'\boldsymbol{y}) = \boldsymbol{w}'\mathrm{var}(\boldsymbol{y})\,\boldsymbol{w} \tag{3.8}$$

where $\mathrm{var}(\boldsymbol{y})$ is again obtained by substituting the assumed model for $\boldsymbol{y}$, and using the probability calculus to derive the variance.

The *Best Estimator* is an estimator with the smallest MSE value. The *Best Linear Estimator* is an estimator with the smallest MSE value among linear estimators. The *Best Linear Unbiased Estimator* (BLUE) is an estimator with the smallest possible variance among linear unbiased estimators. If BLUE is dependent on some covariance parameters that are not known but are estimated, then the obtained estimator is called Estimated BLUE, EBLUE. The above definitions are generalized in a self-evident way to the case where we estimate parameter vector $\boldsymbol{a}$ with a linear estimator:

$$\widehat{\boldsymbol{a}} = \boldsymbol{a}_0 + \boldsymbol{W}\boldsymbol{y} \tag{3.9}$$

where $\boldsymbol{a}_0$ is a fixed vector, $\boldsymbol{W}$ is a fixed matrix having as many rows as there are elements in $\boldsymbol{a}$, and as many columns as there are observations in $\boldsymbol{y}$. When $\boldsymbol{a}$ is a vector, variances are compared using Equation (3.1).

Quite often in statistics, we restrict the consideration to unbiased estimators. The minimum variance for an unbiased estimator of a certain parameter is called *Cramér-Rao lower bound* for the variance (Casella and Berger 2001). An estimator with this variance is said to be an *efficient* or a *minimum-variance unbiased* estimator. However, one might question the routine use of unbiased estimators because, as discussed above, a biased estimator may have a smaller MSE than the best unbiased estimator.

3.4.2 Estimating expected value of any distribution

Consider random variables $X_1, X_2, \ldots, X_n$ on n sampling units. The sample mean is defined as

$$\bar{X} = \frac{X_1 + X_2 + \ldots + X_n}{n} = \frac{1}{n}\sum_{i=1}^{n} X_i\,.$$

The sample mean is a random variable due to the randomness of X_i. Assuming that the random variables have a common mean μ, $\mathrm{E}\,X_1 = \mathrm{E}\,X_2 = \ldots = \mathrm{E}\,X_n = \mu$, we see that the sample mean is an unbiased estimator of the common expected value:

$$\mathrm{E}(\bar{X}) = \mathrm{E}\left(\frac{1}{n}\sum_{i=1}^{n} X_i\right) = \frac{1}{n}\sum_{i=1}^{n}\mathrm{E}(X_i) = \frac{1}{n}n\mu = \mu$$

using rules (2.12) and (2.13) on p. 15. The only assumption needed for this result is that of a common mean. Therefore, the result holds for any variance-covariance structure and joint distribution of variables $X_1, \ldots, X_n$, if they all have the same mean. Box 3.1 illustrates that the sample mean is the least squares estimator of the common expected value. This idea is extended to the case where the mean is a function of fixed predictor variables and regression coefficients (recall Example 2.3, p. 11) in Section 4.4.1.

Box 3.1 Least squares estimation of μ

Let us estimate μ by such a constant that minimizes the sum of squared differences between the random variables $X_1, \ldots, X_n$ and μ:

$$SS = \sum_{i=1}^{n} (X_i - \mu)^2 \,.$$

To find the estimate, we set the first derivative of SS with respect to μ to zero to get $\frac{\partial SS}{\partial \mu} = -\sum 2\,(X_i - \mu)$. Setting it to zero gives

$$\sum X_i - n\mu = 0 \,.$$

Solving for μ gives the ordinary least squares estimator $\widehat{\mu}_{OLS} = \bar{X}$.

If X_i's are uncorrelated and have a common variance σ^2, Equations (2.44) and (2.27) (p. 24) give the variance of the sample mean as

$$\begin{aligned} \text{var}(\bar{X}) &= \text{var}\left(\frac{1}{n}\sum_{i=1}^{n} X_i\right) = \left(\frac{1}{n}\right)^2 \text{var}\left(\sum_{i=1}^{n} X_i\right) \\ &= \frac{1}{n^2} n\sigma^2 = \frac{1}{n}\sigma^2 \,. \end{aligned} \tag{3.10}$$

Taking the square root gives the corresponding standard deviation; the well-known *standard error of the mean* as $\text{sd}(\bar{X}) = \sigma/\sqrt{n}$. Therefore, if one takes several samples of size n from a population where all units have the same expected value μ, the mean of the sample means approaches μ as the number of samples increases. In addition, if the population units are mutually independent and have the same variance σ^2, the sample means vary around μ with standard deviation $\text{sd}(\bar{X}) = \sigma/\sqrt{n}$. Furthermore, the *Central Limit Theorem (CLT)* states that if X_i are independent and identically distributed (and have a finite expected value and variance), the distribution of $\bar{X}$ approaches the Normal distribution as the sample size n gets large. This result holds regardless of the shape of the common distribution of X_i.

Example 3.1. Central Limit Theorem. Assume that observations are taken from a population with underlying *Exponential*(λ) distribution with $\lambda = 0.5$. The mean and variance of the exponential distribution are $\text{E}(X) = 1/\lambda$ and $\text{var}(X) = 1/\lambda^2$. The distribution is highly skewed to the right, as shown in the upper left plot of Figure 3.1. A total of 1000 samples of sizes 5, 20 and 50 were taken from the underlying model. Figure 3.1 shows the distributions of sample means in the 1000 realizations of a sample with the three applied sample sizes. Figure 3.1 shows that the normal approximation improves as the sample size increases, and is fairly close when $n = 50$. Essential parts for the code are shown below.

```
> rate<-1/2; mu<-1/rate; sd<-1/rate
> means5<-means20<-means50<-rep(NA,1000)
> for (i in 1:1000) {
+   s5<-rexp(5,rate)
+   s20<-rexp(20,rate)
+   s50<-rexp(50,rate)
+   means5[i]<-mean(s5)
+   means20[i]<-mean(s20)
+   means50[i]<-mean(s50)
+ }
> # graph for n=5
> hist(means5,freq=FALSE,main="Mean of 5 observations",xlab=expression(bar(x)),ylab="Density")
> lines(x,dnorm(x,mu,sd/sqrt(5)))
```

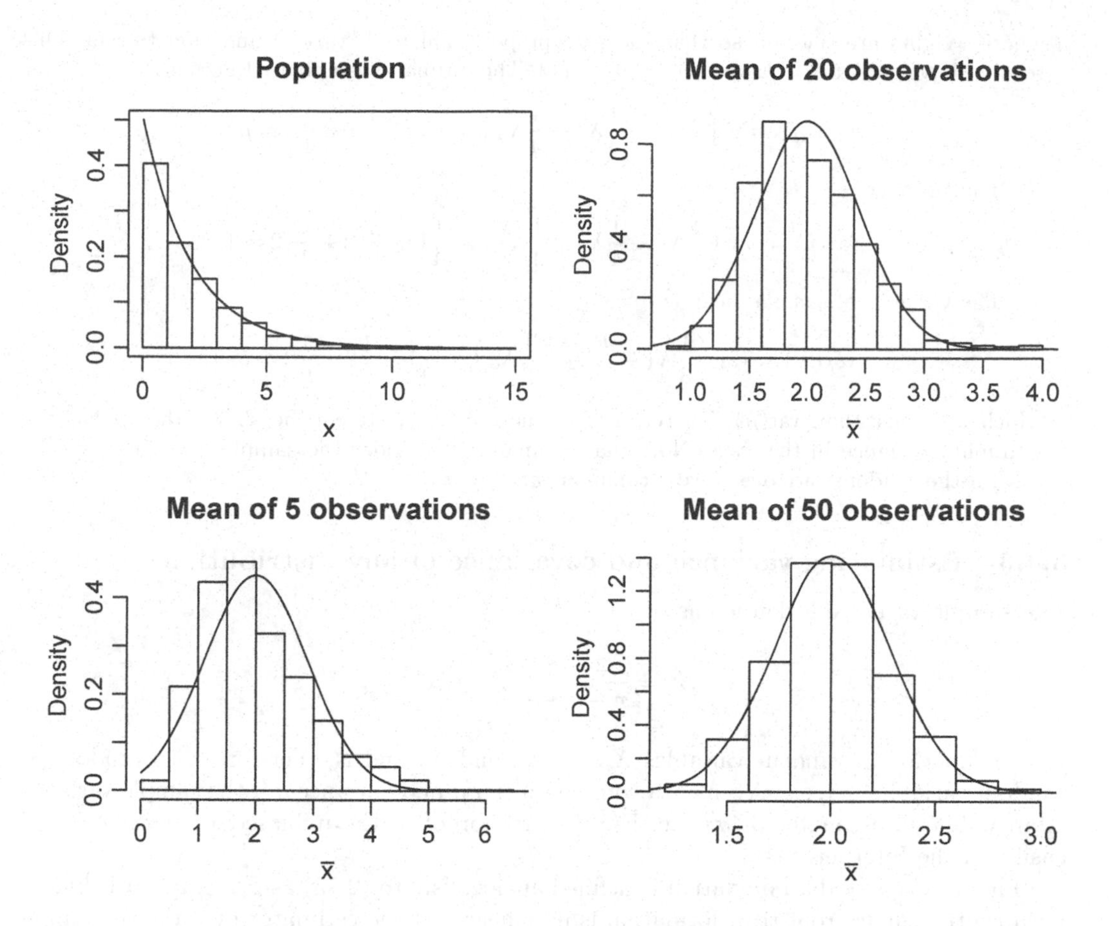

FIGURE 3.1 The distribution of sample mean in the exponentially distributed population of Example 3.1.

Sample mean is not the only alternative as an estimator of population mean. For example, consider $X_1, \ldots, X_n$ with common mean μ and variance σ^2. The weighted average

$$\widehat{\mu} = \sum_{i=1}^{n} w_i X_i$$

where the weights w_i are positive and fulfill $\sum w_i = 1$ provides many alternative unbiased estimators of μ. However, all these estimators have greater variance than the sample mean.

It can be shown that there are no unbiased estimators with a lower standard error than the sample mean, i.e. the variance is as small as possible. Therefore the sample mean is the efficient estimator of μ. It is linear with respect to X_i, therefore it is also BLUE. However, the situation changes if $X_1, \ldots, X_n$ all have a different variance σ_i^2 (but are still uncorrelated). In that case, the weighted mean where weights are proportional to $\frac{1}{\sigma_i^2}$ is efficient.

Example 3.2. Variance of alternative estimators of μ. Consider random variables X_1, X_2, X_3 which all have a common mean μ but the variances are $\text{var}(X_1) = 1$ and $\text{var}(X_1) = \text{var}(X_2) = 2$. We already know that the sample mean is an unbiased estimator of μ. Another estimator of the mean is provided as

$$\hat{\mu} = \sum_{i=1}^{3} w_i X_i \,,$$

where weights are selected so that they are proportional to $1/\operatorname{var}(X)$ and sum to one. This results in weights $w_1 = 1/2$ and $w_2 = w_3 = 1/4$. The estimator is unbiased because

$$E(\hat{\mu}) = E\left(\frac{1}{2}X_1 + \frac{1}{4}X_2 + \frac{1}{4}X_3\right) = \frac{1}{2}\mu + \frac{1}{4}\mu + \frac{1}{4}\mu = \mu$$

The variance is

$$\operatorname{var}(\hat{\mu}) = \operatorname{var}\left(\frac{1}{2}X_1 + \frac{1}{4}X_2 + \frac{1}{4}X_3\right) = \frac{1}{4}1 + \frac{1}{16}2 + \frac{1}{16}2 = 1/2$$

The variance of sample mean is

$$\operatorname{var}(\bar{X}) = \operatorname{var}\left(\frac{1}{3}X_1 + \frac{1}{3}X_2 + \frac{1}{3}X_3\right) = \frac{1}{9}1 + \frac{1}{9}2 + \frac{1}{9}2 = 5/9,$$

which is higher than $\operatorname{var}(\hat{\mu})$. Therefore, $\hat{\mu}$ is more efficient. It can be shown that $\hat{\mu}$ has the minimum variance in this case. Note that estimator $\hat{\mu}$ provides the sample mean as a special case, if the random variables have a common variance σ^2.

3.4.3 Estimating variance and covariance of any distribution

The sample variance is defined as

$$S^2 = \frac{1}{n-1}\sum_{i=1}^{n}(X_i - \bar{X})^2 .$$

It is a function of random variables $X_i, \ldots, X_p$ and $\bar{X}$, and is, therefore, also random. If random variables $X_1, \ldots, X_n$ have a common mean μ and variance σ^2, the sample variance is an unbiased estimator of σ^2, i.e. $E(S^2) = \sigma^2$; proof of the unbiasedness can be found easily on the Internet.

The sample standard deviation is defined analogously to (2.15) as $S = \sqrt{S^2}$. It is biased because the square root transformation is nonlinear and the estimator of variance is unbiased.

A question arises why the denominator of sample variance is $n-1$, not n. An intuitive answer to this result is as follows. The sample mean can be seen as the least squares estimator of the expected value. The value of a that minimizes the sum of squares $\sum_{i=1}^{n}(X_i - a)^2$ is the sample mean $\bar{X}$ (recall Box 3.1 on p. 48). Because the sample mean minimizes the sum of squares and $\bar{X}$ takes exactly the value μ with zero probability, the sum of squared differences around the true mean μ, $\sum_{i=1}^{n}(X_i - \mu)^2$, is necessarily higher than $\sum_{i=1}^{n}(X_i - \bar{X})^2$. Therefore, estimator $\frac{1}{n}\sum_{i=1}^{n}(X_i - \bar{X})^2$ provides an underestimate of the true variance. An unbiased estimator of σ^2 results from adjusting the denominator downwards to $n-1$. In other words, $n-1$ is used in the denominator because one degree of freedom of the data was used for estimating μ as $\bar{X}$. Formally, one could show that $E(\sum_{i=1}^{n}(X_i - \bar{X})^2) = (n-1)\sigma^2$.

Let random variables $Y_1, Y_2, \ldots, Y_n$ and $X_1, X_2, \ldots, X_n$ describe two different properties of the same sampling units $1, \ldots, n$. The sample covariance is defined as

$$C = \frac{1}{n-1}\sum_{i=1}^{n}(X_i - \bar{X})(Y_i - \bar{Y}) .$$

It is an unbiased estimator of the true covariance between X and Y (Equation 2.23, p. 22). The sample correlation is defined using the sample covariance and sample standard deviation in Equation (2.24). The expression reduces to

$$R = \frac{\sum_{i=1}^{n}(X_i - \bar{X})(Y_i - \bar{Y})}{\sqrt{\sum_{i=1}^{n}(X_i - \bar{X})^2 \sum_{i=1}^{n}(Y_i - \bar{Y})^2}} .$$

We have presented the sample mean, standard deviation and correlation as estimators of the true expected value, standard deviation and correlation of random variables. However, they can also be used merely as descriptive summaries of the observed data.

3.4.4 Estimating any parameter of a specific distribution

The previous subsections provided unbiased estimators for the first- and second-order properties of the data. However, the distribution of the data is not always parameterized using them. For example, the variance or covariance may be functions of predictors. The method of Maximum likelihood provides a general approach to estimate any parameters of an assumed distribution.

Empirical collected data provides realizations of a stochastic process, which defines the joint distribution of the data. Based on the assumed stochastic process we may have more or less justified an assumption about the distribution family of the data, such as Poisson, Binomial or Beta-Binomial for random counts of different type, Bernoulli for binary data, or multivariate Normal distribution for a continuous random vector. The parameters of these distributions are of interest. Often, the parameters are functions of predictors, as we illustrated in Examples 2.2 and 2.3 (p. 10 and 11).

A pertinent question is "What are the parameter values for which the observed data are most likely?" The question leads to the method of maximum likelihood, where the parameters are estimated by maximizing the likelihood function for the collected data with respect to parameter $\boldsymbol{\theta}$. The likelihood function is the joint *pdf*/*pmf* of the observed data $\boldsymbol{x}$, as a function of distribution parameters $\boldsymbol{\theta} = (\theta_1, \ldots, \theta_k)'$:

$$L(\boldsymbol{\theta}|\boldsymbol{x}) = f(\boldsymbol{x}|\boldsymbol{\theta}) \,. \tag{3.11}$$

It is mathematically more convenient to use the logarithmic likelihood in estimation. This will not affect the solution because logarithm is a monotonic function. Therefore, $\log L(\boldsymbol{\theta}|\boldsymbol{x}) = l(\boldsymbol{\theta}|\boldsymbol{x})$ and $L(\boldsymbol{\theta}|\boldsymbol{x})$ are maximized at the same value of $\boldsymbol{\theta}$.

Example 3.3. Consider a normally distributed random vector of length 3. All three components have a known common standard deviation of 1.92 and an unknown common mean μ. In addition, the correlation $\text{cor}(X_1, X_2)$ is known to be 0.61. This kind of correlation structure may result when the data consists of two groups (e.g. two sample plots) so that X_3 belongs to one group and X_1 and X_2 to another group. The log-likelihood is the logarithmic multivariate normal density (2.47) (p. 26), where

$$\boldsymbol{\mu} = \begin{pmatrix} \mu \\ \mu \\ \mu \end{pmatrix}, \qquad \boldsymbol{\Sigma} = \begin{pmatrix} 3.69 & 2.25 & 0 \\ 2.25 & 3.69 & 0 \\ 0 & 0 & 3.69 \end{pmatrix}.$$

Because the variance-covariance matrix $\boldsymbol{\Sigma}$ is known, the log-likelihood is a function of μ only. Figure 3.2 shows the likelihood as a function of μ. The likelihood is maximized when $\mu = 4.83$.

The estimate falls between the arithmetic mean of the three observations ($\bar{X} = 5$) and the mean of group-specific means ($(5.5 + 4)/2 = 4.75$). This is justified, because the two observations X_1 and X_2 from group 1 together provide more information about the population mean than the single observation X_3 from group 2. However, the independent observation X_3 alone provides more information than either of the mutually correlated variables X_1 and X_2.

The R-script for Figure 3.2 is shown below. Note that the variance-covariance structure of the data is here defined using s_1^2 and s_2^2, the variances between groups and within group in a variance component model, which will be defined and discussed in Chapter 5.

```
> data<-c(6,5,4)
> s1<-1.5^2; s2<-1.2^2
> Sigma<-matrix(c(s1+s2,s1,0,s1,s1+s2,0,0,0,s1+s2),ncol=3)
> sqrt(diag(Sigma))
```

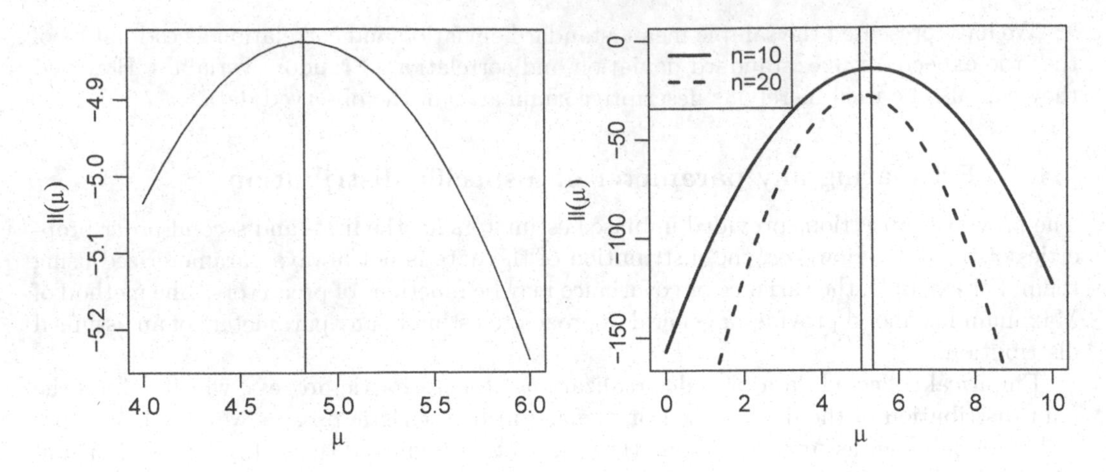

FIGURE 3.2 The normal log-likelihood of Examples 3.3 (left) and 3.6 (right) as a function of μ. The vertical lines show the maximum likelihood estimates.

```
[1] 1.920937 1.920937 1.920937
> cov2cor(Sigma)
          [,1]      [,2] [,3]
[1,] 1.0000000 0.6097561    0
[2,] 0.6097561 1.0000000    0
[3,] 0.0000000 0.0000000    1
>
> library(mnormt)
> ll<-function(mu) log(dmnorm(data, mean = rep(mu,3), varcov=Sigma))
> x<-seq(4,6,0.1)
> ll<-sapply(x,nll)
> plot(x,ll,type="l",xlab=expression(mu),ylab=expression(ll(mu)))
```

Assuming that the log-likelihood function is differentiable with respect to the components of $\boldsymbol{\theta} = (\theta_1, \ldots, \theta_p)$, the likelihood is maximized either at the boundary of the parameter space or at the point where all partial derivatives of l with respect to the components of $\boldsymbol{\theta}$ are zero. This leads to the system of p *likelihood equations*, which are of form

$$\frac{\partial}{\partial \theta_i} l(\boldsymbol{\theta}|\boldsymbol{x}) = 0, \qquad i = 1, \ldots, p$$

The solution of the likelihood equations provides a candidate for the maximum likelihood estimate of $\boldsymbol{\theta}$. Those candidates may be global minima or maxima, local minima or maxima, or inflection points. Other candidates are at the boundary of the parameter space. Of these candidates, the *maximum likelihood estimate* (MLE) is the one that provides the maximum value of likelihood. It is denoted by $\widehat{\boldsymbol{\theta}}$.

The MLE of $\boldsymbol{\theta}$ is invariant to monotonic transformation $g(\theta)$ of the parameters. The property is useful as a constraint for parameters that have a bounded range. For example, the variance must be positive. The normal distribution can be reparameterized in terms of logarithmic variance by defining $\theta = \log(\sigma^2)$. Now, any real-valued estimate of θ will provide a positive estimate of variance as $\sigma^2 = \exp(\theta)$. For parameters that have both an upper and lower bound, such as the probability of success for a *Bernoulli*(p) random variable, function $\text{logit}(p) = \log\left(\frac{p}{1-p}\right)$ can be used to implement the restriction. These restrictions are closely related to the link functions in the context of generalized linear models (see below and Section 8.1.4).

The maximum-likelihood leads to consistent estimators that have the minimum variance if the sample size is sufficiently large. In addition, a CLT for ML-estimates states that the distribution of the parameters is normal in large samples. For distributions that are members of the exponential family (e.g. normal, Poisson, Bernoulli, binomial, exponential and gamma distributions), the estimators of the expected value are the sample mean, which is the minimum-variance unbiased estimator, even in small samples. The (asymptotic) variance of the ML-estimator is defined by the Crámer-Rao lower bound as:

$$\text{var}(\widehat{\boldsymbol{\theta}}) = \left\{ -\text{E}\left[\frac{\partial^2 l(\boldsymbol{\theta}|\boldsymbol{x})}{\partial \boldsymbol{\theta}^2}\right]\right\}^{-1} \tag{3.12}$$

where the $p \times p$ matrix $-\text{E}\left[\frac{\partial^2 l(\boldsymbol{\theta}|\boldsymbol{x})}{\partial \boldsymbol{\theta}}\right]$ is called *the Fisher information.* In applications, the large sample variance is estimated by replacing $\boldsymbol{\theta}$ with the ML estimates $\widehat{\boldsymbol{\theta}}$. Therefore, the variance of an ML-estimator is estimated by differentiating the log-likelihood twice with respect to the parameter vector $\boldsymbol{\theta}$, which results in a $p \times p$ matrix of functions. These functions are evaluated using the ML-estimates in the place of the parameters, and are multiplied by -1 to get an estimate of the Fisher's information. Inversion provides an approximate variance-covariance matrix of the parameter estimates.

The definition of likelihood considerably simplifies if the elements of $\boldsymbol{x} = (x_i, x_2, \ldots, x_n)'$ are independent. In that case, the joint density is the product of marginal densities, and the log-likelihood is, therefore, a sum of logarithmic densities:

$$l(\boldsymbol{\theta}|\boldsymbol{x}) = \sum_{i=1}^{n} \log f(x_i|\boldsymbol{\theta}) \,.$$

An extension to the regression context arises by assuming that the parameters are functions of some predictor variables, as illustrated already in Examples 2.2 (p. 10) and 2.3 (p. 11). Most commonly, the expected value μ is written as a function of a linear predictor $\boldsymbol{x}'\boldsymbol{\beta}$, where $\boldsymbol{x}$ includes predictors and $\boldsymbol{\beta}$ the regression coefficients. If the range of the expected value is bounded, then nonlinear transformation, $\mu = h(\boldsymbol{x}'\boldsymbol{\beta})$ called *the mean function*, is used to ensure that the expected value falls within its range. This formulation leads to the generalized linear models, which will be discussed in Chapter 8.

Example 3.4. Let $\boldsymbol{x} = (x_1, x_2, \ldots, x_n)'$ be observations from a normal distribution with variance 1 and mean μ. The log likelihood function with respect to unknown μ is

$$l(\mu|\boldsymbol{x}) = \sum_{i=1}^{n}\left[\log\left(\frac{1}{\sqrt{2\pi}}\right) - \frac{1}{2}(x_i - \mu)^2\right] = n\log\left(\frac{1}{\sqrt{2\pi}}\right) - \frac{1}{2}\sum_{i=1}^{n}(x_i - \mu)^2$$

Differentiating the log likelihood with respect to μ and setting equal to 0 gives the likelihood equation

$$\sum_{i=1}^{n} x_i - n\mu \quad = \quad 0$$

Solving for μ gives $\mu = \bar{x}$; therefore, the sample mean is a candidate for MLE, and the only solution to the ML equation. To formally verify that it is a global maximum, we should study the behavior of the second derivative. The Fisher information is $\frac{\partial^2 l(\mu|\boldsymbol{x})}{\partial \mu^2} = -n$, which is a constant. Using this in Equation (3.12) gives

$$\text{var}(\hat{\mu}) \quad = \quad 1/n \,.$$

Thus, the standard error of the estimated mean is $1/\sqrt{n}$; the same result that was found in Equation (3.10) without the assumption of normality.

Example 3.5. Let $x_1, x_2, \ldots, x_n$ be observations from a normal distribution with an unknown mean μ and an unknown variance σ^2. The ML-estimators become (show this!)

$$\begin{aligned}\widehat{\mu} &= \bar{x} \\ \widehat{\sigma^2} &= \frac{\sum_{i=1}^{n}(x_i - \bar{x})^2}{n}\end{aligned}$$

We see that the ML estimator of the expected value for the normal distribution is the same as the (least squares) estimation presented in Section 3.4.2. However, the estimator of variance differs from the unbiased estimator presented in Section 3.4.3; the ML-estimator has n in the denominator whereas the unbiased estimator has $n-1$. The ML estimator of variance is, therefore, downward biased. However, it is consistent: the bias vanishes as $n \to \infty$.

Example 3.6. The ML- estimation using numerical algorithms is fast and provides a satisfactory result in most cases. For such estimation, one just needs to define an R-function to compute the negative log-likelihood in the data. Consider independent samples from a common normal distribution below, with a known variance of 1 and an unknown mean.

```
> library(stats4)
> y10<-c(4.99, 4.42, 5.95, 4.49, 5.75, 3.92, 7.74, 5.77, 4.75, 5.61)
> y20<-c(5.80, 6.00, 5.35, 3.98, 4.99, 5.34, 5.14, 5.36, 3.19, 4.62,
+        4.57, 3.23, 5.57, 4.18, 4.66, 6.76, 5.83, 4.74, 6.91, 4.44)
```

We use function `mle` of package `stats4` for estimation. As the algorithm is based on minimization of a function, we need to define a function that evaluates the negative log-likelihood. The function for the smaller sample is:

```
> nll10<-function(mu) -sum(log(dnorm(y10,mu)))
```

The likelihoods in both samples as a function of μ are shown in Figure 3.2. We see that the likelihood based on the larger sample is narrower and has a sharper peak than the likelihood based on the smaller sample. The ML estimates can be found numerically using function `mle`.

```
> sol10<-mle(minuslogl=nll10,start=list(mu=0))
> summary(sol10)

Coefficients:
    Estimate Std. Error
mu     5.339  0.3162278
-2 log L: 28.99266
```

The estimate of μ for the smaller sample is 5.339, with the estimated variance of 0.316^2. For the larger data, we obtain $\widehat{\mu} = 5.033$ and $\widehat{\text{var}}(\widehat{\mu}) = 0.223^2$. The larger sample size leads to more precise estimates. This can also be seen from the plot of likelihoods: standard errors are inversely related to the second derivative of the likelihood at the solution, which is higher with the sharper peak of the likelihood with larger n. The ML estimators from the previous example leads to same estimates:

```
> mean(y10)
[1] 5.339
> sqrt(1/10)
[1] 0.3162278
```

Example 3.7. Consider an observed binary vector that includes 8 successes, coded as 1, and 12 failures, coded as 0.

```
x<-c(1,1,0,1,0,0,1,1,1,0,0,0,0,0,0,0,0,1,0,1)
```

The log-likelihood for independent data $x_1, \ldots, x_n$ from the *Bernoulli*(μ) distribution (equation (2.4), p. 10) is

$$l(\mu|\boldsymbol{x}) = \sum_{i=1}^{n} x_i \ln(\mu) + \left(n - \sum_{i=1}^{n} x_i\right) \ln(1-\mu)$$

Differentiating with respect to μ, setting to 0, and solving for μ gives

$$\widehat{\mu} = \bar{x}\,.$$

In our data, it gives $\widehat{\mu} = 0.4$.

We now illustrate how the MLE of μ is found numerically. A logit transformation $g(\mu) = \ln\left(\frac{\mu}{1-\mu}\right)$ is used to restrict the search space of the parameter μ to the interval $(0,1)$. Therefore, we reparameterize the Bernoulli *pmf* as

$$f(x|\gamma) = \left(\frac{e^\gamma}{1+e^\gamma}\right)^x \left(\frac{1}{1+e^\gamma}\right)^{1-x}$$

where $-\infty < \gamma < \infty$ is the parameter to be estimated. Any finite value of γ now leads to success probability between 0 and 1. The inverse logit function and negative log-likelihood are implemented below. We use value 0.5 as initial guess for μ. MLE solution is searched using function `stats::mle`. We get $\widehat{\gamma} \approx 13.863$, which gives, after back-transformation, the known MLE $\widehat{\mu} = 0.4$.

```
> logit<-function(p) log(p/(1-p))
> ilogit<-function(gamma) exp(gamma)/(1+exp(gamma))
> nll<-function(lp=logit(0.5)) -sum(log(dbinom(x,1,ilogit(lp))))
> sol2<-mle(nll,control=list(trace=1,maxit=1000))
initial  value 13.862944
final  value 13.460233
converged
> ilogit(coef(sol2))
 lp
0.4
```

Reparameterization using the logit function is particularly useful if μ is a function of predictors; recall Example 2.2 (p. 10). Estimation for the corresponding model with count data and log-link is demonstrated in Examples 3.9 and 8.4 (p. 260). Implementation for the data of Example 2.2 is left as an exercise.

Example 3.8. Assume that tree diameters in a forest stand follow *Weibull*(α, β) distribution, and an independent sample of diameters is available. We use this data to find the ML-estimates for parameters α and β in this stand. Both α and β have to be positive. Therefore, we reparameterize the Weibull density using $\theta_1 = \log(\alpha)$ and $\theta_2 = \log(\beta)$, which can have any values within $(-\infty, \infty)$. The following ML-estimates are found numerically: $\widehat{\theta}_1 = 2.03$ and $\widehat{\theta}_2 = 2.74$, with estimated large-sample standard variances $\widehat{\text{var}}(\widehat{\theta}_1) = 0.179^2$ and $\widehat{\text{var}}(\widehat{\theta}_1) = 0.0310^2$. To get estimates of α and β, we use the exponential transformation to θ_1 and θ_2. The estimates are $\widehat{\alpha} = 7.60$ and $\widehat{\beta} = 15.50$.

```
> DBH<-c(15.1, 17.1, 10.0, 13.5, 16.6, 18.1, 11.0, 13.4, 15.2, 17.1,
+        15.9, 17.7, 14.4, 13.8, 14.8, 11.9, 13.0, 14.7, 16.5, 11.0)
>
> nll<-function(theta1,theta2) {
+      shape<-exp(theta1)
+      scale<-exp(theta2)
+ #     cat(shape,scale,"\n")
+      -sum(log(dweibull(DBH,shape,scale)))
+      }
>
> # save ML-estimate into object 'solution'
> solution<-mle(minuslogl=nll,start=list(theta1=log(10),theta2=log(15)))
>
> summary(solution)
Maximum likelihood estimation

Call:
mle(minuslogl = nll, start = list(theta1 = log(10), theta2 = log(15)))

Coefficients:
       Estimate Std. Error
theta1 2.028662 0.17946940
theta2 2.740981 0.03097805
```

```
-2 log L: 88.8954
> exp(coef(solution))
   theta1    theta2
 7.603904 15.502190
```

Web Example 3.9. Maximum likelihood estimation of the parameters that specify the expected value of a Poisson distribution as a function of known predictor variable x, using log link.

3.4.5 Bayesian estimation

Bayesian statistics (see e.g. Gelman *et al.* 2013) provides an alternative to the methods presented in the previous subsections, which are sometimes called *frequentist statistics.* The fundamental difference is that the parameters are assumed to be random variables, not fixed values as we have assumed up to now.

Let $\boldsymbol{\theta}$ be a random vector of model parameters and let $\boldsymbol{y}$ be the vector of data. In Bayesian statistics, we want to estimate the conditional *pdf* $f(\boldsymbol{\theta}|\boldsymbol{y})$. Estimation is based on the Bayes formula (2.19) on p. 20, which is used to express the distribution of interest in terms of the conditional *pdf,* i.e. the likelihood of the data $f(\boldsymbol{y}|\boldsymbol{\theta})$ and the marginal distribution of the parameters $f(\boldsymbol{\theta})$ as

$$f(\boldsymbol{\theta}|\boldsymbol{y}) = \frac{f(\boldsymbol{y}|\boldsymbol{\theta})f(\boldsymbol{\theta})}{\int f(\boldsymbol{y}|\boldsymbol{\theta})f(\boldsymbol{\theta})d\boldsymbol{\theta}} \tag{3.13}$$

The marginal distribution $f(\boldsymbol{\theta})$ is called *the prior distribution* and $f(\boldsymbol{\theta}|\boldsymbol{y})$ *the posterior distribution* of parameter vector $\boldsymbol{\theta}$. The estimate of model parameter vector is, therefore, not a fixed vector but a joint distribution of $\boldsymbol{\theta}$. However, it is often summarized to single values, such as the modes. Furthermore, an interval that includes the (realized) parameter value with given probability, i.e. *the credible interval,* can be constructed based on the quantiles of the distribution.

The prior distribution needs to be specified and can be selected so that it includes such information about $\boldsymbol{\theta}$ that is available before the data are used. The Bayesian methods have been criticized because the researcher can manipulate the results by deliberate selection of priors. The counter-argument is that the Bayesian methodology enables accumulation of knowledge, if information of previous studies is used in setting the prior distribution. In addition, non-informative priors can also be used.

In traditional Bayesian statistics so-called *conjugate priors* are often used (see Section 2.9). If they are used, the prior and posterior distributions are of the same functional form. In addition, the cumbersome integral in the denominator has a closed solution in such cases.

The development and wide use of Bayesian computational methods, such as the Markov Chain Monte Carlo (MCMC), has led to wide use of Bayesian methods with complex models for which maximum likelihood estimators are not easily found. The Bayesian computational methods allow simulation of realizations of the posterior distribution with freely selected priors without evaluating the integral in the denominator. The only cost is increased computing time. To do so, one needs to be able to simulate realizations from the prior distribution and evaluate the likelihood in the observed data using the simulated value of the parameter vector. If the motivation for the Bayesian approach is to enable MCMC computations for complex models but not to include any prior information about them in the analysis, then one can use very flat, non-informative priors.

In this book, the use of Bayesian estimation in the context of linear mixed-effects models is discussed briefly in Section 5.3.5. In addition, the whole concept of mixed-effects models has sometimes been called empirical Bayes methods. The data are used to estimate the prior distribution, which is parameterized in terms of the fixed regression coefficients and the variance-covariance structure of the random effects and residual errors. For the prediction of random effects, the local data from each group is combined with the empirically estimated prior distribution to find the expected value of the posterior distribution of random effects. The `Rstan` package in `R`, provides a widely used tool for Bayesian analyses (Stan Development Team 2018).

Example 3.10. Consider a simple situation where we want to estimate a scalar parameter θ using observed data $\boldsymbol{y}$. The simplest MCMC method is the Metropolis algorithm. Denote the numerator of Equation (3.13), $f(\boldsymbol{y}|\theta)f(\theta)$, by $a(\theta)$. The algorithm includes the following steps:

1. Choose an arbitrary starting point θ_0 from Θ (the sample space of θ). Choose an arbitrary symmetric *pdf* (the proposal density) $g(\theta_{t+1}|\theta_t)$, which suggests the new value of θ (θ_{t+1}) given the previous value θ_t. Usually we use $N(\theta_t|\sigma^2)$ as the proposal density, where the variance σ^2 is given by the user.
2. At each iteration t

   2.1 Generate a candidate of θ: θ' using $g(\theta_{t+1}|\theta_t)$.

   2.2 Compute the acceptance ratio
$$\alpha = \frac{a(\theta')}{a(\theta_t)}$$

   2.3 If $\alpha \geq 1$, then accept the proposal and set $\theta_{t+1} = \theta'$. If $\alpha < 1$, then accept θ' as θ_{t+1} with probability α. If θ' is not accepted, set $\theta_{t+1} = \theta_t$.

3. Repeat steps 2.1 – 2.3 until a desired sample size has been obtained.

Consider the sample $\boldsymbol{y} = (9.76, 9.08, 16.23, 10.28, 10.51)'$ from a normally distributed population with known variance $\sigma^2 = 2^2$. The aim is to estimate the population mean θ. Assume that we know beforehand that the population mean is within the range $(7, 13)$ and utilize that information through an informative prior

$$\theta \sim Uniform(7, 13)\,.$$

We use the standard normal density as the proposal distribution. The iteration is started from a random number from the prior. The size of the generated MCMC sample from the posterior distribution is $M = 1000$. The estimation algorithm has been implemented below.

```
> y<-c(9.76,9.08,16.23,10.28,10.51)
> lb<-7; ub<-13 # bounds of the uniform prior
> # numerator of the posterior distribution
> postProfile<-function(theta) {
+   sapply(theta, function(thet) prod(dnorm(y,thet,2))*dunif(thet,lb,ub))
+ }
>
> # The Metropolis algorithm
> # Use standard normal density as the proposal distribution
> # Start from a random number from the prior
> theta<-runif(1,lb,ub)
> M<-1; nsim<-1000
> thetasim<-rep(NA,nsim)
> while (M<=nsim) {
+       thetanew<-theta+rnorm(1) # Generate a proposal
+       while (thetanew<lb|thetanew>ub) thetanew<-theta+rnorm(1)
+       alpha<-postProfile(thetanew)/postProfile(theta)
+       if (alpha>1|runif(1)<alpha) {
+          theta<-thetasim[M]<-thetanew
+          M<-M+1
+          }
+       }
```

After running the script, vector `thetasim` includes 1000 values from the desired posterior distribution. In an MCMC sample, successive values of the simulated sample are correlated. If that correlation is very strong, then the simulation algorithm does not sample well all parts of the posterior; in that case it is said that the algorithm does not mix the posterior distribution suf-

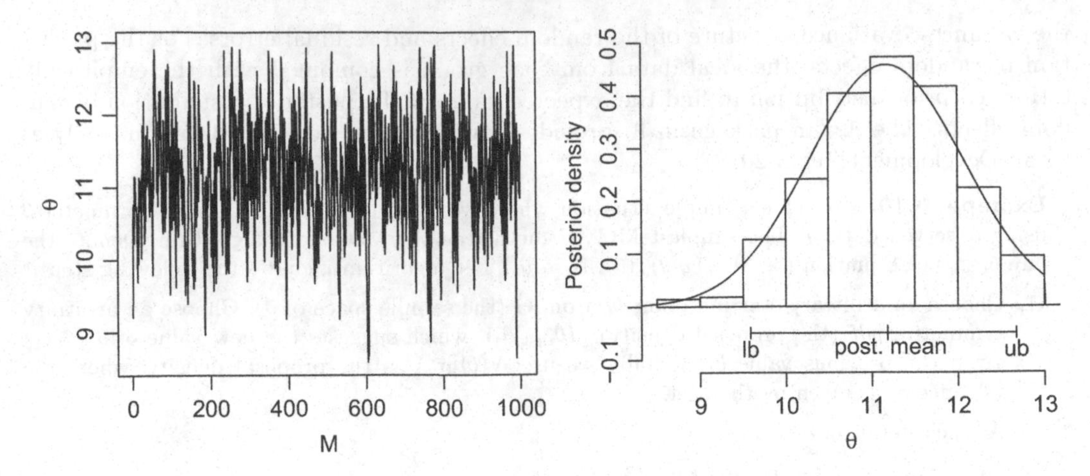

FIGURE 3.3 The trace plot (left) and histogram (right) of the values of θ produced by the Metropolis algorithm of Example 3.10. The smooth line on the histogram illustrates the true posterior distribution and the horizontal line below the histogram the estimated posterior mean and the Bayesian 95% credible interval for population mean.

ficiently well. To explore the mixing, a trace plot of the simulated values can be produced. It is a plot where the simulated values have been plotted in the order that they were generated. The left-hand graph in Figure 3.3 shows the trace plot of the simulated posterior sample. It shows some autocorrelation but the extent of autocorrelation is much shorter than the sample size. Therefore, the algorithm has mixed sufficiently well. The mixing could be improved by increasing the number of iterations and by thinning the sample, e.g. by using $M = 10000$ and only using every 10th simulated value in the final posterior sample. Also, the variance of the proposal density could be changed. The right-hand graph shows the histogram of the generated sample. The smooth line shows the true posterior *pdf*, which was produced using

```
> thetavec<-seq(0,20,0.01)
> postden<-postProfile(thetavec)/integrate(postProfile,lb,ub)$value
> lines(thetavec,postden)
```

The posterior mean is 11.181 and the Bayesian 95% credible interval is [9.58, 12.68]. The traditional estimates (see Section 3.4.2) and confidence intervals are $\widehat{\theta} = 11.172$, and [9.42, 12.92]. The narrower confidence intervals of the Bayesian method result from the assumed prior distribution. Using a wider prior distribution would also make the Bayesian confidence interval wider.

```
> mean(thetasim)
[1] 11.18086
> quantile(thetasim,probs=c(0.025,0.975))
     2.5%     97.5%
 9.584519 12.678903
>
> # Using sample mean, its known standard error and normal distribution
> mean(x)
[1] 11.172
> qnorm(c(0.025,0.975),mean(x),2/sqrt(length(x)))
[1]  9.418955 12.925045
```

3.5 Prediction

Let us then consider predictors, i.e. methods for estimating values of random variables. For different predictor concepts, see e.g. McCulloch *et al.* (2008). The concepts needed for describing predictors are similar to the concepts needed for estimators in Section 3.4.1, but some important differences exist. Note that the term 'predictor' will be used with a different meaning in the context of linear models, where the x-variables can be called the predictors.

3.5.1 Comparing alternative predictors

With respect to a predictor, we need to separate two biases. Let us assume that we predict a random variable or parameter h_1 using an observed random vector $\boldsymbol{h_2}$. In mixed models h_1 could be, for example, a random effect (denoted e.g. with the symbol b). In a regression context, unknown values of the dependent variable y may also be predicted. A predictor $\widetilde{h}_1$ is *marginally unbiased* if:

$$\mathrm{E}\left(\widetilde{h}_1 - h_1\right) = E_{\boldsymbol{h_2}}\left(\widetilde{h}_1\right) - E_{h_1}\left(h_1\right) = 0 \Leftrightarrow E_{\boldsymbol{h_2}}\left(\widetilde{h}_1\right) = E_{h_1}\left(h_1\right) \tag{3.14}$$

Symbol $E_{\boldsymbol{h_2}}$ means that the expectation is taken with respect to the multivariate distribution of $\boldsymbol{h_2}$. E_{h_1} is the expectation with respect to the distribution of h_1. Without further specification 'unbiased predictor' always refers to the marginally unbiased predictor.

Predictor $\widetilde{h}_1$ is *conditionally unbiased* if

$$\mathrm{E}\left(\left(\widetilde{h}_1 - h_1\right)|\boldsymbol{h_2}\right) = \mathrm{E}\left(\widetilde{h}_1|\boldsymbol{h_2}\right) - \mathrm{E}\left(h_1|\boldsymbol{h_2}\right) = 0 \Leftrightarrow \mathrm{E}\left(\widetilde{h}_1|\boldsymbol{h_2}\right) = \mathrm{E}\left(h_1|\boldsymbol{h_2}\right) \tag{3.15}$$

Prediction variance is the variance of the prediction error, i.e. $\mathrm{var}\left(\widetilde{h}_1 - h_1\right)$. This should not be confused with the variance of the predictor $\mathrm{var}\left(\widetilde{h}_1\right)$. To describe the precision of prediction, the former is of interest, the latter is not, even if to fully understand the prediction theory the behavior of $\mathrm{var}\left(\widetilde{h}_1\right)$ also needs to be understood. Note that $\mathrm{var}\left(\widetilde{h}_1 - h_1\right) = \mathrm{var}\left(\widetilde{h}_1\right) + \mathrm{var}\left(h_1\right) - 2\,\mathrm{cov}\left(\widetilde{h}_1, h_1\right)$. In the estimation of fixed α, the variance of the estimation error $\mathrm{var}\left(\widehat{\alpha} - \alpha\right)$ and the variance of the estimator $\mathrm{var}\left(\widehat{\alpha}\right)$ are the same.

Marginal prediction variance, conditional variance and conditional expectation are connected using the law of total variance, Equation (2.46) on page 25:

$$\mathrm{var}\left(\widetilde{h}_1 - h_1\right) = E_{\boldsymbol{h_2}}\,\mathrm{var}\left(\left(\widetilde{h}_1 - h_1\right)|\boldsymbol{h_2}\right) + \mathrm{var}_{\boldsymbol{h_2}}\,\mathrm{E}\left(\left(\widetilde{h}_1 - h_1\right)|\boldsymbol{h_2}\right) \tag{3.16}$$

The bias and variance of a predictor are combined in MSE, either in the marginal MSE:

$$MSE\left(\widetilde{h}_1 - h_1\right) = \left(\mathrm{E}\left(\widetilde{h}_1 - h_1\right)\right)^2 + \mathrm{var}\left(\widetilde{h}_1 - h_1\right) \tag{3.17}$$

or in the conditional MSE:

$$MSE\left(\left(\widetilde{h}_1 - h_1\right)|\boldsymbol{h_2}\right) = E\left(\left(\widetilde{h}_1 - h_1\right)|\boldsymbol{h_2}\right)^2 + \mathrm{var}\left(\left(\widetilde{h}_1 - h_1\right)|\boldsymbol{h_2}\right) \tag{3.18}$$

Best Predictor (BP) is a predictor that has the smallest MSE. The conditional expectation of h_1 given $\boldsymbol{h_2}$, $\mathrm{E}\left(h_1|\boldsymbol{h_2}\right)$ is always the best predictor (Christensen 2011). Thus, in the prediction of a random variable, we are actually searching for the conditional mean given the data.

3.5.2 Linear prediction

A linear predictor is defined similarly as the linear estimator (3.6) or (3.9), i.e. $\widetilde{h}_1 = w_0 + \boldsymbol{w}'\boldsymbol{h}_2$ or for vector $\widetilde{\boldsymbol{h}}_1 = \boldsymbol{w}_0 + \boldsymbol{W}\boldsymbol{h}_2$. Neither w_0, $\boldsymbol{w}$, $\boldsymbol{w}_0$ or $\boldsymbol{W}$ depend on $\boldsymbol{h}_2$.

Let us now assume that we predict a random vector $\boldsymbol{h}_1$ from an observed random vector $\boldsymbol{h}_2$. Let us denote that

$$\boldsymbol{h} = \begin{pmatrix} \boldsymbol{h}_1 \\ \boldsymbol{h}_2 \end{pmatrix}$$

It is assumed that $\mathrm{E}(\boldsymbol{h}_1) = \boldsymbol{\mu}_1$, $\mathrm{E}(\boldsymbol{h}_2) = \boldsymbol{\mu}_2$, $\mathrm{var}(\boldsymbol{h}_1) = \boldsymbol{V}_1$, $\mathrm{var}(\boldsymbol{h}_2) = \boldsymbol{V}_2$, and $\mathrm{cov}(\boldsymbol{h}_1, \boldsymbol{h}_2) = \boldsymbol{V}_{12}$. This can be written as

$$\begin{pmatrix} \boldsymbol{h}_1 \\ \boldsymbol{h}_2 \end{pmatrix} \sim \left[\begin{pmatrix} \boldsymbol{\mu}_1 \\ \boldsymbol{\mu}_2 \end{pmatrix}, \begin{pmatrix} \boldsymbol{V}_1 & \boldsymbol{V}_{12} \\ \boldsymbol{V}_{12}' & \boldsymbol{V}_2 \end{pmatrix} \right]$$

The Best Linear Predictor (BLP) of $\boldsymbol{h}_1$ is (Christensen 2011, p. 35)

$$BLP(\boldsymbol{h}_1) = \widetilde{\boldsymbol{h}_1} = \boldsymbol{\mu}_1 + \boldsymbol{V}_{12}\boldsymbol{V}_2^{-1}(\boldsymbol{h}_2 - \boldsymbol{\mu}_2) \tag{3.19}$$

with the prediction variance

$$\mathrm{var}(\widetilde{\boldsymbol{h}}_1 - \boldsymbol{h}_1) = \boldsymbol{V}_1 - \boldsymbol{V}_{12}\boldsymbol{V}_2^{-1}\boldsymbol{V}_{12}' \tag{3.20}$$

This result means that if the expectations and variance-covariance matrices of two random vectors are known, and either one of them is observed, the other one can be predicted using Equation (3.19). Furthermore, the variance of the prediction error can be calculated using Equation (3.20).

If $\boldsymbol{h}$ follows the multinormal distribution, then BLP is the Best Predictor, i.e. the conditional mean of y given $\boldsymbol{x}$; recall Equation 2.48 on p. 28. BLP can be BP also otherwise.

Best Linear Unbiased Predictor (BLUP) is the marginally unbiased linear predictor that has smallest marginal variance. Typically, BLUPs are obtained using the BLP Equation (3.19), so that the mean vectors $\boldsymbol{\mu}_1$ and $\boldsymbol{\mu}_2$ are computed using BLUE estimates of regression model parameters (Christensen 2011). BLUPs will be considered in Sections 4.7.2, 5.4.1, 6.5, 6.6, 7.3.5, 9.2.4, 9.4.3, 10.1.1 and 11.5.3.

When estimated variance and covariance parameters are used in place of unknown true parameters to compute variances and covariances in Equation (3.19), the obtained best linear unbiased predictor is called an *Estimated BLUP* (EBLUP). If y is predicted using BLP (3.19) from a p dimensional $\boldsymbol{x}$, and var $(\boldsymbol{x})$ is diagonal (the covariances are zero), then

$$\widetilde{y} = \mu_y + \sum_{i=1}^{p} \frac{\mathrm{cov}(y, x_i)}{\mathrm{var}(x_i)}(x_i - \mu_i),$$

$$\mathrm{var}(\widetilde{y} - y) = \mathrm{var}(y) - \sum_{i=1}^{p} \frac{\mathrm{cov}(y, x_i)^2}{\mathrm{var}(x_i)}$$

Example 3.11. Let us assume that it holds for a times series y_t that $\mathrm{E}(y_t) = 5$ and $\mathrm{cov}(y_t, y_{t+r}) = 4\exp(-0.6r)$. We note first that $\mathrm{var}(y_t) = 4\exp(0) = 4$, $\mathrm{cov}(y_t, t_{t+1}) = 2.1952$ and $\mathrm{cov}(y_t, t_{t+2}) = 1.2048$. Let us predict y_3 when we know that $y_2 = 2$. We know in advance that the predicted value is between the observed value at $t = 2$, i.e. 2 and the marginal expected value 5. From Equation (3.19) we get:

$$\widetilde{y}_3 = \mu_y + \frac{\mathrm{cov}(y_2, y_3)}{\mathrm{var}(y_2)}(y_2 - \mu_y) = 5 + \frac{2.1952}{4}(2 - 5) = 3.3535.$$

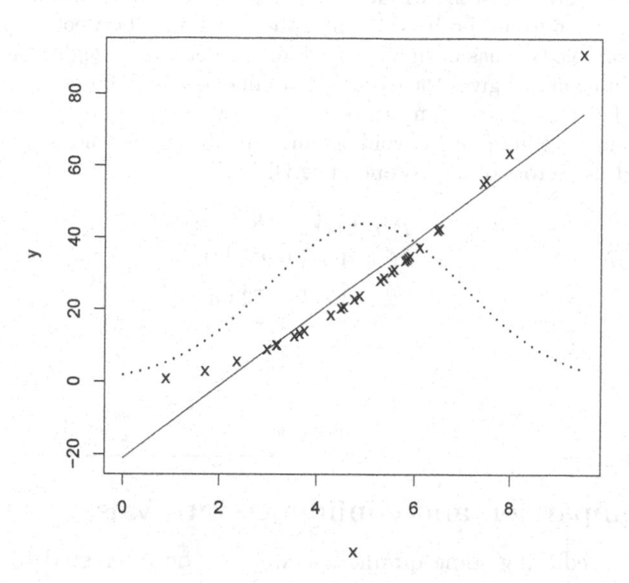

FIGURE 3.4 Illustration of Example 3.12. The solid line is the linear predictor; the dashed line is the probability density function of x. Points are 30 random points generated by the model.

Let us then also utilize the measurement $y_1 = 10$. From Equation (3.19) we get:

$$\widetilde{y}_3 = \mu_y + (\text{cov}(y_1, y_3), \text{cov}(y_2, y_3)) \begin{pmatrix} \text{var}(y_1) & \text{cov}(y_1, y_2) \\ \text{cov}(y_1, y_2) & \text{var}(y_2) \end{pmatrix}^{-1} \begin{pmatrix} y_1 - \mu_y \\ y_2 - \mu_y \end{pmatrix}$$
$$= 5 + (1.2048, 2.1952) \begin{pmatrix} 4 & 2.1952 \\ 2.1952 & 4 \end{pmatrix} \begin{pmatrix} 5 \\ -3 \end{pmatrix} = 3.3535$$

That is, we get exactly the same predicted value when predicting y_3 using y_1 and y_2 as when using only y_2. This is not a coincidence: it follows from the fact that the exponential covariance function implies that BLP has the *Markov*-property: $BLP\,(y_t|y_1, \ldots y_{t-1}) = BLP\,(y_t|y_{t-1})$. In statistics, many models exhibit this property, e.g. the Markov chain. Note that a covariance function $\exp(-cr)$ can be expressed in an equivalent form $(\exp(-c))^r = \gamma^r$

Example 3.12. Let us assume that x is distributed symmetrically with mean μ and variance σ^2. Let $y = x^2 + e$, where e is a random error with mean zero. Let us predict y from x using the BLP (3.19). Now $\text{E}(y) = \text{E}(x^2) = \sigma^2 + \mu^2$. With a symmetrical distribution, the skewness is zero, i.e. $\text{E}(x - \mu)^3 = 0$. This implies that (show it!) $\text{cov}(x, y) = \text{cov}(x, x^2) = 2\mu\sigma^2$. According to Equation (3.19) the BLP is: $\text{E}(y) + \frac{\text{cov}(x,y)}{\text{var}(x)}(x - \mu) = \sigma^2 - \mu^2 + 2\mu x$. In Figure 3.4, the solid line is the linear predictor line and the dashed line shows the distribution of x when x is normally distributed with $\mu = 5$ and $\sigma = 2$. The points are 30 random points when the variance of e is zero (non-zero variance would not affect the predictor). For each value of x the predictor is conditionally biased, except when $\sigma^2 - \mu^2 + 2\mu x = x^2$, i.e. $x = \mu \pm \sigma$, $x = 3$ or $x = 7$ in the figure. However, the expected value of the prediction errors is zero over the distribution of x, i.e. the predictor is marginally unbiased. If $\mu = 0$, the predictor line would be the horizontal line at height σ^2. The best possible predictor would always be x^2, which is not linear with respect to x.

Example 3.13. Assume that a model system is available for prediction of tree volume and tree height using tree diameter only. If we assume that the form of the model system is correct and that the parameters do not include errors, such a model system would give the expected values for height and volume for trees with a known diameter. Furthermore, the residual errors of the model are assumed to have constant variances so that the residual variance of height would be $\boldsymbol{V}_2 = 0.5^2$, that of volume would be $\boldsymbol{V}_1 = 2^2$, and the covariance between them is $\boldsymbol{V}_{12} = 0.5$.

Assume that a sample tree has been measured for diameter and height, the observed height is $\boldsymbol{h}_2 = 14$ m. The volume model gives the expected volume as $\boldsymbol{\mu}_1' = 70$ liters, and the height model gives the expected height as $\boldsymbol{\mu}_2' = 13$ m. In this case, both vectors are of length 1. However, we write them in boldface font in order to avoid the introduction of new notations, and to show that they can be treated as vectors of length one. The BLP is

$$\begin{aligned}\widetilde{\boldsymbol{h}}_1 &= \boldsymbol{\mu} + \boldsymbol{V}_{12}\boldsymbol{V}_2^{-1}(\boldsymbol{h}_2 - \boldsymbol{\mu}_2) \\ &= 70 + 0.5 \times 0.25^{-1}(14 - 13) \\ &= 70 + 2 \times 1 = 72\text{dm}^3\end{aligned}$$

3.6 Model comparison and confidence intervals

After estimating or predicting some quantities, we may be interested to infer what might be the true values of those quantities. More generally, we may be interested if the 'true' model has certain properties. Usually, we want to know if a parameter is zero (or one), or if the difference between two parameters is zero. The questions are analyzed in statistics with statistical hypothesis testing and by constructing confidence intervals. Both testing and confidence intervals are closely related viewpoints to the same problem.

3.6.1 Testing of hypothesis

In statistics, there are two competitive approaches to the testing of a hypothesis. The first approach is based on the work of R. A Fisher, and the second approach centers on the work of J. Neyman and E. S. Pearson. The approaches have been widely compared and discussed, for example, by Lehmann (1993), Berger (2003), Hubbard *et al.* (2003) and Christensen (2005).

Testing has the following components.

1. Assumptions that are anticipated to be true in all cases. This is what we call the *full model.* The estimation is already based on these assumptions. Testing is always done conditional on the full model. The full model cannot be tested without assuming an even fuller model. The researcher is responsible for assuming a reasonable full model. The fit of the full model should be ensured, for example, by careful graphical evaluation.

2. We may be interested to know whether a certain special case of the full model is true. This special case is called the null hypothesis, H_0. The Neyman-Pearson theory of hypotheses testing is based on the notion of comparing the null hypothesis to some alternative *simple hypothesis*, or to a set of hypotheses called a *composite hypothesis.* Fisher merely tested the null hypothesis without considering specific alternatives. Fisherian testing is philosophically based on the idea of proof by contradiction in which the contradiction is not absolute. While we describe the alternative hypothesis in this book, for us it is just the complement of the null hypothesis among models that satisfy the assumptions of the full model. So, in this book, our views are closer to those expressed by Fisher. Sometimes a one-sided test may be reasonable, and then the set of alternative hypotheses is more restricted.

3. We compute from the data a test statistic that is sensitive to deviations from H_0, and we know the distribution or an approximate distribution of the statistic assuming H_0 is true. If the statistic obtains a value that is unlikely when H_0 is true, H_0 will be rejected. How likely the obtained value is, is computed as the probability, *p-value*, to get a value of the test statistics that is at least as weird as the obtained value if H_0 is true. In two-sided tests, the *p*-value is computed using both tails of the distribution of the statistic. In a one-sided test only one tail of the distribution is used.

The *p*-value is a measure of the significance for the evidence for rejecting H_0. For Fisher, the *p*-value is the primary result of a test. A *p*-value may then be classified into significance classes. Usually, a deviation from H_0 is declared significant if $p \leq 0.05$. In earlier times, when the *p*-values could not be easily computed, the quantiles of test statistics were tabulated, and researchers could find the significance class from these tables. Stars were used to indicate to which class the *p*-value belonged.

In the Neyman-Pearson theory, the researcher sets the *significance level* α before the analysis, which is the probability of making a *type I error*, i.e. rejecting H_0 when it is true. The testing is done by defining a rejection area for the test statistic, so that H_0 is rejected if the test statistic falls in the rejection area, and the probability of rejecting true H_0 is α. In practice, the rejection can be done by rejecting H_0 if $p \leq \alpha$. If H_0 is rejected, then the alternative hypothesis is accepted. If $p > \alpha$, then H_0 is (temporarily) accepted. Usually $\alpha = 0.05$ is used. When repeatedly making independent tests where H_0 is true, the relative frequency of falsely rejecting H_0 approaches α. A common misconception is to consider that the *p*-value is the probability of the false rejection of H_0 (α is the probability of false rejection). In the orthodox Neyman-Pearson theory, the exact p is irrelevant: it is only important to ascertain whether $p \leq \alpha$.

If H_0 is accepted, even though it is false, then a *type II error* is made. The *power of the test* is defined as 1- probability of type II error. The power of a test is dependent on the true model, thus it can be computed only for specific alternative hypotheses. Power computations are part of the Neyman-Pearson test theory where the key result is the *Neyman-Pearson lemma.* This lemma states that when testing a simple hypothesis against a simple alternative (e.g. $H_0 : \mu = 0$ against $H_A : \mu = 5$) with a given significance level, the most powerful test is obtained by setting a threshold for the likelihood ratio so that the previously set significance level is obtained.

The Neyman-Pearson theory puts the hypothesis testing into a decision theoretical context where H_0 is either rejected or accepted following rules that lead in the long run to rational behavior. Fisher opposed this decision theoretic framework. For him, hypothesis testing was a part of inductive inference where H_0 is never really accepted, even if its rejection has failed in the study at hand. In general, our attitudes in this book are closer to Fisher, that is, we oppose a mechanical attitude for testing. However, clear cut decisions need to be often made e.g. when selecting explanatory variables to a regression equation.

If H_0 is not rejected, there are two reasons for not saying that it is accepted. First, following Sir Karl Popper, the philosophy of science has recognized that the hypothesis cannot be accepted, just that rejection has failed. Second, we usually can be sure that all relevant factors have some effect, we just do not have enough data to reject the hypothesis of no effect. This is related to the concept of the power of the test. Because the power of the test requires a fully specified alternative hypothesis, it is not usually considered in applications, but the concept should be kept in mind when interpreting why H_0 is not rejected. In science, based on statistical methods, we never make a final decision in regard to the correctness of some hypothesis. Thus, testing can be seen as a method for organizing and presenting the results of a study at hand.

It should be noted that *p*-values are random variables (Murdoch *et al.* 2008). If H_0 is true, then the *p*-value is a uniform random variable between 0 and 1. In cases where an alternative hypothesis is true, the p-value is also a random variable. Its variance is less than under H_0 but anyhow it is larger than most researchers envision (Boos and Stefanski 2011). Owing to the variability of p-values, significant p-values are often not repeated in replicated studies. Therefore, it is misleading to give p-values with several significant digits. A less rigid attitude to p-values is also emphasized by the fact that p-values are often computed using quite inaccurate approximations to the true distribution of the test statistics. These approximations deteriorate as more complex methodology is used.

If the computed p-value is biased upwards (downwards), then the test is called *conservative* (anticonservative). Conservative tests are generally considered to be better (safer) than anticonservative tests when testing whether a regression coefficient deviates from zero. Finding unwarranted causal associations is a more serious problem in science than not finding effects, which will be probably found in future.

Often researchers only consider small p-values. Christensen (2003) points out that attention should also be paid to p-values very close to 1. For instance, when testing deviation of regression coefficients from zero using an F-test, a p-value very close to 1 can be caused by not accounting for negatively correlated data or heteroscedasticity. Similarly, small p-values may also be caused by model misspecification. The direction of the bias in F- tests is determined both by the true correlation structure of residuals and by the values of the predictor variables (see Box 4.3 on p. 95).

Many researchers consider that the purpose of a statistical analysis is to obtain as many significant p-values as possible. Often significant p-values are obtained by making false assumptions, e.g. about the independence of observations or the homogeneity of error variance. For an honest scientist, the purpose is to make valid statistical inferences, not to obtain statistically significant p-values. If two alternative models are used for testing a certain effect, the model that more realistically models the underlying random process should be selected, not the one that provides the lower p-value.

A common misconception is to consider the p-value as the probability that H_0 is true. Such a probability is obtained only in Bayesian statistics. In the Bayesian testing theory, no distinction is made between the null hypothesis and alternative hypotheses; one just computes the posterior probability for each hypothesis using the prior probabilities, the data and the Bayes theorem. The hypothesis with the highest posterior probability is selected. Studies in Bayesian testing theory show that the posterior probability of H_0 can be much larger than the p-value, often by several orders of magnitude (Berger and Sellke 1987). It may be surprisingly high, even if the p-value is significant. In a new approach to statistical hypotheses testing, called Constrained Bayesian Methods (CBM), the observation space contains the regions for making the decision and the regions for not-making the decision (Kachiashvili 2019).

3.6.2 Multiple tests

If we make K tests at significance level α, and in fact H_0 is true for all tests, then the probability that we obtain at least one significant H_0 is $1 - (1-\alpha)^K$. For instance, if $K = 10$ and $\alpha = 0.05$, the probability that we obtain at least one significant test result is 0.40. An important method for avoiding false significant test results is to first test whether a set of effects is jointly zero, and to test individual effects only if the hypothesis that all effects are zero is rejected. Before testing individual correlations, one should test the hypothesis that all correlations are zero. Another method is to control the *family-wise error rate* by correcting the separate p-values so that the family-wise probability of rejecting the hypothesis that all effects are zero is less or equal to α in case all H_0's are true. A simple conservative correction method is to use *Bonferroni correction*, which is obtained by multiplying each individual p-value by K (values greater than 1 are taken to be 1). The

Box 3.2 Publication bias

In many fields, journals publish only papers that present significant test results. This leads to a publication bias. Suppose there is common interest to study whether a certain chemical substance has a positive effect on some disease. In fact, there is no effect at all. One hundred studies are made. As we may expect, about five studies report a significant test for the effect. Only those studies are published, and thus the scientific community starts to believe that the substance has a positive effect. If a well-planned and well-written study shows that there is no significant evidence to support a scientifically interesting hypothesis, these studies should be published, so the publication bias can be avoided.

value of K should be the number of parameters in the original full model, even if some predictors have thereafter been dropped. Less conservative and thus more powerful testing procedures are the sequential Bonferroni (or Holm-Bonferroni) and Sidak (or Holm-Sidak) methods.

Another problem related to making several tests is the case where one first tests if a certain property of the model holds, and further tests are made with the assumption that the first test gave the correct answer. We will discuss this in the context of mixed models where we argue that it is reasonable to keep those random effect terms in the model, which are suggested by the data structure, even though their variances do not significantly differ from zero.

3.6.3 Confidence intervals

If β is a parameter then a $100\gamma\%$ *confidence interval* is the interval $[L, U]$ defined by:

$$P(L \leq \beta \leq U) = \gamma \tag{3.21}$$

Interval $[L, U]$ is also called the *interval estimate* of β. Note that $[L, U]$ varies from sample to sample. Usually, one tries to construct the interval so that $P(\beta < L) = P(\beta > U) = \gamma/2$, even if the shortest possible interval can also be used. The confidence interval is directly related to testing. The hypothesis that $\beta = \beta_0$ can be rejected at significance level $1 - \gamma$ if $\widehat{\beta} < L$ or $\widehat{\beta} > U$.

There are two possible interpretations for the probability statement in Equation (3.21). In the frequentist paradigm generally applied in this book, the probability is γ that the random confidence interval computed using the same method covers the true fixed parameter. In Bayesian statistics, it means that the probability is γ that the random parameter β is included in the interval $[L, U]$, which is considered as fixed after the data have been observed. In Bayesian statistics, these intervals are called *credible intervals.*

A common mistake is to test whether two parameters are different by examining whether their confidence intervals overlap or not (Schenker and Gentleman 2001). If the estimates of two parameters are independent and the 95% confidence intervals do not overlap, then the difference is statistically significant at least at 5% level. However, the difference can be significant even if the confidence intervals do overlap. If the estimates are negatively correlated, then the nonoverlapping of intervals does not necessarily imply a significant difference.

3.6.4 Information theoretic model comparison

The value of the log-likelihood itself does not tell us how well the assumed model fits to the data. However, if there are two alternative models with an equal number of parameters, the model with the highest maximized likelihood fits the best. The *Akaike and Bayesian Information criteria (AIC and BIC)* extend this idea to models with a different number of parameters through penalizing model complexity. As such, they can be used to compare models with different number of parameters.

The aim is to approximate the information lost by using two alternative approximations of the true model. The approach can be used for models fitted using the method of maximum likelihood to a sufficiently large data set. Consider a situation where the true population model f is unknown and is approximated by a model g_1 with p_1 parameters and maximized likelihood L_1. In addition, assume that another approximation is provided by model g_2 with p_2 parameters and maximized likelihood L_2. The difference between the information lost using models g_1, and the information lost using approximation g_2, can be approximated in a large sample by comparing the AIC between the two models (Akaike 1974):

$$AIC = 2p - 2\log(L) \tag{3.22}$$

where p is the total number of parameters in the model, and $\log(L)$ is the maximized log-likelihood of the model in the data. The model with the minimum value of AIC will be selected.

From a user's point of view, the application of BIC is similar to that of AIC, even though the underlying theory differs. The former penalizes model complexity more than the latter, and is defined as follows

$$BIC = -\log(n)p - 2\log(L) \tag{3.23}$$

where n is the number of observations, and the other notations are similar to those used in AIC.

Essentially, both criteria have two components: likelihood and a penalty associated with the number of parameters involved in the model. As the model fit is based on *maximum* likelihood, which is multiplied by a negative number, the model with the lowest AIC or BIC is selected. Without the penalty term, likelihood itself as a model selection criteria would favor the most complex model with the highest number of parameters. The penalty terms ensure that such parameters, which do not sufficiently increase the likelihood, will not be included. Information criteria could also be used in model comparison for nested models, although the theory is very different from the theory of hypothesis tests. An essential difference from a user's point of view is that p-values, or other probability statements, are not associated with this procedure.

4

Linear Model

A very common task in many fields of applied sciences is to find and quantify the factors behind the observed variation in a variable of interest, and to predict the value of the variable of interest for a sampling unit with certain known properties. In the context of forest and environmental data, examples of such tasks include modeling of tree allometry, such as height (e.g. Mehtätalo *et al.* 2015), volume and taper (Schumacher and Hall 1933, Gregoire and Schabenberger 1996, Lappi 2006), prediction of tree and stand growth (Vanclay 1994), modeling site index (e.g. Lappi and Bailey 1988) or other traditional questions of the growth and yield (Burkhart and Tomé 2012, Gregoire *et al.* 1995); estimating the response of trees or ecosystems to different treatments, such as elevated temperature, carbon dioxide concentration or thinning (Kilpeläinen *et al.* 2005, Mehtätalo *et al.* 2014); modeling the habitat preferences of an animal (Melin *et al.* 2016a) and modeling the productivity of stump-lifting machines (Palander *et al.* 2016).

A very general and powerful tool for such tasks is the linear model (LM) and its generalizations: the generalized linear model (GLM), generalized additive model (GAM), nonlinear model (NLM), linear mixed-effects model (LME), generalized linear mixed-effects model (GLMM), generalized additive mixed-effects model (GAMM), nonlinear mixed-effects model (NLME) or a multivariate system of any of these models. To use any of these methods, a basic understanding of the linear model (or linear regression) is essential. For example, the way to express the effects of predictors or treatments is common to all these methods.

This chapter summarizes the essential ideas and properties of the linear model for a researcher in an empirical field. We assume that the reader is already familiar with the basic principles to some degree. The aim is to give a large-scale roadmap for practitioners in a limited space, but still cover most of the essential details using sufficiently rigorous notations. For a more elaborated discussion, there are several excellent texts available about regression. Those textbooks include Christensen (2011), Demidenko (2013), Fahrmeir *et al.* (2013), Gałecki and Burzykowski (2013), Harrell (2001), McCullagh and Nelder (1989), McCulloch *et al.* (2008), Montgomery *et al.* (2012), Muller and Stewart (2006), Pinheiro and Bates (2000), Ratkowsky (1990), Searle *et al.* (1992), Stroup (2013) and Weisberg (2014). A major difference of this chapter to most existing texts is that we discuss the possibility of modeling the variance-covariance structure of residual errors from the very beginning. In addition, we illustrate how that model could be utilized for prediction purposes in different situations.

4.1 Model formulation

4.1.1 Model for an individual observation

Consider a data set with n observations. The variable of interest for observation i is denoted by y_i, and is called the *response variable*. Other commonly used names for y_i are *dependent variable*, *regressand* or just *y-variable*. The response is considered as a random variable, and interest focuses on how strongly the specific variables of interest are associated with it, i.e. how well they explain the variability in the response variable.

The p variables that are used to explain the variability of y_i are denoted by $x_i^{(1)}, x_i^{(2)}, \ldots, x_i^{(p)}$ and are called *predictors*. Other commonly used names include *independent variables*, *regressors*, *covariates*, or just *x-variables*. In some instances, these terms have slightly different meanings; for example, Weisberg (2014) uses the term regressor for the variables that are derived from the original predictors, e.g. via transformations. We do not make this distinction. The term covariate is commonly used for a predictor that is not of primary interest in the analysis but is associated with the response and is, therefore, included for better estimation of the effects of primary interest.

The predictors are usually considered as fixed, with values set by the researcher. For example, the predictor may indicate the fixed treatment applied to observation i, or the researcher may select the sample trees from the forest according to a pre-defined tree diameter criteria. In an observational data set, it is better justified to treat predictors as random. Even in that case, one may condition the inference on the observed x_i's and most results are essentially similar to those when the predictors are fixed. Random regressors are discussed further in Section 10.1.

Consider a data set with n observations. The *linear model* for observation i, $1 = 1, \ldots, n$, is defined as

$$y_i = \beta_1 x_i^{(1)} + \beta_2 x_i^{(2)} + \ldots + \beta_p x_i^{(p)} + e_i \,, \tag{4.1}$$

where y_i is the value of the response, and $x_i^{(1)}, \ldots, x_i^{(p)}$ are the values of the predictors for the ith observation. The *regression coefficients* $\beta_1, \ldots, \beta_p$ are unknown fixed constants, common to all observations. They are the parameters of the assumed submodel for $\mathrm{E}(y_i)$. Term e_i is called the *residual error*. It can also be called the *disturbance* or *model error*. The essential task is to estimate the numerical values of the coefficients using the observed data.

For a more compact definition of the model, it is convenient to pool the predictors of observation i into vector $\boldsymbol{x}_i = \left(x_i^{(1)}, x_i^{(2)}, \ldots, x_i^{(p)}\right)'$ and the coefficients into vector $\boldsymbol{\beta} = (\beta_1, \ldots, \beta_p)'$. The model can then be written as

$$y_i = \boldsymbol{x}_i' \boldsymbol{\beta} + e_i \,. \tag{4.2}$$

A notational benefit over Equation (4.1) is that Equation (4.2) looks the same, regardless of the number of predictors. The *systematic* or *fixed part* $\boldsymbol{x}_i'\boldsymbol{\beta}$ specifies the systematic dependence of the response on the predictors. The residual error can correspondingly be called the *random part*.

The model is said to be linear because the systematic part is linear with respect to coefficients $\boldsymbol{\beta}$. It is also linear in the sense that the standard estimator of regression coefficients, which will be presented later in Section 4.4, is a linear function of response vector $\boldsymbol{y}$. The model is also linear with respect to the predictors. But this is not a restriction, as the predictors in $\boldsymbol{x}_i$ can be nonlinear functions of the original predictors. Also, the response may be a nonlinear function of the original response. Therefore, the model is very flexible and permits the expression of the nonlinear relationship between $\boldsymbol{x}_i$ and y_i through nonlinear transformations. For more discussion about these issues, see Sections 4.2, 7.1 and 10.2.

It is usually assumed that the first predictor $x_i^{(1)} = 1$ for all i. The model then simplifies to

$$y_i = \beta_1 + \beta_2 x_i^{(2)} + \ldots + \beta_p x_i^{(p)} + e_i \,, \tag{4.3}$$

where β_1 is called the *intercept* or *constant* of the model. By including an intercept in the model, the systematic part is allowed to be non-zero when all other predictors are zero, through a level shift in the hyperplane specified by the predictors. The intercept term should be included unless there is some strong theoretical reason to assume that it is zero (see Box 10.1, p. 311 for more discussion). A model that includes only one predictor in addition to the intercept is called a *simple linear model.*

It is assumed that the systematic part gives the expectation of the response given the predictor variables:

$$E(y_i) = \boldsymbol{x}_i'\boldsymbol{\beta}\,.$$

A detailed discussion on the formulation of the systematic part follows in Section 4.2. The residual error e_i expresses the difference between y_i and its expected value. Notice that this expected value is a fixed function of the predictor vector $\boldsymbol{x}_i$. As a difference between a random variable and its (fixed) mean, e_i is also a random variable. The assumption on the systematic part implies that

$$E(e_i) = E(y_i - \boldsymbol{x}_i'\boldsymbol{\beta}) = 0$$

for all $\boldsymbol{x}_i$.

A usual starting assumption of the variance of e_i is that it is constant for all values of i:

$$\mathrm{var}(e_i) = \sigma^2\,. \tag{4.4}$$

In addition, another common starting assumption is that the residual errors of all pairs of observations are uncorrelated:

$$\mathrm{cov}(e_i, e_j) = 0 \tag{4.5}$$

for all pairs (i, j), $i \neq j$.

Because the systematic part is a fixed function and the residual error e_i is a simple linear transformation of y_i, the assumptions on residual error can be equivalently specified for y_i as follows

$$\mathrm{var}(y_i) = \mathrm{var}(y_i|\boldsymbol{x}_i) = \sigma^2$$

and

$$\mathrm{cov}(y_i, y_j) = \mathrm{cov}(y_i|\boldsymbol{x}_i, y_j|\boldsymbol{x}_j) = 0$$

for all pairs (i, j), $i \neq j$. The conditioning on the fixed $\boldsymbol{x}_i$ and $\boldsymbol{x}_j$ is written explicitly to emphasize that the variance and covariance of y_i are conditional on the predictors. The regression coefficients $\beta_1, \ldots, \beta_p$ and the residual variance σ^2 are called model *parameters*. The square root of σ^2 is called *the residual standard error*, and it quantifies how much the values of y_i vary around the expected value $\boldsymbol{x}_i'\boldsymbol{\beta}$.

Normality of e_i is sometimes assumed:

$$e_i \sim N(0, \sigma^2)$$

which can be equivalently stated as:

$$y_i \sim N(\boldsymbol{x}_i'\boldsymbol{\beta}, \sigma^2)\,.$$

To summarize, the model formulation includes the following 3–4 assumptions on the data-generating process:

1. The systematic part is true. Formally $E(y_i) = \boldsymbol{x}_i'\boldsymbol{\beta}$ or $E(e_i) = 0$ for all i.

2. The residual errors (and, therefore, the response as well) have a constant variance. Formally $\mathrm{var}(e_i) = \sigma^2$ or $\mathrm{var}(y_i) = \sigma^2$ for all i.

3. The residual errors (and, therefore, the response as well) are uncorrelated. Formally $\mathrm{cov}(y_i, y_j) = 0$ or $\mathrm{cov}(e_i, e_j) = 0$ for all pairs (i, j), $i \neq j$.

4. [The residual errors (and, therefore, the response as well) have a specific distribution, usually a normal distribution. Formally $e_i \sim N(0, \sigma^2)$ or $y_i \sim N(\boldsymbol{x}_i'\boldsymbol{\beta}, \sigma^2)$.]

The first assumption is related to the first-order properties, assumptions 2 and 3 are related to the second-order properties, and assumption 4 is related to the higher-order properties of the response variable. The first three model assumptions are crucial to anything that will be discussed later in regard to parameter estimation and inference of the model. Assumption 1 is made in all modeling situations. Assumptions 2 and 3 are called here *the OLS assumptions*, because they are the preconditions for the optimality of Ordinary Least Squares (OLS) method for parameter estimation. Assumption 4 may be needed for inference in data sets with small n, for the construction of prediction intervals, and for parameter estimation using likelihood-based methods. It is important to recognize that normality is not a necessary assumption for a linear model.

There is no guarantee that these assumptions are generally met in any empirical data set. Therefore, critical evaluation of the starting assumptions is an important part of the modeling process, and will be discussed later in Section 4.5. Often, the assumptions can be only evaluated after the model parameters have been estimated. Therefore, the modeling process commonly iterates between model formulation, estimation and evaluation, until a satisfactory model formulation has been found.

If the model assumptions are not met, then there is a mismatch between the data and the model. Then, two general principles can be used:

1. *Transform the data to match with the model.* Redefine the systematic part and the variables of the model to meet the assumptions, e.g. by changing the formulation of the systematic part, or by making nonlinear transformations to the response variable.

2. *Change the model to match with the data.* Relax the model assumptions, e.g. by allowing non-constant error variance or correlation among observations.

In practice, these two principles are commonly used together.

The starting assumption $\text{var}(e_i) = \sigma^2$ can be relaxed by allowing inconstant residual variance through parametric variance functions. In addition, the assumption on zero correlation can be relaxed through parametric covariance functions, which model the dependency caused by temporal or spatial dependence, as well as by the grouped structure of the data. In those cases, additional parameters to model these properties will be included in the model. For more discussion, see Section 4.3. When dependent data are modeled, one needs to explicitly express the dependence of data using pairwise covariances or through the complete joint distribution of observations. Such data can no longer be seen as n independent realizations from a common model, but rather as one realization of an n-dimensional random vector, as we discussed in Section 2.12.

The linear model implicitly assumes that y_i is defined at the ratio or interval scale. In addition, the model can be used for variables on "a good ordinal scale", such as the Likert-scale with a sufficient number of non-empty classes. An orthodox model for such data should take into account the non-comparability of the differences between subsequent classes, e.g. through ordinal logistic regression (e.g. Harrell 2001, Chapters 13–14). However, the conclusions based on a linear model may not be very different from conclusions based on a more sophisticated model, which may therefore be unnecessary in many practical situations.

The response variable in a linear model can be either continuous or discrete. However, for many discrete variables, such as binary or count data, the starting assumptions about the mean and variance of y_i are not well justified. The GLM provides better justified models for such cases (Chapter 8). Therefore, the linear model and its extensions provide a very general and widely applicable toolbox for statistical data analysis.

4.1.2 The model for all data

Let us pool the responses y_i of all observations i into a column vector $\boldsymbol{y}_{n\times 1}$, all predictors from all observations into matrix $\boldsymbol{X}_{n\times p}$, and the residual errors into a column vector $\boldsymbol{e}_{n\times 1}$. Model (4.2) for all data becomes

$$\begin{pmatrix} y_1 \\ y_2 \\ \vdots \\ y_n \end{pmatrix} = \begin{pmatrix} \boldsymbol{x}'_1 \\ \boldsymbol{x}'_2 \\ \vdots \\ \boldsymbol{x}'_n \end{pmatrix} \begin{pmatrix} \beta_1 \\ \beta_2 \\ \vdots \\ \beta_p \end{pmatrix} + \begin{pmatrix} e_1 \\ e_2 \\ \vdots \\ e_n \end{pmatrix}$$

or simply

$$\boldsymbol{y} = \boldsymbol{X}\boldsymbol{\beta} + \boldsymbol{e}\,. \tag{4.6}$$

In Equation (4.6), subscripts and superscripts for parameters and observations are not needed; the number of parameters and observations is determined through the dimensions of the matrices and vectors. Additional benefits arise through a convenient definition of the error variance-covariance structure.

The matrix $\boldsymbol{X}$ is commonly called the *model matrix*. It includes the values of predictors for all observations, so that each observation corresponds to a row, and each predictor to a column. If the model has an intercept, then the first column of the model matrix is a vector of ones, $\boldsymbol{1}_{n\times 1}$. The term *design matrix* is often used synonymously with 'model matrix', but it is better to reserve the former for the determination of treatment variable levels in an experimental design.

The previously specified assumptions about residual errors: zero mean, constant variance, and zero covariance between all pairs of errors, are stated by specifying the expected value and variance-covariance matrix of $\boldsymbol{e}$ as follows:

$$\mathrm{E}(\boldsymbol{e}) = \boldsymbol{0} \tag{4.7}$$
$$\mathrm{var}(\boldsymbol{e}) = \sigma^2\boldsymbol{I}\,. \tag{4.8}$$

The constant diagonal elements of var($\boldsymbol{e}$) specify that all n components of the multivariate random variable $\boldsymbol{e}$ have equal variance. All off-diagonal elements of the matrix have an interpretation: they specify the zero correlation separately for all possible pairs of observations.

Following the naming convention of Freedman (2009), we call the above specified model *the OLS model*, because the optimal method to estimate the model parameters for that model is ordinary least squares (see Section 4.4 for details). Section 4.3.3 introduces the *GLS model*, which is such a generalization of the OLS model, where the weighted and generalized least squares are the optimal estimation methods. We recognize the drawback of naming the models according to the parameter estimation methods because model formulation and parameter estimation are two separate steps. For example, OLS may sometimes be a satisfactory parameter estimation method for the GLS model, and ridge regression or LASSO (Box 4.2, p. 91) may be satisfactory for the OLS model. An alternative naming convention uses *the classical linear model* for the OLS model, and *the general linear model* for the GLS model (e.g. Fahrmeir *et al.* 2013). However, the latter is easily confused with the generalized linear model (GLM, Chapter 8). Therefore, we use the terms OLS and GLS model.

4.2 The systematic part

The systematic part describes the expected value of the response as a function of predictors. The systematic part should be as *parsimonious* as possible, i.e. it should have a minimal number of parameters. The possible flexibility of the systematic part is also restricted by the available data: the number of coefficients in the systematic part should be much lower than the number of observations used for parameter estimation. A total of 20 observations per regression coefficient is usually a sufficient number, and 10 observations per coefficient are the absolute minimum for a linear model; see Section 4.4 of Harrell (2001) for a more elaborated discussion and suggestions for generalized linear models. If the systematic part is too flexible, then it starts to model also the noise in the data; this phenomenon is called *overfitting*.

The data provides information about the trends of $\mathrm{E}(y_i)$ only within the range of x-variables in the data, and it is, therefore, usually risky to assume that the model is valid far outside the range of the x−variables of the model fitting data. The use of a model for prediction outside the range of modeling data is called *extrapolation.*

4.2.1 Identifiability and multicollinearity

A technical restriction on the systematic part is that the predictors should be *linearly independent* in order to make the model *identifiable*. This means that no column of the model matrix $\boldsymbol{X}$ should be a linear function of the other columns. In matrix calculus, this means that the rank of the model matrix $\boldsymbol{X}$ should be p, i.e. $\boldsymbol{X}$ should be of *full rank.* For a model that is not identifiable, there are many parameter vectors $\boldsymbol{\beta}$ that provide the same value of $\boldsymbol{X}'\boldsymbol{\beta}$ and therefore, an equally good fit to the data. The model is necessarily unidentifiable if the number of predictors is greater than the number of observations, if the indicator variables of all factor levels are included in a model that includes the intercept (see Section 4.2.4), or if the absolute correlation between some predictors is 1. A model that is not identifiable is *overparameterized.*

There are mathematical methods to deal with unidentifiable models using *generalized inverses* of matrices. In addition, many linear model textbooks define unidentifiable models but then the estimation considers *estimable functions* of the parameters of an unidentifiable model. However, we prefer to keep things more easily understandable and will discuss only identifiable models.

Multicollinearity of predictors means that some predictors $x_i^{(k)}$ and $x_i^{(l)}$ are correlated. If the multicollinearity is very strong, the model almost becomes unidentifiable. Strong multicollinearity may be a problem in observational data sets, where the researcher cannot control the values of predictors. A controlled experiment should be planned so that the predictors are not correlated. See Sections 4.2.3 and 4.4.1 for more discussion.

4.2.2 Formulating the systematic part

The mathematical formula for the systematic part should be based on subject-matter theory or previous knowledge of the relationship whenever possible. Quite often, the theory suggests a function that is nonlinear with respect to the parameters of interest. Sometimes a linear model can be found based on such a nonlinear relationship, but that is not the case with all nonlinear equations. In those cases, either a nonlinear model or a linear approximation of the nonlinear function can be used. These situations are illustrated in the following examples.

Example 4.1. Tree stem is a three-dimensional object, which suggests that a model for stem biomass could be based on

$$\text{volume} = \text{form factor} \times \text{diameter}^{\beta_1} \times \text{height}^{\beta_2} .$$

The model for biomass could be, for example

$$\begin{aligned} \text{biomass} &= \text{mean wood density} \times \text{volume} \\ &= \text{mean wood density} \times \text{form factor} \times \text{diameter}^{\beta_1} \times \text{height}^{\beta_2} , \end{aligned}$$

where β_1 is close to 2 and β_2 is close to 1. A logarithmic transformation to both sides of the equation leads to a linear systematic part

$$\log(\text{biomass}) = \beta_1 + \beta_2 \log(\text{diameter}) + \beta_3 \log(\text{height}) , \quad (4.9)$$

which has meaningful interpretations of regression coefficients as the sum of logarithmic form factor and wood density, and the powers of diameter and height. Therefore, the biomass model can be conveniently handled in the context of the linear model. For more discussion of logarithmic regression, see Section 10.2.1, p. 314.

Example 4.2. Consider the effect of Photosynthetically Active Radiation (PAR_i) on the net photosynthesis, which is measured in terms of net carbon dioxide uptake ($NCO2_i$). For such a process, the hyperbolic model has been suggested in literature:

$$NCO2_i = -\phi_1 + \frac{\phi_2 PAR_i}{\phi_3 + PAR_i} , \quad \phi_1, \phi_2, \phi_3 > 0, \quad PAR_i > 0 .$$

This function cannot be linearized with respect to all of its parameters.

In most practical cases, theory of the studied phenomena does not exist. In those cases, the systematic part should be formulated so that it is sufficiently flexible to be interpreted as an approximate of $\mathrm{E}(y_i)$. Such function is often found by trial and error. In addition, the formulation should allow evaluation of possible hypotheses of interest. It is generally desirable that the mathematical formulation of the systematic part is based on prior hypotheses on the process being modeled, not on the collected empirical data, even though that condition is seldom met in practice. Data-driven formulation of the systematic part may lead to overestimation of the effects of the selected predictors. That is because one tends to keep those predictors in the model for which the coefficients are overestimated and drop those predictors for which the coefficients are underestimated. In practice, compromises are inevitable, but the risks should be recognized, and the model building process should be clearly described in research reports. For more discussion on the topic, see Chapter 4 of (Harrell 2001) and the related Box 3.2 about publication bias on p. 65.

One of the most important decisions about the systematic part of a linear model is the selection of the predictors. The model that has all p predictors and shows a good fit to the data after a graphical evaluation of the fit is called the *full model*. Restricted versions of this model are obtained, e.g. by dropping some predictors (i.e. restricting the associated coefficients to 0), restricting the coefficients to some other fixed value, or making other linear restrictions to the relationships among a group of predictors. In hypothesis testing, the main approach is to compare the full model with a restricted model.

The relationship between a predictor and the response is *nonlinear*, or *curvilinear*, if $\mathrm{E}(y)$ does not follow a straight line as a function of x. A curvilinear relationship can be modeled with a linear model by applying nonlinear transformations, either to the predictors, to the response, or to both of these. However, transformations to y are not necessary for good fit of the systematic part, because any nonlinear relationship can be well approximated by using a sufficient number of wisely chosen transformations of the predictors. The resulting model is sometimes called a *curvilinear* model.

Including the same predictor several times as different nonlinear transformations permits modeling of the curvilinear relationship between the response variable and the predictor. A common example is the *polynomial regression*, where the model includes first k powers of the predictor. As a high-degree polynomial can approximate any nonlinear relationship with arbitrarily high accuracy, the polynomial regression can fit any data if the degree of polynomial is high enough. In particular, one can obtain a perfect fit if the degree is equal to $n-1$. However, the behavior of a high-degree polynomial for such values of x-variables that were not present in the data may be very surprising. *Regression splines* provide another means to obtain a very flexible systematic part, where the fit is improved by the increased number of knots in the spline (Harrell 2001); see Section 10.4 for more discussion.

Even though a nonlinear transformation of the response variable may not be necessary for good fit of the systematic part, there may be other reasons to use such one. It may lead to:

1. A more parsimonious systematic part,
2. Meaningful interpretations to the parameters, as we have illustrated above in Equation (4.9), or
3. More simple/valid assumptions in regard to the random part.

Especially, if the variance of y_i changes as a function of $\mathrm{E}(y_i)$ or any of the predictors, a wisely selected nonlinear transformation $g(y_i)$ may have a constant variance. However, if a nonlinear transformation $g(y_i)$ is modeled on predictors $\boldsymbol{x}_i$, the systematic part models the expected value of the transformed response $\mathrm{E}(g(y_i))$, not the transformed expected value $g(\mathrm{E}(y_i))$. Therefore, a model fitted for $g(y_i)$ gives unbiased predictions of $g(y_i)$, but direct back-transformation leads to biased predictions for y_i. Back-transformation bias is discussed in more detail in Section 10.2. Transformation of the y-variable is also linked with the GLM (see Box 8.1, p. 246).

4.2.3 Interpretation of regression coefficients

It is important to understand the interpretations of the regression coefficients. For example, in hypothesis testing the hypotheses are statements about the regression coefficients or other parameters of the model. Ability to interpret the regression coefficients is therefore a prerequisite for hypothesis testing.

A simple computation gives an interpretation to the regression coefficients. Consider a situation where the value of the kth predictor is increased from a fixed value x_0 by one unit to x_0+1, and the other predictors are kept unchanged. The resulting difference to $\mathrm{E}(y_i)$ is

$$\beta_1 x_i^{(1)} + \ldots + \beta_k(x_0+1) + \ldots + \beta_p x_i^{(p)} - \left(\beta_1 x_i^{(1)} + \ldots + \beta_k(x_0) + \ldots + \beta_p x_i^{(p)}\right) = \beta_k \,.$$

Therefore, each regression coefficient can be interpreted as the change in $\mathrm{E}(y_i)$ as the predictor in question is increased by one unit and the others are kept constant. An implicit assumption behind that interpretation is, however, that the change in $\mathrm{E}(y)$ is not associated by the value of x_0, and the other predictors remain unchanged when x_0 is increased.

The interpretation is more complicated if the predictors of the model are dependent, because the condition to keep the other predictors unchanged may not make sense. There may be either a deterministic mathematical relationship, or a linear association among the predictors, which is called *(multi)collinearity*. A deterministic mathematical relationship exists, for example, between the predictors of the second-order polynomial model

$$\mathrm{E}(y_i) = \beta_1 + \beta_2 x_i^{(2)} + \beta_3 x_i^{(3)}$$

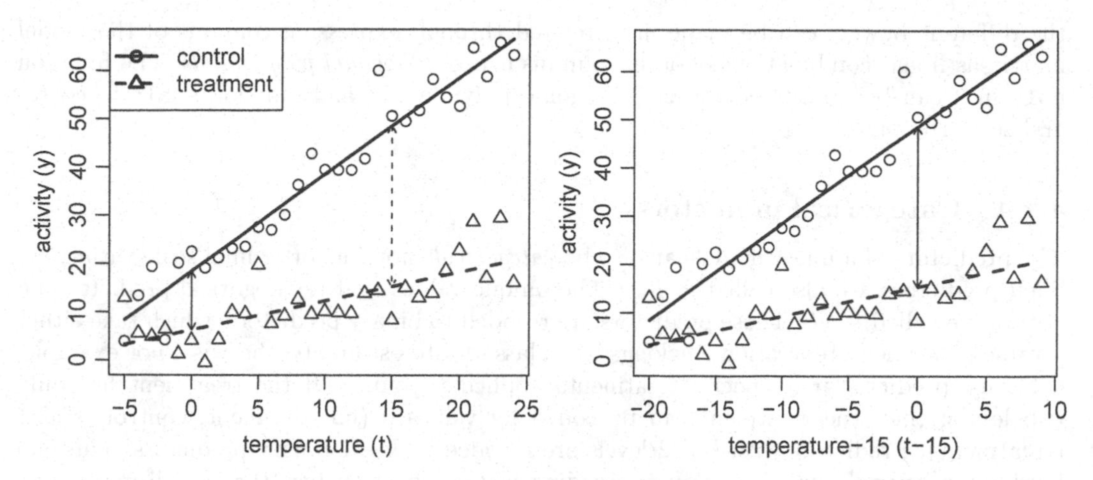

FIGURE 4.1 The variability in variable "activity" is explained by a model, where $E(y_i) = \beta_1 + \beta_2 T_i + \beta_3 t_i + \beta_4 T_i t_i$; T_i is an indicator for treated category and t_i is the temperature. Now, β_1 specifies the expected activity for the control class when temperature is 0, and β_2 specifies the effect of treatment when temperature is 0 (the solid arrow in the left-hand graph). An equivalent model uses $E(y_i) = \gamma_1 + \gamma_2 T_i + \gamma_2(t_i - 15) + \gamma_4 T_i(t_i - 15)$. Now γ_1 and γ_2 specify the expected activity for the control class and the effect of treatment, respectively, when $t_i - 15 = 0$ (i.e. $t_i = 15$), as illustrated by the solid arrow in the right-hand graph and dashed arrow in the left-hand graph. Note also that $\beta_3 = \gamma_3$ and $\beta_4 = \gamma_4$.

where $x_i^{(3)} = \left(x_i^{(2)}\right)^2$. A one-unit increase in either $x_i^{(2)}$ or $x_i^{(3)}$ is always associated with an increase in the other predictor as well. As an example of correlated predictors, consider a model for stem biomass as a function of tree diameter and height. A one-unit increase in diameter is commonly associated with an increase in height in forest tree data. Therefore, direct interpretation of the regression coefficient for diameter is misleading because one-centimeter diameter growth without growth in height is not biologically meaningful. The multicollinearity does not mean that there are any problems in model formulation, parameter estimation or predictive power of the model; it just needs to be taken into account in interpretation of the regression coefficients and in the related inference.

Linear transformations in the predictors and response do not affect the model fit. However, they may lead to benefits in interpretation, reporting and numerical computation of parameter estimates. Therefore, such transformations are generally recommended if any of these benefits can be achieved. For example, a certain predictor $x_i^{(k)}$ can be rescaled through multiplication with a constant. This may help in reporting the results as a similar scale of regression coefficients is employed, provide benefits in interpretation and lead to more stable estimation procedures. Value $1/\operatorname{sd}(x_i^{(k)})$ is often useful in scaling. Another widely applicable linear transformation results from the subtraction of a fixed constant c_k from predictor $x_i^{(k)}$. The constant may be the mean of the predictor in the modeling data, or any otherwise justified fixed value. A benefit from such transformation is that the intercept now has the interpretation as the expected value of y when $x_i^{(k)} = c_i$. In a model that includes interactions (see the next subsection), such a transformation changes the interpretation of the main effect terms, from effects when $x_i^{(k)} = 0$ to effects when $x_i^{(k)} = c_k$, which is often convenient (see Figure 4.1). If the mean of $x_i^{(k)}$ is used as c_k, the resulting predictor is said to be *centered*. A predictor that is first centered by using the mean and thereafter scaled using $1/\operatorname{sd}(x_i^{(k)})$ is called *standardized predictor*. In polynomial regression, the different powers can be made uncorrelated through expressing columns of the model matrix as linear combinations of other columns using *orthogonal polynomials*. The function `poly` in `R` can be used to construct orthogonal polynomials, and function `scale` to center and scale the predictors.

4.2.4 Categorical predictors

The predictors of a linear model can also be *categorical* (nominal or ordinal scale variables). Such predictors are also called *factors*. The unique values that a categorical predictor can obtain are called *levels*. Those predictors are recoded to binary predictors, which take either a value of 1 when observation i belongs to a class of interest and 0 otherwise. For example, a binary predictor may specify treatments applied to unit i. If the treatment has only two levels, then the treatment can be coded by values 0 (no treatment, control) and 1 (treatment). Predictors with $k > 2$ levels are recoded to $k - 1$ binary predictors. This can be done in several ways. The way of recoding has no effect on how the overall model fits, but it changes the interpretation of the coefficients. The researcher is free to choose the recoding approach in the way that best serves the research questions of the analysis.

The most common approach is to select one level as the default and recode the other levels so that the associated coefficients express the difference to the default level in a model that includes the intercept. This is implemented by including separate indicator variables for each level of the categorical predictor, except for the default level. For example, the "control", "moderate treatment" and "heavy treatment" levels can be recoded by including the binary predictors "moderate" (which takes the value of 1 for an observation treated with the moderate treatment and 0 otherwise) and "heavy" (1 for heavy and 0 otherwise). This parameterization is specified by recoding the three treatments in the two columns of the model matrix as follows:

Original	*Recoded*	
Treatment	moderate	heavy
control	0	0
moderate	1	0
heavy	0	1

One might be tempted to also include a third column for the control treatment using:

Original	*Recoded*		
Treatment	control	moderate	heavy
control	1	0	0
moderate	0	1	0
heavy	0	0	1

However, this would overparameterize the model: the sum of these binary predictors would be 1 for all observations, and the first predictor (1 for intercept) would be the direct sum of the three binary predictors. However, the model becomes identifiable if the intercept is not included in the model.

The interpretation of the regression coefficients of binary predictors is similar to that of continuous predictors. However, because only 0 and 1 are possible values, there is only one possible 1-unit increase in the value of the predictor: the change from 0 to 1. In the first example above, the regression coefficients for the two binary variables quantify the difference in $\mathrm{E}(y_i)$ between the "moderate" and "control" levels and between "heavy" and "control" levels. Finding interpretations for the parameters of the latter definition is left as an exercise.

A third alternative recoding of the treatments to the model matrix for the same variable may be specified by

Original	*Recoded*	
Treatment	moderate or heavy	heavy
control	0	0
moderate	1	0
heavy	1	1

in a model that includes an intercept. The regression coefficient of the first binary variable now expresses the difference in $E(y_i)$ between the "moderate" and "control" levels, and the coefficient of the second predictor expresses the difference between the "heavy" and "moderate" levels. For example, Melin *et al.* (2016b) used the third formulation above to quantify the differences in the habitat preferences of moose animals between each successive pair of months over the year.

The tables shown above are commonly (but slightly confusingly) called *contrast matrices* (Venables and Ripley 2002, Section 6.2), even though the rows of the matrix are not contrasts in their exact meaning in mathematical statistics (see Section 11.2 of (Casella and Berger 2001)).

Predictors that are the products of two other predictors are called *interactions*. Consider the systematic part

$$E(y_i) = \beta_1 + \beta_2 x_i^{(2)} + \beta_3 x_i^{(3)} + \beta_4 x_i^{(2)} x_i^{(3)}$$

where $x_i^{(2)}$ is binary and $x_i^{(3)}$ is continuous. The last term is now an interaction of the two predictors. The second and third terms are called *main effects* in this context. Consider first the case $x_i^{(2)} = 0$, i.e. the systematic part for non-members of the class indicated by the binary predictor. The second and fourth terms vanish, and the coefficients of the remaining terms, β_1 and β_3 specify the intercept and slope of a simple linear model of a non-member. The model for a member results by specifying $x_i^{(2)} = 1$. Simple algebra shows that the intercept and slope for a member become $\beta_1 + \beta_2$ and $\beta_3 + \beta_4$, respectively. Therefore, β_2 and β_4 can be interpreted as differences in the intercept and slope between a member and a non-member.

Interpretation of interactions of two categorical predictors are developed in a similar way. It is also possible to formulate models with more than two predictors and higher-order interactions. However, considerable care is needed in interpreting the coefficients of such models.

4.2.5 Examples

This section presents two examples that describe the formulation of the systematic part of a linear model.

Example 4.3. Two-way ANOVA in agriculture. Data `immer` of package `MASS` includes observations of barley yield of five varieties on six farms (locations) in 1931 and 1932 (Immer *et al.* 1934). The variables in the data are: location(`Loc`), variety (`Var`), yield in 1931 (`Y1`) and yield in 1932 (`Y2`). We want to study whether the varieties differ in terms of mean yield during these two years. We also allow a possible effect of location.

```
> library(MASS) # The library of book "Modern applied statistics with S and S-Plus"
> data(immer)
```

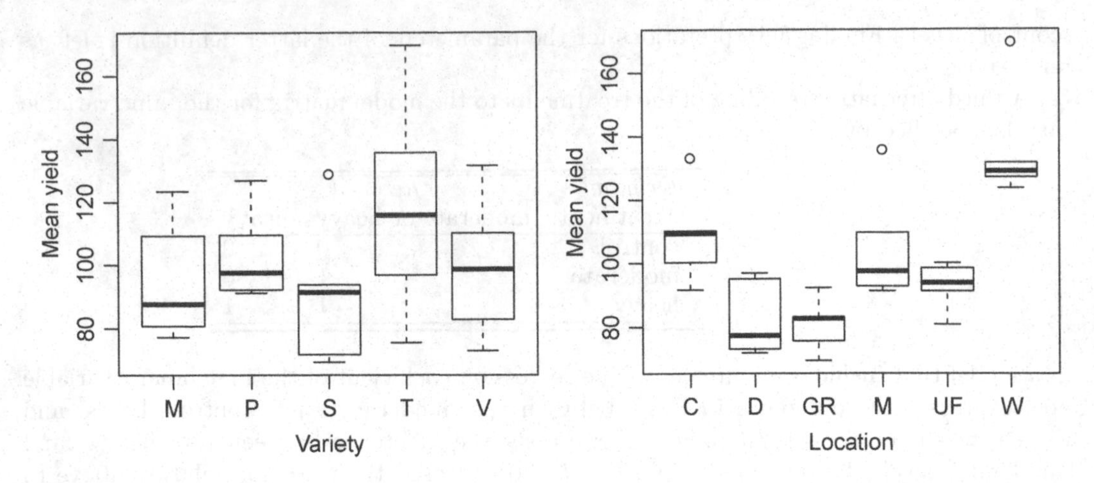

FIGURE 4.2 A boxplot of the mean barley yield (bushels per acre) on variety (left) and location (right) in the `immer` data set of package `MASS`.

Id	Loc	Var	Y	Id	Loc	Var	Y	Id	Loc	Var	Y
1	UF	M	80.85	11	M	M	92.70	21	GR	M	82.65
2	UF	S	93.85	12	M	S	91.20	22	GR	S	69.45
3	UF	V	100.05	13	M	V	97.45	23	GR	V	82.90
4	UF	T	98.45	14	M	T	135.60	24	GR	T	75.60
5	UF	P	91.25	15	M	P	109.60	25	GR	P	92.20
6	W	M	123.50	16	C	M	109.35	26	D	M	77.30
7	W	S	128.75	17	C	S	91.65	27	D	S	71.90
8	W	V	131.45	18	C	V	110.10	28	D	V	73.15
9	W	T	169.60	19	C	T	133.15	29	D	T	96.80
10	W	P	126.90	20	C	P	100.25	30	D	P	95.05

TABLE 4.1 Data set `immer` of package `MASS`

If the class of a variable is a factor, `R` automatically treats the different levels as separate categories. If needed, one could define a variable as a factor by using `immer$Var<-as.factor(immer$Var)`. In this data set, the `Loc` and `Var` have already been defined as factors.

```
> class(immer$Var)
[1] "factor"
> class(immer$Loc)
[1] "factor"
```

We compute the mean yield of the two years into a new variable called `Y` using:

```
> immer$Y<-(immer$Y1+immer$Y2)/2
```

Figure 4.2 shows the mean yield as a function of variety and location. The complete data set is shown in Table 4.1. The two-way analysis of variance model for this data is

$$\begin{aligned} y_i &= \beta_1 + \beta_2 VP_i + \beta_3 VS_i + \beta_4 VT_i + \beta_5 VV_i + \\ &\quad \beta_6 LD_i + \beta_7 LGR_i + \beta_8 LM_i + \beta_9 LUF_i + \beta_{10} LW_i + e_i \end{aligned} \tag{4.10}$$

where $VP_i, \ldots, VV_i$ are binary indicator variables for varieties P,...,V and $LD_i, \ldots, LW_i$ are binary indicators for locations D,...,W. Variety M and location C are not included in the model, which means that they are the default levels of the factors. Now the intercept β_1 is the expected mean yield of variety M at location C. Coefficient β_2 expresses the expected difference between varieties P and M, which is assumed to be equal at all locations. Coefficients $\beta_3, \ldots, \beta_5$ express the corresponding expected differences between the other varieties and variety M. Coefficient β_6 expresses the expected difference between locations D and C, which is assumed to be equal for all varieties. Coefficients $\beta_7, \ldots, \beta_{10}$ express the corresponding expected differences between other locations and location C. It is assumed that $\mathrm{E}(e_i) = 0$ and $\mathrm{var}(e_i) = \sigma^2$ for all i, and $\mathrm{cov}(e_i, e_j) = 0$ for all pairs (i, j), $i \neq j$.

The above-described coding of varieties and locations was made automatically in R, which selects the level that is first in alphabetical order as the default. It can be seen in the following output:

```
> contrasts(immer$Var)
  P S T V
M 0 0 0 0
P 1 0 0 0
S 0 1 0 0
T 0 0 1 0
V 0 0 0 1
```

The default level of a categorical predictor is the level with zeroes in the corresponding row. The default level can be changed as follows:

```
> levels(immer$Var)
[1] "M" "P" "S" "T" "V"
# For example, set P as the default
> contrasts(immer$Var)<-contr.treatment(levels(immer$Var),base=2)
> contrasts(immer$Var)
  M S T V
M 1 0 0 0
P 0 0 0 0
S 0 1 0 0
T 0 0 1 0
V 0 0 0 1
```

The contrasts can be changed using `contrasts(immer$Var)<-M`, where `M` is the new matrix of appropriate dimensions. There are also different contrast functions (in addition to `contr.treatment`), which automatically generate matrices with the correct dimensions and convenient names of rows and columns.

Model (4.10) can be presented as $\boldsymbol{y} = \boldsymbol{X\beta} + \boldsymbol{e}$, where $\mathrm{E}(\boldsymbol{e}) = \boldsymbol{0}$ and $\mathrm{var}(\boldsymbol{e}) = \sigma^2\boldsymbol{I}$, by defining the following matrices and vectors:

$$\boldsymbol{y} = \begin{pmatrix} 80.85 \\ 93.85 \\ 100.05 \\ 98.45 \\ 91.25 \\ 123.50 \\ 128.75 \\ 131.45 \\ \vdots \\ 96.80 \\ 95.05 \end{pmatrix}, \boldsymbol{X} = \begin{pmatrix} 1 & 0 & 0 & 0 & 0 & 0 & 0 & 0 & 1 & 0 \\ 1 & 0 & 1 & 0 & 0 & 0 & 0 & 0 & 1 & 0 \\ 1 & 0 & 0 & 0 & 1 & 0 & 0 & 0 & 1 & 0 \\ 1 & 0 & 0 & 1 & 0 & 0 & 0 & 0 & 1 & 0 \\ 1 & 1 & 0 & 0 & 0 & 0 & 0 & 0 & 1 & 0 \\ 1 & 0 & 0 & 0 & 0 & 0 & 0 & 0 & 0 & 1 \\ 1 & 0 & 1 & 0 & 0 & 0 & 0 & 0 & 0 & 1 \\ 1 & 0 & 0 & 0 & 1 & 0 & 0 & 0 & 0 & 1 \\ \vdots & \vdots & \vdots & \vdots & \vdots & \vdots & \vdots & \vdots & \vdots & \vdots \\ 1 & 0 & 0 & 1 & 0 & 1 & 0 & 0 & 0 & 0 \\ 1 & 1 & 0 & 0 & 0 & 1 & 0 & 0 & 0 & 0 \end{pmatrix},$$

$$\boldsymbol{\beta} = \begin{pmatrix} \beta_1 \\ \beta_2 \\ \vdots \\ \beta_{10} \end{pmatrix}, \boldsymbol{e} = \begin{pmatrix} e_1 \\ e_2 \\ \vdots \\ e_{30} \end{pmatrix}, \mathrm{var}(\boldsymbol{e}) = \begin{pmatrix} \sigma^2 & 0 & 0 & \cdots & 0 \\ 0 & \sigma^2 & 0 & \cdots & 0 \\ 0 & 0 & \sigma^2 & \cdots & 0 \\ \vdots & \vdots & \vdots & \ddots & \vdots \\ 0 & 0 & 0 & \cdots & \sigma^2 \end{pmatrix}.$$

Example 4.4. Linear model in forest work study. Consider a random process where three machines lift stumps of different diameters on a clearcut area for use as a bioenergy crop. We assume machine-specific 2nd order polynomial relationships between mean processing time and stump diameter:

$$\begin{aligned} t_i =& \beta_1 + \beta_2 d_i + \beta_3 d_i^2 + \\ & \beta_4 M2_i + \beta_5 M2_i d_i + \beta_6 M2_i d_i^2 + \\ & \beta_7 M3_i + \beta_8 M3_i d_i + \beta_9 M3_i d_i^2 + e_i \end{aligned} \tag{4.11}$$

where t_i is the processing time for stump i (seconds), d_i is the diameter of stump i (cm), $M2_i$ is a binary indicator specified as follows:

$$M2_i = \begin{cases} 1 & \text{if stump } i \text{ was processed using machine 2} \\ 0 & \text{otherwise}, \end{cases}$$

and

$$M3_i = \begin{cases} 1 & \text{if stump } i \text{ was processed using machine 3} \\ 0 & \text{otherwise}. \end{cases}$$

For illustration, assume that the model parameters are known to have the following values

$\beta_1 = -30$	$\beta_2 = 3$	$\beta_3 = 0$	$\beta_4 = 100$	$\beta_5 = -5$
$\beta_6 = 0.1$	$\beta_7 = 50$	$\beta_8 = -5$	$\beta_9 = 0.1$	$\sigma^2 = 25^2$

To illustrate the vector formulation of the systematic part, consider the following three stumps:

Stump id	Diameter	Machine
1	16	1
2	18	3
3	29	2

Vector $\boldsymbol{x}_i$ for these stumps, as well as the common vector $\boldsymbol{\beta}$ are defined as:

$$\boldsymbol{x}_1 = \begin{pmatrix} 1 \\ 16 \\ 256 \\ 0 \\ 0 \\ 0 \\ 0 \\ 0 \\ 0 \end{pmatrix}, \boldsymbol{x}_2 = \begin{pmatrix} 1 \\ 18 \\ 324 \\ 0 \\ 0 \\ 0 \\ 1 \\ 18 \\ 324 \end{pmatrix}, \boldsymbol{x}_3 = \begin{pmatrix} 1 \\ 29 \\ 841 \\ 1 \\ 29 \\ 841 \\ 0 \\ 0 \\ 0 \end{pmatrix}, \boldsymbol{\beta} = \begin{pmatrix} \beta_1 \\ \beta_2 \\ \beta_3 \\ \beta_4 \\ \beta_5 \\ \beta_6 \\ \beta_7 \\ \beta_8 \\ \beta_9 \end{pmatrix}.$$

The machine-specific functions for processing time can be constructed. For machine 1, the function results by specifying $M2 = 0$ and $M3 = 0$:

$$t = -30 + 3d.$$

For machine 2, we specify $M2 = 1$ and $M3 = 0$ to get

$$t = -30 + 100 + (3 - 5)d + (0 + 0.10)d^2 = 70 - 2d + 0.10d^2.$$

For machine 3, a similar logic yields

$$20 - 2d + 0.10d^2.$$

Data set `stumplift` could have been generated by model (4.11). The data includes information on the productivity of three stump lifting machines on three Norway Spruce (*Picea abies*) clearcut areas in Central Finland, where stumps were lifted for bioenergy purposes. A different machine and driver operated at each site, so that the differences between machines are the joint effects of site, machine and driver, but they are called effects of machine for simplicity. There is a total of 485 observations in the data set: 144, 180 and 161 stumps were lifted by machines 1, 2 and 3, respectively.

For matrix formulation of model (4.11) for the machine-diameter combinations of the **stumplift** data set, we define $\boldsymbol{y} = (t_1, t_2, \ldots, t_{485})'$ and $\boldsymbol{e} = (e_1, e_2, \ldots, e_{485})'$ Matrices and vectors $\boldsymbol{X}$, $\boldsymbol{\beta}$, and $\mathrm{var}(\boldsymbol{e})$ take the form

$$\boldsymbol{X}_{485\times 9} = \begin{pmatrix} 1 & 37 & 1369 & 0 & 0 & 0 & 0 & 0 & 0 \\ 1 & 50 & 2500 & 0 & 0 & 0 & 0 & 0 & 0 \\ \vdots & \vdots & \vdots & \vdots & \vdots & \vdots & \vdots & \vdots & \vdots \\ 1 & 24.5 & 600.25 & 0 & 0 & 0 & 0 & 0 & 0 \\ 1 & 27.0 & 729 & 1 & 27.0 & 729 & 0 & 0 & 0 \\ \vdots & \vdots & \vdots & \vdots & \vdots & \vdots & \vdots & \vdots & \vdots \\ 1 & 50 & 2500 & 1 & 50 & 2500 & 0 & 0 & 0 \\ 1 & 29 & 841 & 0 & 0 & 0 & 1 & 29 & 841 \\ \vdots & \vdots & \vdots & \vdots & \vdots & \vdots & \vdots & \vdots & \vdots \\ 1 & 19.5 & 380.25 & 0 & 0 & 0 & 1 & 19.5 & 380.25 \end{pmatrix}$$

$$= \begin{pmatrix} \boldsymbol{X}_1 & \boldsymbol{0} & \boldsymbol{0} \\ \boldsymbol{X}_2 & \boldsymbol{X}_2 & \boldsymbol{0} \\ \boldsymbol{X}_3 & \boldsymbol{0} & \boldsymbol{X}_3 \end{pmatrix},$$

$$\boldsymbol{\beta}_{9\times 1} = \begin{pmatrix} \beta_1 \\ \beta_2 \\ \vdots \\ \beta_9 \end{pmatrix}, \ \mathrm{var}(\boldsymbol{e})_{485\times 485} = \begin{pmatrix} \sigma^2 & 0 & 0 & \cdots & 0 \\ 0 & \sigma^2 & 0 & \cdots & 0 \\ 0 & 0 & \sigma^2 & \cdots & 0 \\ \vdots & \vdots & \vdots & \ddots & \vdots \\ 0 & 0 & 0 & \cdots & \sigma^2 \end{pmatrix}$$

where $\boldsymbol{X}_1$, $\boldsymbol{X}_2$ and $\boldsymbol{X}_3$ are the model matrices for the machine-specific models for machines 1, 2 and 3, respectively.

Assuming normality of e_i, we simulate one possible realization of the assumed model for the machine-diameter combinations of the data set. The simulated data are shown in Figure 4.3 and the essential R-scripts are below.

```
> library(lmfor)
> data(stumplift)

> # True parameter values
> beta<-c(-30, 3, 0, 100, -5, 0.1, 50, -5, 0.1)
> sigma<-25

> # Construct the model matrix for diameter-machine combinations
> # of dataset stumplift
> x<-stumplift$Diameter
> M<-stumplift$Machine
> M2<-as.numeric(M==2)
> M3<-as.numeric(M==3)
> X<-cbind(1,x,x^2,M2,M2*x,M2*x^2,M3,M3*x,M3*x^2)

> # Compute the mean processing time and the simulated values
> Et<-X%*%beta
> tsim<-Et+rnorm(length(x),mean=0,sd=sigma)
```

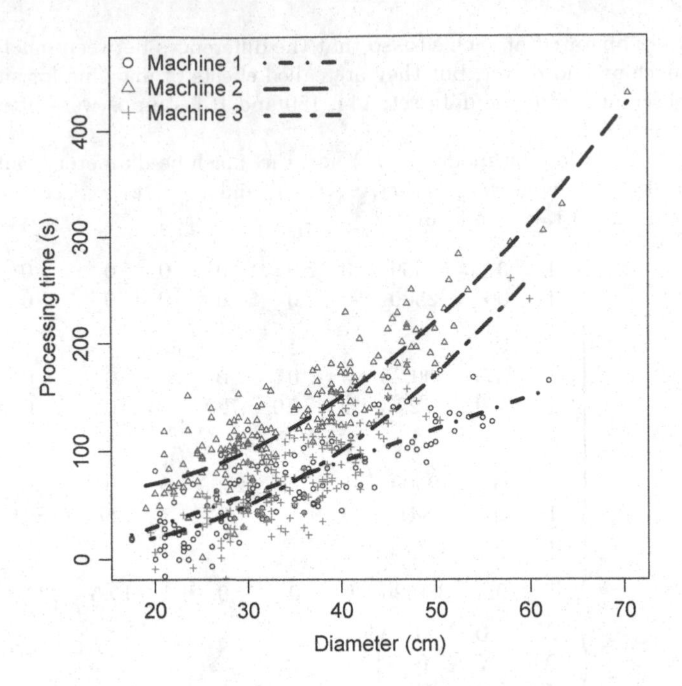

FIGURE 4.3 Mean processing time against stump diameter (lines) and simulated processing times (dots) for the diameter-machine combinations of `stumplift` data set under the model described in Example 4.4, assuming a normal distribution for the residual errors.

4.3 The random part

A researcher often faces a situation where the starting assumptions about the error do not hold. In particular, the assumption of constant error variance, or the assumption of zero correlation among residual errors may not be justified. A common approach to fix the problem of non-constant error variance is the use of nonlinear transformations to the response. This solution simultaneously changes the assumption on the systematic part of the model, as we have already discussed in Section 4.2. In the case of dependency, the situation is not as simple. One might obtain independent data by computing means of y in independent subsets of the original data, especially if the data have a grouped structure. However, this solution might be fully acceptable only in a balanced grouped data set with group-specific predictors only. If the dependence is caused by temporal or spatial dependency, one might consider the removal of observations that are too close to each other in space or time. However, such an approach would be a waste of collected data.

The next subsections present the main principles, and some simple but widely applicable models, for modeling the error variance and correlation. When those models are used, the data is not manipulated to meet the OLS-assumptions. Instead, the model is changed to more realistically describe the random process behind the data. More alternatives and discussion about modeling the error variance-covariance structure in the context of linear models with R can be found in Pinheiro and Bates (2000) and Gałecki and Burzykowski (2013, Chapters 10–12).

4.3.1 Modeling inconstant error variance

Quite often the assumption $\text{var}(e_i) = \sigma^2$ about a constant variance is not met: the residual errors are *heteroscedastic*, i.e. they have *heterogeneous variance*. In those cases, the assumption can be replaced with

$$\text{var}(e_i) = f(z_i|\boldsymbol{\theta}), \tag{4.12}$$

where function f models the error variance, z_i is a predictor of the variance function, and $\boldsymbol{\theta}$ includes the variance function parameters. Vector $\boldsymbol{\theta}$ includes a scaling parameter σ^2 and possibly other parameter(s) that describe the change of variance as a function of z_i. The variance varies because z_i varies. Therefore, a single scalar-valued error variance or residual standard error does not exist. The most common choice as z_i is the unknown $\text{E}(y_i)$, which is replaced by its estimate once estimates of $\boldsymbol{\beta}$ are available. However, any scalar valued function of the predictors $\boldsymbol{x}_i$ can be used as z_i as well, to make the variance depend on a vector of predictors instead of one scalar predictor.

A widely applicable function for modeling the error variance is the power function:

$$\text{var}(e_i) = \sigma^2 |z_i|^{2\delta}, \tag{4.13}$$

where σ^2 and δ are the scale and shape parameters to be estimated. Depending on the value of δ, the error variance may be decreasing ($\delta < 0$), constant ($\delta = 0$, implying the OLS model, Section 4.1.2) or increasing ($\delta > 0$) as a function of z_i. A drawback of the power function is that it approaches zero when $z_i \to 0$. Therefore, for models where z_i may be close to zero and variance does not go to zero when z_i goes to zero, one may instead use

$$\text{var}(e_i) = \sigma^2 \left(\delta_1 + |z_i|^{\delta_2}\right)^2 .$$

It is important to recognize that parameter σ^2 now has a different interpretation than in the model where the residual variance is constant. It is a scaling factor of the variance function, and nothing more. The error variance for each observation is obtained from the variance function, and the value depends on σ^2, z_i and on the other parameters of the variance function.

4.3.2 Modeling dependence using an autocorrelation function

Both the model with constant variance and the model with heterogeneous variance have assumed uncorrelated residual errors: $\text{cov}(e_i, e_j) = 0$ when $i \neq j$. This assumption may not be true: there may be serial (temporal) autocorrelation in a data set of repeated measurements, spatial correlation in a spatial data set, or correlation among members of the same group in a grouped data set. Such correlation can be modeled by assuming a specific structure for the dependence among observations.

As an example, consider data of repeated measurements, where the data are ordered according to the measurement time. Let k be the integer-valued distance in time between two measurements, which is expressed in terms of measurement intervals and call it the *time lag*. The correlation between observations with time lag k can be modeled, e.g. by the widely used and relatively simple *autoregressive process* of order 1, AR(1) (recall Equation (2.55) p. 42):

$$\text{cor}(e_i, e_{i+k}) = \phi^k, \tag{4.14}$$

where ϕ, the correlation between successive observations, is a parameter to be estimated. This model assumes that the correlation is only a function of the time lag between the two observations. The extension introduces one additional parameter ϕ compared to the OLS model. The implicit assumption is that the time lag between all observations is constant, which is commonly met if the observations are taken at fixed time intervals (such as hourly, daily, weekly or annually) and there is no missing data. A simple extension of the model is the continuous-time AR process (CAR), where the time points t_j belong to a continuous index set and $\text{cor}(e_i, e_j) = \phi^{|t_j - t_i|}$. If additionally non-constant error variance is assumed through the power-type variance function (4.13), the covariance between each pair becomes the product of the assumed correlation and the non-equal residual standard errors of the two observations:

$$\text{cov}(e_i, e_j) = \text{sd}(e_i)\,\text{sd}(e_j)\,\text{cor}(e_i, e_j) = \sigma^2\,|z_i z_j|^\delta\,\phi^{|i-j|}.$$

Alternatives and extensions to the AR(1) model are AR(p), MA(q) and ARMA(p, q) models (recall Section 2.12).

For spatially dependent data, an *isotropic* spatial correlation structure may be assumed. It is an extension of the temporal correlation structure for continuous time, where the time lag between observations is replaced with a spatial distance $d(i, j)$, most commonly the euclidean distance between the locations of observations i and j. The term isotropic means that the correlation is a function of distance between the two observations only, i.e. it is not affected by the direction. A nonlinear function that best models the decrease of the spatial correlation as a function of increasing distance is selected as the parametric *correlogram*. A simple example is the exponential correlogram

$$\text{cor}(e_i, e_j) = \rho(d) = \begin{cases} (1 - \psi_1)\exp(-d/\psi_2) & 0 < \psi_1 < 1; \psi_2 > 0; d > 0 \\ 1 & d = 0\,, \end{cases} \tag{4.15}$$

where d is the (euclidian) distance between observations i and j. The upper function is a decreasing correlation function, approaching the maximum value of $1 - \psi_1$ as $d \to 0$, and the minimum value of 0 as distance d approaches infinity. Parameter ψ_2 is called *the range* of the spatial correlation. When $d = \psi_2$, the correlation is $100\exp(-1) = 36\%$ of the maximal correlation, and the correlation is 14% of the maximum when $d = 2\psi_2$. Parameter ψ_1, which is restricted to range $[0, 1]$, allows non-zero variability in e_i even for infinitely small, non-zero distances d, and is called the *nugget effect* due to early applications of this model in the mining industry. Equivalently, the spatial dependency can be expressed through the variogram

$$\gamma(d) = \frac{1}{2}\,\text{var}(e_i - e_j)\,.$$

The correlogram and variogram are related through $\rho(d) = 1 - \frac{1}{\sigma^2}\gamma(d)$, if $\text{var}(e_i) = \text{var}(e_j) = \sigma^2$ (show this!).

For grouped data, the simplest correlation structure results from assuming a constant correlation among all members of the same group:

$$\text{cor}(y_i, y_j) = \begin{cases} \rho & \text{if } y_i \text{ and } y_j \text{ belong to the same group} \\ 0 & \text{if } y_i \text{ and } y_j \text{ belong to different groups.} \end{cases}$$

This correlation structure, called the *compound symmetry* (e.g. Pinheiro and Bates 2000, p. 161, 227), covers the mixed-effects model with a random intercept as the special case when $\rho > 0$. More discussion in regard to the interpretation and formulation of this model, as well as more advanced variance-covariance structures for grouped data, can be found in Chapter 5.

4.3.3 Variance-covariance structures in the model for all data

The assumptions of non-constant error variance and non-zero correlation among residual errors are implemented in model (4.6) by replacing the assumption (4.8) with

$$\text{var}\,(\boldsymbol{e}) = \sigma^2 \boldsymbol{V} \tag{4.16}$$

where $\boldsymbol{V}$ is any positive definite matrix of size $n \times n$. The resulting extended model is called the *GLS model*, because the optimal parameter estimation method for the model is the Generalized Least Squares (GLS). The OLS model is such a special case of the GLS model where $\boldsymbol{V} = \boldsymbol{I}$.

Because the model matrix $\boldsymbol{X}$ and coefficients $\boldsymbol{\beta}$ are treated as fixed, the assumptions on the residual error can be equivalently stated as

$$\begin{aligned} \mathrm{E}(\boldsymbol{y}) &= \boldsymbol{X\beta} \\ \mathrm{var}(\boldsymbol{y}) &= \sigma^2 \boldsymbol{V} . \end{aligned}$$

If the assumption about normality of residual errors is needed, it is specified as

$$\boldsymbol{e} \sim N(\boldsymbol{0}, \sigma^2 \boldsymbol{V})$$

or equivalently

$$\boldsymbol{y} \sim N(\boldsymbol{X\beta}, \sigma^2 \boldsymbol{V}) .$$

The matrix $\sigma^2 \boldsymbol{V}$ specifies the variance-covariance structure for all pairs of observations of the data so that the diagonal elements separately specify the variance for each observation and the off-diagonal elements separately specify the covariances for all possible pairs of observations. The following matrices show two examples of the resulting variance-covariance matrix under some previously specified models for error variance and correlation.

If a non-constant error variance is assumed but the assumption of zero correlation is retained, only the diagonal elements of $\boldsymbol{V}$ are non-zero. The difference to the OLS model is that these elements are no longer the same. For example, the power-type variance function (4.13) with zero correlation among observations yields the following diagonal variance-covariance matrix of residual errors:

$$\mathrm{var}(\boldsymbol{e}) = \sigma^2 \begin{pmatrix} |z_1|^{2\delta} & 0 & 0 & \cdots & 0 \\ 0 & |z_2|^{2\delta} & 0 & \cdots & 0 \\ 0 & 0 & |z_3|^{2\delta} & \cdots & 0 \\ \vdots & \vdots & \vdots & \ddots & \vdots \\ 0 & 0 & 0 & \cdots & |z_n|^{2\delta} \end{pmatrix} . \tag{4.17}$$

If dependence of observations is assumed, then the off-diagonal elements of $\boldsymbol{V}$ are no longer zeroes. For example, if the AR(1) model (4.14) is used to model the temporal dependence, and constant error variance (σ^2) is assumed, we get

$$\mathrm{var}(\boldsymbol{e}) = \sigma^2 \begin{pmatrix} 1 & \phi & \phi^2 & \cdots & \phi^{n-1} \\ \phi & 1 & \phi & \cdots & \phi^{n-2} \\ \phi^2 & \phi & 1 & \cdots & \phi^{n-3} \\ \vdots & \vdots & \vdots & \ddots & \vdots \\ \phi^{n-1} & \phi^{n-2} & \phi^{n-3} & \cdots & 1 \end{pmatrix} . \tag{4.18}$$

To define a model where both heteroscedasticity and dependence is assumed, let $\sigma^2 \boldsymbol{\Gamma}$ be a matrix that includes the residual error variances on the diagonal, and zeroes elsewhere (e.g. Equation (4.17)), and let $\boldsymbol{C}$ be the correlation matrix of residuals (e.g. Equation (4.18) without the scaling factor σ^2). Now

$$\mathrm{var}(\boldsymbol{e}) = \sigma^2 \boldsymbol{\Gamma}^{1/2} \boldsymbol{C} \boldsymbol{\Gamma}^{1/2}$$

where the diagonal matrix $\boldsymbol{\Gamma}^{1/2}$ is defined so that $\boldsymbol{\Gamma} = \boldsymbol{\Gamma}^{1/2}\boldsymbol{\Gamma}^{1/2}$, i.e. $\sigma\boldsymbol{\Gamma}^{1/2}$ includes the residual standard errors on the diagonal, and zeroes elsewhere. The above-specified variance-covariance-structures and the compound symmetric structure for grouped data are illustrated in the following example.

4.3.4 Example

We illustrate alternative variance-covariance structures that might be justified in the `stumplift` data set.

Example 4.5. Variance-covariance structures in a forest work study. Let us replace the assumption of constant error variance in Example 4.4 with the assumption

$$\mathrm{var}(e_i) = \sigma^2 \left| x_i^{(2)} \right|^{2\delta} ,$$

where $\sigma^2 = 0.5^2$ and $\delta = 1.1$; the other assumptions are kept unchanged. One realization of this model is shown in the upper left-hand graph in Figure 4.4. The graph shows that the observations with a greater diameter vary around the machine specific lines more than observations with small diameters. The phenomena could be seen even more clearly by plotting the residual errors $y_i - \boldsymbol{x}_i'\boldsymbol{\beta}$ on diameter.

The stumps in the data set within each clearcut area are saved in the order of processing. It may be possible that the errors of successive stumps are correlated, e.g. because of variation in the alertness of the operator over time, or location-related effects in the processing speed (e.g. small obstacles that have temporary effects on the processing speed but then affect the processing of several successive stumps). This temporal autocorrelation could be described by an AR(1) process within each machine.

For the purpose of demonstration, assume that the lag-1 correlation of the process is known to be $\phi = 0.9$ within all three machines. The other assumptions were kept unchanged from Example 4.4. The assumed correlation structure is illustrated in the lower left-hand graph in Figure 4.4, so that black indicates a correlation of 1 and white indicates zero correlation. We see that the correlation between any pair of stumps harvested by the same machine is non-zero for short lags and gets weaker as the lag increases. The shift from machine 1 to 2 is visible at observation number 145, and from machine 2 to 3 at observation number 325.

The upper right-hand graph in Figure 4.4 shows one realization of the data set from the model with constant error variance and the AR(1) model for the autocorrelation. A simulated realization from another model, where a non-constant variance according to the above-specified power-type variance function was assumed in addition to the AR(1) process with $\phi = 0.9$, is shown in the middle left-hand graph. The presence of very strong temporal autocorrelation cannot be detected from these scatter plots, and more advanced analytic tools are needed to detect it. Such tools will be discussed later in Section 4.5.

The compound symmetric correlation structure is demonstrated by specifying $\mathrm{cor}(e_i, e_j) = 0.5$ for all pairs of the same machine. The resulting correlation matrix is illustrated in the lower right-hand graph in Figure 4.4. A realized data set from such a model, assuming the constant error variance with $\sigma^2 = 25^2$ is shown in the middle right-hand graph. The constant correlation among the observations of the same machine causes level shifts in the data sets. For example, almost all observations of machine 2 are below the machine-specific curve that shows the expected value for the machine. This level shift would not be recognized in an empirical data set, where the true machine-specific models are not known, and it could not be separated from the true differences between machines, if the data set does not include replicates of clearcut areas within machines. In addition, the random variability of observations around the area-specific line (which is not shown) is smaller than in the data set where observations are independent. This means that inference about the differences between machines would be misleading if the data were analyzed with the OLS model. If replicates of clearcut areas within machines were available, the differences between machines could be separated from the random variability between clearcuts. A natural way to analyze such a grouped data set is provided by the mixed-effect models, which will be formulated in Chapter 5.

Selected parts of the code:

```
> # 1. Non-constant error variance with zero correlation.
> # Matrix V
> V1<-diag(x^(2*1.1))

> # 2. AR(1) with constant error variance.
```

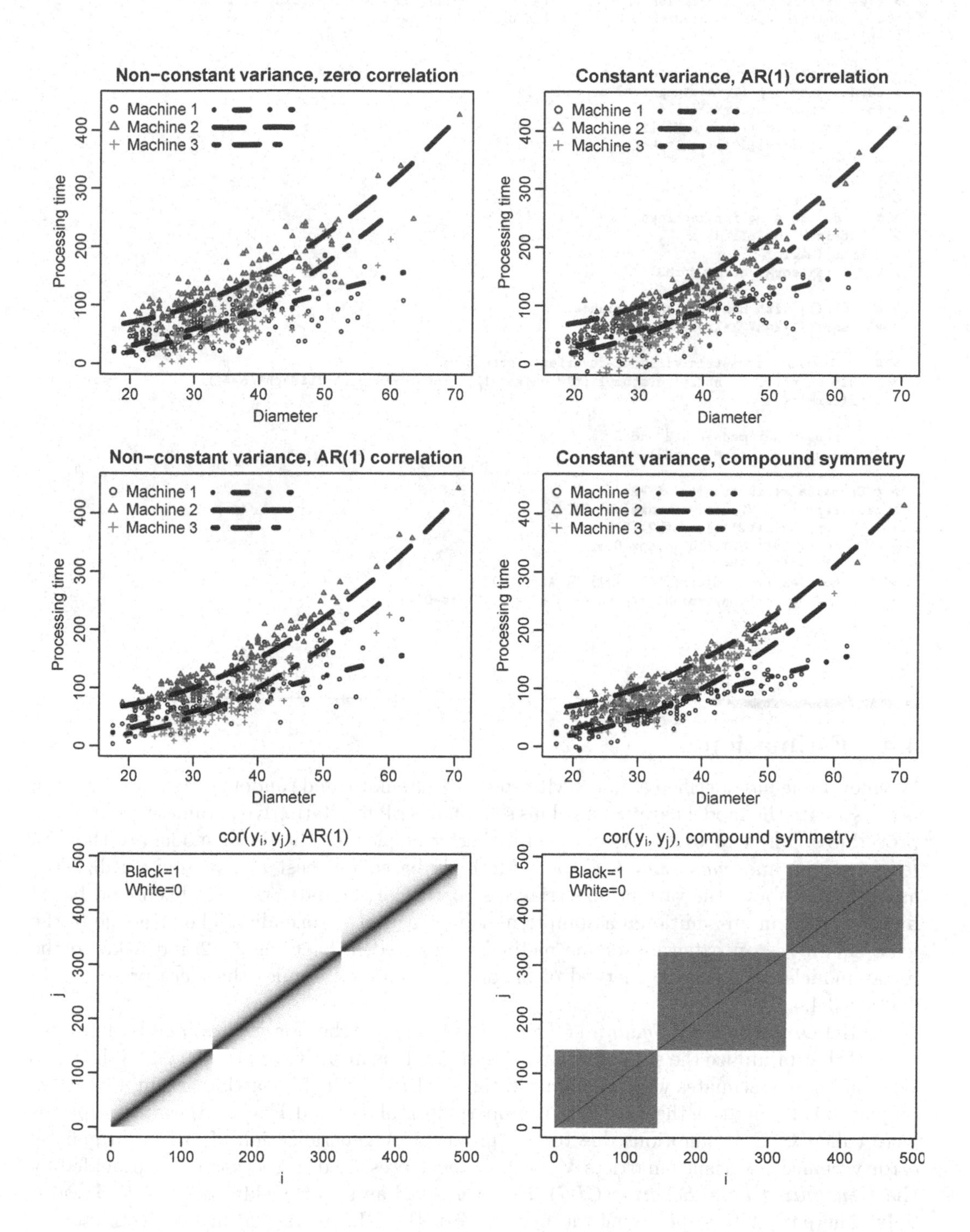

FIGURE 4.4 Realizations of four different models for the `stumplift` data set (upper and middle), and illustration of the two applied correlation structures (lower).

```
> # We utilize the information that the data are primarily ordered according to machine,
> # and within each machine, according to the order of processing.
> phi<-0.9

> # Correlation matrix for machine 1
> corM1<-diag(rep(1,sum(stumplift$Machine==1)))
> for (i in 1:dim(corM1)[1]){
+      for (j in 1:dim(corM1)[2]) {
+           corM1[i,j]<-phi^abs(i-j)
+      }
+ }

> # .. do the same for machines 2 and 3 ..
> # Matrix V for all data
> library(magic)
> V2<-adiag(corM1,corM2,corM3)

> # 3. AR(1) with non-constant variance
> V3<-sqrt(V1)%*%V2%*%sqrt(V1)

> # 4. Compound symmetric variance-covariance structure
> corM1<-matrix(0.5, ncol=sum(stumplift$Machine==1), nrow=sum(stumplift$Machine==1))
> diag(corM1)<-1

> # .. repeat for machines 2 and 3 ..
> V4<-adiag(corM1,corM2,corM3)

> # Graphical illustration of V2
> plot(rep(1:dim(V2)[1],each=dim(V2)[2]),
+      rep(1:dim(V2)[1],dim(V2)[2]),
+      col=gray(1-V2),pch=15,cex=0.1,
+      xlab="i",ylab="j",
+      main=expression(cor(y[i],y[j])*", AR(1)"))
> legend("topleft",box.col=NA,legend=c("Black=1","White=0"))
```

4.4 Estimation

Whenever one has specified a model with specific systematic and random parts, the next step is to estimate the model parameters. This section describes alternative parameter estimation procedures. The two widely used model fitting principles for the linear model are those of *least squares* and *maximum likelihood*. Methods based on least squares only utilize the assumptions about the variance-covariance structure of the data, whereas likelihood-based methods rely on an additional assumption of (multivariate) normality. The principle of the least squares is an extension of the methods presented in Sections 3.4.2 and 3.4.3 to the linear models, whereas the method of maximum likelihood applies the ideas presented in Section 3.4.4.

In the *Ordinary Least Squares* (*OLS*) method, one searches for such estimates of parameters $\boldsymbol{\beta}$ that minimize the sum of squared residuals. This intuitive approach can be shown to yield unbiased estimates with minimum variance (BLUE), if the starting assumptions 1–3 on page 69 are met (i.e. the fixed part is properly formulated and $\boldsymbol{V} = \boldsymbol{I}$); these assumptions were called OLS assumptions due to this property. A generalization of OLS to a general error variance-covariance matrices $\boldsymbol{V}$, such as those presented in Section 4.3, is provided by the *Generalized Least Squares* (*GLS*). However, OLS and GLS yield estimates of $\boldsymbol{\beta}$ and σ only. The parameters specifying the matrix $\boldsymbol{V}$ of the GLS model, or at least estimates of them, need to be known before GLS can be applied.

In the methods based on maximum likelihood, one specifies the multivariate normal likelihood and maximizes it with respect to the model parameters. Two variants of the maximum likelihood are used: the maximum likelihood (ML) and *restricted* (or *residual*) *maximum likelihood* (*REML*). The maximum likelihood can be used for estimation of both the $\boldsymbol{\beta}$ and the parameters specifying $\boldsymbol{V}$. However, at least some of the estimators of the parameters of $\boldsymbol{V}$ are biased in data sets with small n (but they are consistent). The REML method provides estimators that are unbiased, at least in certain special cases, are generally less biased than the ML estimators, and seem to be less sensitive to outliers than ML estimators (McCulloch *et al.* 2008, p. 178–179). REML provides estimators only for the parameters of the error variance-covariance structure $\boldsymbol{V}$. Once these have been estimated, one can use GLS to estimate parameters $\boldsymbol{\beta}$, treating the estimated matrix $\boldsymbol{V}$ as if it were true. This is usually the default estimation method in statistical packages. There are also some simpler methods than REML and ML for estimation of the variance-covariance structure $\boldsymbol{V}$ in some specific cases. For example, *iteratively reweighted GLS* can be applied.

4.4.1 Ordinary least squares (OLS)

The OLS finds parameter estimates $\widehat{\boldsymbol{\beta}}$ that minimize the residual sum of squares $\sum_{i=1}^{n}\left(y_i - \boldsymbol{x}_i\widehat{\boldsymbol{\beta}}\right)^2$. The problem definition yields a system of p equations, which are solved for the regression coefficients. For a general solution and simple presentation, it is convenient to use the matrix notation. The sum of squared residuals can be written in a matrix form as $(\boldsymbol{y} - \boldsymbol{X}\widehat{\boldsymbol{\beta}})'(\boldsymbol{y} - \boldsymbol{X}\widehat{\boldsymbol{\beta}}) = \widehat{\boldsymbol{e}}'\widehat{\boldsymbol{e}}$. Minimizing this with respect to $\widehat{\boldsymbol{\beta}}$ yields the OLS estimator of $\boldsymbol{\beta}$:

$$\widehat{\boldsymbol{\beta}}_{OLS} = (\boldsymbol{X}'\boldsymbol{X})^{-1}\boldsymbol{X}'\boldsymbol{y}\,. \tag{4.19}$$

The estimator is unbiased:

$$\begin{aligned} \mathrm{E}(\widehat{\boldsymbol{\beta}}_{OLS}) &= \mathrm{E}\left[(\boldsymbol{X}'\boldsymbol{X})^{-1}\boldsymbol{X}'\boldsymbol{y}\right] \\ &= (\boldsymbol{X}'\boldsymbol{X})^{-1}\boldsymbol{X}'\mathrm{E}(\boldsymbol{y}) \\ &= (\boldsymbol{X}'\boldsymbol{X})^{-1}\boldsymbol{X}'\boldsymbol{X}\boldsymbol{\beta} \\ &= \boldsymbol{\beta} \end{aligned} \tag{4.20}$$

and has the variance-covariance matrix

$$\begin{aligned} \mathrm{var}(\widehat{\boldsymbol{\beta}}_{OLS}) &= \mathrm{var}\left[(\boldsymbol{X}'\boldsymbol{X})^{-1}\boldsymbol{X}'\boldsymbol{y}\right] \\ &= (\boldsymbol{X}'\boldsymbol{X})^{-1}\boldsymbol{X}'\mathrm{var}\,(\boldsymbol{y})\,\boldsymbol{X}\,(\boldsymbol{X}'\boldsymbol{X})^{-1} \\ &= \sigma^2\,(\boldsymbol{X}'\boldsymbol{X})^{-1}\boldsymbol{X}'\boldsymbol{I}\boldsymbol{X}\,(\boldsymbol{X}'\boldsymbol{X})^{-1} \\ &= \sigma^2\,(\boldsymbol{X}'\boldsymbol{X})^{-1}\,. \end{aligned} \tag{4.21}$$

Computing the OLS estimator and its variance requires inversion of matrix $\boldsymbol{X}'\boldsymbol{X}$. The inverse exists only if the columns of $\boldsymbol{X}$ are linearly independent. If that is not the case, a unique value of $\boldsymbol{\beta}$ that minimizes the sum of squares cannot be found and the model is unidentifiable, as we discussed in Section 4.2.1. If the predictors have strong multicollinearity, then the inverse exists but the inaccuracy in the estimated regression coefficients (visible in the diagonal elements of $(\boldsymbol{X}'\boldsymbol{X})^{-1}$) becomes very high. However, the correlations (the off-diagonal elements of $(\boldsymbol{X}'\boldsymbol{X})^{-1}$) also become high in absolute value, and so the variance of the fitted values, $\mathrm{var}(\boldsymbol{X}\widehat{\boldsymbol{\beta}})$, is not increased by inclusion of a correlated predictor (recall Equation (2.44) on p. 24).

The estimation variances of regression coefficients depend only on the scalar-valued error variance σ^2, and the values of predictors in $\boldsymbol{X}$. Therefore, with wise selection of the values of x-variables during the experimental design, a researcher can crucially affect the accuracy of $\widehat{\boldsymbol{\beta}}_{OLS}$ and the covariance among the parameter estimates. For more discussion, see Chapter 14. To compute the numerical values of the estimation errors, an estimate of the error

Box 4.1 Stein's paradox

In this box, we briefly consider the James-Stein estimator which has caused much debate among statisticians. Let us assume that $\boldsymbol{Y}$ is an m dimensional random vector distributed as $\boldsymbol{y} \sim N(\boldsymbol{\theta}, \sigma^2 \boldsymbol{I})$, with known σ^2 and unknown mean vector $\boldsymbol{\theta}$. We want to estimate $\boldsymbol{\theta}$ using a single observation $\boldsymbol{y}$ of $\boldsymbol{Y}$. The least squares estimator and the maximum likelihood estimator is $\widehat{\boldsymbol{\theta}} = \boldsymbol{y}$. However, Stein showed that if we want to minimize the mean square error $\mathrm{E}\left(\left\|\boldsymbol{\theta} - \widehat{\boldsymbol{\theta}}\right\|^2\right) = \mathrm{E}\left(\sum_{i=1}^m \left(\theta_i - \widehat{\theta}_i\right)^2\right)$, the ML estimate is not optimal. The result is known as Stein's paradox. An estimator with lower MSE for $m \geq 3$ is the James-Stein estimator:

$$\boldsymbol{\theta}_{JS} = \left(1 - \frac{(m-2)\,\sigma^2}{\|\boldsymbol{y}\|^2}\right) \boldsymbol{y}$$

The James-Stein estimator shrinks the natural estimator towards zero, and therefore, is called a *shrinkage estimator*. Shrinkage estimators are biased but can have lower MSEs than any unbiased estimator. The estimator can also be modified so that the shrinkage is towards any fixed vector. Of course, the estimator is better the closer the fixed vector is to the true value. The estimator can also be modified to handle unknown variance, several observations, correlations and unequal variances. What makes the James-Stein estimator paradoxical is that the different dimensions of the $\boldsymbol{Y}$ vector need not have anything in common, for example, one variable may come from astronomy, the second from sociology and the third from forestry. Stein's paradox is mathematically interesting, but we recommend that James-Stein estimators should not be used in forest science. They would be better with respect to MSE, but we do not consider mean square error such an absolute criterion that it would support the use of illogical estimators.

The BLUP-predictors in the context of mixed models (see Section 5.4.1), Bayesian estimates of any model parameters, ridge and LASSO estimators of regression coefficients (see Box 4.2 on page 91) are examples of such shrinkage estimators that have a sound theoretical and practical justification.

variance is needed. An unbiased estimator is

$$\widehat{\sigma}^2_{OLS} = \frac{\widehat{\boldsymbol{e}}'\widehat{\boldsymbol{e}}}{n-p}, \tag{4.22}$$

where $\widehat{\boldsymbol{e}} = \boldsymbol{y} - \boldsymbol{X}\widehat{\boldsymbol{\beta}}_{OLS}$ includes the (empirical) residuals. Because $\boldsymbol{X}$ is fixed, using the unbiased estimator of σ^2 in Equation (4.21) also leads to an unbiased estimator of $\mathrm{var}(\widehat{\boldsymbol{\beta}}_{OLS})$, $\widehat{\mathrm{var}}(\widehat{\boldsymbol{\beta}}_{OLS})$. The estimate of the residual standard error is obtained as the square root of $\widehat{\sigma}^2_{OLS}$, and it is biased (recall Section 3.4.3).

The OLS estimator of $\boldsymbol{\beta}$ is the BLUE of the parameters of model (4.6). This results from the Gauss-Markov theorem (Christensen 2011, Harville 1981). That is, from among estimators that are unbiased and linear with respect to $\boldsymbol{y}$, the OLS estimator has the smallest variance. However, there are biased estimators that have smaller variance, and also smaller MSE than the OLS estimator (see Boxes 4.1 and 4.2).

The only assumption that was used to prove the zero bias of OLS estimator $\boldsymbol{\beta}_{OLS}$ was the assumption that the systematic part is correct: $\mathrm{E}(\boldsymbol{y}) = \boldsymbol{X}\boldsymbol{\beta}$. Therefore, even if the OLS assumptions of the residual errors do not hold, the OLS estimator of $\boldsymbol{\beta}$ is unbiased. However, the OLS estimator is no more BLUE, and estimators with lower variance can be found. In addition, the estimator (4.22) is either biased (if the residual errors are correlated but have a constant variance) or does not make sense (if the true variance is not a single number). In those cases, $\widehat{\mathrm{var}}(\widehat{\boldsymbol{\beta}}_{OLS})$ is also biased.

If the model includes an intercept, then the average of fitted values $\widehat{y}_i = \boldsymbol{x}_i'\widehat{\boldsymbol{\beta}}_{OLS}$ is equal to the average of the y-variable in the data, and the mean of residuals is zero. If all

Box 4.2 Ridge regression and LASSO

Ridge regression is a shrinkage estimator of regression coefficients (See Box 4.1 on page 90). More specifically, it is a modification of the OLS estimator, where parameter estimates are shrunken towards zero so as to reduce the MSE. The cost of reduced MSE is a non-zero bias. Equation (4.19) described the OLS estimator in a linear regression model. It is unbiased but it can have large variance under multicollinearity. In ridge regression (see e.g. Fahrmeir *et al.* (2013)), the diagonal of the moment matrix is increased:

$$\widehat{\beta}_{ridge} = \left(\boldsymbol{X}'\boldsymbol{X} + \lambda \boldsymbol{I}\right)^{-1}\boldsymbol{X}'\boldsymbol{y} \tag{4.23}$$

Note the similarity of Equation (4.23) to Equation (10.30) on p. 324. The ridge estimator is biased, but for suitable values of λ the MSE of the parameter estimates is smaller than in OLS regression. The ridge estimator minimizes the penalized sum of squares

$$\text{PLS}(\boldsymbol{\beta}) = (\boldsymbol{y} - \boldsymbol{X}\boldsymbol{\beta})'(\boldsymbol{y} - \boldsymbol{X}\boldsymbol{\beta}) + \lambda\boldsymbol{\beta}'\boldsymbol{\beta} \tag{4.24}$$

where the first term is the sum of squared residuals and the second term is the sum of squared parameter estimates. Compare this to Equation (10.29). In order to make the scales of different parameters comparable, the predictor variables are usually standardized. So as not to penalize the intercept, the predictor variables are also centered so that the intercept will be zero. A problem in ridge regression is the determination of λ. A common procedure is to draw the *ridge trace*, i.e. the parameter estimates with respect to λ and examine where the estimates begin to stabilize (they will finally stabilize at zero).

Another method to shrink the parameter estimates is to use the sum of absolute values of parameters as the penalty term, instead of sum of squares of the parameters. This method is called LASSO (Least Absolute Shrinkage and Selection Operator, see e.g. Fahrmeir *et al.* (2013) or Hastie *et al.* (2009)). The LASSO estimates cannot be expressed in a closed form. LASSO and other methods based on similar ideas are commonly used in modern data science for variable selection in regression models.

x-variables take their mean value, then the predicted value is equal to the average of y in the data. If all predictors are centered, then the estimated intercept is equal to the average value of y.

Example 4.6. Consider model (4.10) defined in Example 4.3 (p. 77) for the `immer` data set. The model is fitted below using function `lm` of the base `R`:

```
> imod1<-lm(Y~Var+Loc,data=immer)
> summary(imod1)

Coefficients:
             Estimate Std. Error t value Pr(>|t|)
(Intercept)  102.202      6.078  16.815 2.88e-13 ***
VarP           8.150      6.078   1.341 0.194983
VarS          -3.258      6.078  -0.536 0.597810
VarT          23.808      6.078   3.917 0.000854 ***
VarV           4.792      6.078   0.788 0.439728
LocD         -26.060      6.658  -3.914 0.000860 ***
LocGR        -28.340      6.658  -4.256 0.000386 ***
LocM          -3.590      6.658  -0.539 0.595705
LocUF        -16.010      6.658  -2.405 0.025996 *
LocW          27.140      6.658   4.076 0.000589 ***

Residual standard error: 10.53 on 20 degrees of freedom
```

The estimated expected yield for variety M in location C is 102 bushels per acre. The expected yields for varieties P, S, T and V are estimated to be 8.2, -3.3, 23.8 and 4.8 kg higher than for variety M, respectively. The estimated expected differences between the other locations and location C are -26.1, -28.3, -3.6, -16.0 and 27.1 bushels. The estimated residual variance is $\widehat{\sigma}^2 = 10.53^2$ bushels^2.

Example 4.7. The OLS estimation of the model defined in Example 4.4 (p. 80) is carried out below using function `nlme::gls`.

```
> library(nlme)
> mod1<-gls(Time~Diameter+Machine*Diameter+Machine*I(Diameter^2),
+ data=stumplift)

> summary(mod1)
Generalized least squares fit by REML
  Model: Time ~ Diameter + Machine * Diameter + Machine * I(Diameter^2)
  Data: stumplift
       AIC      BIC   logLik
  4513.619 4555.273 -2246.81

Coefficients:
                           Value Std.Error   t-value p-value
(Intercept)            -26.82633  24.48858 -1.095462  0.2739
Diameter                 2.65388   1.40829  1.884470  0.0601
Machine2               116.54694  32.80309  3.552925  0.0004
Machine3                77.40427  42.51924  1.820453  0.0693
I(Diameter^2)           -0.01415   0.01910 -0.740899  0.4591
Diameter:Machine2       -6.74999   1.83028 -3.687952  0.0003
Diameter:Machine3       -5.02022   2.37284 -2.115697  0.0349
Machine2:I(Diameter^2)   0.11895   0.02428  4.898340  0.0000
Machine3:I(Diameter^2)   0.09463   0.03209  2.948914  0.0033

Residual standard error: 24.79195
Degrees of freedom: 485 total; 476 residual
```

The OLS estimates are reported in column `Value` and their standard errors (the square roots of $\widehat{\text{var}}(\widehat{\boldsymbol{\beta}})$) are shown in column `Std.Error`. The correlations of fixed effects are based on the off-diagonal elements of that matrix. The estimate of σ^2 is 24.79^2. Function `lm` could be used for model fitting as well, but we use function `nlme::gls` here to ensure consistency with the later examples.

Web Example 4.8. Implementing the OLS estimation of the previous example manually by using matrix operations.

4.4.2 Generalized least squares (GLS)

If the OLS assumptions do not hold and matrix $\boldsymbol{V}$ is known, the efficient and unbiased estimator $\widehat{\boldsymbol{\beta}}$ is provided by GLS, where the sum of weighted squared residuals $(\boldsymbol{y} - \boldsymbol{X}\widehat{\boldsymbol{\beta}})'\boldsymbol{V}^{-1}(\boldsymbol{y} - \boldsymbol{X}\widehat{\boldsymbol{\beta}})$ is minimized with respect to $\widehat{\boldsymbol{\beta}}$. The solution is:

$$\widehat{\boldsymbol{\beta}}_{GLS} = \left(\boldsymbol{X}'\boldsymbol{V}^{-1}\boldsymbol{X}\right)^{-1}\boldsymbol{X}'\boldsymbol{V}^{-1}\boldsymbol{y}\,. \tag{4.25}$$

The estimator is unbiased and has variance

$$\text{var}(\widehat{\boldsymbol{\beta}}_{GLS}) = \sigma^2\left(\boldsymbol{X}'\boldsymbol{V}^{-1}\boldsymbol{X}\right)^{-1}\,. \tag{4.26}$$

An unbiased estimator of the scaling factor σ^2 is given by

$$\widehat{\sigma}^2_{GLS} = \frac{\widehat{\boldsymbol{e}}'\boldsymbol{V}^{-1}\widehat{\boldsymbol{e}}}{n-p}\,. \tag{4.27}$$

Obviously, all results with OLS are obtained by replacing $\boldsymbol{V}$ with $\boldsymbol{I}$.

Interestingly, GLS can be implemented as OLS by using appropriately transformed data and predictors. In particular, such matrix $\boldsymbol{V}^{-1/2}$ exists that fulfills $\boldsymbol{V}^{-1/2}\left(\boldsymbol{V}^{-1/2}\right)' = \boldsymbol{V}^{-1}$ and $\left(\boldsymbol{V}^{1/2}\right)'\boldsymbol{V}^{1/2} = \boldsymbol{V}$. The matrix is found using the so-called Cholesky decomposition of

$\widehat{V}$ (see Searle 1982). Using this matrix, the GLS model can be written as the following OLS model:

$$y^* = X^*\beta + e^* , \tag{4.28}$$

where $y^* = \left(V^{-1/2}\right)' y$, $X^* = \left(V^{-1/2}\right)' X$, and $e^* = \left(V^{-1/2}\right)' e$. Now, $\operatorname{var} e^* = \sigma^2 I$. Therefore, the GLS estimate of β can be found as

$$\widehat{\beta}_{GLS} = \left((X^*)' X^*\right)^{-1} (X^*)' y^* , \tag{4.29}$$

i.e. by estimating the regression coefficients of the transformed data using OLS. See Example 4.13 (p. 104) for an R implementation of this procedure.

Example 4.9. Coefficients β of the model in Example 4.4 (p. 80) were estimated using GLS by assuming (recall Example 4.5 on p. 86):

1. No correlation among observations; error variance is proportional to diameter$^{2.2}$.
2. Dependence according to AR(1) within each machine with $\phi = 0.9$; constant error variance.
3. Dependence according to AR(1) within each machine with $\phi = 0.9$; error variance is proportional to diameter$^{2.2}$.
4. Dependence according to the compound symmetry structure with an intra-clearcut correlation of 0.5; constant error variance.

Table 4.2 shows the estimates using these specifications of V and their standard errors, together with the results from the OLS fit in Example 4.7. The applied assumption on error variance-covariance structure has a large effect especially on the estimates of β_1, β_4 and β_7, which specify the level of the machine-specific curves. Changes regarding error variance have an effect on the regression coefficients, whereas changes in the correlation structure have effects on the estimated standard errors of the regression coefficients.

```
> # Example of GLS using V1.
> betaGLS1<-solve(t(X)%*%solve(V1)%*%X)%*%t(X)%*%solve(V1)%*%y
> ehat<-y-X%*%betaGLS1
> sigma2GLS1<-t(ehat)%*%solve(V1)%*%ehat/(n-p)
> varbetaGLS1<-as.numeric(sigma2GLS1)*solve(t(X)%*%solve(V1)%*%X)
```

In practice, V is never known exactly, and it needs to be replaced by an estimated value $\widehat{V}$. One may have a justified knowledge of its structure, but the values of parameters are, in most cases, unknown. For example, we may have a good justification for the use of the structures presented in Section 4.3, but the parameters ϕ, ρ, or δ are unknown. The common solution is to estimate matrix V using the data and, thereafter, use the estimated V in the GLS estimator as if it were true. This approach ignores the estimation error of V. The resulting estimator is not BLUE, but if V is estimated consistently (e.g. using REML or ML), it approaches BLUE as the sample size increases, and has lower variance than $\widehat{\beta}_{OLS}$, at least if the data set is sufficiently large.

One alternative for the estimation of V is the iterative generalized least squares (IGLS). Let θ include the parameters that specify the assumed structure of V based on the data-generating process. IGLS has the following steps:

1. Assuming that $V = I$, estimate the regression coefficients.
2. Using the residuals of the current model, estimate θ and then use it to update the estimate of V.
3. Using the current estimate of V, re-estimate β and σ^2 using GLS.
4. Repeat steps 2 and 3 until there is no change in the estimates.

	Parameter estimates				
	OLS	GLS, V1	GLS, V2	GLS, V3	GLS, V4
β_1	-26,83	-33,33	-3,60	-11,20	-26,83
β_2	2,65	3,05	1,32	1,74	2,65
β_3	-0,01	-0,02	0,00	0,00	-0,01
β_4	116,55	80,04	115,97	98,50	116,55
β_5	-6,75	-4,67	-6,93	-5,96	-6,75
β_6	0,12	0,09	0,13	0,11	0,12
β_7	77,40	58,70	56,74	22,19	77,40
β_8	-5,02	-3,95	-3,42	-1,25	-5,02
β_9	0,09	0,08	0,07	0,03	0,09
	Standard errors				
	OLS	GLS, V1	GLS, V2	GLS, V3	GLS, V4
β_1	24,49	19,29	33,08	18,46	34,85
β_2	1,41	1,23	1,31	1,35	1,41
β_3	0,02	0,02	0,02	0,02	0,02
β_4	32,80	28,24	44,83	26,65	48,01
β_5	1,83	1,75	1,75	1,87	1,83
β_6	0,02	0,03	0,02	0,02	0,02
β_7	42,52	35,45	51,47	34,20	55,11
β_8	2,37	2,18	2,22	2,29	2,37
β_9	0,03	0,03	0,03	0,03	0,03

TABLE 4.2 Estimates of the regression coefficients and their standard errors in Example 4.4 in the `stumplift` data set using ordinary least squares (OLS) and generalized least squares (GLS) with the four variance-covariance structures specified in Example 4.5.

The special case where only heteroscedasticity is modeled is called Iteratively Reweighted Least Squares (IRLS). The variance function can be estimated by modeling the squared residuals using OLS (see Lappi (1997) for an example). For a detailed presentation of the methods, see Gałecki and Burzykowski (2013, p. 142–144). The benefit of IGLS is that it does not make assumptions on the distribution of the data. Another approach to estimate $\boldsymbol{\theta}$ is provided by maximum likelihood, which is presented in the next subsections. If step 2 of the procedure is properly implemented, IGLS leads to the ML estimates (Nelder and Wedderburn 1972, del Pino 1989). An example of an iterative procedure for estimating error variance of a mixed-effects model is given in Example 6.5 on page 193.

4.4.3 Maximum likelihood (ML)

The maximum likelihood method is based on an assumption about the distribution of $\boldsymbol{y}$. The likelihood is the joint *pdf* of $\boldsymbol{y}$ as a function of model parameters. Parameters values are searched for that maximize the value of the likelihood, i.e. the joint density given the observed data and assumed model. The method provides estimators for all parameters: $\boldsymbol{\beta}$, σ^2 and parameters that specify $\boldsymbol{V}$ in the GLS-model, although the chief benefit results from the estimation of the parameters that describe the variance-covariance structure. For this reason, we formulate the method directly for the GLS model under normality.

Consider the following model:

$$\boldsymbol{y} \sim N\left(\boldsymbol{X\beta}, \sigma^2 \boldsymbol{V}(\boldsymbol{\theta})\right) ,$$

where $\boldsymbol{V}(\boldsymbol{\theta})$ is a positive definite matrix parameterized by $\boldsymbol{\theta}$. This is a special case of model (4.6) (p. 71) because it includes an assumption on multivariate normality of $\boldsymbol{y}$. The log-

> **Box 4.3** BLUE or minimum RMSE?
>
> The use of GLS, instead of OLS, makes a difference between a statistical modeler and a curve-fitter. The latter seeks good results, whereas the former seeks realistic ones. GLS always provides a higher mean square error and a lower R-square (see Section 4.5.3) than OLS. The minimum RMSE cannot be used as a general model selection criterion. If the OLS assumptions are not met and OLS is used, then the obtained estimates for variances of the regression parameters and the obtained p-values are generally biased. Depending on both the correlation structure and the model matrix, the bias can occur in both directions (Christensen 2003). If it happens that OLS provides, in addition to smaller RMSE and higher R^2, also smaller estimates for the variances of the parameter estimates and smaller p-values, it is not a valid justification for the use of OLS.
>
> Assume that we are estimating simple linear regression
>
> $$y_{ij} = \beta_1 + \beta_2 x_{ij} + e_{ij}$$
>
> where $i = 1, \ldots, M$, $j = 1, \ldots, n$. Denote that $N = Mn$, $r_e = \text{cor}(e_{ij}, e_{ij'})$ is the within-group correlation of residual errors and $R_x = \sum_{i=1}^{M} \sum_{j=1}^{n} \sum_{j' \neq j} (x_{ij} - \bar{x})(x_{ij'} - \bar{x}) / (n-1) \sum_{i=1}^{M} \sum_{j=1}^{n} (x_{ij} - \bar{x})^2$ is the empirical within-group correlation of x. Then it can be shown that if $(N-1) r_e R_x + r_e > 0$, the p-value and variance for the slope parameter β_2 provided by OLS are too small. If the inequality is reversed, the p-value provided by OLS is too large.
>
> When GLS is used, then the effect of the correlation structure on the variances of the parameter estimates also depends on the model matrix. In the simple linear regression case described above, a non-zero r_e makes the variance of $\widehat{\beta}_2$ smaller when ${r_e}^2 < R_x r_e$.

likelihood becomes:

$$l(\boldsymbol{\beta}, \boldsymbol{\theta}, \sigma^2 | \boldsymbol{y}) = -\frac{n}{2} \log 2\pi - \frac{1}{2} \log \left| \sigma^2 \boldsymbol{V}(\boldsymbol{\theta}) \right| - \frac{1}{2\sigma^2} (\boldsymbol{y} - \boldsymbol{X\beta})' \boldsymbol{V}(\boldsymbol{\theta})^{-1} (\boldsymbol{y} - \boldsymbol{X\beta}) . \quad (4.30)$$

The derivation for the ML estimators requires differentiating Equation (4.30) with respect to $\boldsymbol{\beta}$, $\boldsymbol{\theta}$ and σ^2, setting the derivatives to zero, and solving for the parameters (see e.g. McCulloch *et al.* (2008, Chapter 6 Appendix)). The resulting estimator of $\boldsymbol{\beta}$ is the GLS estimator (4.25), which is unbiased, and BLUE when $\boldsymbol{V}$ is known. The ML estimator of the scaling factor σ^2 is

$$\widehat{\sigma}^2_{ML} = \frac{1}{n} \widehat{\boldsymbol{e}} \widehat{\boldsymbol{V}}^{-1} \widehat{\boldsymbol{e}} .$$

Under the OLS assumptions, we get $\widehat{\sigma}^2_{ML} = \frac{\widehat{\boldsymbol{e}}' \widehat{\boldsymbol{e}}}{n}$ (compare with Example 3.5, p. 54). These estimators differ from the unbiased estimators (4.22) and (4.27) by factor $\frac{n-p}{n}$ and are, therefore, downward biased. However, the estimators are consistent and the bias becomes negligible, if n is large and if p is small compared to n.

In contrast to OLS and GLS, the method also provides an estimator of $\boldsymbol{\theta}$. That estimator is usually biased but consistent. An implementation of the maximum likelihood for one special case: the linear mixed-effects model with a random intercept is given in Web Example 5.15 (p. 155).

4.4.4 Restricted maximum likelihood (REML)

The *restricted* or *residual ML* (*REML*) approach is a version of the ML that fixes the bias problem in the estimation of σ^2 and $\boldsymbol{\theta}$. That is why REML is the most widely applied estimation method for linear mixed models and variance component models. However, it can also be applied to other GLS models. The method provides estimates only for the parameters related to matrix $\boldsymbol{V}$. The parameters $\boldsymbol{\beta}$ are estimated using GLS after an REML-estimate of $\boldsymbol{V}$ has been found.

REML applies the maximum likelihood method to vector $\boldsymbol{K}'\boldsymbol{y}$ of length $n-p$ instead of the original vector of observations, $\boldsymbol{y}$ of length n. The contrast matrix $\boldsymbol{K}$ is selected wisely, so that it removes all such variation from vector $\boldsymbol{y}$ that can be explained by the fixed effects specified by a model matrix $\boldsymbol{X}$ but retains all other variation.

How then to find the matrix $\boldsymbol{K}$? The key condition is to define each column of matrix $\boldsymbol{K} = (\boldsymbol{k}_1, \ldots, \boldsymbol{k}_{n-p})$, such that $\boldsymbol{k}_i'\boldsymbol{X} = \boldsymbol{0}$ for $i = 1, \ldots, n-p$; such a column is called a *contrast*. The estimation is possible only if $\boldsymbol{K}$ has a full column rank. A matrix result (Searle 1982, Section 9.7 a) states that the maximum number of *linearly independent* columns that fulfill the above condition is $n-p$. However, there are infinitely many possible sets of $n-p$ contrasts, of which only one needs to be selected to formulate matrix $\boldsymbol{K}$. The choice of $\boldsymbol{K}$ has no effect on the final parameter estimates, which simplifies the task considerably. However, the choice of $\boldsymbol{K}$ has an effect on the value of the log-likelihood function at the solution.

One way to find columns $\boldsymbol{k}_i$ for a given model matrix $\boldsymbol{X}$ is to use (McCulloch *et al.* 2008)

$$\boldsymbol{k}_i' = \boldsymbol{c}_i' \left(\boldsymbol{I} - \boldsymbol{X}\boldsymbol{X}^-\right) \tag{4.31}$$

where $\boldsymbol{X}^-$ is a generalized inverse of matrix $\boldsymbol{X}$ (there are many possible $\boldsymbol{X}^-$ matrices) and $\boldsymbol{c}_i$ is a non-zero vector of length n. Any choices of $\boldsymbol{c}_i'$ and $\boldsymbol{X}^-$ will give $\boldsymbol{k}_i$ such that $\boldsymbol{k}_i'\boldsymbol{X} = \boldsymbol{0}$, but they should be selected so that matrix $\boldsymbol{K}$ is of full rank. The final REML estimates are not affected either by the choice of $\boldsymbol{X}^-$ or by the choice of $\boldsymbol{c}_i$.

Pre-multiplication of

$$\boldsymbol{y} = \boldsymbol{X}\boldsymbol{\beta} + \boldsymbol{e}$$

by $\boldsymbol{K}'$ yields

$$\boldsymbol{K}'\boldsymbol{y} = \boldsymbol{K}'\boldsymbol{X}\boldsymbol{\beta} + \boldsymbol{K}'\boldsymbol{e}\,.$$

The property $\boldsymbol{K}'\boldsymbol{X} = \boldsymbol{0}$ yields

$$\boldsymbol{K}'\boldsymbol{y} = \boldsymbol{K}'\boldsymbol{e}\,.$$

If $\text{var}(\boldsymbol{e}) = \sigma^2\boldsymbol{V}(\boldsymbol{\theta})$, then $\text{var}(\boldsymbol{K}'\boldsymbol{e}) = \sigma^2\boldsymbol{K}'\boldsymbol{V}(\boldsymbol{\theta})\boldsymbol{K}$. Furthermore, if $\boldsymbol{y} \sim N(\boldsymbol{X}\boldsymbol{\beta}, \sigma^2\boldsymbol{V}(\boldsymbol{\theta}))$, then $\boldsymbol{K}'\boldsymbol{y} \sim N(\boldsymbol{0}, \sigma^2\boldsymbol{K}'\boldsymbol{V}(\boldsymbol{\theta})\boldsymbol{K})$. The estimation can be based on this likelihood, which is called *restricted likelihood.* Essential differences between the likelihood of $\boldsymbol{K}'\boldsymbol{y}$ and the likelihood of $\boldsymbol{y}$ are (1) difference in the lengths of $\boldsymbol{y}$ (n) and $\boldsymbol{K}'\boldsymbol{y}$ ($n-p$), and (2) restricted likelihood does not include $\boldsymbol{\beta}$. The former ensures adjustment of the estimators of σ^2 and $\boldsymbol{\theta}$ for the degrees of freedom needed for the estimation of $\boldsymbol{\beta}$. The latter makes the likelihood much simpler compared to the full likelihood of $\boldsymbol{y}$, but also implies that REML does not provide estimators for $\boldsymbol{\beta}$. However, once $\boldsymbol{\theta}$ has been estimated, $\boldsymbol{\beta}$ can be estimated using the GLS estimator

$$\widehat{\boldsymbol{\beta}} = \left(\boldsymbol{X}'\widehat{\boldsymbol{V}}^{-1}\boldsymbol{X}\right)^{-1}\boldsymbol{X}'\widehat{\boldsymbol{V}}^{-1}\boldsymbol{y}\,.$$

This estimator is believed to yield at least nearly optimal estimators under different variance-covariance structures. In the literature and statistical software, the REML method refers to the use of REML for estimation of $\boldsymbol{\theta}$ and σ^2 and GLS for estimation of $\boldsymbol{\beta}$. In `R`, function `nlme::gls` can be used in fitting GLS models with many useful structures of $\boldsymbol{V}$. Box 4.4 summarizes two issues that should be kept in mind when using REML in model comparisons.

Web Example 4.10. Finding the REML estimator of σ^2 for the OLS-model in a data set of five observations using matrix operations in `R`.

Example 4.11. Let us replace the assumption of constant error variance on Example 4.7 with the power-type variance function

$$\text{var}(e_i) = \sigma^2\,|\widehat{y}_i|^{2\delta}\,.$$

Box 4.4 Warnings about model comparison using REML fits

Restricted maximum likelihood is the standard method for the estimation of variance-covariance parameters of the GLS model, including the linear mixed-effects model. However, there are two properties of the method that should be recognized by the user.

1. The likelihood is based on such a likelihood where the effect of part $\boldsymbol{X\beta}$ has been removed. Therefore, likelihood-based inference, such as likelihood ratio tests and the information criteria, should not be based on REML fits, if the models that have different fixed predictors.
2. There are infinitely many $\boldsymbol{K}$ matrices that can be used in REML fitting, and they all lead to identical estimates of the parameters. However, the value of the likelihood function at these estimates is affected by the choice of $\boldsymbol{K}$. Therefore, the REML likelihoods provided by different software packages are not necessarily comparable, even if the fixed parts are identical.

Therefore, likelihood-based comparisons can be based on REML only if the models have identical fixed parts and have been fitted using the same $\boldsymbol{K}$ matrix.

We further assume that the data are normally distributed and estimate parameter δ using the REML procedure, as implemented in function `nlme::gls`:

```
> mod2<-update(mod1,weights=varPower())
> summary(mod2)
Generalized least squares fit by REML
  Model: Time ~ Diameter + Machine * Diameter + Machine * I(Diameter^2)
  Data: stumplift
       AIC      BIC    logLik
  4389.066 4434.886 -2183.533

Variance function:
 Structure: Power of variance covariate
 Formula: ~fitted(.)
 Parameter estimates:
    power
0.8806234

Coefficients:
                            Value Std.Error   t-value p-value
(Intercept)             -33.07694  12.59961 -2.625235  0.0089
Diameter                  3.02842   0.81987  3.693788  0.0002
Machine2                 68.97624  28.93235  2.384053  0.0175
Machine3                 52.47329  33.11648  1.584507  0.1137
I(Diameter^2)            -0.01929   0.01237 -1.559887  0.1195
Diameter:Machine2        -3.92545   1.78586 -2.198071  0.0284
Diameter:Machine3        -3.53743   2.06793 -1.710609  0.0878
Machine2:I(Diameter^2)    0.07968   0.02623  3.037388  0.0025
Machine3:I(Diameter^2)    0.07317   0.03135  2.333951  0.0200

Residual standard error: 0.5738837
Degrees of freedom: 485 total; 476 residual
```

Example 4.12. Let us further assume an AR(1) process (4.14) (p. 83) for the errors within each machine in the model described in Example 4.11 (recall Example 4.5 for the definition of the AR(1) structure).

```
> mod3<-update(mod2,cor=corAR1(form=~1|Machine))
> summary(mod3)
Generalized least squares fit by REML
  Model: Time ~ Diameter + Machine * Diameter + Machine * I(Diameter^2)
  Data: stumplift
       AIC      BIC    logLik
  4383.002 4432.987 -2179.501

Correlation Structure: AR(1)
 Formula: ~1 | Machine
 Parameter estimate(s):
```

```
        Phi
0.1445364
Variance function:
 Structure: Power of variance covariate
 Formula: ~fitted(.)
 Parameter estimates:
    power
0.8617775

Coefficients:
                          Value Std.Error   t-value p-value
(Intercept)           -27.52623  12.97412 -2.121626  0.0344
Diameter                2.70023   0.84099  3.210787  0.0014
Machine2               75.48322  29.03036  2.600148  0.0096
Machine3               42.90643  33.04667  1.298359  0.1948
I(Diameter^2)          -0.01491   0.01264 -1.179603  0.2387
Diameter:Machine2      -4.36785   1.79223 -2.437108  0.0152
Diameter:Machine3      -2.81333   2.05299 -1.370359  0.1712
Machine2:I(Diameter^2)  0.08686   0.02637  3.294615  0.0011
Machine3:I(Diameter^2)  0.06081   0.03094  1.965252  0.0500

Residual standard error: 0.6220353
Degrees of freedom: 485 total; 476 residual
```

The estimated lag-1 autocorrelation coefficient was 0.145, which is not very high.

4.5 Does the model fit?

Model fit should be explored in various ways, including wisely selected graphics and numerical criteria that measure the goodness of fit. Practitioners often seek a simple, general and objective model evaluation procedure using simple number-valued criteria. Unfortunately, such criteria do not exist. Therefore, graphs and numerical model fitting criteria are not alternatives to each other, but instead should be used together. Graphical displays of various kinds provide an overall picture about the model fit and may show potential model misspecifications. Numerical criteria provide an objective summary of the most essential features and may be useful especially if two alternative models for the same response variable are compared in the same data. They can answer the question whether one model fits better than the other, but they do not necessarily tell us whether any of the models fits sufficiently well, except in some special cases discussed on page 115 and in Section 8.2.6.

4.5.1 Model residuals

The residual errors e_i would be a perfect tool to graphically evaluate whether the model assumptions 1–4 in Section 4.1.1 have been met. Unfortunately, the residuals errors are not available. However, after the model parameters have been estimated, the *residuals* can be computed by subtracting the fitted values from the observed values of y_i. This gives the *raw residuals*

$$\widehat{e}_i = y_i - \widehat{y}_i \,.$$

It is straightforward to show that the expected value of the residuals is zero:

$$\begin{aligned} \mathrm{E}\left(\widehat{e}_i\right) &= \mathrm{E}\left(y_i - \boldsymbol{x}_i'\widehat{\boldsymbol{\beta}}\right) \\ &= \mathrm{E}(y_i) - \boldsymbol{x}_i'\,\mathrm{E}(\widehat{\boldsymbol{\beta}}) \\ &= \boldsymbol{x}_i'\boldsymbol{\beta} - \boldsymbol{x}_i'\boldsymbol{\beta} = 0\,. \end{aligned}$$

To derive the variance-covariance matrix of residuals, we first note that the fitted value of $\boldsymbol{y}$ can be written in terms of true value as follows:

$$\begin{aligned}\widehat{\boldsymbol{y}} &= \boldsymbol{X}\widehat{\boldsymbol{\beta}}_{GLS} \\ &= \boldsymbol{X}\left(\boldsymbol{X}'\boldsymbol{V}^{-1}\boldsymbol{X}\right)^{-1}\boldsymbol{X}\boldsymbol{V}^{-1}\boldsymbol{y} \\ &= \boldsymbol{H}\boldsymbol{y},\end{aligned}$$

where matrix $\boldsymbol{H} = \boldsymbol{X}\left(\boldsymbol{X}'\boldsymbol{V}^{-1}\boldsymbol{X}\right)^{-1}\boldsymbol{X}\boldsymbol{V}^{-1}$ is called the *hat matrix* because pre-multiplying $\boldsymbol{y}$ by $\boldsymbol{H}$ gives $\widehat{\boldsymbol{y}}$, i.e. adds the hat on $\boldsymbol{y}$. The vector of residuals can now be written as

$$\begin{aligned}\widehat{\boldsymbol{e}} &= \boldsymbol{y} - \widehat{\boldsymbol{y}} \\ &= (\boldsymbol{I} - \boldsymbol{H})\boldsymbol{y},\end{aligned}$$

and the variance becomes

$$\begin{aligned}\text{var}(\widehat{\boldsymbol{e}}) &= (\boldsymbol{I} - \boldsymbol{H})\,\text{var}(\boldsymbol{y})\,(\boldsymbol{I} - \boldsymbol{H})' \\ &= \sigma^2\,(\boldsymbol{I} - \boldsymbol{H})\,\boldsymbol{V}\,(\boldsymbol{I} - \boldsymbol{H})'.\end{aligned}$$

The matrix $\text{var}(\widehat{\boldsymbol{e}})$ differs from the variance of the (true) residual errors by taking into account the estimation error of the regression coefficients. If the residual errors $\boldsymbol{e}$ are normally distributed, then the residuals $\widehat{\boldsymbol{e}}$ are also normally distributed because they are linear combinations of the residual errors.

Under the OLS model, the variance of residuals simplifies to

$$\text{var}(\widehat{\boldsymbol{e}}) = \sigma^2\,(\boldsymbol{I} - \boldsymbol{H}),$$

where $\boldsymbol{H} = \boldsymbol{X}(\boldsymbol{X}'\boldsymbol{X})^{-1}\boldsymbol{X}'$. The diagonal elements are

$$\begin{aligned}\text{var}(\widehat{e}_i) &= \sigma^2\left(1 - \boldsymbol{x}_i'\left(\boldsymbol{X}'\boldsymbol{X}\right)^{-1}\boldsymbol{x}_i\right) \\ &= \sigma^2 - \boldsymbol{x}_i'\,\text{var}(\widehat{\boldsymbol{\beta}}_{OLS})\boldsymbol{x}_i,\end{aligned} \tag{4.32}$$

which shows that $\text{var}(\widehat{e}_i) < \text{var}(e_i)$. The inequality approaches equality as the data set gets large and the estimation error of $\widehat{\boldsymbol{\beta}}$ vanishes. This means that even though the residual errors have a constant variance, this does not hold for the residuals. The more extreme the observations are in terms of the x-variables, the lower is the variance of the residuals.

The *standardized residuals* are obtained by dividing the raw residuals by $\sqrt{\widehat{\text{var}}(\widehat{e}_i)}$, where $\widehat{\text{var}}(\widehat{e}_i)$ is obtained by replacing σ^2 and $\text{var}(\widehat{\boldsymbol{\beta}}_{OLS})$ with their OLS estimates in Equation (4.32). Such residuals have a zero mean and unit variance. They are useful in finding outlying observations in small data sets. In large data sets, Pearson residuals (see below) are satisfactory for the same purpose. Computation of standardized residuals are commonly implemented in statistical software for the OLS model only.

For exploring the fit of a GLS model, *Pearson* and *normalized* residuals are useful. Pearson residuals are obtained by standardizing the raw residuals by the estimated residual standard errors:

$$\widetilde{e}_i = \frac{\widehat{e}_i}{\sqrt{f(z_i, \widehat{\boldsymbol{\theta}})}},$$

where f is the assumed variance function ((4.12), p. 83) and $\widehat{\boldsymbol{\theta}}$ includes the estimated variance function parameters. These residuals are useful in exploring whether a fitted variance function properly models the non-constant error variance. If the variance function fits well, the Pearson residuals should have an approximately constant variance of one.

Normalized residuals can be used to explore whether the applied variance-covariance structure properly models the dependence among observations. They are residuals that are "de-correlated" according to the estimated variance-covariance matrix of residual errors. Recall the previous definition of the Cholesky factor $(\boldsymbol{V}^{-1/2})$ from Section 4.4.2. The normalized residuals are the residuals of model (4.28), which are found as

$$\dot{e} = \frac{1}{\widehat{\sigma}}\left(\boldsymbol{V}^{-1/2}\right)'\widehat{e}.$$

If the residuals are multivariate normal and $\text{var}(\widehat{e}) = \sigma^2\boldsymbol{V}$, then $\text{var}(\dot{e}) = \boldsymbol{I}$. Therefore, if the estimated covariance structure properly models the dependence among the observations and the error variance, the normalized residuals $\dot{e}$ show zero-correlation (and unit variance). Notice that the ith normalized residual is a linear combination of the entire vector $\widehat{e}$. Therefore, it is not only associated with observation i of the original data, but with other observations of the data that are correlated with e_i.

Notice also that unlike the standardized residuals, the Pearson and normalized residuals do not take into account the estimation errors of regression coefficients. However, the effect is usually marginal if the data set is not very small.

The terminology used in the standardization varies slightly, and readers should be careful as to the meaning (e.g. Gałecki and Burzykowski (2013, Sections 4.5, 7.5 and 10.5)). In particular, the term "standardized residuals" is sometimes used for the Pearson residuals.

4.5.2 Graphical evaluation of the fit

The specific model assumptions to be explored are

1. Is the systematic part properly formulated so that $\text{E}(y_i) = \boldsymbol{x}'\boldsymbol{\beta}$?

2. Does the residual variance follow the model we assumed?

3. Do the residual errors have the correlation structure we assumed?

4. Do the residuals have (multivariate) normal distribution?

In addition, one should check and recognize

5. The presence and effect of outliers.

6. The presence and effect of influential observations.

7. The multicollinearity of predictors.

We point out that items 1–4 are *model assumptions* that have been made for model formulation. Item 5 is related to assumption 4. The influential observations (item 6) and multicollinear predictors (item 7) do not violate the model assumptions, but they should be taken into account in interpretation of the estimated model.

If the model includes an intercept, either explicitly so that the model matrix includes a unit vector as a column, or implicitly so that the requested unit vector can be constructed as a linear combination of the columns in the model matrix, the mean of residuals following an OLS fit is exactly zero. That is not the case under a GLS fit, but it is not an acceptable argument against the use of GLS. If the mean of standardized or raw residuals is too far from zero, it may indicate a poor fit of the fixed part. For example, the intercept term might be needed if such one is not included, or more flexible mean function should be used.

Assumption 1 can be evaluated by looking at the raw, Pearson or standardized residuals with respect to *all predictors of the model* and to the *fitted values* $\widehat{y}_i$. If the local *mean of residuals* shows trends over the predictor range, the systematic part of the model may need improvement. The problem of graphical evaluation is always how to recognize the real trends behind the random noise. One possibility is to classify the data into a small number of classes and see whether the mean in each class is significantly different from zero. Practical tools for this purpose are provided, e.g. in R-packages `lmfor` and `Hmisc`. Other nonparametric smoothing methods can be used as well, such as the *locally weighted average smoothing* (loess), implementation of which is available in R-function `lowess`. In large data sets, one can find trends in the residuals using these methods, even if the conventional residual plots do not show any trends. A common mistake for a novice modeler to make is to look at the trends in the residuals on y_i. Such a graph always has a trend, but that trend does not indicate problems in model fit. It just underlines the fact that small values of y are commonly associated with negative residuals and high values of y with positive residuals.

If the dependent variable is assigned only discrete values (e.g. if preference classes are interpreted to be measured almost at the interval scale), the residuals are on the lines with slope -1 in a plot with respect to $\widehat{y}$. However, this does not indicate problems in model formulation (Searle 1988). When the dependent variable receives a value d, then $y - \widehat{y} = d - \widehat{y}$, so residuals of observations with $y = d$ are on the line with slope -1 and intercept d, see Figure 4.5. Similar lines are visible whenever y is discrete, such as with count and binary data, or when the response is rounded to a specified accuracy. However, if there are many unique values of the response in the data, the lines may not be clearly visible. Figure 8.4 on p. 264 also shows similar 'suspicious' lines in residual plots for a generalized linear model.

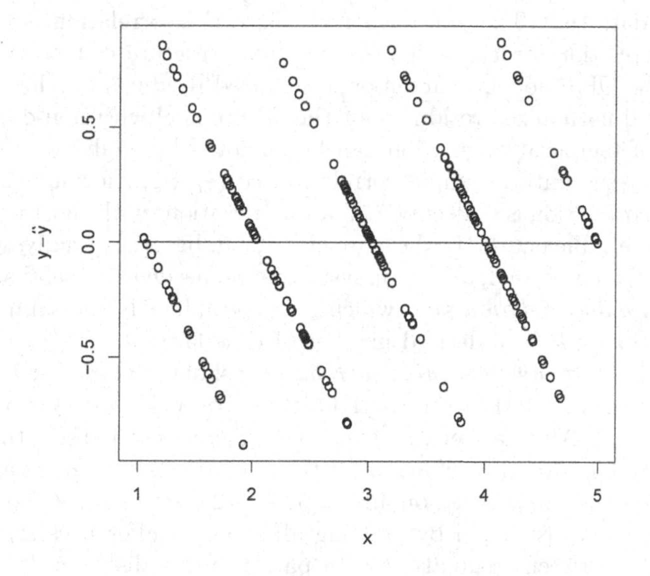

FIGURE 4.5 Residuals $y - \widehat{y}$ plotted on $\widehat{y}$ when y is assigned values 1,2,3,4,5 only. The x variable has a uniform distribution, the y variable was first generated by adding a uniform random error to x, scaled to interval $[0.5, 5.5]$ and then rounded to the nearest integer. Then y was regressed on x. The regression is statistically valid OLS regression without normality assumptions.

Non-constant error variance and correlation structure are properties of the data, or more specifically, of the process that has generated the data. They cannot be removed or eliminated, they can just be modeled (but recall what was discussed at the beginning of

Section 4.3). This means that even for a sophisticated variance-covariance structure, the residual graphs based on raw residuals are not essentially changed by the use of a model for the dependence structure. A proper question in this context is, whether the structure was realistically modeled.

Assumption 2 (error variance) can be evaluated by exploring the *variability of the residuals around zero.* Any residuals can be used for the OLS model, and Pearson residuals can be used for the GLS model. If the plot does not show any trend in the variability of residuals within the range of predicted values, then one could believe that the assumption about error variance has been met. In a model with several predictors, one should also look at the behavior of the residuals on all predictors of the model. If any of them shows non-constant variability within the range of the predictor, the variance function should probably be included or reformulated to better model the variability in the error variance. Again, it may be hard to discern the real trends in the variability from the random noise, especially if the data are not uniformly distributed over the x-axis. The interpretation may be easier if one explores the standard deviation or the variance of residuals in classes of predictors/predicted values to see the possible trends in the variance; `lmfor::mywhiskers` with argument `se=FALSE` illustrates those lines by vertical lines around the class means. One could also directly plot the squared residuals on fitted values or predictors; or their means by classes with additional error bars based on the standard error of the mean of squared residuals in each class.

The starting point in assumption 3 is usually the zero correlation of residual errors. It should be questioned whenever there is some reason for correlation. We have discussed in Section 4.3.2 modeling of (i) temporal autocorrelation of repeated measurements, (ii) spatial autocorrelation in spatial data sets, or (iii) dependence of observations of the same group in grouped data sets. The approach to explore the correlation is chosen based on the anticipated correlation structure. The correlation structure can be explored by using any residuals for the OLS model, Pearson or normalized residuals for the GLS model with non-diagonal $\boldsymbol{V}$, and normalized residuals for the GLS model with non-diagonal $\boldsymbol{V}$.

The existence of temporal correlation can be explored by analyzing the correlation of *lagged residuals*, i.e. computing sample correlation $\text{cor}(\dot{e}_i, \dot{e}_{i-k})$ for sufficiently many lags $k = 1, 2, \ldots$. If the correlation is only expected for observations with short time intervals, the first few time lags are sufficient. If the data are cyclic, e.g. because of daily, weekly or annual cycles in the data, then a more detailed analysis may be needed. A useful summary of such computations is an *autocorrelation plot*, which shows graphically the sample correlation of residuals as a function of k for a desired number of time lags.

In a spatial data set, a *sample semivariogram* of residuals can be used to show the correlation of residuals with respect to the spatial distance between observations (e.g. Pinheiro and Bates 2000, p. 242). Alternatively, *a sample correlogram* can express the same information. Recall Section 4.3.2 for the definition of the theoretical variogram and correlogram. The empirical semivariogram is based on all $n \times (n-1)/2$ distinct pairs of observations. The semivariogram cloud is constructed by plotting all squared differences $\Delta_k = 0.5(\dot{e}_i - \dot{e}_j)^2$ $i, j = 1, \ldots, n$, $i \neq j$ between residuals of each pair k on the distance d_k between the two observations. Such a plot is usually very noisy. Therefore, it is more informative to classify the pairs according to the spatial distance between the points, and plot the mean of Δ_k in each class. A trend line (e.g. based on a lowess smoother) can be added to the plot to help in the interpretation.

An increasing trend in the variogram of the Pearson residuals indicates increasing spatial correlation with decreasing distance. If spatial correlation has been properly modeled, the variogram of normalized residuals should not show any trend as a function of d. See Example 4.14 for implementation.

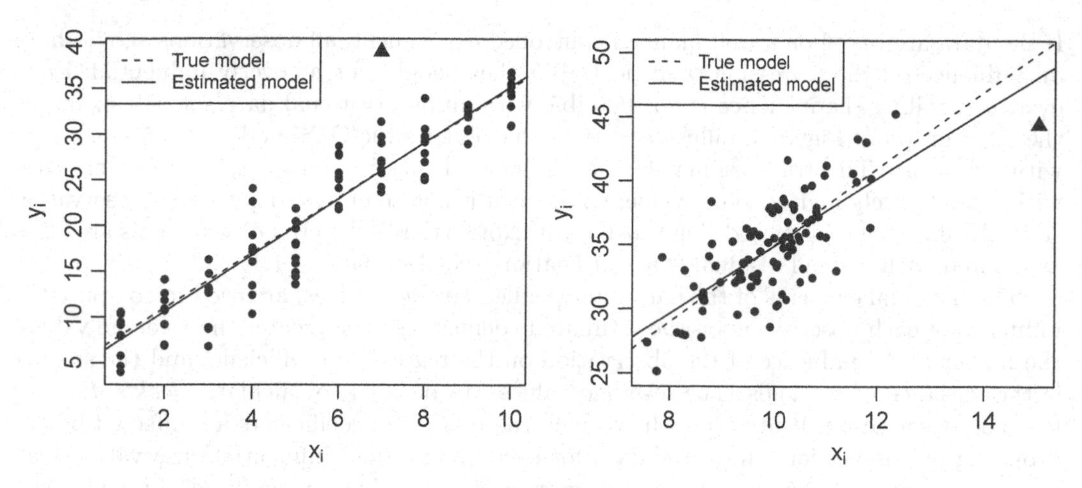

FIGURE 4.6 An outlier (left) and an influential observation (right) illustrated by triangles in two simulated data sets.

In grouped data sets, mixed-effects models (see Chapter 5) provide a framework for the analysis of the correlation structure.

The situation becomes problematic if there is dependence among observations, but the necessary information for modeling such dependence is not available. For example, the temporal order of repeated measurements, the spatial locations or group membership information may be missing from the data set. There are no good solutions for such a situation.

Assumption 4 (normality of residual errors) can be evaluated using a *normal quantile-quantile plot* (*normal q-q plot*) of the residuals, where n systematically selected quantiles of the standard normal distribution are plotted on the residuals of the model. Any residuals can be used for the OLS-model, and Pearson or normalized residuals can be used for the GLS model. If the assumption about normality is met, the points in the plot should formulate a straight line.

Outliers (item 5) are observations for which the residual is exceptionally large because y_i has been generated by a different random process than the other data points (see left-hand graph in Figure 4.6). For example, there may be coding errors or significant measurement errors in some observations. Observations with a high absolute value of the residual are potential outliers, but are not necessarily outliers. To detect candidate outliers, the residual plot is useful. They are also seen as heavy tails in the normal q-q plot. After the candidate outliers have been found, a closer look should be taken (e.g. from the original measurement records) to see whether something exceptional is associated with their measurement. If something exceptional is found, the observation can be removed from the data. However, one should not routinely remove all candidate outliers because they may just be exceptional in the population of interest. Removing such an observation will decrease the variability in the sample and, therefore, lead to downward biased estimates of the standard errors and to anti-conservative test results. If the residuals have heavy tails due to the properties of the process being modeled, one could evaluate whether their exclusion has an effect on the main results; if it does, then both models should probably be reported or at least discussed. A formal test to find potential outliers when the OLS model is valid is described in many textbooks; an R-implementation is `car::outlierTest`.

Influential observations (item 6) are observations that have a large effect on the regression coefficients. In general, the farther away other observations from an observation in terms of x, the larger is the influence of that observation on the estimates of regression coefficients.

If the data are based on a well-planned controlled experiment, all observations should have an influence on the regression coefficients but they should not be overly influential. However, controlling the influence is not possible with an observational data set. For example, the right graph of Figure 4.6 illustrates data that follows the OLS-model $y_i = 5 + 3x_i + e_i$, with $\sigma^2 = 3^2$. The true residual for the influential observation at $x_i = 15$ is negative with a moderately high absolute value; however, it is not an outlier. That single observation strongly affects the estimated slope coefficient. Notice that influential observations are often associated with a small residual (for justification, recall Equation (4.32)).

The diagonal elements of the hat matrix, called *leverage values*, are used to compute the influence of each observation on the estimated coefficients. The greater the leverage values, the higher is the influence of the observation on the regression coefficients and the smaller is the variance of residuals. The leverage values can be used to calculate *Cook's distance* for each observation. It illustrates how much the regression coefficients are affected by the dropping of observation i from the data for each observation. Influential observations can be analyzed using a graph that shows the standardized residuals on leverage. If the absolute values of standardized residuals for observations with large leverage are high, it means that they are potentially overly influential. Cook's distance can be used to define the limit of "large" standardized residual for each leverage value. Such a plot is routinely produced by the `plot` method of objects fitted using function `lm`. Function `influence.measures` can be used for more advanced analysis of influential observations. Formal analyses of outliers and influential observations are only well developed for the OLS model. However, they can be directly applied with the GLS-model formulation (4.28) (p. 93) when $\boldsymbol{V}$ is diagonal (see also Martin (1992)).

The *multicollinearity* of predictors can be identified, e.g. by plotting the predictors against each other, by computing the sample correlation matrix of the predictors, or by using so-called *variance inflation factors* (Fahrmeir *et al.* 2013). As we have already discussed in Section 4.2, the multicollinearity does not violate the model assumptions, but leads to problems in the interpretation of the regression coefficients, and to high variance of the estimated coefficients. In a very extreme case, the correlation is ± 1 and the model becomes unidentifiable. In experimental data sets, the x-variables are often under the control of the researcher, and the values can be selected so that multicollinearity is avoided. Also, it may be wise to avoid multicollinearity among predictors in observational data sets so as to avoid misleading conclusions (see, e.g. Harrell 2001, for alternative procedures). However, one should not routinely follow this instruction, but select the procedure that is best justified for the research question. For example, tree diameter and height are strongly correlated, but they both should, of course, be included as predictors in a model for stem volume. Shrinkage estimators (see Box 4.2 on p. 91) can also be useful in the case of multicollinear predictors (Fahrmeir *et al.* 2013, Chapter 4).

Example 4.13. The fit of model (4.11) on p. 80 (also see Example 4.5 on p. 86) was explored by the diagnostic plots shown in Figure 4.7. The upper left-hand graph shows the box plot of residuals for the three machines, the upper right-hand graph the residuals on stump diameter, and the lower left-hand graph the residuals on fitted value. The lower right-hand graph shows the autocorrelation plot of the residuals.

To help in interpretation, the data are divided into a small number of classes of equal width using the variable on the x-axis (either diameter or the fitted value), and the mean of residuals m_k and standard deviation of residuals s_k were computed for each class k of n_k observations. An approximate 95% prediction interval for an individual residual was computed as $m_k \pm 1.96s_k$. In addition, the standard error of the mean, $se_k = s_k/\sqrt{n_k}$ was computed for each class and used to construct a 95% confidence interval of the mean of each class as $m_k \pm 1.96se_k$. The thin vertical lines show the prediction intervals and the thick lines show the confidence intervals of the mean in each class; gray is used to highlight those intervals that do not cross the line $y = 0$.

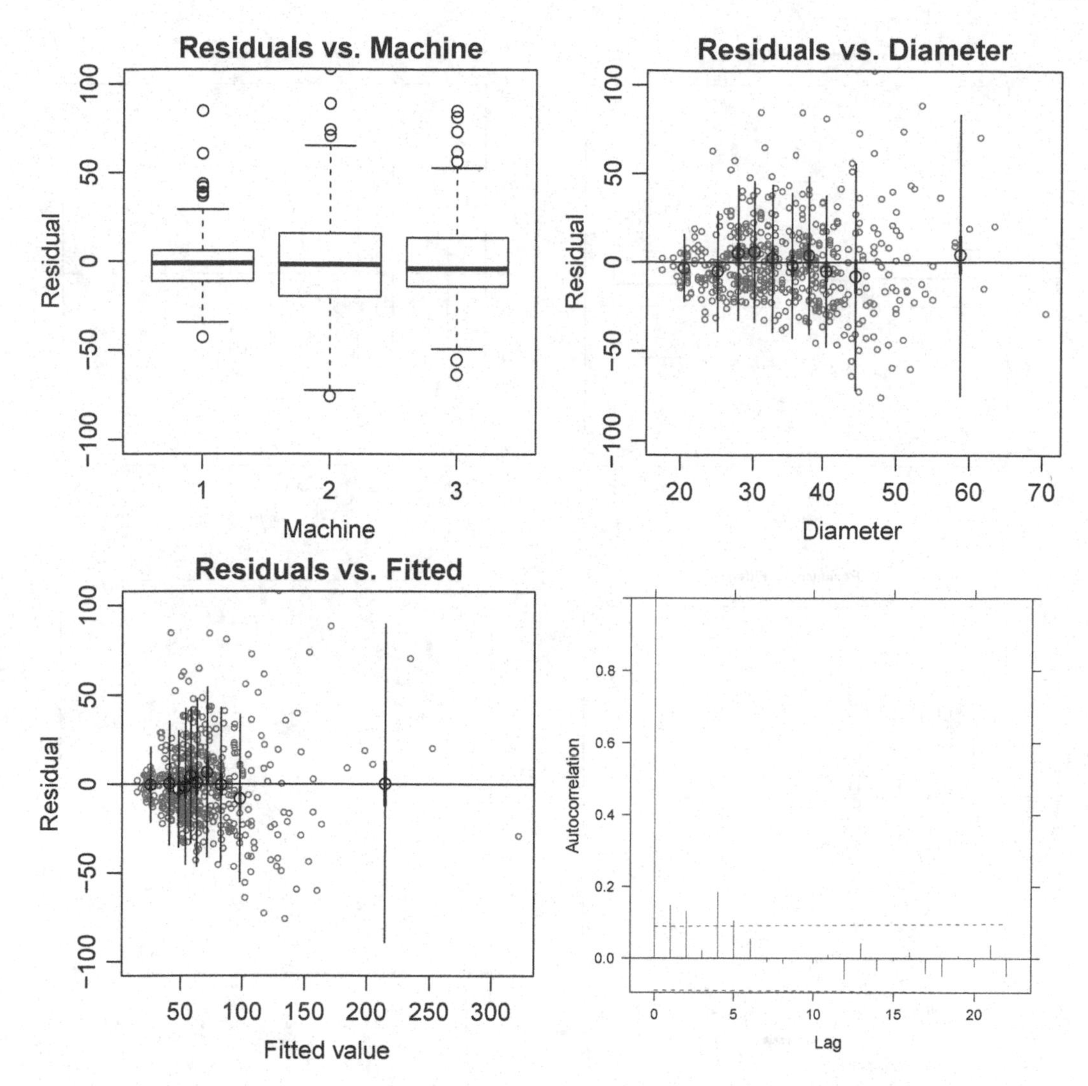

FIGURE 4.7 Diagnostic plots on the ordinary least squares fit of model (4.11).

The plot of residuals against diameter does not show a clear nonlinear trend in the mean of residuals, even though there might be a lack of fit for smallest diameters in the systematic part. The variance of residuals clearly increases as a function of stump diameter and as a function of predicted value. The variance also seems to be lower for machine 1 than for the other machines.

The observations in the data set are in the order of processing. There might be some temporal autocorrelation because successive trees have been lifted at the same locations with common working conditions, or the alertness of the driver may have gradually changed during the processing of the site. The autocorrelation plot indicates a slight temporal autocorrelation for the shortest time lags $k = 1, \ldots, 5$.

The R-script for the upper graphs and the lower right-hand graph of Figure 4.7 is below:

```
> plot(stumplift$Machine,resid(mod1),main="Residuals vs. Machine",
+      xlab="Machine",ylab="Residual")
> plot(stumplift$Diameter, resid(mod1), main="Residuals vs. Diameter",
+      xlab="Diameter",ylab="Residual",cex=0.5,col=gray(0.5))
> library(lmfor)
> mywhiskers(stumplift$Diameter,resid(mod1),add=TRUE,lwd=2)
> mywhiskers(stumplift$Diameter,resid(mod1),add=TRUE,lwd=1,se=FALSE)
> abline(0,0)
```

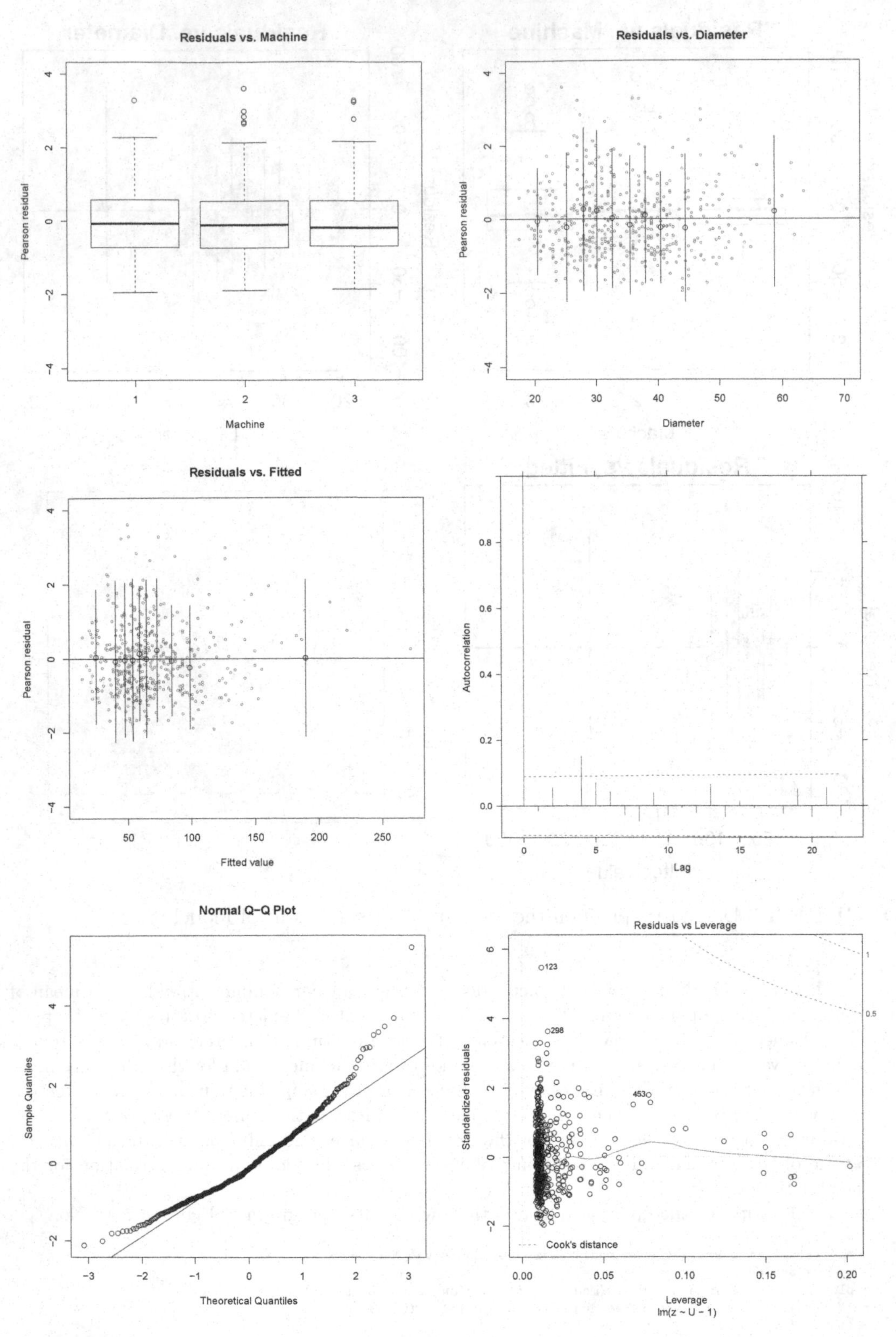

FIGURE 4.8 The Pearson residuals of the model `mod2` on the predicted value, machine and stump diameter. The autocorrelation and normal-q-q plot of the normalized residuals of model `mod3`. Standardized residuals on leverage for model `mod1` and the Cook's distance contours.

```
> plot(ACF(mod1,form=~1|Machine),alpha=0.05)

> # Compute lagged correlations manually.
> # lag 1 correlation (not absolutely correct because of the effect of machines,
> # think why?)
> cor(resid(mod1)[-1],resid(mod1)[-485])
[1] 0.1438193
> # lag 2 correlation
> cor(resid(mod1)[-c(1,2)],resid(mod1)[-c(484,485)])
[1] 0.1290783
> cor(resid(mod1)[-c(1:3)],resid(mod1)[-c(483:485)])
[1] 0.027211
> cor(resid(mod1)[-c(1:4)],resid(mod1)[-c(482:485)])
[1] 0.1808821
```

Model mod2, which assumed a power-type variance function for residual errors, was fitted in Example 4.11. The AR(1) process was further assumed in mod3 of Example 4.12. To explore whether these extensions adequately modeled the variance, the Pearson residuals of model mod2 were plotted against fitted values, machines and stump diameter in Figure 4.8. All graphs show a rather constant variability. The autocorrelation plot based on the normalized residuals of model mod3 does not have significant autocorrelation, even at the shortest time lag. This indicates that the assumed AR(1) process satisfactorily modeled the slight autocorrelation of the residual errors. The normal q-q plot of the normalized residuals of model mod3 do not form a straight line. The largest observed residuals are clearly larger than the corresponding theoretical quantiles of the normal distribution, which indicates a much heavier right tail for the distribution than would be expected by the normal distribution. For the left tail, the situation is the opposite: the residuals are smaller in absolute value than would be expected by the normal distribution. However, our data set has almost 500 observations, and so this lack of normality hardly causes any problem and we can trust the Central Limit theorem in regards to the normality of parameter estimates. For confirmation, a more detailed analysis could be done, e.g. by using bootstrapping (Hastie *et al.* 2009). The R-script of the upper-right, middle-right and lower-left graphs of Figure 4.8 is shown below:

```
> plot(stumplift$Diameter, resid(mod2,type="p"))
> plot(ACF(mod3,resType="n"),alpha=0.05)
> qqnorm(resid(mod3,type="n"))
> qqline(resid(mod3,type="n"))
```

The observation with the highest normalized residual might be an outlier. A careful modeler should explore the effect of dropping that observation on the overall results of the analysis.

The overly influential observations were checked with model mod2 that was refitted in the form (4.28) (p. 93) using OLS. The parameter δ of the variance function was first extracted using:

```
> delta<-coef(mod2$modelStruct)[1]
```

and matrix $\boldsymbol{V}$ was, thereafter, constructed and saved as object V1. Thereafter, model mod2 was re-estimated using function lm and saved to object mod2b.

```
> delta<-coef(mod2$modelStruct)[1]
> V1<-diag(fitted(mod2)^(2*delta))
> Vinvsqrt<-chol(solve(V1))
> z<-Vinvsqrt%*%stumplift$Time
> U<-Vinvsqrt%*%model.matrix(mod3,data=stumplift)
> mod2b<-lm(z~U-1)
> plot(mod2b,which=5)
```

The summary of mod2b showed exactly the same estimates of $\boldsymbol{\beta}$ and σ^2 as mod2. The final command of the script produced the lower-right graph in Figure 4.8. If the standardized residuals with high leverage values are within Cook's ellipsoids (shown using dashed lines), the observations would be overly influential. The most influential observations are the stumps on rows 123, 298 and 453, but they are not classified as overly influential, as they do not fall within the Cook's distance contours. Model mod3 was not used in this analysis because matrix $\boldsymbol{V}$ would be non-diagonal and the residuals of model (4.28) would not be associated with the original observations.

Example 4.14. Data set sp::meuse (Rikken and Van Rijn 1993, Pebesma and Bivand 2005) includes measurements of topsoil heavy metal concentrations at the observation locations in a flood plain of the Meuse river. The upper-left graph in Figure 4.9 shows the mapped data, so that the symbol size is proportional to zinc concentration. Zinc concentration has been plotted against distance to the river in the middle left graph, which shows that the mean concentration decreases as the distance from the river increases, and the decrease is nonlinear. In addition, it seems that the variance is higher at smaller distances.

We explain zinc concentration with the distance to the river. The shortest distance from each point to the river was computed using the functions of package **spatstat** (Baddeley and Turner 2005). The nonlinearity in the relationship could be taken into account in various ways, including polynomial regression, splines, and transformations to response, which could also stabilize the error variance. To avoid the need for bias correction in predictions, we do not want to make transformations to the response. Instead, we model the nonlinear trend using a spline, and the heteroscedastic variance parametrically with a variance function. In addition, we explore the spatial dependence of the residuals and also model it parametrically.

After some initial analysis, we find that the following model is suitable for the data:

$$z_i = \beta_1 + \beta_2 r_i + \beta_2 r_i^{(1)} + e_i$$

where z_i is the topsoil zinc concentration at location (x_i, y_1) (mg/kg), r_i is the shortest distance from location (x_i, y_1) to the Meuse river (km), and

$$r_i^{(1)} = (r_i - t_1)_+^3 - \frac{t_3 - t_1}{t_3 - t_2}(r_i - t_2)_+^3 \frac{t_2 - t_1}{t_3 - t_2}(r_i - t_3)_+^3 .$$

is such a transformation of distance r_i that the systematic part of the model becomes a restricted cubic spline based on three knots (t_1, t_2 and t_3) (see Section 10.4.1). The subscript + indicates operation $y_+ = \max(y, 0)$. The spline function is continuous and has continuous first and second derivatives at the knots. It is based on two third-order polynomials within $t_1 \le r_i < t_3$ and is restricted to be linear for $r_i < t_1$ and $r_i \ge t_3$. Following the suggestion by Harrell (2001), the knots were placed on the 10th, 50th and 90th percentiles of r_i to get $t_1 = 0.02$ km, $t_2 = 0.27$ km and $t_3 = 0.63$ km. Furthermore, we assume that

$$\mathrm{var}(e_i) = \sigma^2 r_i^{2\delta} .$$

The model was fitted using

```
> library(sp)
> library(spatstat)
> data(meuse)
> natspline<-function(X,t,j){
+   k<-length(t)
+   pmax(0,X-t[j])^3-
+        pmax(0,X-t[k-1])^3*(t[k]-t[j])/(t[k]-t[k-1])+
+        pmax(0,X-t[k])^3*(t[k-1]-t[j])/(t[k]-t[k-1])
+ }
> # recalcute the distance from river for compatibility with latter example
> meuse.ppp<-ppp(meuse$x,meuse$y,win)
> meuse.riv.psp<-psp(meuse.riv[-1,1],meuse.riv[-1,2], meuse.riv[-n,1], meuse.riv[-n,2], win)
> dist.m<-nncross(meuse.ppp,meuse.riv.psp)$dist+10
> meuse$r<-dist.m/1000 # to get realistic scale for coefficients
> t<-quantile(meuse$r,probs=c(0.1,0.5,0.9))
> meuse$a1<-natspline(meuse$r,t,1)
> mod1.meuse<-gls(zinc~r+a1,data=meuse,weights=varPower(-0.5, ~r))
```

The systematic part of this model has been illustrated by the line in the middle-left graph in Figure 4.9. The function seems to sufficiently model the relationship between distance and expected zinc concentration. The Power-type variance function satisfactorily models the error variance, as one can see in the lower-left graph in Figure 4.9.

To explore the spatial autocorrelation among the residuals, the empirical semivariogram of normalized residuals was constructed for distances [0, 1000] meters between observations. The

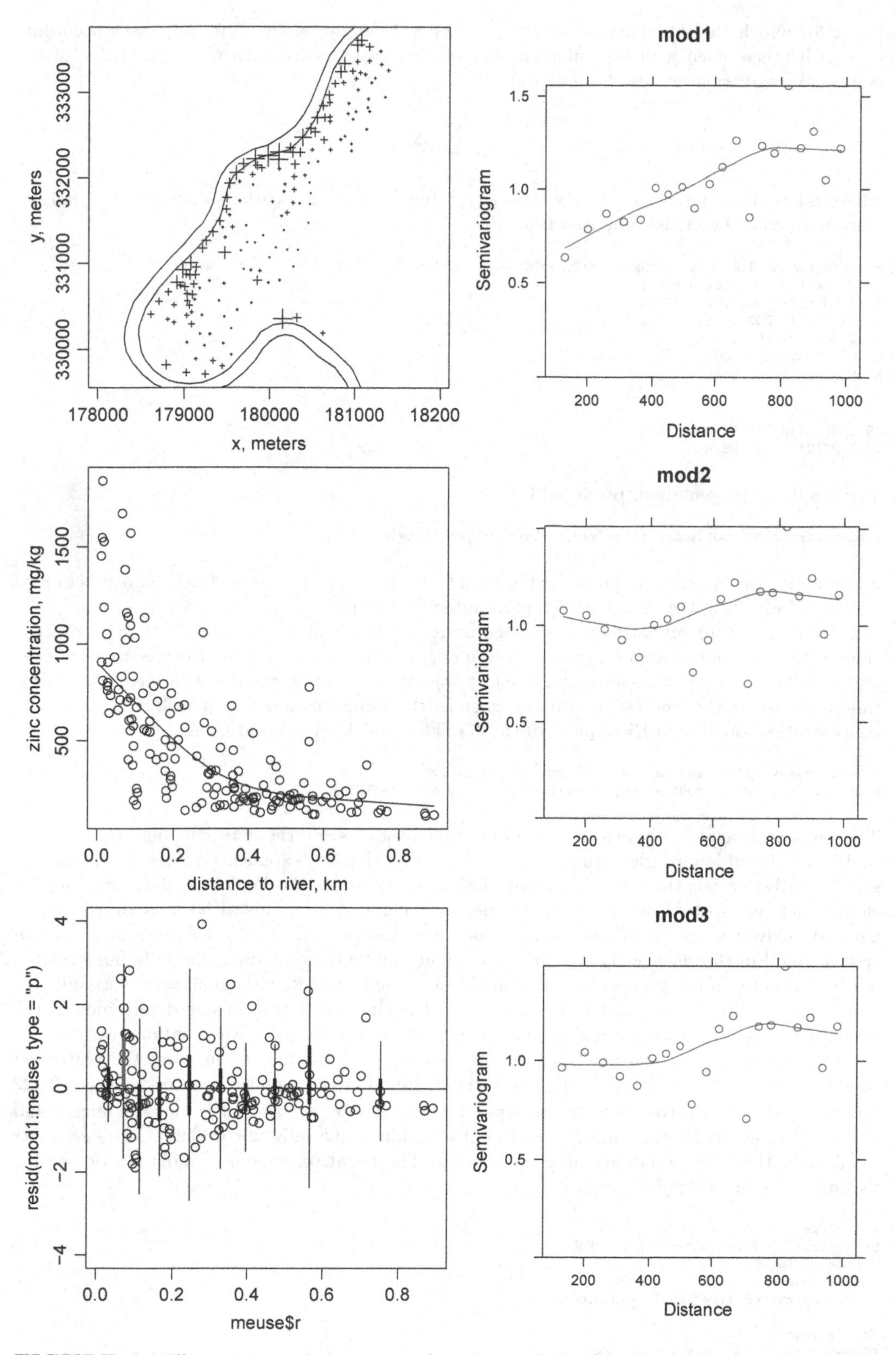

FIGURE 4.9 Illustrations of the `meuse` data set and the fitted models. See Example 4.14 for details.

pairs, for which the inter-location distance is below 1000, are assigned to 20 classes according to the distance, each with a similar number of observations. For each class, the value of the empirical semivariogram was computed as

$$\frac{1}{2m}\sum_{k=1}^{m}\Delta_k^2$$

where Δ_k is the difference in the normalized residuals $\dot{e}_i - \dot{e}_j$ at locations i and j. The semivariogram for model `mod1` is computed below:

```
> Variogram(mod1.meuse,form=~x+y,resType="n",maxDist=1000)
      variog     dist n.pairs
1  0.6358282 130.9752     212
2  0.7873041 202.2004     213
3  0.8695124 259.2393     213
4  0.8260057 311.6809     213
5  0.8369303 363.7705     213
.
.
19 1.0509446 937.4375     213
20 1.2210705 982.9878     213
```

The graphical presentation, produced by

```
> plot(Variogram(mod1.meuse,form=~x+y,resType="n",maxDist=1000))
```

is shown in the upper-right graph in Figure 4.9. It shows an increasing trend as a function of distance, indicating the existence of spatial autocorrelation.

The spatial autocorrelation was modeled using the exponential model (4.15) (p. 84). We considered two alternative models: one where the nugget effect was restricted to zero (`mod2.meuse`) and a second where a non-zero nugget effect was allowed (`mod3.meuse`). The latter allows for random noise in the zinc concentration, whereas the former assumes a smooth surface for the concentration, such as in Example 2.19 (p. 27). The models were fitted using:

```
> mod2.meuse<-update(mod1.meuse,cor=corExp(form=~x+y))
> mod3.meuse<-update(mod1.meuse,cor=corExp(form=~x+y,nugget=TRUE))
```

The empirical semivariograms of the normalized residuals for these two models (Figure 4.9, middle-right and lower-right) are rather flat, indicating that the exponential correlation function satisfactorily models the spatial autocorrelation of the residuals. Which of these two models should then be selected? Because the data does not include observations with very short distances, the data-driven selection of the model is not well justified. Even though there is a smooth spatial trend in the zinc concentration, there might still be some random, spatially uncorrelated additional noise in the y-variable, which might be associated, e.g. with small-scale variability in zinc concentration, or measurement errors when the zinc concentration was determined in the laboratory. Therefore, we selected the model with the nugget effect as the final model.

The summary of model `mod3.meuse` below shows the following estimates of the regression coefficients: $\widehat{\boldsymbol{\beta}} = (899, -2210, 5452)'$. The variance function parameter estimates are $\widehat{\delta} = -0.422$ and $\widehat{\sigma}^2 = 105^2$. The correlation structure parameter estimates are $\widehat{\psi}_1 = 0.382$ for the nugget and $\widehat{\psi}_2 = 199$ (meters) for the range. Therefore, the random, spatially uncorrelated noise accounts for 38% of the total variability at each location. The negative value of δ indicates decreasing variance as a function of r_i.

```
> mod3.meuse
Generalized least squares fit by REML
  Model: zinc ~ r + a1
  Data: meuse
  Log-restricted-likelihood: -1012.027

Coefficients:
(Intercept)           r          a1
   1017.083   -2497.955    7426.976
```

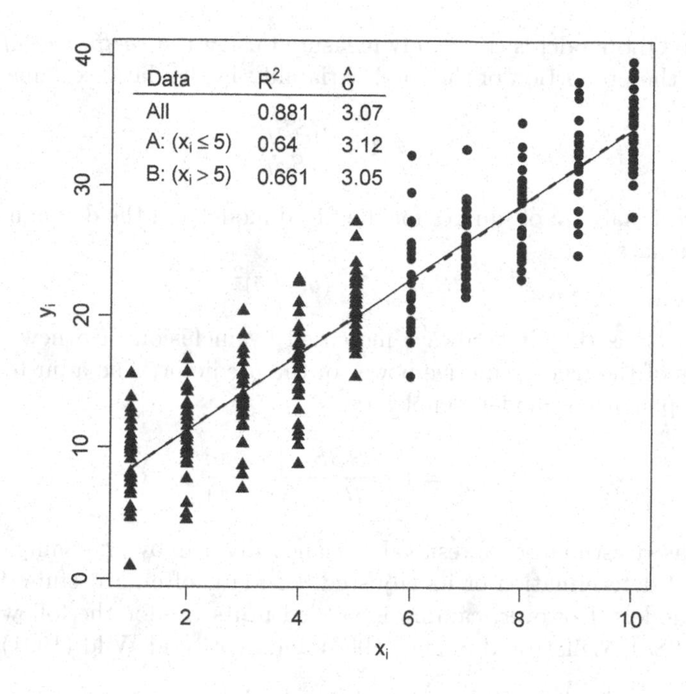

FIGURE 4.10 Illustration of R^2 and $\widehat{\sigma}$ using a simple linear regression in a simulated data set that meets the OLS assumptions. The data were split into two subsets: subset A with small values of x_i and subset B with large values of x_i. The simple linear regression model was fitted to all data and both subsets. The figure shows the data, and the fitted regression lines for all data (solid line) and for the two subsets (dashed lines). The tabulated values show R^2 and $\widehat{\sigma}$ for the three models. Even though the fitted models are practically identical, the values of R^2 in the subsets are much lower than for the full data set because the total variation is different. The estimates of residual standard error are similar for all three models.

```
Correlation Structure: Exponential spatial correlation
 Formula: ~x + y
 Parameter estimate(s):
       range      nugget
149.3274067   0.2311708
Variance function:
 Structure: Power of variance covariate
 Formula: ~r
 Parameter estimates:
     power
-0.5960734
Degrees of freedom: 155 total; 152 residual
Residual standard error: 88.68876
```

4.5.3 Numerical criteria for model fit

For the OLS model, the estimate of residual standard error $\widehat{\sigma}$ measures the variability that was not explained by the model. Because it is expressed in the same unit as the y-variable, it can be compared to the marginal standard deviation of y in the population, and to the differences between treatments or effect of other predictors. Unlike the R^2 (see below), it is also comparable between two similar models fitted in different data sets (see Figure 4.10). However, the residual standard error is not as such a useful model fitting criterion.

The fit of the OLS model is commonly measured using the *coefficient of determination*, which estimates the proportion of the total variability in the data explained by the model:

$$R^2 = 1 - \frac{RSS_f}{RSS_t}, \tag{4.33}$$

where RSS_f is residual sum of squares for the fitted model and the denominator is the total variability in the data

$$RSS_t = \sum (y_i - \bar{y})^2 .$$

A problem with R^2 is that it is always increased by inclusion of a new predictor in the model regardless of the true predictive power of the predictor. The adjusted R^2 value fixes this problem by penalizing model complexity:

$$R^2_{adj} = 1 - \frac{RSS_f/(n-p)}{RSS_t/(n-1)}, \tag{4.34}$$

i.e. 1−the unbiased estimator of residual variance divided by the sample variance of y. The coefficient of determination or its adjusted version is often a useful criterion for quick comparison of models. However, caution is needed in its use for the following reasons; see also Kvålseth (1985), Willett and Singer (1988) and Scott and Wild (1991):

- R^2 is sensitive to the data. If the residual standard error in a population is constant, R^2 is only affected by the spread of the x-variables. By extending the range of x-variables, one can artificially increase R^2 in an experiment without improving the model itself. For different data sets, comparison using the estimated residual standard error might be better justified, as illustrated in Figure 4.10.

- Neither R^2 nor $\widehat{\sigma}^2$ are comparable for models where different transformations of the response are used. For example, models fitted for y_i and $\log(y_i)$ should not be compared using R^2, as is sometimes done.

- R^2 should not be used for comparison of parameter estimation methods. It does not take into account the possible non-constant variance and dependence of the data. It uses equal weighting for all observations regardless of their correlation structure or variance. If it is used, it will select the OLS method from the methods used for otherwise similar linear models. This happens because the minimized sum of squares under the assumed model (RSS_f), which is used in the numerator of R^2, is the model fitting criterion in OLS (see Box 4.3, p. 95).

- Two equivalent models can have dramatically different R^2 values depending on the selection of the dependent variable. For example, by modeling height growth using total height as the dependent variable and height of the previous year as an independent variable, one can obtain very high R^2. The equivalent model with height increment as the dependent variable has much lower R^2.

- By using dummy variables to 'explain' exceptional years in a time series (or outliers in any kind of data, in general), one can get misleadingly high R^2.

In general, $\widehat{\sigma}^2$ and R^2 provide a rough summary of a model fitted by OLS to a data set where the OLS assumptions are met. However, they do not provide an alternative to a graphical evaluation of model fit. They are not very useful and are often misleading in data sets where the errors are correlated and/or have a non-constant variance. Better and more widely applicable model comparison criteria are discussed in the next subsection.

4.6 Inference

4.6.1 Hypothesis tests of nested models

Once a model has been fitted, one might want to test hypotheses about its parameters, recall Section 3.6 for a general discussion about hypothesis testing. Examples of such questions are:

1. Does the model significantly explain the variation in $\boldsymbol{y}$?
2. Is a specific predictor or group of predictors needed?
3. Do some parameters have a certain relationship? For example, is the effect similar at two levels of multilevel treatment? Or is the effect similar between all successive levels of an ordinal categorical predictor?
4. Could some of the parameters be restricted to a fixed value supported by subject-matter theory or a previous study? Often the parameters are restricted to 0.
5. Is the assumed residual variance and autocorrelation structure justified?

All these questions lead to the comparison of two models: a *full model* and a *constrained model*. The latter is also called a *null model* or *restricted model*. The constrained model is formulated by making restrictions to the regression coefficients $\boldsymbol{\beta}$ and variance-covariance parameters $\boldsymbol{\theta}$ of the full model; therefore the models are said to be *nested*: the constrained model is a special case of the full model. Once the full model is found, a restricted version of it is formulated. Formally, we specify the following null and alternative hypotheses:

$$H_0 : \boldsymbol{\theta} = \boldsymbol{\theta}_0$$
$$H_1 : \boldsymbol{\theta} \neq \boldsymbol{\theta}_0$$

where $\boldsymbol{\theta}$ includes those model parameters that are constrained to a fixed value $\boldsymbol{\theta}_0$ by the null hypothesis H_0. Parameter $\boldsymbol{\theta}$ may also include some functions of the original model parameters. For example, constraint $\beta_1 = \beta_2$ is implemented by specifying $\theta = \beta_1 - \beta_2$ and $H_0 : \theta = 0$.

When the value of $\boldsymbol{\theta}$ is estimated without restrictions, we find the best-fitting model, the full model, from among the set of models defined by H_1. The selection of the full model is, therefore, a critical step. The full model should be carefully formulated and its fit graphically evaluated before switching from the model formulation stage to the hypothesis testing stage. If the selected full model fits poorly to the data, then the null model will also fit poorly. This leads to comparison of two poorly fitting models, and may lead to erroneous conclusions.

Hypothesis testing is a tool to evaluate whether an effect exists in the population from which the data has been sampled. However, significance of an effect does not necessarily mean that the effect is strong. In modern large data sets, especially, very small and practically meaningless effect, can be statistically significant. For more discussion, see Section 10.7.

Some testing procedures in R require that both full and restricted models are fitted. If the H_0 sets some parameters to zero, the constrained model can be fitted by dropping the corresponding predictors from the model. If the null hypothesis is not that simple, such as $H_0 : \beta_1 = \beta_2$, implementation is a little more complicated. Recall that constraint can be implemented by specifying $\theta = \beta_1 - \beta_2$ and using $H_0 : \theta = 0$. To fit the restricted model, the model needs to reparameterized so that the new model is equivalent to the original model, but θ is one parameter of the model. The constrained model is then obtained by dropping that predictor from the model. If β_1 and β_2 are coefficients of two binary predictors T_1 and T_2, the constrained model can be implemented by defining a new binary predictor T_{12}, which takes a value of 1 when $T_1 = 1$ or $T_2 = 1$, and using it as a predictor in the constrained model instead of T_1 and T_2 of the full model. However, fitting of the constrained model in many cases does not need to be explicit, as will be discussed in the next subsection.

4.6.2 Wald's F- and χ^2 tests of β

When testing hypotheses about the regression coefficients, we set the constraint $\boldsymbol{L\beta}_F = \boldsymbol{k}$ to the regression coefficient vector $\boldsymbol{\beta}_F$ of the full model. Matrix $\boldsymbol{L}_{r\times p}$ specifies a linear combination of regression coefficients that is set to $\boldsymbol{k}$ by the null hypothesis. That is, it specifies $r = p - q$ constraints to $\boldsymbol{\beta}_F$ of length p. The null and alternative hypotheses are:

$$H_0 : \boldsymbol{L\beta}_F = \boldsymbol{k}$$
$$H_1 : \boldsymbol{L\beta}_F \neq \boldsymbol{k}$$

Often, $\boldsymbol{k} = \boldsymbol{0}$. In the most common situation, one of the regression coefficients (say, the jth one) is set to zero by the null hypothesis. In that case, $\boldsymbol{L}$ has only one row, with zeroes in all elements except for the jth one, and $\boldsymbol{k} = \boldsymbol{0}$.

The hypothesis test is based on the weighted sums of squares for the constrained (RSS_C) and full (RSS_F) model, which are defined as follows:

$$RSS_C = (\boldsymbol{y} - \boldsymbol{X}_C\widehat{\boldsymbol{\beta}}_C)'\boldsymbol{V}^{-1}(\boldsymbol{y} - \boldsymbol{X}_C\widehat{\boldsymbol{\beta}}_C)$$
$$RSS_F = (\boldsymbol{y} - \boldsymbol{X}_F\widehat{\boldsymbol{\beta}}_F)'\boldsymbol{V}^{-1}(\boldsymbol{y} - \boldsymbol{X}_F\widehat{\boldsymbol{\beta}}_F)$$

where $\widehat{\boldsymbol{\beta}}_C$ and $\widehat{\boldsymbol{\beta}}_F$ are the GLS estimates of the regression coefficients of the two nested models, and $\boldsymbol{X}_C$ and $\boldsymbol{X}_F$ are the corresponding model matrices. The test statistic is

$$F_{obs} = \frac{(RSS_C - RSS_F)/r}{RSS_F/(n-p)} \tag{4.35}$$

where p is the number of fixed predictors (including the intercept) in the full model, r is the number of constraints in the restricted model (i.e. the difference between the number of parameters in the full model and in the restricted model), and n is the number of observations. Note that the denominator is the GLS estimator of σ^2 (4.27) in the full model; we denote it by $\widehat{\sigma}^2_F$.

The formulation in Equation (4.35) points out that the hypothesis test is based on a comparison of two models. However, for the linear model with known $\boldsymbol{V}$, the test can be equivalently presented as the *Wald F-test*. Such formulation does not require estimation of the restricted model. Therefore, it is usually presented in literature and implemented in computer software:

$$F_{obs} = \frac{\left(\boldsymbol{L}\widehat{\boldsymbol{\beta}}_F - \boldsymbol{k}\right)'\left(\boldsymbol{L}\left(\boldsymbol{X}'\boldsymbol{V}^{-1}\boldsymbol{X}\right)^{-1}\boldsymbol{L}'\right)^{-1}\left(\boldsymbol{L}\widehat{\boldsymbol{\beta}}_F - \boldsymbol{k}\right)/r}{\widehat{\sigma}^2_F}\,. \tag{4.36}$$

Note that $\left(\boldsymbol{X}'\boldsymbol{V}^{-1}\boldsymbol{X}\right)^{-1}$ is the variance of regression coefficients in the full model without the scaling constant $\widehat{\sigma}^2_F$ and the matrix product in the numerator quantifies how much it is increased by the constraints specified by H_0. If the null model is true and $\boldsymbol{e} \sim N(\boldsymbol{0}, \sigma^2\boldsymbol{V})$ with known $\boldsymbol{V}$, we have

$$F_{obs} \sim F(r, n-p)\,. \tag{4.37}$$

Equation (4.37) is asymptotically true even without normality, as we will discuss in Section 4.6.3. This test is commonly suggested for regression coefficients estimated using OLS or GLS.

The justification of the statistic (4.36) is as follows: if $\boldsymbol{V}$ and σ^2 are known, the residual errors are normally distributed and H_0 is true, we have

$$W = \frac{1}{\sigma^2}\left(\boldsymbol{L}\widehat{\boldsymbol{\beta}}_F - \boldsymbol{k}\right)'\left(\boldsymbol{L}\left(\boldsymbol{X}'\boldsymbol{V}^{-1}\boldsymbol{X}\right)^{-1}\boldsymbol{L}'\right)^{-1}\left(\boldsymbol{L}\widehat{\boldsymbol{\beta}}_F - \boldsymbol{k}\right) \sim \chi^2(r)\,. \tag{4.38}$$

In addition, $S = \frac{1}{\sigma^2}\widehat{\boldsymbol{e}}'\boldsymbol{V}^{-1}\widehat{\boldsymbol{e}} \sim \chi^2(n-p)$, and it is independent of W. We can write $F_{obs} = (W/r)\,/\,(S/\,(n-p))$, which has the $F(r, n-p)$ distribution (recall Section 2.11.3). The true σ^2 cancels out because it occurs both in the numerator and in the denominator, and the remaining term in the denominator is the unbiased estimate $\widehat{\sigma}^2_F$.

An alternative test statistic, Wald's χ^2, is obtained by replacing σ^2 in Equation (4.38) by its unbiased estimator $\widehat{\sigma}^2_F$. The resulting statistic can also be written as

$$W_{obs} = rF_{obs}\,. \tag{4.39}$$

The statistic has asymptotically a $\chi^2(r)$ distribution. This result is sometimes used for testing in large data sets. The test is equivalent to the F-test when n approaches infinity. Unlike the F-test, this test is asymptotic even under normality of the residual errors because of the estimation error of $\widehat{\sigma}^2_F$. It is not suggested for general use but is the only possibility, e.g. when a certain composite hypothesis is tested in grouped data sets (see Example 5.23, p. 176).

One widely used application of the F-test is the overall test of the regression, which is formulated by specifying such a null model that includes no other predictors beyond the intercept, i.e. $\boldsymbol{L} = \begin{pmatrix} \boldsymbol{0} & \boldsymbol{I}_{p-1} \end{pmatrix}$. This provides, e.g. the F-test for one-way ANOVA. In general, the F-test of two nested models provides a framework for a very wide class of hypothesis tests of regression coefficients on nested models. One just needs to formulate the model appropriately so that the hypothesis of interest can be specified as $\boldsymbol{L}\widehat{\boldsymbol{\beta}}_F = \boldsymbol{k}$, as illustrated in Example 4.15. Another specific application is the outlier test procedure of `car::outlierTest`.

If the linear model has only one categorical predictor with two levels, the overall test of the regression is equivalent to the two-sample t-test. If the only predictor has more than two levels, then the test is equivalent to the one-way analysis of variance (ANOVA). A model with two categorical predictors specifies the two-way ANOVA and a model with both categorical and continuous predictors is sometimes called an analysis of covariance (ANCOVA) model. These classifications were important in the days before modern computers when different specialized formulas were used for different model types.

In some cases, the F-test can be used for *the lack-of-fit test* of the systematic part of the model. Assume a single-predictor regression, where the data includes several observations for each single value of the x-variables. The full model in this case is such a model where each unique value of the x-variable provides a level of a categorical predictor. The constrained model is a model where x is treated as a continuous predictor. Comparison of these models using the F-test provides a formal evaluation of the goodness of fit. If the data does not include replicates for each value of x, a modified version uses classified values of x in the full model. For more details, see Weisberg (2014). Note that other tests on the regression coefficients do not test whether a model fits *well*; they just test whether a model fits *better* than another model. Goodness-of-fit tests for GLM are encountered in Section 8.2.6.

4.6.3 Normality of residual errors in testing

The F tests statistic has the exact distribution specified in Equation (4.37), if the null hypothesis is true and the residual errors are normally distributed. The same also holds for the $t-$tests of regression coefficients, which will be presented later. If normality is not met, the result is still valid asymptotically, i.e. if the number of observations is sufficiently large, due to an extension of the Central Limit Theorem, which was demonstrated in Example 3.1, (p. 48). However, the assumptions about the model form and variance-covariance structure of residual errors should be met regardless of the number of observations (see, e.g., Casella and Berger 2001, Section 11.2.1).

The Wald's χ^2 test is valid only asymptotically, even under normality, but it is also valid for non-normal data. However, to achieve a certain level of accuracy for non-normal data, a larger sample size may be needed than for normal data

What then is "a sufficiently large data set"? Ultimately, it depends on the shape of the distribution of residual errors. The closer the distribution is to a normal distribution, the smaller the number of observations are required. Usually, the test statistics converge to the null distribution surprisingly fast, and small or moderate violations of the normality assumption are often meaningless. However, very heavy tails of the distribution and the presence of outliers may lead to a slow convergence. Nowadays it is easy to explore such issues through simulation.

In some cases, the assumption of normality of residuals is evaluated through formal tests. However, such tests are not useful: they seldom reject the null hypothesis about normality in a small data set, where normality is important, and they easily reject it in a large data set, where it is unimportant. Graphical evaluation of normality is much more informative in the regression context.

4.6.4 Marginal and sequential testing procedures

In a multiple regression model, one usually needs to test several sets of possible predictors. For example, in our example with the `stumplift` data, we might want to test the effects of machine, stump diameter and the interaction of the stump diameter and machine. In which order should these tests be done, and does the order have an effect on the conclusions? The answer is not always clear, and unfortunately, different procedures may lead to different conclusions.

The model usually has a lower-order term and higher-order terms. The higher-order terms are derived from the lower-order terms, and are most commonly interactions or powers of them. Weisberg (2014, Section 6.2) suggests the *marginality principle* for hypothesis testing. In that approach, one starts from the full model and tests the highest-order terms first. If the highest-order term is judged to be non-zero, then none of the corresponding lower-order terms are tested but they are kept in the model. If the highest-order term is judged to be zero and dropped from the model, then one proceeds to test the terms of the second-highest order. The marginality principle leads to the use of so-called *type II sums of squares*.

Alternative approaches to this procedure are the *type I sums of squares*, also called *sequential sums of squares* and *type III sums of squares*, also called *marginal sums of squares*. In the sequential approach, one starts from an empty model and adds terms to the model in the order specified by the model equation. That is, one first evaluates a full model including the first predictor against the null model without any predictor. Thereafter, a full model with two first predictors is tested against the model including the first predictor alone. This procedure is continued until all predictors are included. However, even though the residual sums of squares in the numerator of the test statistic (4.35) are based on a different full model in every test, the estimated standard error in the denominator is always based on the full model that includes all the predictors. In the sequential procedure, the order in which the predictors are included makes a difference. When marginal sums of squares are used, the full model always contains all the predictors. The null model is obtained from the full model by dropping each of the predictors in turn. This principle leads to testing of the presence of lower-order terms in a model that includes higher-order terms, which seldom makes sense.

To summarize, types I, II and III sums of squares specify a different ordering of individual tests based on the test statistics (4.35) and (4.36). The results from these different procedures are similar in some special cases (e.g. in balanced experimental data sets) but not generally. In addition, the reports from these procedures do not necessarily include the tests that are needed for the research question at hand, and all tests do not necessarily make sense. One should, therefore, understand what effects are actually tested in each of the reports produced by the statistical software. This in turn requires a full understanding of the model equations and interpretations of the model parameters. The researcher should never rely on an output of an automatic testing procedure if one does not fully understand what tests it is based on. Indeed, it should be the researcher who decides what tests are needed and carried out, not the computer software.

4.6.5 The effect of estimation error in V

Equations (4.35) and (4.39) require that $\boldsymbol{V}$ is known. If we believe that the OLS model is correct, we have $\boldsymbol{V} = \boldsymbol{I}$, which is known. In that case, $\widehat{\boldsymbol{\beta}}_C$ and $\widehat{\boldsymbol{\beta}}_F$ are also the OLS estimates, and the test leads to an F-test on the OLS model. Under the GLS model, $\boldsymbol{V}$ is usually not known: it includes inaccuracy caused by the model selection and the estimation of unknown parameter $\boldsymbol{\theta}$, which may include, e.g. the parameters of the assumed variance and autocorrelation functions. In the mixed-effects models (Chapter 5), $\boldsymbol{\theta}$ will also include the variances and covariances of the random effects. There are at least three solutions to this problem, which are presented below.

One possibility is to use a *naive approach*, replacing the known $\boldsymbol{V}$ in Equation (4.35) with its estimate, $\widehat{\boldsymbol{V}}$, based on the full model fitting. The resulting test procedure is called *the conditional F-test* (Pinheiro and Bates 2000) to emphasize that it is performed conditional to the estimated variance-covariance structure, or *approximate Wald/F -testing procedure* (Stroup 2013), so as to emphasize that the result is only approximate due to the estimation errors of $\boldsymbol{V}$.

When an estimate of $\boldsymbol{V}$ is used in Equation (4.25) (p. 92), Equation (4.26) underestimates the true variance of regression coefficients (Kackar and Harville 1984). Therefore, the confidence intervals of $\boldsymbol{\beta}$ using the naive approach are too narrow, and tests based on Equation (4.35) with estimated $\boldsymbol{V}$ may be strongly anti-conservative with small sample sizes (i.e. p-values are too small). One solution to this problem is to use $\widehat{\boldsymbol{V}}$ in the place of $\boldsymbol{V}$, and to additionally adjust the denominator degrees of freedom in the theoretical distribution of the test statistic in order to take into account the estimation errors of $\boldsymbol{V}$. This is useful especially in the context of linear mixed-effects models.

Another possibility is to use the *Kenward-Roger* approximation of $\widehat{\mathrm{var}}(\widehat{\boldsymbol{\beta}})$ and of the denominator degrees of freedom (Kenward and Roger 1997). The formula is based on a second-order Taylor approximation. See McCulloch *et al.* (2008, p. 165–168) or Stroup (2013, p. 168–170) for a summary of the procedure. In `R`, the approximation is available in package `pbkrtest` for mixed-effect models fitted using package `lme4`.

In `R` function `nlme::gls`, only the naive approach with degrees of freedom defined as in Equation (4.35) is available. That approach is used in Example 4.15 below. Approaches available for mixed-effects models are discussed in Section 5.6, where an example is presented that compares these three approximations in different situations.

4.6.6 Examples

This section illustrates the testing procedures described in the previous subsections.

Example 4.15. The conditional F-tests for a model fitted using `nlme::gls` can be conducted using the one-argument form of `anova`. By default, the sequential (type I) sums of squares are used. Such a test for model `mod3` of the stumplift data is performed below.

```
> anova(mod3)
Denom. DF: 476
                       numDF   F-value p-value
(Intercept)                1 2292.1899  <.0001
Diameter                   1  501.9529  <.0001
Machine                    2   59.5533  <.0001
I(Diameter^2)              1    1.8318  0.1766
Diameter:Machine           2   17.2478  <.0001
Machine:I(Diameter^2)      2    6.3329  0.0019
```

Row `Diameter` presents an F-test comparing a full model that includes stump diameter as the only predictor to a constrained model that does not include any predictors. The p-value is below 0.0001, indicating that stump diameter significantly explains the variation in the processing time. Row `Machine` compares a full model that includes stump diameter and machine with a constrained model that includes diameter as the only predictor. As the p-value for this row is also small, the inclusion of machine after diameter to the model significantly improves the fit. However, adding the second power of diameter to the model, which already includes diameter and machine as predictors, does not significantly improve the fit (p=0.177). Furthermore, adding the interaction of diameter and machine (i.e. allowing machine-specific coefficients for the first order terms of the polynomial but keeping the coefficients for the second-order term common to all machines - although this model may not make much sense) significantly improves the model fit, compared to the model with machine, diameter and squared diameter as predictors. Finally, allowing machine-specific second-order terms of the polynomial further improves the model fit significantly. This indicates that even though the main effect for the squared diameter was non-significant, there is a possible nonlinear response of processing time to stump diameter, at least for some of the machines.

Tests based on the marginal (type III) sums of squares are carried out below:

```
> anova(mod3,type="m")
Denom. DF: 476
                      numDF   F-value p-value
(Intercept)               1  4.501297  0.0344
Diameter                  1 10.309156  0.0014
Machine                   2  3.746246  0.0243
I(Diameter^2)             1  1.391464  0.2387
Diameter:Machine          2  3.392062  0.0345
Machine:I(Diameter^2)     2  6.332938  0.0019
```

In this test it is extremely important to understand what effects were actually tested in each of the tests, as many of the conducted tests do not make sense. The full model was of the following form:

$$\begin{aligned} t_i =& \beta_1 + \beta_2 d_i + \beta_3 d_i^2 + \\ & \beta_4 M2_i + \beta_5 M2_i d_i + \beta_6 M2_i d_i^2 + \\ & \beta_7 M3_i + \beta_8 M3_i d_i + \beta_9 M3_i d_i^2 + e_i \,. \end{aligned} \tag{4.40}$$

1. Row `(Intercept)` tests hypothesis $H_0 : \beta_1 = 0$, i.e. whether the processing time of machine 1 for a zero-diameter stump significantly differs from zero. Such a test may not be of much interest. In general, a high p-value for the intercept should not on its own be used as an argument to drop it from the model (see Box 10.1, p. 311).
2. Row `Diameter` tests hypothesis $H_0 : \beta_2 = 0$ when all other coefficients are unrestricted. Therefore, it tests whether the parabola $\beta_1 + \beta_3 d^2$, which has the minimum/maximum at zero diameter, is not significantly worse than the general parabola $\beta_1 + \beta_2 d + \beta_3 d^2$ for machine 1. This test might not be well justified either.

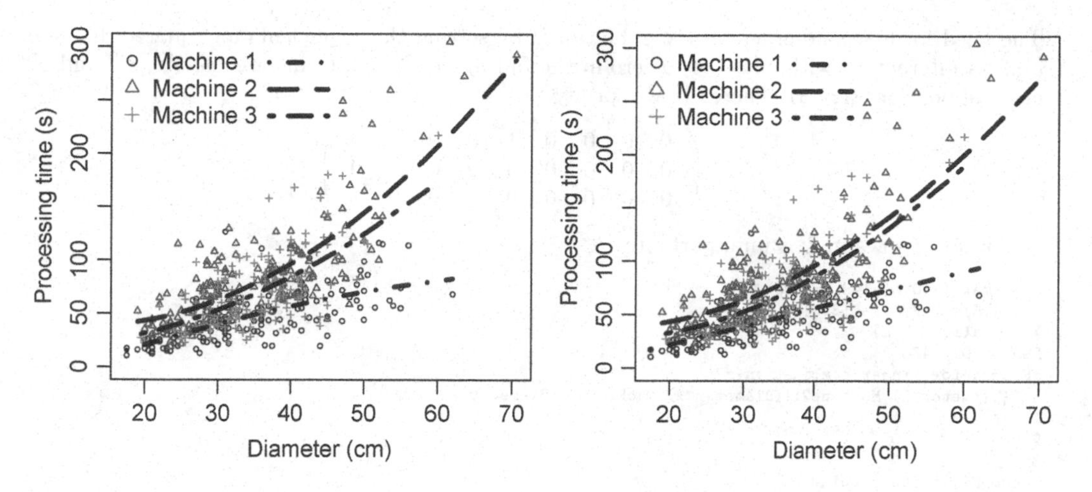

FIGURE 4.11 Stump processing time (s) and predicted values of models (4.40) (left) and (4.41) (right) against stump diameter (cm) (see Example 4.15).

3. Row `Machine` evaluates, hypothesis $H_0 : \beta_4 = \beta_7 = 0$. If these effects are set to zero, it means that the processing time for a zero-diameter stump is equal for all machines. Again, the test may not be of much interest, as interactions $\beta_5 M2_i d_i$, $\beta_8 M3_i d_i$, $\beta_6 M2_i d_i^2$ and $\beta_9 M3_i d_i^2$ are also included in the model.
4. Row `I(Diameter^2)` tests hypothesis $H_0 : \beta_3 = 0$, i.e. whether a linear function $\beta_1 + \beta_2 d_i$ is sufficient for machine 1. The p-value is 0.24 and the null hypothesis is not rejected. It indicates that the second-order term is not necessarily needed for machine 1.
5. Row `Diameter:Machine` tests hypothesis $H_0 : \beta_5 = \beta_8 = 0$, i.e. whether the polynomials for all machines have a common first-order term β_2, given that the second-order term is machine-specific. We do not see any justified reason to assume such a restriction.
6. Row `Machine:I(Diameter^2)` tests hypothesis $H_0 : \beta_6 = \beta_9 = 0$, i.e. whether the polynomials for all machines have a common second-order term β_3. This means that the parabolas for all machines are of the same shape. The low p-value suggests rejection of the null hypothesis, and keeping the machine-specific second-order terms in the model.

The sequential and marginal tests suggest the inclusion of machine-specific second-order terms, but the sequential test for `I(Diameter^2)` was insignificant. However, the full model in the sequential test for `I(Diameter^2)` did not include interactions, and fits rather poorly to our data. Furthermore, as the tests suggest the inclusion of the interaction of machine and the second power of diameter in the model, the marginality principle suggests that the result about the effects of `I(Diameter^2)` in the sequential test should be ignored.

The relationship for machine 1 might be linear. This is supported by the left-hand panel of Figure 4.11 and by the marginal test for `I(Diameter^2)`. This may be related to the fact that very few large stumps were operated by machine 1. However, the data clearly shows nonlinear curves for machines 2 and 3.

The sequential and marginal tests described above did not make a comparison where all the machine-specific second-order terms are dropped from the full model. Such a test can be done by specifying an appropriate hypothesis matrix $\boldsymbol{L}$ for a general test $H_0 : \boldsymbol{L\beta} = \boldsymbol{0}$. To define $\boldsymbol{L}$, we need to know the order of fixed effects in the model. Each corresponds to one column in $\boldsymbol{L}$.:

```
> names(coef(mod3))
[1] "(Intercept)"             "Diameter"                 "Machine2"
[4] "Machine3"                "I(Diameter^2)"            "Diameter:Machine2"
[7] "Diameter:Machine3"       "Machine2:I(Diameter^2)" "Machine3:I(Diameter^2)"
```

The need for a second-order term can be tested by setting the coefficients for squared diameter and its interaction with machine to zero in the full model. These terms are the 5th, 8th and 9th elements of `coef(mod3)`. Therefore we specify

$$\boldsymbol{L} = \begin{pmatrix} 0 & 0 & 0 & 0 & 1 & 0 & 0 & 0 & 0 \\ 0 & 0 & 0 & 0 & 0 & 0 & 0 & 1 & 0 \\ 0 & 0 & 0 & 0 & 0 & 0 & 0 & 0 & 1 \end{pmatrix};$$

the order of rows in $\boldsymbol{L}$ is unimportant.

```
> L<-matrix(0,ncol=9,nrow=3)
> L[1,5]<-L[2,8]<-L[3,9]<-1
> anova(mod3,L=L)
Denom. DF: 476
 F-test for linear combination(s)
  I(Diameter^2) Machine2:I(Diameter^2) Machine3:I(Diameter^2)
1             1                      0                      0
2             0                      1                      0
3             0                      0                      1
  numDF F-value p-value
1     3 4.56783  0.0036
```

The null hypothesis is rejected; the machine-specific second-order terms should be kept in the model. The same test can also be done using argument **Terms** in **anova** as follows (numbers 4 and 6 refer to the rows of the output from **anova(mod3)**):

```
> anova(mod3,Terms=c(4,6))
Denom. DF: 476
 F-test for: I(Diameter^2), Machine:I(Diameter^2)
  numDF F-value p-value
1     3 4.56783  0.0036
```

We might also want to test whether the shapes of the curves for machines 2 and 3 are similar. If they are similar, then $\beta_5 = \beta_8$ and $\beta_6 = \beta_9$. This leads to null hypothesis $H_0 : \beta_5 - \beta_8 = \beta_6 - \beta_9 = 0$, which is implemented as $H_0 : \boldsymbol{L\beta} = \boldsymbol{0}$ by defining

$$\boldsymbol{L} = \begin{pmatrix} 0 & 0 & 0 & 0 & 0 & 1 & -1 & 0 & 0 \\ 0 & 0 & 0 & 0 & 0 & 0 & 0 & 1 & -1 \end{pmatrix}.$$

The test below gives a p-value 0.69, indicating that the null hypothesis is not rejected. It is possible that the first and second powers of the diameter have common coefficients for machines 2 and 3, and there is only a level shift in the curves:

```
> L<-matrix(0,ncol=9,nrow=2)
> L[1,6]<-L[2,8]<-1
> L[1,7]<-L[2,9]<--1
> anova(mod3,L=L)
Denom. DF: 476
 F-test for linear combination(s)
  Diameter:Machine2 Diameter:Machine3 Machine2:I(Diameter^2) Machine3:I(Diameter^2)
1                 1                -1                      0                      0
2                 0                 0                      1                     -1

  numDF   F-value p-value
1     2 0.3736314  0.6884
```

A test for a common curve for machines 2 and 3 ($H_0 : \beta_4 - \beta_7 = \beta_5 - \beta_8 = \beta_6 - \beta_9$) gave a p-value of 0.0028 (code not shown), indicating that the level of the curve differs between these two machines. We could also ask whether all three machines only differ in terms of the intercept. Such a test gave a p-value below 0.0001 (code not shown), indicating that interactions between stump diameter and machine are needed.

To conclude, a restricted model assuming linear dependence for machine 1 and common first- and second-order terms for machines 2 and 3 is sufficient. A test for such a model is based on hypothesis $H_0 : \beta_3 = \beta_5 - \beta_8 = \beta_6 - \beta_9 = 0$, and gave a p-value 0.5446. Such a restricted version of the model has the formula

$$t_i = \beta_1 + \beta_2 d_i + \beta_3 M2_i + \beta_4 M3_i + \beta_5 M23_i d_i + \beta_6 M23_i d_i^2 + e_i \,, \tag{4.41}$$

where $M23_i$ takes a value 1 for stumps processed using machine 2 or 3, and 0 for machine 1. The assumptions in regard to the error term are the same as previously described for model `mod3`. The model is fitted below, and the fitted values from the model are shown in the right-hand panel of Figure 4.11.

```
> mod4<-update(mod3,model=Time~Diameter+Machine+I((Machine!=1)*(Diameter))+I((Machine!=1)*(Diameter^2)))
```

Example 4.16. For the two-way ANOVA model of data set `immer` (recall Example 4.6), the F-test for the overall fit of the model is defined as $H_0 : \beta_2 = \beta_3 = \ldots = \beta_{10} = 0$ (see model (4.10)). The result of the test is shown at the end of the model summary:

```
> summary(imod1)
...
F-statistic:  13.3 on 9 and 20 DF,  p-value: 1.216e-06
```

The variation in the mean barley yield is significantly explained by variety and location.

The sequential F-tests for variety and location can be conducted using the one-argument form of `anova`. The results from the marginal tests would be identical because the data are balanced. Notice that the arguments `Terms` and `L` are not available in method `anova.lm`, which is used when method `anova` is applied to a model fitted using `lm`.

```
> anova(imod1)
Analysis of Variance Table

Response: Y
          Df  Sum Sq Mean Sq F value    Pr(>F)
Var        4  2655.0  663.75  5.9891  0.002453 **
Loc        5 10610.5 2122.09 19.1480 5.212e-07 ***
Residuals 20  2216.5  110.83
---
Signif. codes:  0 '***' 0.001 '**' 0.01 '*' 0.05 '.' 0.1 ' ' 1
```

More sophisticated hypothesis tests for models fitted using `lm` could be carried out by using the two-argument form of `anova`, which requires that both full and constrained models are fitted. Even more flexible alternatives are available, e.g. in package `car` (see Fox and Weisberg (2011)). Function `Anova` (note the capital `A`) of that package provides anova-tables using type II and III sums of squares. Function `linearHypothesis` (or just `lht`) provides a tool for testing the general hypothesis of form $H_0 : \boldsymbol{L\beta} = \boldsymbol{k}$. There are also other packages for similar purposes, such as package `multcomp` Bretz *et al.* (2011), which makes adjustments for multiple comparisons, but does not allow as general hypotheses as `car::lht` does.

4.6.7 t-tests of $\boldsymbol{\beta}$

As illustrated in the previous section, all two-sided tests for regression coefficients $\boldsymbol{\beta}$ can be formulated through the F-tests. Equivalently, they can be formulated as t-tests. The formal *t-test of regression coefficients* is based on the statistic

$$t_{obs} = \frac{\boldsymbol{l}'\widehat{\boldsymbol{\beta}} - \boldsymbol{l}'\boldsymbol{\beta}}{\hat{\sigma}\sqrt{\boldsymbol{l}'(\boldsymbol{X}'\boldsymbol{V}^{-1}\boldsymbol{X})^{-1}\boldsymbol{l}}} \tag{4.42}$$

where the row vector $\boldsymbol{l}'_{p\times 1}$ implements a restriction on the parameter vector in the same way that matrix $\boldsymbol{L}$ did in the F-test. If the residual errors are normally distributed, $\boldsymbol{V}$ is known and H_0 is true, then

$$t_{obs} \sim t(n-p) .$$

The properties of the t-test are the same as those of F-test: the t-distribution is exact only under normality and for known $\boldsymbol{V}$; the effect of non-normality vanishes as n gets large; and the result is only approximate if an estimate of $\boldsymbol{V}$ is used, which is commonly called the *conditional t-test* or *approximate Wald test.* If $\boldsymbol{V}$ is estimated, the tests are anti-conservative.

The squared t-statistic with $n-p$ degrees of freedom equals the F-statistic with 1 and $n-p$ degrees of freedom. Therefore, the two-sided (conditional) t-test is equivalent to the (conditional) F-test where the null model is formulated by restricting a regression coefficient to zero using row vector $\boldsymbol{l}$ as one-row matrix $\boldsymbol{L}$ in the F-test. The t-distribution of regression coefficients is useful as a basis for confidence intervals of regression coefficients, and in the construction of one-sided tests for regression coefficients.

The most commonly used application of the t-test is to examine whether a certain coefficient is zero, which leads to

$$\frac{\widehat{\beta}_i}{\mathrm{se}(\widehat{\beta}_i)} \sim t(n-p) \tag{4.43}$$

where $\mathrm{se}(\widehat{\beta}_i)$ is the square root of the ith diagonal element of matrix $\widehat{\sigma}^2(\boldsymbol{X}'\boldsymbol{V}^{-1}\boldsymbol{X})^{-1}$. Tests for the null hypothesis $H_0 : \beta_i = 0$ for all coefficients separately are commonly reported by statistical packages in the model summary. In practice, the formal $t-$test can often be replaced by the following rule of thumb: such coefficients whose absolute estimates are higher than twice the standard error are significant at the 0.05 significance level.

It should also be recognized that the tests for a certain predictor are carried out by assuming that all other predictors are included in the model. Therefore, the t-tables routinely produced by statistical packages are equivalent to the corresponding F-tests based on type III sum of squares. The warnings that were given previously about the interpretation of such F-tests also apply for these t-tables. The t-tests are commonly used as so-called *post-hoc* tests in the analysis of variance, where all pairs of treatment levels are tested for equality. Such a procedure leads to a high number of tests, and the resulting p-values are, therefore, commonly adjusted for multiple tests (see Section 3.6.2 and Example 4.18). Post-hoc tests should be done only if the categorical predictor with multiple levels is found to be significant in an F-test. If it is not found to be significant, post-hoc tests should not be done.

Example 4.17. Consider the model `mod4`, which was fitted in Example 4.15. The t-tests for $H_0 : \beta_i = 0$ for all individual regression coefficients are reported in the model summary:

```
> summary(mod4)
Coefficients:
                                    Value Std.Error   t-value p-value
(Intercept)                     -12.67112  4.190869 -3.023506  0.0026
Diameter                          1.71661  0.145419 11.804584  0.0000
Machine2                         55.26731 20.046643  2.756936  0.0061
Machine3                         44.85483 20.451428  2.193237  0.0288
I((Machine != 1) * (Diameter))   -2.98056  1.211771 -2.459675  0.0143
I((Machine != 1) * (Diameter^2))  0.06481  0.017765  3.648293  0.0003
```

The interpretation for these tests is similar to the F-tests based on marginal sums of squares. Therefore, the test on `Machine2`, for example, tests the null hypothesis $H_0 : \beta_3 = 0$ (using notations of Equation (4.41)). That is, the null hypothesis assumes that the processing time for a zero-diameter stump with machine 2 is different from that of machine 1 (which is unrealistically estimated to be -12.6 seconds in this model).

Example 4.18. In Example 4.16 the analysis of `immer` data indicated that variety is a statistically significant predictor of mean barley yield. To find which varieties differ from each other (post-hoc test), we use function `glht` of package `multcomp`. The general hypotheses are defined in matrix $\boldsymbol{K}$, where each line specifies a linear hypothesis for a t-test $H_0 : \boldsymbol{l}'\boldsymbol{\beta} = k$. For variable `Var`, there are five levels leading to 10 possible pairs. The comparisons of pairs are implemented below in object `K`:

```
> library(multcomp)
> K<-matrix(0,ncol=10,nrow=10)
> K[1,2]<-K[2,3]<-K[3,4]<-K[4,5]<-1
> K[5:7,2]<-K[8:9,3]<-K[10,4]<--1
```

```
> K[5,3]<-K[6,4]<-K[7,5]<-K[8,4]<-K[9,5]<-K[10,5]<-1
> colnames(K)<-names(coef(imod1))
> rownames(K)<-c("P-M","S-M","T-M","V-M","S-P","T-P","V-P","T-S","V-S","V-T")
> K
    (Intercept) VarP VarS VarT VarV LocD LocGR LocM LocUF LocW
P-M           0    1    0    0    0    0     0    0     0    0
S-M           0    0    1    0    0    0     0    0     0    0
T-M           0    0    0    1    0    0     0    0     0    0
V-M           0    0    0    0    1    0     0    0     0    0
S-P           0   -1    1    0    0    0     0    0     0    0
T-P           0   -1    0    1    0    0     0    0     0    0
V-P           0   -1    0    0    1    0     0    0     0    0
T-S           0    0   -1    1    0    0     0    0     0    0
V-S           0    0   -1    0    1    0     0    0     0    0
V-T           0    0    0   -1    1    0     0    0     0    0
```

The post-hoc tests without adjustment for multiple comparisons are shown below:

```
> summary(glht(imod1,K),test=adjusted("none"))

	 Simultaneous Tests for General Linear Hypotheses

Fit: lm(formula = Y ~ Var + Loc, data = immer)

Linear Hypotheses:
         Estimate Std. Error t value Pr(>|t|)
P-M == 0    8.150      6.078   1.341 0.194983
S-M == 0   -3.258      6.078  -0.536 0.597810
T-M == 0   23.808      6.078   3.917 0.000854 ***
V-M == 0    4.792      6.078   0.788 0.439728
S-P == 0  -11.408      6.078  -1.877 0.075185 .
T-P == 0   15.658      6.078   2.576 0.018029 *
V-P == 0   -3.358      6.078  -0.553 0.586700
T-S == 0   27.067      6.078   4.453 0.000244 ***
V-S == 0    8.050      6.078   1.324 0.200289
V-T == 0  -19.017      6.078  -3.129 0.005288 **
---
Signif. codes:  0 '***' 0.001 '**' 0.01 '*' 0.05 '.' 0.1 ' ' 1
(Adjusted p values reported -- none method)
```

A easier-to-use but less transparent and flexible way to define the pairwise comparisons is to use function `mcp`. In addition, we use the Bonferroni method to adjust the p-values for multiple comparisons (recall Section 3.6.2).

```
> summary(glht(imod1, linfct = mcp(Var = "Tukey")),test=adjusted("bonferroni"))

	 Simultaneous Tests for General Linear Hypotheses

Multiple Comparisons of Means: Tukey Contrasts


Fit: lm(formula = Y ~ Var + Loc, data = immer)

Linear Hypotheses:
           Estimate Std. Error t value Pr(>|t|)
P - M == 0    8.150      6.078   1.341  1.00000
S - M == 0   -3.258      6.078  -0.536  1.00000
T - M == 0   23.808      6.078   3.917  0.00854 **
V - M == 0    4.792      6.078   0.788  1.00000
S - P == 0  -11.408      6.078  -1.877  0.75185
T - P == 0   15.658      6.078   2.576  0.18029
V - P == 0   -3.358      6.078  -0.553  1.00000
T - S == 0   27.067      6.078   4.453  0.00244 **
V - S == 0    8.050      6.078   1.324  1.00000
V - T == 0  -19.017      6.078  -3.129  0.05288 .
---
Signif. codes:  0 '***' 0.001 '**' 0.01 '*' 0.05 '.' 0.1 ' ' 1
(Adjusted p values reported -- bonferroni method)
```

The results are identical to the previous ones, except that the adjusted p-values are 10 times higher than the non-adjusted values because there are 10 pairwise comparisons; the values above 1 are just shrunk to 1. The conclusion from the adjusted p-values is that the differences in the mean yield between varieties T and M (23 bushels per acre) and T and S (27 bushels per acre) are statistically significant.

In some cases, it might be interesting to test whether the mean of the effects in some classes differs from the mean of other classes. In particular, one of the classes may be a control and then one might be willing to test the mean effect of treatments. Such a test is demonstrated here. Assume that variety M is regarded as the control (e.g. because it is an old traditionally grown variety) and we want to test whether the mean of other varieties differs from the mean of M. Our hypothesis is $H_0 : \beta_1 = \frac{(\beta_1+\beta_2)+(\beta_1+\beta_3)+(\beta_1+\beta_4)+(\beta_1+\beta_5)}{4}$, which simplifies to $H_0 : \frac{1}{4}\beta_2 + \frac{1}{4}\beta_3 + \frac{1}{4}\beta_4 + \frac{1}{4}\beta_5 = 0$. This has been implemented as $H_0 : \boldsymbol{L\beta} = \boldsymbol{0}$ below. We use **car::lht** for this purpose, but **multcomp::glht** would do the work as well.

```
> library(car)
> L<-matrix(0,ncol=10,nrow=1)
> L[,2]<-L[,3]<-L[,4]<-L[,5]<-1/4
> L
     [,1] [,2] [,3] [,4] [,5] [,6] [,7] [,8] [,9] [,10]
[1,]    0 0.25 0.25 0.25 0.25    0    0    0    0     0
> lht(imod1,L)
Linear hypothesis test

Hypothesis:
0.25 VarP  + 0.25 VarS  + 0.25 VarT  + 0.25 VarV = 0

Model 1: restricted model
Model 2: Y ~ Var + Loc

  Res.Df    RSS Df Sum of Sq      F  Pr(>F)
1     21 2553.0
2     20 2216.5  1    336.51 3.0364 0.09677 .
---
Signif. codes:  0 '***' 0.001 '**' 0.01 '*' 0.05 '.' 0.1 ' ' 1
```

The test indicates that the mean of the other varieties does not significantly differ from the mean of variety M.

4.6.8 Likelihood ratio - a general large-sample test procedure for nested models

An alternative to testing two nested models in large data sets is the *likelihood ratio* (LR) test:

$$LR = 2\log\left(\frac{L_F}{L_C}\right) = 2\left[\log\left(L_F\right) - \log\left(L_C\right)\right],$$

where L_F is the maximized likelihood in the full model and L_C is the maximized likelihood of the constrained model. The statistic LR is necessarily positive because the constrained model is a special case of the full model, and is zero in the special case where the restriction sets the parameter of interest exactly at the ML estimate of the full model. Under normality in a large sample, for a null model that sets the parameter in the interior of the parameter space, we have by the Wilk's theorem, (Stuart and Ord 1991, Section 23.7)

$$LR \sim \chi^2(p-q).$$

The likelihood ratio tests can be applied for all parameters of the model, not only for the regression coefficients $\boldsymbol{\beta}$. It is based on the asymptotic distribution of the test statistic, and is, therefore, approximate even under exact normality, except for some special cases. If the test is used for regression coefficients $\boldsymbol{\beta}$, the regular likelihood should be used instead of the restricted likelihood, because the latter does not involve $\boldsymbol{\beta}$. For models with general $\boldsymbol{V}$, the LR is not suggested for regression coefficients; see Pinheiro and Bates (2000, Section 2.4) for justification. The conditional t- and F-tests are recommended instead. The LR test would be exact and identical to the previously presented F-tests, if $\boldsymbol{V}$ were known.

For more discussion on alternative tests see Christensen (2011, Section 3), Fahrmeir *et al.* (2013, Section 3.3), Stroup (2013) and Weisberg (2014).

The LR test is the only alternative to test parameters $\boldsymbol{\theta}$ that specify $\boldsymbol{V}$. Even in this case, these tests have been reported to be conservative (see Pinheiro and Bates (2000, p. 83–87)). Furthermore, the asymptotic results on the likelihood ratio do not hold when the null hypothesis sets the parameter value at the boundary of its support (e.g. when variance is set to 0). We will return to this problem in Chapter 5.

Example 4.19. Earlier in Examples 4.7, 4.11 and 4.12, we fitted models for the processing time of stumps: `mod1` with constant error variance, `mod2` with non-constant error variance (Power-type variance function) and `mod3` with non-constant error variance (Power-type variance function) and correlated errors (AR(1)). The models are nested so that `mod1` is a special case of `mod2`, which is further a special case of `mod3`. They all have the same fixed part, so an LR test can be done on the REML fits. The LR test carried out below suggests `mod2` over `mod1`, and `mod3` over `mod2`. These tests confirm the conclusions in Example 4.13 to select model `mod3`.

```
> anova(mod1,mod2,mod3)
     Model df      AIC      BIC    logLik   Test   L.Ratio p-value
mod1     1 10 4513.619 4555.273 -2246.809
mod2     2 11 4389.066 4434.886 -2183.533 1 vs 2 126.55271  <.0001
mod3     3 12 4383.002 4432.987 -2179.501 2 vs 3   8.06424  0.0045
```

4.6.9 Comparing non-nested models

The AIC and BIC (see Section 3.6.4) can be used in comparing non-nested models. The model summaries in `R` usually report the values of these information criteria for fitted models. Functions `AIC` and `BIC` can also be used for computing these criteria for models fitted using different functions. In Example 4.19, the values of AIC and BIC are lowest for model `mod3`, leading to the same conclusion found in the LR tests. The information criteria are not comparable if models differ in the fixed part and are fitted using REML. For a discussion of other model selection criteria, see (Fahrmeir *et al.* 2013, Section 3.4).

4.6.10 Confidence intervals for model parameters

The confidence intervals of regression coefficients are constructed based on the normality of regression coefficients, which is justified either by normality of residual errors or by the large sample properties of the regression coefficient estimates. The formulation of confidence intervals of any linear combination of the regression coefficients is based on Equation (4.42). The $100(1-\alpha)\%$ symmetric two-sided confidence interval of the linear combination $\boldsymbol{l}'\boldsymbol{\beta}$ is given by

$$\boldsymbol{l}'\widehat{\boldsymbol{\beta}} \pm \widehat{\sigma} t_{n-p,\alpha/2} \sqrt{\boldsymbol{l}' \left(\boldsymbol{X}'\boldsymbol{V}^{-1}\boldsymbol{X}\right)^{-1} \boldsymbol{l}} \,. \tag{4.44}$$

The most commonly used application uses Equation (4.43), which produces marginal confidence intervals over the distribution of other parameter estimates. This is justified if the estimation errors of the regression coefficients are not strongly correlated. However, these intervals do not reveal the entire picture if the estimates are correlated. For correlated estimation errors, it is better justified to formulate the joint confidence ellipsoids for the entire group of related coefficients. At least the effect of possible correlation of regression coefficients should be recognized in the interpretation. Most `R`-functions have the method `intervals` that produces the marginal confidence intervals of parameters from models of different type.

The confidence intervals of variance-covariance parameters can be constructed based on the asymptotic multivariate normality of the variance-covariance parameters. The confidence intervals are computed in the natural parametrization used in the definition of the likelihood. For example, the AR(1) coefficient is parameterized through a transformation that does not allow values outside the range $[-1, 1]$, and the normality-based intervals are constructed for that transformation. The endpoints of the resulting confidence interval are, thereafter, transformed to the scale where they are reported. Such intervals will never include the boundary of the parameter space, such as a zero variance.

4.7 Prediction

We consider prediction using a linear model in two situations: in the case where information about the residual error for the sampling unit in question is not available and in the case where such information is available. In both cases, the prediction is the conditional expectation of the y-variable given the available data.

4.7.1 Prediction of an uncorrelated observation

With the linear model, prediction for a new individual with predictors $\boldsymbol{x}_0$ is obtained as $\tilde{y}_0 = \boldsymbol{x}_0'\widehat{\boldsymbol{\beta}}$. The prediction includes errors caused by

- model uncertainty,
- estimation errors of the parameters, $\text{var}(\widehat{\boldsymbol{\beta}})$, and
- residual errors.

The prediction variance for an observation with predictors $\boldsymbol{x}_0$ accounts for the residual error and uncertainty of parameter estimates, assuming that the systematic part is correctly formulated. Direct calculation gives

$$\begin{aligned} \text{var}(\tilde{y}_0 - y_0) &= \text{var}\left(\boldsymbol{x}_0'\widehat{\boldsymbol{\beta}} - (\boldsymbol{x}_0'\boldsymbol{\beta} + e_0)\right) \\ &= \boldsymbol{x}_0' \, \text{var}(\widehat{\boldsymbol{\beta}} - \boldsymbol{\beta})\boldsymbol{x}_0 + \text{var}(e_0) \\ &= \sigma^2 \boldsymbol{x}_0' \left(\boldsymbol{X}'\boldsymbol{V}^{-1}\boldsymbol{X}\right)^{-1} \boldsymbol{x}_0 + \text{var}(e_0) \,. \end{aligned} \tag{4.45}$$

Here, $\text{var}(e_0)$ is the residual error variance when the predictors take value $\boldsymbol{x}_0$. The middle equation is derived from the upper equation because residual errors are uncorrelated with the estimation errors of regression coefficients. In the OLS model, the first term of Equation (4.45) above and, thus, the total prediction error is smallest when $\boldsymbol{x}_0 = \bar{\boldsymbol{x}}$, where $\bar{\boldsymbol{x}}$ includes the averages of the x variables in the data.

The $100(1-\alpha)\%$ *prediction interval* of a new observation can be formulated based on this prediction variance, and the $(\alpha/2)^{th}$ and $(1-\alpha/2)^{th}$ quantiles of the $t(n-p)$-distribution. In large data sets, the normal distribution leads to practically the same results. If we are predicting the mean of n y-values for a fixed $\boldsymbol{x}_0$, the prediction error variance is the same as (4.45), although the last term is replaced by $\text{var}(e_0)/n$. The term vanishes as n approaches infinity. If the resulting variance is used in the construction of the interval, it is called the $100(1-\alpha)\%$ *confidence interval* of the mean.

The prediction interval for an individual observation takes into account both the estimation errors of the regression coefficients and the residual uncertainty. Normality is assumed for the sum of these components. Therefore, the interval may be misleading if the normality of errors is not met, regardless of the size of the model fitting data set. An approach, such as bootstrapping, which takes into account the shape of the distribution of the residual errors, might be better in this case. The confidence interval of the mean is well approximated by the normality-based interval if the estimation data set is sufficiently large.

Equation (4.45) can be generalized to prediction for several values of $\boldsymbol{x}_0$ jointly. Let us pool them to a matrix $\boldsymbol{X}_0$, which has similar structure to the model matrix $\boldsymbol{X}$, but now each of the n_0 rows represents an observation for which the prediction is to be made. Denote the corresponding variance-covariance matrix of residual errors by $\sigma^2 \boldsymbol{V}_0$. The prediction is obtained as $\widetilde{\boldsymbol{y}}_0 = \boldsymbol{X}_0 \widehat{\boldsymbol{\beta}}$ and the variance of prediction errors is

$$\begin{aligned} \text{var}(\widetilde{\boldsymbol{y}}_0 - \boldsymbol{y}_0) &= \boldsymbol{X}_0 \, \text{var}(\widehat{\boldsymbol{\beta}}) \boldsymbol{X}_0' + \sigma^2 \boldsymbol{V}_0 \\ &= \sigma^2 \boldsymbol{X}_0 \left(\boldsymbol{X}' \boldsymbol{V}^{-1} \boldsymbol{X} \right)^{-1} \boldsymbol{X}_0' + \sigma^2 \boldsymbol{V}_0 \, . \end{aligned} \tag{4.46}$$

For the evaluation, σ^2 and $\boldsymbol{V}$ need to be replaced by their estimates. The resulting object is a $n_0 \times n_0$ matrix, where the diagonal elements are the same as one could get using (4.45) separately for each row of $\boldsymbol{X}_0$, and the off-diagonal elements are non-zero even for an OLS model. The non-zero covariances result from common estimation errors of fixed effects. Evidently, the covariances are even higher if the covariances in $\boldsymbol{V}_0$ are non-zero. Note the similarity between the first term and the hat matrix (Section 4.5.1).

When repeated predictions with an estimated regression model are carried out, the effects of the residual errors of the new observations cancel each other out, although the effects of the estimation errors of the parameters do not. Therefore, estimation errors of regression coefficients behave like bias in that instance. An example is the use of volume or biomass models, which are routinely applied once they have been estimated for a certain region and tree species. When such models are fitted, it is extremely important that the modeling data set is sufficiently large, so that $\text{var}(\widehat{\boldsymbol{\beta}})$ is negligibly small.

Example 4.20. Assume that we want to predict the mean barley yield in 2017 and 2018 for all varieties at location D of the `immer` data set. We assume (probably unrealistically) that growth conditions have not changed since data collection. The prediction and the 95% confidence intervals are computed below using function `predict`. For instance, we are 95% confident that the mean yield of variety M in these two years will be 50.8–101.5 bushels per acre.

```
> immer1718<-immer[immer$Loc=="D",]
> pred<-predict(imod1,newdata=immer1718,interval="predict")
> rownames(pred)<-immer1718$Var
> pred
       fit      lwr       upr
M 76.14167 50.78479 101.49855
S 72.88333 47.52645  98.24021
V 80.93333 55.57645 106.29021
T 99.95000 74.59312 125.30688
P 84.29167 58.93479 109.64855
```

Example 4.21. Consider model `mod4`, which was fitted in Example 4.15. The script below computes the variance of prediction error for stumps of diameters $20, 21, \ldots, 60$ cm operated by machine 2. Because of the AR(1) structure of the residual errors, the order of new trees is important. It is assumed that the trees in vector `d` are given in the order of harvesting. In addition, we compute the 95% confidence interval for the mean processing time.

```
> # Prediction error variances for some new stumps using machine 2:
> d<-20:60
> X0<-cbind(1,d,1,0,d,d^2)
> tFit<-X0%*%coef(mod4)
> delta<-0.8386222
> phi<-0.1417738
> sigma<-0.6856451
> V1<-diag(as.vector(tFit^(2*delta)))
> corMat<-diag(rep(1,length(d)))
> for (i in 1:dim(corMat)[1]){
```

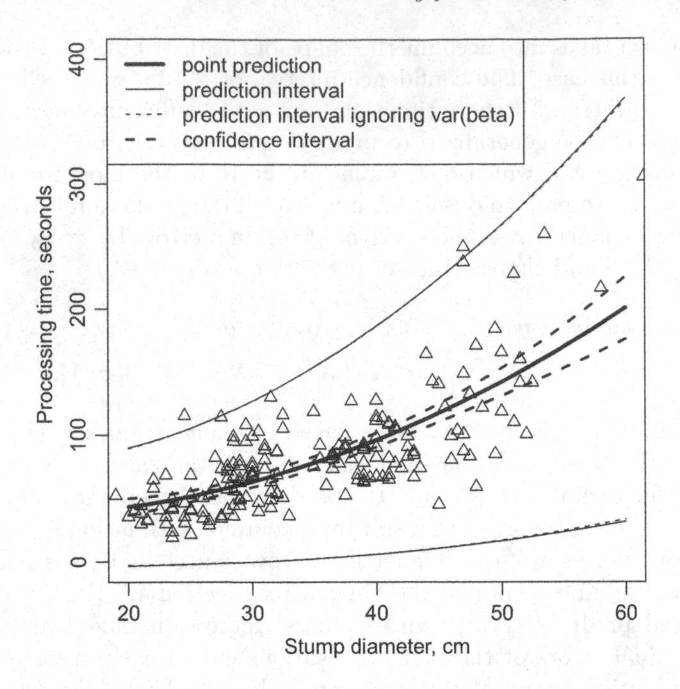

FIGURE 4.12 Prediction and confidence intervals for machine 2 in the `stumplift` data set. The dots show the original data, the thick line is the prediction, and the thinner lines show different prediction and confidence intervals computed in Example 4.21.

```
+      for (j in 1:dim(corMat)[2]){
+          corMat[i,j]<-phi^abs(i-j)
+      }
+ }
> V0<-sqrt(V1)%*%corMat%*%sqrt(V1)
> X<-model.matrix(mod4,data=stumplift)
> varBeta<-mod4$varBeta # matrix sigma^2*V
> varMean<-X0%*%varBeta%*%t(X0)
> varPred<-varMean+V0
> varPred[1:5,1:5]
           [,1]       [,2]      [,3]       [,4]       [,5]
[1,] 565.886080  90.960455  20.82431   9.655063   7.143146
[2,]  90.960455 593.885834  93.51740  19.805341   8.327641
[3,]  20.824313  93.517398 626.01580  97.117691  19.342219
[4,]   9.655063  19.805341  97.11769 662.373190 101.701926
[5,]   7.143146   8.327641  19.34222 101.701926 703.075295
```

The above matrix shows the variance-covariance matrix of prediction errors for the first five trees (i.e. those of diameters 20, ..., 24 cm). The successive trees are most strongly correlated.

To construct the prediction interval, we assume that the data are normally distributed, and as the data set is large, it is sufficient to use the normal distribution instead of the t-distribution to construct the prediction intervals. The resulting 95% prediction interval is shown in Figure 4.12 with solid thin lines. The confidence band is quite wide, especially at the lower bound, and includes more than 95% of the original data points. This can be explained by the skewed distribution of the residuals. To construct more realistic prediction intervals, we should use some other distribution to the normal distribution. In particular, a right-skewed distribution would be better suited to this situation. The thin dashed lines show the prediction intervals by ignoring the estimation errors of fixed effects.

The thick dashed lines in Figure 4.12 show the 95% confidence interval for the mean processing time. We have 95% confidence that the true mean processing time for stumps of a given diameter are within these bounds. Due to the asymptotic normality of $\widehat{\beta}$, this interval is well approximated by the normal distribution.

```
> pred<-data.frame(Diamm=d,Prediction=tFit,
+                  Lbound=tFit-qnorm(0.975)*sqrt(diag(varPred)),
+                  Lbound2=tFit-qnorm(0.975)*sqrt(diag(V0)),
+                  Lbound3=tFit-qnorm(0.975)*sqrt(diag(varMean)),
+                  Ubound=tFit+qnorm(0.975)*sqrt(diag(varPred)),
+                  Ubound2=tFit+qnorm(0.975)*sqrt(diag(V0)),
+                  Ubound3=tFit+qnorm(0.975)*sqrt(diag(varMean)))
>
```

4.7.2 Prediction of a correlated observation

In the above derivations we have assumed that the new observation to be predicted is uncorrelated with the observations in the estimation data. However, if such correlations exist, better predictions can be obtained. Let us assume GLS model formulation $\boldsymbol{y} = \boldsymbol{X\beta}+\boldsymbol{e}$ with $\mathrm{E}(\boldsymbol{e}) = \boldsymbol{0}$ and $\mathrm{var}(\boldsymbol{e}) = \boldsymbol{V}$. We predict a new observation $y_0 = \boldsymbol{x}_0'\boldsymbol{\beta} + e_0$ which is correlated with the observations in the estimation data. Let $\boldsymbol{c} = \mathrm{cov}\,(\boldsymbol{e}, e_0)$. Goldberger (1962) has shown that the best linear unbiased predictor, BLUP, of y_0 is

$$\widetilde{y}_0 = \boldsymbol{x}_0'\widehat{\boldsymbol{\beta}} + \boldsymbol{c}'\boldsymbol{V}^{-1}\left(\boldsymbol{y} - \boldsymbol{X}\widehat{\boldsymbol{\beta}}\right) \tag{4.47}$$

where $\widehat{\boldsymbol{\beta}}$ is the GLS estimator of $\boldsymbol{\beta}$. The prediction variance is:

$$\mathrm{var}\,(\widetilde{y}_0 - y_0) = \left(\boldsymbol{x}_0' - \boldsymbol{c}'\boldsymbol{V}^{-1}\boldsymbol{X}\right)\mathrm{var}\left(\widehat{\boldsymbol{\beta}}\right)\left(\boldsymbol{x}_0 - \boldsymbol{X}'\boldsymbol{V}^{-1}\boldsymbol{c}\right) + \mathrm{var}\,(e_0) - \boldsymbol{c}'\boldsymbol{V}^{-1}\boldsymbol{c} \tag{4.48}$$

The first term of Equation (4.48) represents the error variance due to estimation errors of $\boldsymbol{\beta}$, and the last two terms represent the error variance when e_0 is predicted from known $\boldsymbol{e}$. Equation (4.45) is obtained when $\boldsymbol{c} = \boldsymbol{0}$. The BLUP and its variance in mixed models are also obtained from the important results presented in Goldberger (1962).

The method where Equations (4.47) and (4.48) are utilized for prediction in a spatial context, is called *kriging* in the context of geostatistics; see e.g. Diggle and Ribeiro (2007) for the theory, and Bivand *et al.* (2008) for applications in R. The following example illustrates that those methods can also be seen as applications of the GLS model where the spatial dependence has been modeled by using an autocorrelation function.

Example 4.22. In Example 4.14 (p. 108), the following model was fitted to the topsoil zinc concentration data set `sp::meuse`

$$\begin{aligned} z_i &= \beta_1 + \beta_2 r_i + \beta_2 r_i^{(1)} + e_i \\ \mathrm{var}(e_i) &= \sigma^2 r_i^{2\delta} \\ \mathrm{cor}(e_i, e_j) &= \begin{cases} (1-\psi_1)\exp(-d/\psi_2) & 0 < \delta_1 < 1; \delta_2 > 0; d > 0 \\ 1 & d = 0\,, \end{cases} \end{aligned}$$

The parameter estimates of model `mod3.meuse` were $\widehat{\boldsymbol{\beta}} = (899, -2210, 5452)'$, $\widehat{\delta} = -0.422$, $\widehat{\sigma}^2 = 105^2$, $\widehat{\psi}_1 = 0.382$, and $\widehat{\psi}_2 = 199$ (meters). Based on this model and the original modeling data, we now make predictions for a total of 34 new regularly spaced locations within the area covered by the original data (see Figure 4.13).

The black circles on the left-hand graph show the predictions that utilize only the fixed part of the model. They are large at locations close to the riverside. However, the residual errors at the new locations are correlated with the original measurements, and the correlation is stronger the closer the new locations are to the observations of the original data set. This correlation is utilized by predicting the residual error of the model at each new location, and adding it to the prediction based on the fixed part of the model using Equation (4.47). Such adjusted predictions are illustrated by the gray shaded circles. This method adjusts the predictions downwards if the nearby observed concentrations are smaller than expected by distance to the river, and upwards in the opposite case. If no observations are sufficiently close, or if the measurements are similar to

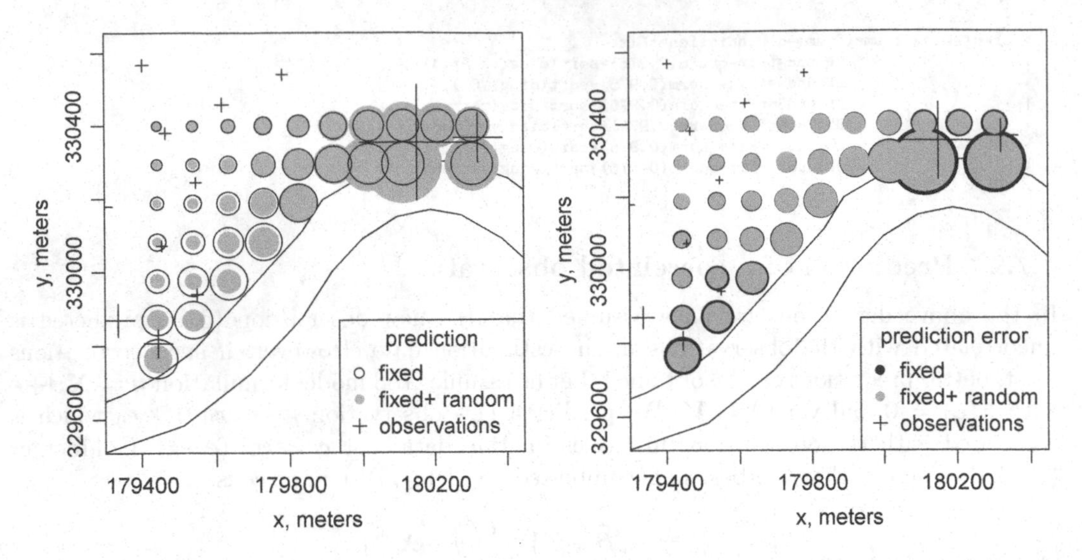

FIGURE 4.13 Illustration of predictions and standard deviation of prediction errors of topsoil zinc concentration in a subset of the area covered by data set `meuse` in Example 4.22. The symbol widths are proportional to the concentrations or standard deviations. The gray symbols on the right-hand graph are plotted on the black symbols and overlap them completely whenever the two prediction error variances are equal.

the predictions based on the fixed part, the predicted residual is close to zero and both predictions are similar.

The prediction errors based on the fixed part of the model were computed by applying Equation (4.45) separately for each new location. The prediction error variances based on the fixed and random parts were computed using Equation (4.48). We could also have used Equation (4.46); Equations (4.47) and (4.48) could be correspondingly written to produce the predictions for all new points simultaneously. Such formulas would also produce the full variance-covariance matrix of prediction errors. However, because computation is heavier and those covariances are seldom utilized in practice, we implemented the prediction for each new point separately. The R-script for the matrix formulation is included, however, for those readers who are interested.

The right-hand graph illustrates the prediction errors so that the symbol radius is proportional to the standard deviation, and symbol area to the variance. The prediction errors based on the fixed part were first plotted using solid black symbols, and they were overlaid with gray symbols that illustrate the prediction errors based on the fixed and random parts, which are always lower or equal to the prediction error based on the fixed part only. Because of the inconstant error variance, the prediction errors nearer the river are very high for both predictions. The errors for the prediction that utilizes the random part are lower than the errors for the prediction based on the fixed part only, especially at locations that are close to the original measurements. The large prediction errors near the river are decreased more in absolute terms than the small errors at locations farther from the river. The R-script is available on the book website.

5

Linear Mixed-Effects Models

The mixed-effects model is a model for a *grouped data set.* Examples of grouped data sets are individuals within spatially explicit subgroups, such as sample trees within forest sample plots; observations of a split-plot or a block in a controlled experiment; and repeated measurements of any subjects, such as plants or sample plots.

The effects of groups can be modeled either as fixed or random effects. In the *fixed-effects model,* categorical predictors are used as a predictor to describe the group membership. The coefficients are then estimated for each group in the data. A fixed-effects model is justified if all the groups are represented in the modeling data, or at least the interest lies in those particular groups that are included. They may also be justified if the group effects are not of primary interest in the analysis, such as block effects in a controlled experiment. A drawback of the fixed-effects model is that group-specific predictors cannot be used as additional fixed predictors, because their use would lead to an unidentifiable model. This prevents the use of fixed plot effects in a split-plot experiment. Furthermore, a sufficiently large number of observations per group should be available for a satisfactorily accurate estimation of the fixed group effects. The fixed-effects model does not allow prediction of the response variable for groups that are not included in the data. However, inference can be carried out for other regression coefficients, such as the differences between treatments in a randomized block design where the block effects are treated as fixed (see Chapter 14).

It is quite common that the groups of the data represent only a sample from a population of groups, while one is interested in making the inference about the entire population of groups. For example, all possible sample plots would not be included in forest tree data, and the data, therefore, represents a sample from a population of sample plots. In those situations, the group effects are naturally treated as random variables, which leads to *the mixed-effects model,* also called the *mixed model* and *random-effects model.* Additional benefits of random group effects over the fixed effects are that the number of observations per group can be small; even one observation per group may be enough for satisfactory estimation accuracy, provided some groups have several observations. As groups get larger, the estimated group effects from the mixed-effects model approach those of the fixed-effects model. Mixed-effects models also enable modeling of variability between groups using fixed predictors that are defined at the group level. Furthermore, prediction is also possible for groups that were not present in the modeling data.

The simplest mixed-effects model is the *variance component model.* It is comparable to the one-way analysis of variance model, dividing the total variation among the observations into variation within groups and between groups. After estimating these *variance components*, we can use the observations from the group in question to predict the realized value of the random effect; the *group effect* for that group. Mixed-effects models develop these ideas further to more general regression models. The models allow prediction of group effects even when the number of observations per group is smaller than the number of group effects. It can also be done afterwards for groups that were not included in the original modeling data. In addition, the variance components have an interpretation related to the covariance between observations of the same group. Therefore, the models parameterize the dependency of observations, leading to a well-justified GLS model for grouped data. The

Box 5.1 Pseudo-replication

Pseudo-replication refers to an invalid statistical inference where the correlated *measurement units* are analyzed as if they were independent *treatment units* or *sampling units*. This concept is usually applied in designed experiments, but it applies equally well to observational studies. A forestry example could be an experiment where treatments are applied randomly to plots according to an experimental design, and then the data are analyzed as if seedlings within the plots would be independent replicates. In that case, the seedlings are pseudo-replicates as the plots would be the true replicates. Temporal pseudo-replication occurs when repeated measurements are analyzed as if they were independent. The pseudo-replicates can also have different levels, e.g. needles within seedlings within plots.

In earlier times there were two ways to carry valid statistical inference when the measurement units were not independent. First, one could use treatment or sampling unit averages as observations. But if there were different numbers of measurements per treatment or sampling unit, then the averages have different variances. Moreover, it would not be possible to use measurement unit level variables as covariates. Second, it would be possible to pick such MSE terms from the ANOVA tables that the F tests are valid. However, this requires a lot of input from the analyst.

Fortunately, there is now a simple solution to the pseudo-replication problem: use a mixed model, which contains random effects for different grouping factors. In this way the correlations between different measurement units will automatically be taken into account.

mixed-effect models, therefore, provide a statistically sound framework for analysis of multiple, usually correlated, observations within independent sampling units. Such observations are sometimes called *pseudoreplicates* (see Box 5.1).

A grouped data set may have either a *single level of grouping* or *multiple levels of grouping*. In the case of multiple levels, the levels may either be *nested* or *crossed* (Table 5.1). In a nested grouping, the observations of a certain subgroup always belong to the same main group, so such data sets are commonly called *hierarchical*. In a crossed grouping, the groups do not have hierarchical ordering, instead members of a certain group at one level of grouping can belong to many groups in the other levels of grouping. The nested and crossed grouping structures can also be mixed in various ways. Examples of data sets with a nested grouping structure are:

- Sample trees within the sample plots within clusters of sample plots in national forest inventories.
- Split-plot or blocked experiments with repeated measurements for each split plot.

Examples of data sets with a crossed grouping structure are:

- Repeated measurements of sample trees that have been observed in the same years within a rather small area. The observations at a given point share a common year effect because of shared weather conditions. In addition, the repeated measurements of a tree have a common tree effect.
- Observations of sample trees on overlapping aerial images, where all trees of a given image share a common image effect and repeated observations of a certain tree on different images share a common tree effect (e.g. Korpela *et al.* 2014).
- Ecological experiments where bird's eggs are moved from the nest of the biological mother to another nest.
- Heritage data of any polygamous species.

		Level 1 groups		
		1	2	3
	1	x		
	2	x		
Level 2	3	x		
groups	4		x	
	5		x	
	6		x	
	7		x	
	8			x
	9			x

		Level 1 groups		
		1	2	3
	1	x	x	x
	2	x	x	x
Level 2	3	x	x	x
groups	4	x	x	x
	5	x	x	x
	6	x	x	x
	7	x	x	x
	8	x	x	x
	9	x	x	x

TABLE 5.1 An example of possible combinations of groups at levels 1 and 2 in a data set where grouping level 2 is nested within grouping level 1 (left), and in a data set where the levels of grouping are crossed (right).

- Controlled experiments with clones or origins of plants, where each experimental unit includes (at least partially) the same clones/origins.

5.1 Model formulation

5.1.1 The variance component model

A variance component model is the simplest mixed-effects model. It has no other predictors beyond the group indicator. The model partitions the variation of y into variation within groups and between groups.

We start by defining a fixed-effects model for grouped data. Denote the y-variable of individual j, $j = 1, \ldots, n_i$, of group i, $i = 1, \ldots, M$, by y_{ij}. A fixed-effects model for the data is

$$y_{ij} = \beta_0 + \beta_i + e_{ij} \tag{5.1}$$

where β_0 is the intercept, β_i ($i = 1, \ldots, M$) is the fixed regression coefficient for level i of the categorical predictor "Group", and e_{ij} is a random residual error with $\mathrm{E}(e_{ij}) = 0$, $\mathrm{var}(e_{ij}) = \sigma^2$, and $\mathrm{cov}(e_{ij}, e_{i'j'}) = 0$ for all $ij \neq i'j'$. The model is a special case of a linear model, more specifically, the model of one-way analysis of variance. This model is not identifiable because it includes the indicator variables for all levels of the variable "Group" and the intercept. It becomes identifiable if one of the $M + 1$ predictors is dropped from the model. If the indicator of level k is dropped, then the intercept β_0 quantifies the mean of the (default) group k, and β_i ($i = 1, \ldots, M; i \neq k$) specifies the differences between the means of the other groups and group k. It is convenient to drop the intercept from the model and include the effects of all groups. This gives

$$y_{ij} = \beta_i + e_{ij}\,, \tag{5.2}$$

where β_i is now directly the mean of group i.

The variance component model is the corresponding linear mixed-effects model. It differs from the fixed-effects model only by the assumption that the group effect is a random variable, not a fixed parameter. The model is specified as

$$y_{ij} = \mu + b_i + \epsilon_{ij}\,, \tag{5.3}$$

where μ is a fixed population *mean of group means.* It can also be interpreted as the *mean for an average group of the data* or, if the mean and mode of the random effects are equal, the *mean for a typical group.* An average group refers to such group where the group effects b_i takes its expected value.

The assumptions about the random group effect and residual error are specified as

$$b_i \sim N(0, \sigma_b^2)$$

and

$$\epsilon_{ij} \sim N(0, \sigma^2),$$

where b_i are independent among groups, ϵ_{ij} are independent among observations, and b_i and ϵ_{ij} are independent of each other.

Part $b_i + \epsilon_{ij}$ is called the *random part*. It divides the difference $y_{ij} - \mu$ into two independent parts: a random group effect for group i, and a random residual error for individual j of group i. Each observation y_{ij} is a sum of a fixed constant μ and two independent, normally distributed, random variables: the group effect b_i and residual error ϵ_{ij}. The grouped data includes information related to the M realizations of the group effects and $N = \sum_{i=1}^{M} n_i$ realizations of the residual errors. The model is identifiable if the modeling data contains at least one group with more than one observation. Of course, a reasonable estimation precision requires several groups with more than one observation.

The parameters of the model (5.3) are μ, σ_b^2 and σ^2. They quantify the mean of group means, the variance of the random effects and the variance of residual errors. The latter two are *the variance components*: the *between-group variance* and *within group variance*, which together quantify the *total variability* in y_{ij}.

A grouped data set is said to be *balanced* if all groups have the same number of observations. Estimation for balanced data is much simpler than for unbalanced data, but modern model fitting procedures do not require balanced data. However, greater care is needed in analysis and in the interpretation of the results based on unbalanced data.

The above-specified assumption in regard to the normality of random effects and residual errors is not really necessary for the mixed models. The normality assumptions can be replaced with weaker assumptions, such that the expected values are zero and the variances are constant. The variances can then be estimated using quadratic forms (see Section 5.3.1). Interestingly, these methods often lead to the same estimates as the estimators that are based on normality.

Using the observed data, fixed parameters are *estimated*, and random variables are *predicted.* For the variance component model, the model parameters are μ, σ_b^2, and σ^2. After finding the estimates of them, we can compute predictions $\widetilde{b}_i$ for all groups $i = 1, \ldots, M$ and further use them to compute the group-level predictions $\widetilde{y}_{ij} = \widehat{\mu} + \widetilde{b}_i$. We will demonstrate these estimates and predictions in the next example, even though details about the estimation and prediction procedures will be given later in Section 5.3.

Example 5.1. Data `lmfor::spati2` includes 1678 measurements of individual tree height from 56 Scots pine (*Pinus sylvestris*) stands in North Karelia. Start with the fixed-effects model without predictors,

$$y_{ij} = \mu + e_{ij},$$

which is fitted using

```
> library(lmfor)
> data(spati2)
> lm1<-gls(h~1,data=spati2)
> lm1
Coefficients:
(Intercept)
   9.565673
```

```
Residual standard error: 5.288785
```

The estimates of model parameters μ and σ^2 are the sample mean and sample variance of y_{ij}

```
> mean(spati2$h)
[1] 9.565673
> sd(spati2$h)
[1] 5.288785
```

The model describes how the individual tree heights vary around the mean over all the data. The grouping to sample plots is ignored.

The inclusion of fixed plot effects leads to model (5.2):

```
> spati2$plot<-as.factor(spati2$plot)
> lm2<-gls(h~plot-1,data=spati2)
```

The total number of fixed coefficients in the model is 56 (the number of plots). The estimates for the first two plots are shown below. The estimates are equal to the plot-specific means of tree heights.

```
> coef(lm2)[1:2]
plot1    plot2
12.56000 14.40000
> mean(spati2$h[spati2$plot==1])
[1] 12.56
> mean(spati2$h[spati2$plot==2])
[1] 14.4
> mean(coef(lm2))
[1] 12.32486
```

The mean of regression coefficients is 12.3 m, which differs from the mean height estimated by model `lm1` because the data are not balanced. If the data were balanced, these two means would be equal. The residual standard error

```
> lm2$sigma
[1] 2.74074
```

quantifies the variation that is not explained by the plot means and can, therefore, be called *within-plot variation.* It is much lower than the residual standard error of model `lm1`; the residual of which accrues from the variability among plots and within plots.

Assuming that the plot effects are random, we fit the variance component model (5.3) using:

```
> lmm1<-lme(h~1,random=~1|plot,data=spati2)
> summary(lmm1)
Linear mixed-effects model fit by REML

Random effects:
 Formula: ~1 | plot
        (Intercept) Residual
StdDev:    4.943386 2.740718

Fixed effects: h ~ 1
               Value Std.Error   DF  t-value p-value
(Intercept) 12.30439 0.6664204 1622 18.46341       0

Number of Observations: 1678
Number of Groups: 56
```

The output shows that the estimated grand mean of group (plot) means is 12.30 m. This differs slightly from the mean of `lm2` coefficients due to different weighting in the computation of the mean due to the varying group size. The estimates of variance components are $\widehat{\sigma}_b^2 = 4.94^2$ and $\widehat{\sigma}^2 = 2.74^2$.

Even though the plot effects are not estimated as the parameters of the variance component model, they can be predicted afterwards using the estimated variance components and observed values of y_{ij} from the group in question. The script below shows these predictions: `ranef(lmm1)` gives the predicted random effects $\widetilde{b}_i$ and `coef(lmm1)` gives the predicted plot-level coefficients $\widehat{\mu} + \widetilde{b}_i$. These are shown for the first five plots below.

```
> ranef(lmm1)[1:5,]
[1]  0.25173869  2.05838925 -1.21629399 -0.90035267  0.01984519
> coef(lmm1)[1:5,]
[1] 12.55613 14.36278 11.08810 11.40404 12.32424
```

Comparing these values to the previously computed fixed plot effects of model lm2 shows only small differences.

The parameters of the two fixed-effects models can be compared to the parameters of the mixed-effects model:

- The population mean $\widehat{\mu} = 12.30$ of lmm1 is comparable to $\bar{\beta} = 12.32$ of lm2. They both quantify the mean of group means, which can also be understood as the mean of an average group. These means would also be comparable to the mean of observations $\widehat{\mu} = 9.57$ of lm1 if the date were balanced.
- The residual variance $\widehat{\sigma}^2 = 2.74^2$ of lmm1 is equal to $\widehat{\sigma}^2 = 2.74^2$ of lm2. They both quantify the within-group variance.
- The *total variability* $\widehat{\sigma}_b^2 + \widehat{\sigma}^2 = 4.94^2 + 2.74^2 = 5.65^2$ of lmm1 approximately compares to $\widehat{\sigma}^2 = 5.29^2$ of lm1. However, the estimated error variance of the fixed-effects model without group effects is always lower than the estimated total variability in a mixed-effects model. The difference gets larger as the number of groups decreases. In the very extreme case of $M = 1$, the two fixed-effects models lm1 and lm2 are equivalent. In that case, the residual variance still quantifies the within-group variance but evidently not the total variance, as there is no information about the variability between groups in the data. Therefore, the simple fixed-effects model lm1 is not useful for serious estimation of total variability in grouped data.
- The predictions $\widehat{\mu} + \widetilde{b}_i$ of lmm1 are comparable to $\widehat{\beta}_i$ of lm2. For example, $\widehat{\mu} + \widetilde{b}_1 = 12.556$ compares to $\widehat{\beta}_1 = 12.560$ but is slightly closer to the population mean 12.30 than the latter. Explanation for this shrinkage towards population mean is given later in Section 5.3.

5.1.2 Mixed-effects model with a random intercept

The variance component model can be generalized to a situation where μ in Equation (5.3) is replaced by a linear function of predictors. The resulting linear mixed-effects model with random intercept and single level of grouping is defined as

$$\begin{aligned} y_{ij} &= \beta_1 + \beta_2 x_{ij}^{(2)} + \ldots + \beta_p x_{ij}^{(p)} + b_i + \epsilon_{ij} \\ &= \boldsymbol{x}_{ij}' \boldsymbol{\beta} + b_i + \epsilon_{ij} \, . \end{aligned} \tag{5.4}$$

Part $\boldsymbol{x}_{ij}'\boldsymbol{\beta}$ is the *fixed part* of the model and $b_i + \epsilon_{ij}$ is the *random part.* The fixed part specifies the dependence of y on the predictors in an average group of the data. However, it may also have other interpretations, which will be discussed later in Section 6.4.

The assumptions about the random effects are similar to those made for the variance component model:

$$b_i \sim N(0, \sigma_b^2)$$

and

$$\epsilon_{ij} \sim N(0, \sigma^2) \, .$$

The random effects b_i are independent among groups, and the residual errors ϵ_{ij} are independent among observations. Also, b_i are independent of ϵ_{ij} for all i, j. Notice that the assumption in regards to the zero mean for random effects and residuals implicitly means that the random effects $\boldsymbol{b}_i$ and the residual errors ϵ_{ij} do not depend on $\boldsymbol{x}_i$; this will be discussed more in Section 6.3.

Reorganizing the terms of Equation (5.4) as follows:

$$y_{ij} = (\beta_1 + b_i) + \beta_2 x_{ij}^{(2)} + \ldots + \beta_p x_{ij}^{(p)} + \epsilon_{ij} \, .$$

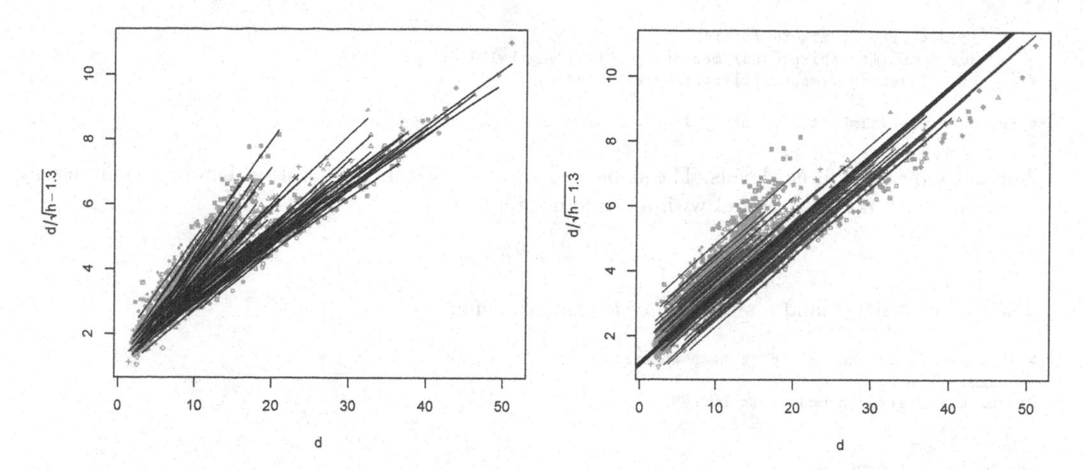

FIGURE 5.1 Plot of $y_{ij} = \frac{d_{ij}}{\sqrt{h_{ij}-1.3}}$ against d_{ij} in data set `spati`, overlaid with fitted plot-specific regression lines (left) and plot-specific fits using a mixed-effects model with a random intercept (right). The right-hand graph also shows the fitted value based on the fixed part only using (thick line).

emphasizes that the model assumes a varying intercept between groups, whereas the other coefficients are common to all groups of the data. The group-specific intercepts vary around the intercept of an average (or typical) group with common variance σ_b^2. Individual observations vary around the group-specific hyper plane with common variance σ^2. The model is suitable for data sets where the group-specific hyper planes differ from each other only through a level shift. If there is only one predictor in the data, then the group-specific y-x relationships are parallel lines when plotted on the predictors, as illustrated in the following example.

Example 5.2. The dependence of tree height on diameter is often satisfactorily modeled using the Näslunds function

$$h = 1.3 + \frac{d^2}{(\beta_1 + \beta_2 d)^2}$$

where h is tree height, d is tree diameter, 1.3 is the level of the diameter measurement (breast height, m), and $\beta^{(1)}$ and $\beta^{(2)}$ are model parameters. Reorganizing the terms shows (recall the discussion of Section 4.2.2, see also Section 10.2.2) that the assumed relationship can be written in a linear form

$$y = \beta_1 + \beta_2 d$$

by defining $y = \frac{d}{\sqrt{h-1.3}}$. We consider this model for the association between tree height and diameter in data `spati2`.

We start by computing the applied transformation of height to the data. Figure 5.1 (left) shows the observations and plot-specific linear regressions of y_{ij} against d_{ij}. Such fits can be carried out in our example data set because we have approximately 30 trees per plot on average, but it would not be possible if the number of observations per groups were small, as is often the case.

```
> spati2$y<-d/sqrt(spati2$h-1.3)
>
> plot(spati2$d,spati2$y,pch=as.numeric(spati2$plot)%%25,cex=0.5,
+      xlab="d",ylab=expression(d/sqrt(h-1.3)),col=gray(0.7))
>
> plots<-unique(spati2$plot)
> for (i in 1:length(plots)) {
+     thisplot<-spati2[spati2$plot==plots[i],]
```

```
+       model<-lm(y~d,data=thisplot)
+       dvec<-seq(min(thisplot$d),max(thisplot$d),length=10)
+       lines(dvec,coef(model)[1]+coef(model)[2]*dvec)
+       }
# try also nlme::lmList and lmfor::linesplot for the same purposes.
```

The lines are at different levels. The slope also varies between plots, but we ignore it temporarily and start with a mixed model with a random intercept

$$y_{ij} = \beta^{(1)} + \beta^{(2)} d_{ij} + b_i + \epsilon_{ij} \,.$$

The model is fitted and a summary of it printed using:

```
> lmm2<-lme(y~d,random=~1|plot,data=spati2)
> summary(lmm2)
Linear mixed-effects model fit by REML

Random effects:
 Formula: ~1 | plot
        (Intercept)  Residual
StdDev:   0.6143678 0.3763636

Fixed effects: y ~ d
                Value  Std.Error   DF   t-value p-value
(Intercept) 1.1455777 0.08667368 1621  13.21713       0
d           0.2132666 0.00172630 1621 123.54000       0
```

Assuming that the model fits (which is not true here because we assumed common slopes for all plots), we would interpret the estimates as follows. The association between y and d on an average (or typical) sample plot is

$$y_{ij} = 1.146 + 0.213 d_{ij}$$

However, the intercept of the model varies around the value of 1.146 with a standard deviation of 0.614, implying that the level of the line varies among plots. If the intercepts are normally distributed, 95% of the plot-specific intercepts in the population of plots where the data were sampled should fall within the range $1.146 \pm 1.96 \times 0.614 = (-0.057, 2.349)$. The standard deviation of the random intercept can be compared to the residual standard error, because they are both given in the unit of the y-variable. The residual standard error of the model is 0.376. The total unexplained variability is $0.614^2 + 0.376^2 = 0.720^2$, of which 73% is between plots, and 27% within plots.

The plot-specific fits can also be produced, after predicting the random effects b_i of the model, for the plots of the modeling data (Figure 5.1, right).

```
> linesplot(spati2$d,predict(lmm2),spati2$plot,add=TRUE,cex=0,col.lin=gray(0.2))    # plot-specific
> abline(fixef(lmm2),lwd=3)                                                         # fixed part
```

The estimates of coefficients $\boldsymbol{\beta}$ of the model can be extracted using `fixef`. The plot-specific predictions of coefficients can be extracted using `coef`, as shown below for plots 1, 2 and 3:

```
> coef(lmm2)[1:3,]
  (Intercept)         d
1    1.106268 0.2132666
2    1.015863 0.2132666
3    1.102746 0.2132666
```

The estimates of intercepts $\beta_1 + b_i$ vary among groups, whereas the coefficients of diameter (β_2) are the same for all groups (because we assumed so), even though the data does not support this (Figure 5.1, left).

The mixed-effects model can be seen as a special case of the fixed-effects model (4.2) (p. 68), where the residual error e is partitioned into two parts: the group effect b and the observation-level residual error ϵ. The random group effect models the correlation among observations of the same group. It can be seen by computing the correlation of two observations j and j' of the same group i:

$$\begin{aligned} \text{cov}(y_{ij}, y_{ij'}) &= \text{cov}(\boldsymbol{x}'_{ij}\boldsymbol{\beta} + b_i + \epsilon_{ij}, \boldsymbol{x}'_{ij'}\boldsymbol{\beta} + b_i + \epsilon_{ij'}) \\ &= \text{cov}(b_i + \epsilon_{ij}, b_i + \epsilon_{ij'}) \\ &= \text{cov}(b_i, b_i) + \text{cov}(b_i, \epsilon_{ij'}) + \text{cov}(b_i, \epsilon_{ij}) + \text{cov}(\epsilon_{ij}, \epsilon_{ij'}) \\ &= \text{var}(b_i) = \sigma_b^2 \end{aligned} \tag{5.5}$$

where the second row results from the first row because $\boldsymbol{x}'_{ij}\boldsymbol{\beta}$ is fixed. The last three terms on the third row are zero due to the model assumptions. Interestingly, parameter σ_b^2 has a double interpretation: on one hand, it is the variance of the group-specific intercepts and on the other hand, it is the covariance between any two observations from the same group.

The variance of observation y_{ij} is

$$\begin{aligned} \text{var}(y_{ij}) &= \text{var}(\boldsymbol{x}'_{ij}\boldsymbol{\beta} + b_i + \epsilon_{ij}) \\ &= \sigma_b^2 + \sigma^2 . \end{aligned}$$

The second row results from the first because $\boldsymbol{x}'_{ij}\boldsymbol{\beta}$ is fixed and $\text{cov}(b_i, \epsilon_{ij}) = 0$. Therefore, the correlation between two observations of the same group becomes

$$\text{cor}(y_{ij}, y_{ij'}) = \frac{\text{cov}(y_{ij}, y_{ij'})}{\text{sd}(y_{ij})\,\text{sd}(y_{ij'})} = \frac{\sigma_b^2}{\sigma_b^2 + \sigma^2} .$$

because $\text{sd}(y_{ij}) = \text{sd}(y_{ij'}) = \sigma_b^2 + \sigma^2$.

The above-specified interpretation of the variance between groups as intra-class covariance leads to a compound symmetric variance-covariance structure for the data (recall Section 4.3). More specifically, the mixed model with random intercept specifies such a compound symmetric variance-covariance structure, where the covariance between the observations of the same group is restricted to be positive. Situations where negative within-group covariances are justified are discussed in Box 5.2 (p. 153).

5.1.3 Multiple random effects

There is no reason to restrict the between-group variation to the intercept only. Associating random effects with other coefficients than the intercept leads to the following, a more general linear mixed-effects model, which is sometimes called the *random coefficient model*:

$$\begin{aligned} y_{ij} &= \beta_1 + \beta_2 x_{ij}^{(2)} + \ldots + \beta_p x_{ij}^{(p)} + b_i^{(1)} + b_i^{(2)} z_{ij}^{(2)} + \ldots + b_i^{(s)} z_{ij}^{(s)} + \epsilon_{ij} \\ &= \boldsymbol{x}'_{ij}\boldsymbol{\beta} + \boldsymbol{z}'_{ij}\boldsymbol{b}_i + \epsilon_{ij} , \end{aligned} \tag{5.6}$$

where $\boldsymbol{x}'_{ij}\boldsymbol{\beta}$ is the fixed part, and $\boldsymbol{z}'_{ij}\boldsymbol{b}_i + \epsilon_{ij}$ is the random part. The coefficients of the random part, $b_i^{(1)}, b_i^{(2)}, \ldots, b_i^{(s)}$, are called *random effects* or *random coefficients*. However, the predictors in $\boldsymbol{z}'_{ij}$ are not random predictors; random predictors refer to the situation where the predictors of a model are considered random variables (see Section 10.1).

We assume that

$$\boldsymbol{b}_i \sim N(\boldsymbol{0}, \boldsymbol{D}_i) ,$$

and $\boldsymbol{b}_i$ are independent among groups. Matrix $\boldsymbol{D}_i$ is a positive semidefinite $s \times s$ variance-covariance matrix of random effects. The random effects are assumed to be identically distributed; therefore, $\boldsymbol{D}_i$ is identical for all groups. This is emphasized, hereafter, by replacing index i with $*$. We assume that a vector of random effects has been independently drawn from the above-specified multivariate normal distribution for each group of the data. The vector of random effects is also independent of the residual error ϵ_{ij}. The other assumptions, in regard to the systematic part and the residual error ϵ_{ij}, are as they were before.

The parameters specifying the random part of the model include σ^2 and the elements of matrix $\boldsymbol{D}_*$. An essential difference to the previous formulation is that $\boldsymbol{D}_*$ also includes covariances. For example, if the length of $\boldsymbol{b}_i$ is 2, then $\boldsymbol{D}_*$ includes three parameters: $\text{var}(b_i^{(1)})$, $\text{var}(b_i^{(2)})$ and $\text{cov}(b_i^{(1)}, b_i^{(2)})$.

It may be assumed that the predictors in the fixed and random parts are the same. In that case, $\boldsymbol{x}_{ij} = \boldsymbol{z}_{ij}$, and model (5.6) can be written as

$$y_{ij} = \beta_1 + b_i^{(1)} + (\beta_2 + b_i^{(2)})x_{ij}^{(2)} + \ldots + (\beta_p + b_i^{(p)})x_{ij}^{(p)} + \epsilon_{ij}\,, \tag{5.7}$$

which emphasizes that we assume a random effect for each fixed coefficient of the linear model. Model (5.7) can be also specified as

$$y_{ij} = \boldsymbol{x}_{ij}'\boldsymbol{c}_i + \epsilon_{ij}\,, \tag{5.8}$$

where

$$\boldsymbol{c}_i \sim N(\boldsymbol{\beta}, \boldsymbol{D}_*)\,.$$

This formulation shows that the regression coefficients in the model can be seen as a random vector specified at the group level, with an unknown expected value $\boldsymbol{\beta}$ and an unknown variance-covariance matrix $\boldsymbol{D}_*$, which is common to all groups. A simple parameter estimation method for this specific model is discussed in Box 5.3 (p. 154).

The number of parameters in $\boldsymbol{D}_*$ is $s + \frac{1}{2}s(s-1)$, which increases rapidly as $\boldsymbol{b}_i$ gets longer. If the length of $\boldsymbol{b}_i$ is 3, then $\boldsymbol{D}_*$ includes six parameters (three variances and three covariances), and if the length is 4, the number of parameters is 10. The number of parameters can be reduced by implementing restrictions to matrix $\boldsymbol{D}_*$, for example, by assuming that the covariances between some random effects are zero, or that $\boldsymbol{D}_*$ is diagonal in the most extreme case. These are not reasonable default starting assumptions, but in some cases may be justified, e.g. to obtain convergence when fitting a nonlinear mixed-effects models (Pinheiro and Bates 2000, Section 8.2). If the predictor variables have been centered, it is often justified to assume that the correlations between the random intercept and random coefficients are zero.

There is seldom a reason to assume that the expected value of a certain parameter $\boldsymbol{c}_i$ is zero. Therefore, a fixed effect corresponding to every random effect should usually be kept in the model. However, it may not be justified or technically possible to assume that all parameters vary among groups. For these reasons, the elements of $\boldsymbol{z}_{ij}$ are also elements of $\boldsymbol{x}_{ij}$, but all elements of $\boldsymbol{x}_{ij}$ are not necessarily included in $\boldsymbol{z}_{ij}$. The mixed-effects model specifies the expected value of y_{ij}, as well as the residual errors and coefficients at different levels of grouping. The coefficients of the fixed part are called *population-level coefficients*. They can also be seen as group-level coefficients that result from the replacement of the random effects with their expected value of zero. Therefore, they are the coefficients for an average group. The *population-level (marginal)* mean of y_{ij} is

$$\text{E}(y_{ij}) = \boldsymbol{x}_{ij}'\boldsymbol{\beta}$$

It gives the mean for an average group of the data, i.e. for a group where $\boldsymbol{b}_i = \text{E}(\boldsymbol{b}_i)$. It can also be interpreted as the marginal mean over the random effects. The corresponding population-level residual errors are

$$\begin{aligned} e_{ij} &= y_{ij} - \text{E}(y_{ij}) \\ &= \boldsymbol{z}_{ij}'\boldsymbol{b}_i + \epsilon_{ij} \end{aligned}$$

The *group-level coefficients* are the coefficients of model (5.8) and they also include the random effects. The resulting *group-level (conditional) mean* of y_{ij} is

$$\mathrm{E}(y_{ij}|\boldsymbol{b}_i) = \boldsymbol{x}'_{ij}\boldsymbol{\beta} + \boldsymbol{z}'_{ij}\boldsymbol{b}_i$$

The corresponding residual errors are $\epsilon_{ij} = y_{ij} - \mathrm{E}(y_{ij}|\boldsymbol{b}_i)$. *The population-level (population-averaged)* and *group-level predictions* are obtained by replacing the parameters and random effects in the expressions by their estimates and predictions; detailed discussion of these predictions follows in Sections 5.4.2 and 6.4. To summarize, depending on the level of coefficients, predictions and residuals, the random effects can be considered to belong either to the residual errors or to the systematic part of the model. If not explicitly stated, the residuals and predictions from the mixed-effects models usually refer to those based on group-level coefficients.

In R, the population-level coefficients in the models fitted using package `nlme` are given by function `fixef` and the group-level coefficients by `coef`. The predictions (fitted values) of the model are computed using function `predict` and `fitted`. Argument `level=0` provides population-level predictions and `level=1` provides the group-level predictions. Corresponding residuals result from using a similar `level` argument in function `residuals`. The default level in functions `predict` and `resid` is `level=1`, thus providing group-level predictions and residuals.

If the model has two fixed parameters and a random effect is assigned to both, we get a model with random intercept and slope.

$$y_{ij} = (\beta_1 + b_i^{(1)}) + (\beta_2 + b_i^{(2)})x_{ij} + \epsilon_{ij}\,, \tag{5.9}$$

where $\boldsymbol{b}_i = \left(b_i^{(1)}, b_i^{(2)}\right)' \sim N\left(\boldsymbol{0}, \boldsymbol{D}_*\right)$ with

$$\boldsymbol{D}_* = \begin{pmatrix} \mathrm{var}\left(b_i^{(1)}\right) & \mathrm{cov}\left(b_i^{(1)}, b_i^{(2)}\right) \\ \mathrm{cov}\left(b_i^{(1)}, b_i^{(2)}\right) & \mathrm{var}\left(b_i^{(2)}\right) \end{pmatrix}. \tag{5.10}$$

Example 5.3. The left-hand graph of Figure 5.1 suggests that the slope also varies among plots. Therefore, model (5.9) (with $x_{ij} = d_{ij}$) was fitted to data set `spati2`:

```
> lmm3<-lme(y~d,random=~d|plot,data=spati2)
> summary(lmm3)
Linear mixed-effects model fit by REML

Random effects:
 Formula: ~d | plot
 Structure: General positive-definite, Log-Cholesky parametrization
            StdDev     Corr
(Intercept) 0.28851536 (Intr)
d           0.05973479 -0.405
Residual    0.24476916

Fixed effects: y ~ d
                Value  Std.Error   DF  t-value p-value
(Intercept) 1.0978967 0.04365712 1621 25.14817       0
d           0.2332299 0.00811378 1621 28.74490       0
```

The mean of group-specific intercepts and slopes is given by the fixed effects $\widehat{\beta}^{(1)} = 1.098$ and $\widehat{\beta}^{(2)} = 0.233$. These values represent the $y - d$ relationship on a typical sample plot, which is estimated to be $1.098 + 0.233d$. The estimate of matrix $\boldsymbol{D}_*$ was

$$\widehat{\boldsymbol{D}}_* = \begin{pmatrix} 0.289^2 & -0.405 \times 0.289 \times 0.0597 \\ -0.405 \times 0.289 \times 0.0597 & 0.0597^2 \end{pmatrix}$$
$$= \begin{pmatrix} 0.0832 & -0.00699 \\ -0.00699 & 0.00357 \end{pmatrix}.$$

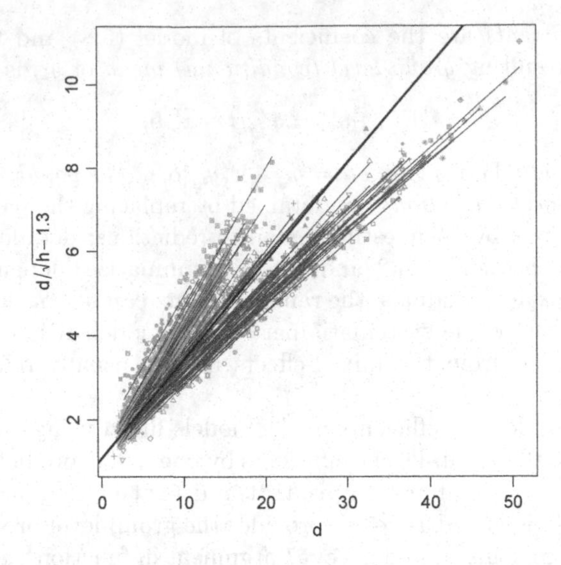

FIGURE 5.2 The group-level predictions (thin lines) and the prediction based on the fixed part (thick line) of model `lmm3` in data set `spati2`.

These parameters show that the plot-specific slopes vary around the slope of the typical plot, 0.233, with a standard deviation of 0.0597. Therefore, approximately 95% of the plots in the underlying population should have a slope within the interval $(0.116, 0.350)$ (mean $\pm$ 1.96×sd). The intercept varies around the mean of 1.098 with a standard deviation of 0.288, which yields a corresponding interval for intercepts in the population of plots as $(0.532, 1.663)$. There is a moderately strong negative correlation (-0.405) between $b_i^{(1)}$ and $b_i^{(2)}$. This means that a plot with a steep slope usually has a low intercept, and a plot with a gentle slope has a high intercept. This correlation is partially an artifact and could be substantially diminished if the centralized predictors $d_{ij} - \bar{d}$ were used. This would lead to an equally well-fitting model, although the interpretation of the intercept and related random effect would be changed. By permitting a random slope, the estimated residual standard error decreased considerably, from 0.376 of model `lmm2` (see Example 5.2) to 0.245.

The group-specific intercepts and slopes both include the group-specific random effect as seen below:

```
> coef(lmm3)[1:3,]
  (Intercept)         d
1   1.5573251 0.1820843
2   0.9252729 0.2186277
3   1.1501239 0.2093376
```

The plot-specific relationships for all plots, and for a typical plot, are shown in Figure 5.2. The model fits much better to the data than the previous model with random intercept (see the right panel of Figure 5.1, p. 137).

5.1.4 Random effects for categorical predictors

Model (5.6) can also be applied to situations where the variables in z'_{ij} are binary. For example, they may be indicator variables that recode a multilevel categorical predictor with s levels to s binary variables. If we include the random intercept and random coefficients for all these binary variables, the model is overparameterized. The situation is similar to the case with fixed categorical predictor, which was discussed earlier in Section 4.2. One possible approach to solve this problem is to drop the indicator variable of one level (the default level) from the model. Let $z_{ij}^{(1)}, \ldots, z_{ij}^{(s)}$ be the indicator variables for levels $1, \ldots, s$ for observation j of group i. If $z_{ij}^{(1)}$ is dropped, an identifiable model is:

$$y_{ij} = \boldsymbol{x}_{ij}\boldsymbol{\beta} + b_i^{(1)} + b_i^{(2)} z_{ij}^{(2)} + \ldots + b_i^{(s)} z_{ij}^{(s)} + \epsilon_{ij} . \tag{5.11}$$

For an observation at level 1, $z_{ij}^{(k)} = 0$ for $k = 2, \ldots, s$. The random part becomes $b_i^{(1)} + \epsilon_{ij}$ and

$$\mathrm{var}\left(y_{ij}\right) = \mathrm{var}\left(b_i^{(1)} + \epsilon_{ij}\right) = \mathrm{var}\left(b_i^{(1)}\right) + \mathrm{var}\left(\epsilon_{ij}\right) .$$

For an observation at level 2, $z_{ij}^{(k)} = 1$ for $k = 2$ and $z_{ij}^{(k)} = 0$ otherwise, leading to the random part $b_i^{(1)} + b_i^{(2)} + \epsilon_{ij}$. Now

$$\begin{aligned}\mathrm{var}\left(y_{ij}\right) &= \mathrm{var}\left(b_i^{(1)} + b_i^{(2)} + \epsilon_{ij}\right) \\ &= \mathrm{var}\left(b_i^{(1)}\right) + \mathrm{var}\left(b_i^{(2)}\right) + 2\,\mathrm{cov}\left(b_i^{(1)}, b_i^{(2)}\right) + \mathrm{var}\left(\epsilon_{ij}\right) .\end{aligned}$$

The formula also looks similar to that of level 2 for levels $3, \ldots, s$. Therefore, the model assumes level-specific variances, which are obtained using the above formulas with the variances and covariances in $\boldsymbol{D}_*$. The above-specified variances can also be interpreted as the covariances between such observations j and j' from the same group i that have the same values of the categorical predictors $\boldsymbol{z}_i$. If two observations of the same group i belong to different levels of the categorical predictors, the covariance is also non-zero; finding it is left as an exercise.

An equivalent model with a more straightforward interpretation of the random parameters is a model that does not include the random intercept, but includes random effects for all levels of the categorical predictor:

$$y_{ij} = \boldsymbol{x}_{ij}\boldsymbol{\beta} + b_i^{(1)} z_{ij}^{(1)} + b_i^{(2)} z_{ij}^{(2)} + \ldots + b_i^{(s)} z_{ij}^{(s)} + \epsilon_{ij} . \tag{5.12}$$

For an observation at level k, $(k = 1, \ldots, s)$, the random part is $b_i^{(k)} + \epsilon_{ij}$ and

$$\mathrm{var}\left(y_{ij}\right) = \mathrm{var}\left(b_i^{(k)}\right) + \mathrm{var}\left(\epsilon_{ij}\right) .$$

Thus, the variances on the diagonal of $\boldsymbol{D}_*$ directly specify the variances for different groups. For two observations from the same group i in level k^{th} of the categorical predictor, $\mathrm{cov}\left(y_{ij}, y_{ij'}\right) = \mathrm{var}\left(b_i^{(k)}\right)$. For two observations from group i in levels k and l of the categorical predictor, $\mathrm{cov}\left(y_{ij}, y_{ij'}\right) = \mathrm{cov}\left(b_i^{(k)}, b_i^{(l)}\right)$. This covariance can be estimated only if different levels of the categorical predictor occur in the same groups. It cannot be estimated if all observations of a group belong in the same level. However, the assumption of a diagonal $\boldsymbol{D}_*$ leads to an identifiable model in such case.

Example 5.4. Data set `BrkRes` includes measurements of the breaking resistance of wood samples from a total of 118 silver birch (*Betula pendula*) trees. A total of 1–3 wood samples per tree were collected. The samples were classified to three ring classes according to the distance of the sample from the pith (class 1 is closest to the pith). Each sample was measured destructively for breaking resistance in the laboratory. The measured variables also included wood density. Figure 5.3 illustrates the variables of interest in the data set. Breaking resistance seems to increase from pith to bark, and with increasing wood density.

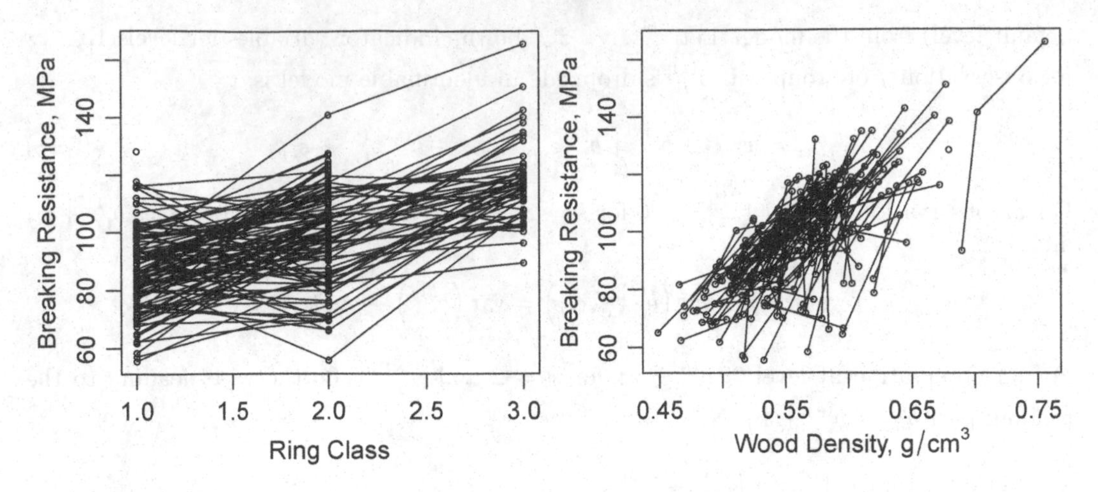

FIGURE 5.3 The breaking resistance as a function of ring class (left) and wood density (right) in data set `BrkRes` of Example 5.4. The lines connect observations from the same tree.

We analyze the association of wood density and ring class with the breaking resistance of the sample using a mixed-effects model with tree-level random effects. To allow different variance for each ring class, we use model (5.12) for the breaking resistance of sample j of tree i:

$$\begin{aligned} BR_{ij} =& \beta_1 + \beta_2 RC2_{ij} + \beta_3 RC3_{ij} + \beta_4 WD_{ij} \\ &+ b_i^{(1)} RC1_{ij} + b_i^{(2)} RC2_{ij} + b_i^{(3)} RC3_{ij} + \epsilon_{ij} \,, \end{aligned}$$

where $RC1_{ij}$, $RC2_{ij}$, and $RC3_{ij}$ are indicator variables for ring classes 1, 2 and 3, respectively from pith to bark, and WD_{ij} is the wood density. The assumptions in regard to the random effects and residual errors are as specified earlier for model (5.6).

```
> data(BrkRes)
> brmod1 <- lme(Resistance~RingClass+Density,
+               random=~RingClass-1|Tree,
+               data=BrkRes)

> summary(brmod1)
Random effects:
 Formula: ~RingClass - 1 | Tree
 Structure: General positive-definite, Log-Cholesky parametrization
           StdDev    Corr
RingClass1  8.633182 RngCl1 RngCl2
RingClass2 12.386222 0.085
RingClass3  5.098128 0.505  0.536
Residual    5.769907
```

The estimated variances of tree effects vary among the ring classes; 8.633^2 in ring class 1, 12.386^2 in ring class 2, and 5.098^2 in ring class 3. Random effects in class 3 are positively correlated with the random effects in classes 1 and 2, but the correlation between random effects in classes 1 and 2 is low. The residual variance is 5.770^2.

For illustration, we also fit model (5.11), where the random part includes an intercept and the effects for ring classes 2 and 3. The estimated variance of $\boldsymbol{z}_{ij}\boldsymbol{b}_i$ for ring class 1 is 8.633^2, and it is seen directly in the output of the model. For ring classes 2 and 3, the variances need to be computed using Equation (2.28), which gives 12.389^2 and 5.100^2, respectively. These values differ only slightly from the estimates based on model `brmod1`. Residual standard errors of the two models are also very similar. The fixed effects of these two models (not shown) are equal up to the 5th decimal. The small differences in the estimates result from rounding of the estimates, and from small differences in the numerical estimation methods when different parameterization of the same variance-covariance structure is used.

```
> brmod2 <- lme(Resistance~RingClass+Density,
+               random=~RingClass|Tree,
+               data=BrkRes)

> summary(brmod2)
Random effects:
 Formula: ~RingClass | Tree
 Structure: General positive-definite, Log-Cholesky parametrization
            StdDev    Corr
(Intercept)  8.633052 (Intr) RngCl2
RingClass2  14.482018 -0.523
RingClass3   7.485481 -0.809  0.711
Residual     5.770086

> sqrt(8.633052^2+14.482018^2-2*0.523*8.633052*14.482018)
[1] 12.38884
> sqrt(8.633052^2+7.485481^2-2*0.809*8.633052*7.485481)
[1] 5.099287
```

5.2 Matrix formulation of the model

5.2.1 Matrix formulation for single group

Model (5.6) for group i can be written in matrix form as

$$\boldsymbol{y}_i = \boldsymbol{X}_i\boldsymbol{\beta} + \boldsymbol{Z}_i\boldsymbol{b}_i + \boldsymbol{\epsilon}_i \tag{5.13}$$

by defining

$$\boldsymbol{y}_i = \begin{pmatrix} y_{i1} \\ y_{i2} \\ \vdots \\ y_{in_i} \end{pmatrix}, \boldsymbol{X}_i = \begin{pmatrix} 1 & x_{i1}^{(2)} & \dots & x_{i1}^{(p)} \\ 1 & x_{i2}^{(2)} & \dots & x_{i2}^{(p)} \\ \vdots & \vdots & \ddots & \vdots \\ 1 & x_{in_i}^{(2)} & \dots & x_{in_i}^{(p)} \end{pmatrix}, \boldsymbol{\beta} = \begin{pmatrix} \beta_1 \\ \beta_2 \\ \vdots \\ \beta_p \end{pmatrix},$$

$$\boldsymbol{Z}_i = \begin{pmatrix} 1 & z_{i1}^{(2)} & \dots & z_{i1}^{(s)} \\ 1 & z_{i2}^{(2)} & \dots & z_{i2}^{(s)} \\ \vdots & \vdots & \ddots & \vdots \\ 1 & z_{in_i}^{(2)} & \dots & z_{in_i}^{(s)} \end{pmatrix}, \boldsymbol{b}_i = \begin{pmatrix} b_i^{(1)} \\ b_i^{(2)} \\ \vdots \\ b_i^{(s)} \end{pmatrix}, \boldsymbol{\epsilon}_i = \begin{pmatrix} \epsilon_{i1} \\ \epsilon_{i2} \\ \vdots \\ \epsilon_{in_i} \end{pmatrix},$$

$$\boldsymbol{D}_* = \begin{pmatrix} \mathrm{var}(b_i^{(1)}) & \mathrm{cov}(b_i^{(1)}, b_i^{(2)}) & \dots & \mathrm{cov}(b_i^{(1)}, b_i^{(s)}) \\ \mathrm{cov}(b_i^{(1)}, b_i^{(2)}) & \mathrm{var}(b_i^{(2)}) & \dots & \mathrm{cov}(b_i^{(2)}, b_i^{(s)}) \\ \vdots & \vdots & \ddots & \vdots \\ \mathrm{cov}(b_i^{(1)}, b_i^{(s)}) & \mathrm{cov}(b_i^{(2)}, b_i^{(s)}) & \dots & \mathrm{var}(b_i^{(s)}) \end{pmatrix},$$

$$\mathrm{var}(\boldsymbol{\epsilon}_i) = \boldsymbol{R}_i = \sigma^2 \boldsymbol{I}_{n_i \times n_i}.$$

Furthermore,

$$\boldsymbol{b}_i \sim N(\boldsymbol{0}, \boldsymbol{D}_*) \quad \text{and} \quad \boldsymbol{\epsilon}_i \sim N(\boldsymbol{0}, \sigma^2 \boldsymbol{I}_{n_i \times n_i}).$$

If a random effect is associated with each of the fixed effects, matrices $\boldsymbol{X}_i$ and $\boldsymbol{Z}_i$ are identical. If random effects are associated only for a subset of the fixed effects, then the corresponding elements in $\boldsymbol{b}_i$, the corresponding columns in $\boldsymbol{Z}_i$, and the corresponding rows and columns in $\boldsymbol{D}_*$ are all retained.

Example 5.5. The model with a random intercept is specified by defining

$$Z_i = \mathbf{1}_{n_i \times 1},\ \boldsymbol{b}_i = \left(b_i^{(1)} \right),\ \boldsymbol{D}_* = \sigma_b^2 ,$$

where $\mathbf{1}_{n_i \times 1}$ is a column vector of ones.

Example 5.6. A model with a random intercept and a random coefficient of primary predictor $x^{(2)}$ is specified by defining

$$\boldsymbol{Z}_i = \begin{pmatrix} 1 & x_{i1}^{(2)} \\ \vdots & \vdots \\ 1 & x_{in_i}^{(2)} \end{pmatrix},\ \boldsymbol{b}_i = \begin{pmatrix} b_i^{(1)} \\ b_i^{(2)} \end{pmatrix},\ \boldsymbol{D}_* = \begin{pmatrix} \mathrm{var}(b_i^{(1)}) & \mathrm{cov}(b_i^{(1)}, b_i^{(2)}) \\ \mathrm{cov}(b_i^{(1)}, b_i^{(2)}) & \mathrm{var}(b_i^{(2)}) \end{pmatrix}.$$

The expected value and variance-covariance matrix of the data from a single group i are

$$\begin{aligned} \mathrm{E}(\boldsymbol{y}_i) &= \mathrm{E}(\boldsymbol{X}_i\boldsymbol{\beta} + \boldsymbol{Z}_i\boldsymbol{b}_i + \boldsymbol{\epsilon}_i) \\ &= \mathrm{E}(\boldsymbol{X}_i\boldsymbol{\beta}) + \mathrm{E}(\boldsymbol{Z}_i\boldsymbol{b}_i) + \mathrm{E}(\boldsymbol{\epsilon}_i) \\ &= \boldsymbol{X}_i\boldsymbol{\beta} \end{aligned} \tag{5.14}$$

$$\begin{aligned} \mathrm{var}(\boldsymbol{y}_i) &= \mathrm{var}(\boldsymbol{X}_i\boldsymbol{\beta} + \boldsymbol{Z}_i\boldsymbol{b}_i + \boldsymbol{\epsilon}_i) \\ &= \mathrm{var}(\boldsymbol{X}_i\boldsymbol{\beta}) + \mathrm{var}(\boldsymbol{Z}_i\boldsymbol{b}_i) + \mathrm{var}(\boldsymbol{\epsilon}_i) \\ &= \boldsymbol{Z}_i\mathrm{var}(\boldsymbol{b}_i)\boldsymbol{Z}_i' + \mathrm{var}(\boldsymbol{\epsilon}_i) \\ &= \boldsymbol{Z}_i\boldsymbol{D}_*\boldsymbol{Z}_i' + \boldsymbol{R}_i . \end{aligned}$$

The second-row follows from the first-row because there is zero correlation between the random effects and residual errors, and the first term on the second-row is zero because $\boldsymbol{X}_i\boldsymbol{\beta}$ is fixed. If normality of random effects and residual errors is assumed, then

$$\boldsymbol{y}_i \sim N(\boldsymbol{X}_i\boldsymbol{\beta}, \boldsymbol{Z}_i\boldsymbol{D}_*\boldsymbol{Z}_i' + \boldsymbol{R}_i).$$

The covariance between the random effects and observations of group i are

$$\begin{aligned} \mathrm{cov}(\boldsymbol{b}_i, \boldsymbol{y}_i') &= \mathrm{cov}(\boldsymbol{b}_i, (\boldsymbol{X}_i\boldsymbol{\beta} + \boldsymbol{Z}_i\boldsymbol{b}_i + \boldsymbol{\epsilon}_i)') \\ &= \mathrm{cov}(\boldsymbol{b}_i, (\boldsymbol{X}_i\boldsymbol{\beta})') + \mathrm{cov}(\boldsymbol{b}_i, (\boldsymbol{Z}_i\boldsymbol{b}_i)') + \mathrm{cov}(\boldsymbol{b}_i, \boldsymbol{\epsilon}_i') \\ &= \mathbf{0} + \mathrm{cov}(\boldsymbol{b}_i, \boldsymbol{b}_i')\boldsymbol{Z}_i' + \mathbf{0} \\ &= \boldsymbol{D}_*\boldsymbol{Z}_i' . \end{aligned}$$

Example 5.7. In the model with a random intercept, the matrices defined in Example 5.5 yield

$$\mathrm{var}(\boldsymbol{y}_i) = \sigma_b^2\mathbf{1}\mathbf{1}' + \sigma^2\boldsymbol{I} = \begin{pmatrix} \sigma_b^2 + \sigma^2 & \sigma_b^2 & \cdots & \sigma_b^2 \\ \sigma_b^2 & \sigma_b^2 + \sigma^2 & \cdots & \sigma_b^2 \\ \vdots & \vdots & \ddots & \vdots \\ \sigma_b^2 & \sigma_b^2 & \cdots & \sigma_b^2 + \sigma^2 \end{pmatrix}.$$

Therefore, the covariance between two observations from the same group is equal to the variance of random effects, as already shown in Equation (5.5) (p. 139).

Example 5.8. In the model with random a intercept and slope, the matrices were defined in Example 5.6. For simplicity, we denote $x_{ij}^{(2)} = x_j$, $\mathrm{var}(b_i^{(1)}) = \sigma_1^2$, $\mathrm{var}(b_i^{(2)}) = \sigma_2^2$ and $\mathrm{cov}(b_i^{(1)}, b_i^{(2)}) = \sigma_{12}$. The variance-covariance matrix for the data of group i becomes (show this!)

$$\begin{aligned} \mathrm{var}(\boldsymbol{y}_i) &= \boldsymbol{Z}_i\boldsymbol{D}_*\boldsymbol{Z}_i' + \boldsymbol{R}_i \\ &= \begin{pmatrix} \sigma_1^2 + 2\sigma_{12}x_1 + \sigma_2^2x_1^2 + \sigma^2 & \cdots & \sigma_1^2 + \sigma_{12}(x_1 + x_{n_i}) + \sigma_2^2x_1x_{n_i} \\ \vdots & \ddots & \vdots \\ \sigma_1^2 + \sigma_{12}(x_1 + x_{n_i}) + \sigma_2^2x_1x_{n_i} & \cdots & \sigma_1^2 + 2\sigma_{12}x_{n_i} + \sigma_2^2x_{n_i}^2 + \sigma^2 \end{pmatrix} \end{aligned}$$

5.2.2 Relaxing the assumptions on residual variance

The assumption $\boldsymbol{R}_i = \sigma^2 \boldsymbol{I}$, can be relaxed by allowing inconstant variance among the residual errors of group i or by allowing (e.g. temporal or spatial) autocorrelation among them. Furthermore, one might assume different variances within each group by including group-specific values of σ_i^2 in $\boldsymbol{R}_i = \sigma_i^2 \boldsymbol{I}$. Whatever general structure is assumed for error variance, it can be specified as $\boldsymbol{\epsilon}_i \sim N(\boldsymbol{0}, \boldsymbol{R}_i)$, where $\boldsymbol{R}_i$ parameterizes the relaxed assumptions on the residuals using a parsimonious set of parameters. The possible structures of $\boldsymbol{R}_i$ are similar to those discussed in Section 4.3. The diagnostic tools needed to detect whether a parametric model is required for the variance-covariance structure, and to analyze whether the assumed model fits to the data, are the same as discussed earlier in the context of linear models (see Sections 4.5.2 and 4.6.8).

Web Example 5.9. The structure of matrix $\text{var}(\boldsymbol{y}_i)$ for a model with the Power-type variance function and random intercept.

Example 5.10. The upper panel in Figure 5.4 shows the residuals of model `lmm3` against the fitted values (left) and against diameter (right). A quick look at the residual plot on the right-hand side of Figure 5.4 indicates decreasing variance as a function of tree diameter. Therefore, a new model using variance function

$$\text{var}(\epsilon_{ij}) = \sigma^2 |d_{ij}|^{2\delta} \tag{5.15}$$

is considered, where δ should be negative to model the decrease in the variance as a function of tree diameter d_{ij}. The model was fitted using:

```
> lmm4<-update(lmm3,weights=varPower(form=~d))
```

However, there is only a small difference between the standardized residual plot (Figure 5.4, lower left) and the raw residual plot of `lmm3`. The model output below also shows that the estimate of parameter δ was very small, which indicates a practically constant variance as a function of tree diameter. To see the complete picture, the lower right-hand graph in Figure 5.4 shows the residuals of model `lmm3` overlaid by a whiskers plot, where the vertical lines are proportional to the sample standard deviation of the residuals in 10 diameter classes. All the whiskers are approximately the same length, which indicates that the assumption of a constant error variance was well met. Therefore, the impression about heteroscedastic variance was a visual artifact, caused by a non-uniform, skewed distribution of tree diameters: most large residuals are associated with small trees because the majority of the trees in the data set are small.

```
> lmm4
Variance function:
 Structure: Power of variance covariate
 Formula: ~d
 Parameter estimates:
     power
0.04490039
```

5.2.3 Matrix formulation for all data

Consider a complete data set of M independent groups drawn from a population of groups, and assume that the data of each group follows model (5.13). The model for all data is defined as

$$\boldsymbol{y} = \boldsymbol{X\beta} + \boldsymbol{Zb} + \boldsymbol{\epsilon} \tag{5.16}$$

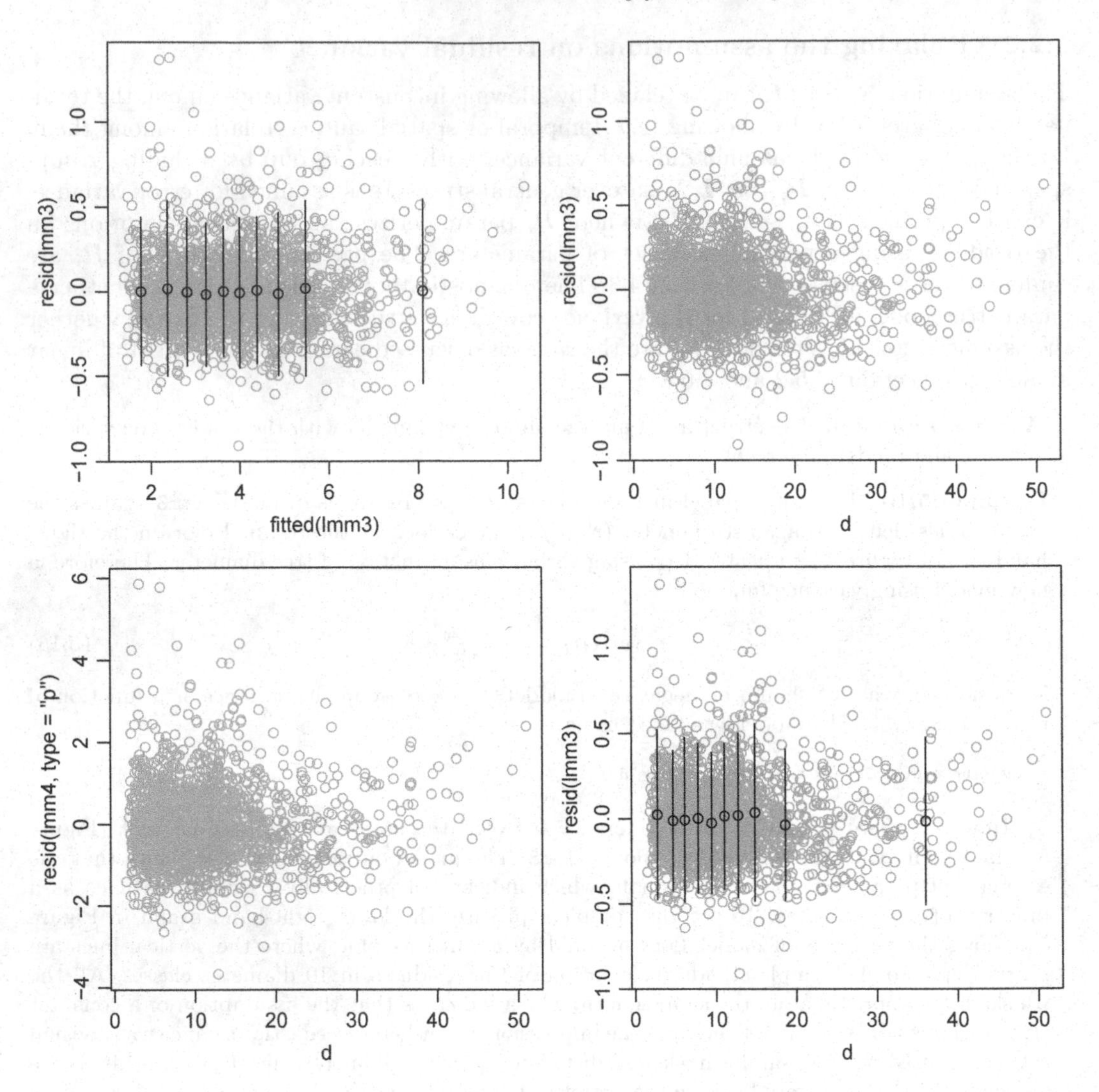

FIGURE 5.4 Residual plots of models `lmm3` and `lmm4`. See Example 5.10 for explanation.

by defining

$$\boldsymbol{y} = \begin{pmatrix} \boldsymbol{y}_1 \\ \boldsymbol{y}_2 \\ \vdots \\ \boldsymbol{y}_M \end{pmatrix}, \boldsymbol{X} = \begin{pmatrix} \boldsymbol{X}_1 \\ \boldsymbol{X}_2 \\ \vdots \\ \boldsymbol{X}_M \end{pmatrix}, \boldsymbol{Z} = \begin{pmatrix} \boldsymbol{Z}_1 & \boldsymbol{0} & \cdots & \boldsymbol{0} \\ \boldsymbol{0} & \boldsymbol{Z}_2 & \cdots & \boldsymbol{0} \\ \vdots & \vdots & \ddots & \vdots \\ \boldsymbol{0} & \boldsymbol{0} & \cdots & \boldsymbol{Z}_M \end{pmatrix},$$

$$\boldsymbol{b} = \begin{pmatrix} \boldsymbol{b}_1 & \boldsymbol{b}_2 & \ldots & \boldsymbol{b}_M \end{pmatrix}', \boldsymbol{\epsilon} = \begin{pmatrix} \boldsymbol{\epsilon}_1 & \boldsymbol{\epsilon}_2 & \ldots & \boldsymbol{\epsilon}_M \end{pmatrix}'.$$

The vector of fixed parameters, $\boldsymbol{\beta}$, is the same as before. Therefore, to formulate a mixed-effects model for all the data, it is essential to formulate the models for each group in the matrix form and use them to define the above-specified matrices for the entire data set.

Group	y	x	Group	y	x	Group	y	x
1	2	10	2	2	11	2	5	15
1	4	13	2	4	14	3	4	12

TABLE 5.2 A small grouped data.

In model (5.16), we have

$$\text{var}(\boldsymbol{b}) = \text{var}\begin{pmatrix} \boldsymbol{b}_1 \\ \boldsymbol{b}_2 \\ \vdots \\ \boldsymbol{b}_M \end{pmatrix} = \begin{pmatrix} \boldsymbol{D}_1 & \boldsymbol{0} & \dots & \boldsymbol{0} \\ \boldsymbol{0} & \boldsymbol{D}_2 & \dots & \boldsymbol{0} \\ \vdots & \vdots & \ddots & \vdots \\ \boldsymbol{0} & \boldsymbol{0} & \dots & \boldsymbol{D}_M \end{pmatrix}.$$

Because $\text{var}(\boldsymbol{b}_i) = \boldsymbol{D}_*$ for all i, this simplifies to

$$\boldsymbol{D} = \boldsymbol{I}_{M\times M} \otimes \boldsymbol{D}_*,$$

where $\otimes$ is the Kronecker product of matrices. The error variance-covariance matrix is

$$\text{var}(\boldsymbol{\epsilon}) = \text{var}\begin{pmatrix} \boldsymbol{\epsilon}_1 \\ \boldsymbol{\epsilon}_2 \\ \vdots \\ \boldsymbol{\epsilon}_M \end{pmatrix} = \begin{pmatrix} \boldsymbol{R}_1 & \boldsymbol{0} & \dots & \boldsymbol{0} \\ \boldsymbol{0} & \boldsymbol{R}_2 & \dots & \boldsymbol{0} \\ \vdots & \vdots & \ddots & \vdots \\ \boldsymbol{0} & \boldsymbol{0} & \dots & \boldsymbol{R}_M \end{pmatrix}.$$

Here, the blocks are of dimension $n_i \times n_i$ and, as such, are not identical. However, the parameters that specify them (σ^2 and possible additional parameters for residual variance-covariance structure) are usually common to all groups.

Example 5.11. Consider the model

$$y_{ij} = \beta_1 + \beta_2 x_{ij} + b_i + \epsilon_{ij},$$

where $b_i \sim N(0, \sigma_b^2)$ and $\epsilon_{ij} \sim N(0, \sigma^2)$. The model can be written in the form of Equation (5.16) for the data set shown in Table 5.2 by defining

$$\boldsymbol{y} = \begin{pmatrix} 2 \\ 4 \\ 2 \\ 4 \\ 5 \\ 4 \end{pmatrix}, \boldsymbol{X} = \begin{pmatrix} 1 & 10 \\ 1 & 13 \\ 1 & 11 \\ 1 & 14 \\ 1 & 15 \\ 1 & 12 \end{pmatrix}, \boldsymbol{Z} = \begin{pmatrix} 1 & 0 & 0 \\ 1 & 0 & 0 \\ 0 & 1 & 0 \\ 0 & 1 & 0 \\ 0 & 1 & 0 \\ 0 & 0 & 1 \end{pmatrix}, \boldsymbol{b} = \begin{pmatrix} b_1 \\ b_2 \\ b_3 \end{pmatrix}, \boldsymbol{\beta} = \begin{pmatrix} \beta_1 \\ \beta_2 \end{pmatrix},$$

$$\text{var}(\boldsymbol{b}) = \sigma_b^2 \boldsymbol{I}_{3\times 3}, \text{var}(\boldsymbol{\epsilon}) = \sigma^2 \boldsymbol{I}_{6\times 6}.$$

Web Example 5.12. Specification of the model with a random intercept and slope for the data set shown in Table 5.2.

Model (5.16) can be written equivalently as a *marginal* or *population-level* model

$$\boldsymbol{y} = \boldsymbol{X\beta} + \boldsymbol{e} \tag{5.17}$$

by defining $\boldsymbol{e} = \boldsymbol{Zb} + \boldsymbol{\epsilon}$. The variance-covariance matrix $\boldsymbol{W}$ of the residual error $\boldsymbol{e}$ equals that of $\boldsymbol{y}$, and is

$$\begin{aligned} \boldsymbol{W} &= \text{var}(\boldsymbol{e}) \\ &= \text{var}(\boldsymbol{Zb} + \boldsymbol{\epsilon}) \\ &= \boldsymbol{ZDZ}' + \boldsymbol{R}. \end{aligned}$$

Therefore, the linear mixed-effects model is a special case of the linear model (4.6) (p. 71), where $\text{var}(\boldsymbol{e}) = \boldsymbol{W} = \boldsymbol{Z}\boldsymbol{D}\boldsymbol{Z}' + \boldsymbol{R}$, and $\boldsymbol{Z}$, $\boldsymbol{D}$, and $\boldsymbol{R}$ are as defined in Equation (5.16). The specified structure of $\boldsymbol{W}$ properly takes the within-group correlation of the data into account. For use in the next subsection and for compatibility with Chapter 4, we define $\boldsymbol{W} = \sigma^2\boldsymbol{V}$, where $\boldsymbol{V}$ defines the variance-covariance structure of the data without the scaling factor σ^2 (see Section 5.3.2). If $\boldsymbol{R} = \sigma^2\boldsymbol{I}$, then the scaling factor σ^2 is the variance of group-level errors ϵ_{ij}.

Example 5.13. In Example 5.11, the model with a random intercept can be written for the entire data set as

$$\boldsymbol{y} = \boldsymbol{X}\boldsymbol{\beta} + \boldsymbol{e}$$

by defining

$$\boldsymbol{e} = \begin{pmatrix} Z_1 b_1 + \epsilon_1 \\ Z_2 b_2 + \epsilon_2 \\ Z_3 b_3 + \epsilon_3 \end{pmatrix} = \begin{pmatrix} b_1 + \epsilon_{11} \\ b_1 + \epsilon_{12} \\ b_2 + \epsilon_{21} \\ b_2 + \epsilon_{22} \\ b_2 + \epsilon_{23} \\ b_3 + \epsilon_{31} \end{pmatrix} \quad \text{and}$$

$$\begin{aligned}
\boldsymbol{W} &= \begin{pmatrix} \sigma_b^2 + \sigma^2 & \sigma_b^2 & 0 & 0 & 0 & 0 \\ \sigma_b^2 & \sigma_b^2 + \sigma^2 & 0 & 0 & 0 & 0 \\ 0 & 0 & \sigma_b^2 + \sigma^2 & \sigma_b^2 & \sigma_b^2 & 0 \\ 0 & 0 & \sigma_b^2 & \sigma_b^2 + \sigma^2 & \sigma_b^2 & 0 \\ 0 & 0 & \sigma_b^2 & \sigma_b^2 & \sigma_b^2 + \sigma^2 & 0 \\ 0 & 0 & 0 & 0 & 0 & \sigma_b^2 + \sigma^2 \end{pmatrix} \\
&= \sigma^2 \begin{pmatrix} \sigma_b^2/\sigma^2 + 1 & \sigma_b^2/\sigma^2 & \dots & 0 \\ \sigma_b^2/\sigma^2 & \sigma_b^2/\sigma^2 + 1 & \dots & 0 \\ \vdots & \vdots & \ddots & \vdots \\ 0 & 0 & \dots & \sigma_b^2/\sigma^2 + 1 \end{pmatrix} .
\end{aligned}$$

The compound symmetric structure of $\boldsymbol{W}$ is determined by the grouping of the data set and is parameterized by the variance between groups σ_b^2 and the variance within groups σ^2.

5.3 Estimation of model parameters

This section covers two topics related to the estimation of a mixed-effects model:

1. Estimation of the fixed effects $\boldsymbol{\beta}$.
2. Estimation of the variance-covariance parameters, including the scaling factor σ^2, the variances and covariances of random effects in matrix $\boldsymbol{D}$, and possible additional parameters for the inconstant error variance and dependence of observations in matrix $\boldsymbol{R}$.

The estimation of $\boldsymbol{D}_*$, σ^2 and $\boldsymbol{\beta}$ can be based on analysis of variance, (RE)ML, or Bayesian methods. The RE(ML) and Bayesian methods can also provide estimates of the other parameters. The analysis of variance approach mimics the traditional analysis of variance and does not make specific assumptions about the distribution of the data. It was the earliest estimation method in the context of mixed-effects models. The general principles of the methods based on least squares and maximum likelihood were presented in detail in Section 4.4. Currently, they are the most widely used estimation methods. Bayesian methods can also be used for parameter estimation; they are attractive tools for models with a complex (e.g. spatio-temporal) variance-covariance structure. They also allow easier computation of variances of the parameter estimates than the methods based on LS or ML.

After the model parameters have been estimated, the random effects $\boldsymbol{b}_i$ for $i = 1, \ldots, M$ can be predicted using methods that will be presented in Section 5.4. The predicted random effects are needed, e.g. for group-specific predictions and model diagnostics. The fixed-effects and random effects can also be estimated jointly, using Henderson's mixed model equations (see Box 5.5, p. 162).

5.3.1 Estimating variance components with ANOVA methods

Estimation of variances and covariances in mixed models with ML or REML can be computationally demanding. Before the time of fast computers, variance components were usually estimated with ANOVA methods. For a history of the methods used to estimate variance components, see Searle *et al.* (1992). To describe the principles of ANOVA methods in brief, let us consider again the simplest possible mixed model:

$$y_{ij} = \mu + b_i + \epsilon_{ij}\,. \tag{5.18}$$

Let us consider a balanced data set with M groups, n observations in each group, and a total of $N = Mn$ observations, $\sigma^2 = \mathrm{var}\,(\epsilon_{ij})$, and $\sigma_b^2 = \mathrm{var}\,(b_i)$. Then, σ^2 can be first estimated in an unbiased way with:

$$\widehat{\sigma}^2 = SSE/(N-M) = \sum_{i=1}^{M}\sum_{j=1}^{n}(y_{ij}-\bar{y}_{i.})^2/(N-M)\,,$$

where $\bar{y}_{i.}$ is the average in group i. Here, SSE is the residual sum of squares if Equation (5.18) is estimated assuming that group effects b_i are fixed, see Equation (5.2). Let us then compute the residual sum of squares ($=SSE_1$) for the restricted model where the intercept is the only predictor:

$$SSE_1 = \sum_{i=1}^{M}\sum_{j=1}^{n}(y_{ij}-\bar{y})^2\,,$$

where $\bar{y}$ is the overall mean. The corresponding mean square is

$$MS_b = (SSE_1 - SSE)/(M-1)\,, \tag{5.19}$$

which has an expected value:

$$\mathrm{E}\,(MS_b) = n\sigma_b^2 + \sigma^2\,.$$

Thus, an unbiased estimate σ_b^2 is (show this!):

$$\widehat{\sigma}_b^2 = \left(MS_b - \widehat{\sigma}^2\right)/n\,. \tag{5.20}$$

Another way to derive the same estimator for σ_b^2 is, perhaps, easier to understand. Notice that $\bar{y}_{i.}$ is an unbiased estimator for $\mu + b_i$. A naive, biased way to estimate σ_b^2 is to use sample variance

$$s_b^2 = \frac{\sum_{i=1}^{M}(\bar{y}_{i.}-\bar{y})}{M-1}\,. \tag{5.21}$$

To correct for bias, we compute the variance of $\bar{y}_{i.}$:

$$\mathrm{var}(\bar{y}_{i.}) = \mathrm{var}(\mu + b_i + \bar{e}_i) = \sigma_b^2 + \frac{\sigma^2}{n}\,. \tag{5.22}$$

The sample variance of $\bar{y}_i$ (5.21) is an unbiased estimate of the variance of $\bar{y}_i$ (5.22). Thus, writing that the sample variance is equal to the variance of $\bar{y}_{i.}$ and solving for σ_b^2, we get an unbiased estimate for σ_b^2:

$$\widehat{\sigma}_b^2 = s_b^2 - \frac{\widehat{\sigma}^2}{n} . \tag{5.23}$$

It is left as an exercise to show that Equations (5.20) and (5.23) are equivalent.

Lappi and Bailey (1988), Lappi (1991a) and Lappi and Malinen (1994) applied similar approaches to estimate variances and covariances of random parameters in more complicated models. The method of Lappi and Bailey (1988) was slightly biased and it was corrected by Lappi and Malinen (1994). The general approach is also described by Rao (1987) whose evident error was corrected by Lappi (1991a).

The general principles of ANOVA methods for estimating variances of random parameters can be formulated as follows. Let r be the number of unknown variances. Then

1. Form r mean squares (or generally quadratic forms).
2. Derive the expected value of each mean square as a linear function of the unknown variances.
3. Write each observed mean square equal to its expected value.
4. Solve the obtained set of equations for the unknown variances.

Historically, Henderson's three methods (see Searle *et al.* (1992)) have been important; an early application in forestry is in Lappi (1986).

In general, the ANOVA methods are not used nowadays. First, as the variance estimates are unbiased, they often produce negative variances if the true variances are small. Some consider this to be illogical, even if a practical solution would be to assign negative estimates a value zero. In Box 5.2 (p. 153) we discuss the possibility to interpret negative variances in terms of negative within-group correlations. Notice that in the maximum likelihood estimation as well, negative estimates would equally often produce the maximum of the likelihood function. However, ML estimates are just defined to be the estimates that produce the maximum of the likelihood in the feasible parameter space. Another problem with ANOVA methods is that there is a limitless range of quadratic forms that can be used for unbalanced data.

An advantage of the ANOVA methods is that they are not based on the assumptions of normality. However, the REML estimates of variance components are usually equal to the ANOVA estimates in balanced data, if ANOVA estimates are in the feasible parameter space. Harville (1977) gave arguments for assuming that, in general, 'the maximum likelihood estimators derived on the basis of normality may be suitable even when the form of distribution is not specified'.

Another advantage in ANOVA estimators is that they can be expressed in closed form. They are generally much faster to compute than ML and REML estimates, which require iterative computations. So, ANOVA estimators may be reasonable in large data sets or when a large number of estimates need to be computed. When comparing ANOVA methods to (RE)ML, it should also be kept in mind that standard statistical software does not necessarily find the global maximum of the (restricted) likelihood when the likelihood surface is multi-modal (e.g. Dixon (2016)).

5.3.2 Maximum likelihood

The parameter estimation of a mixed-effects models can be based on maximum likelihood under normality. The justification for the wide use of likelihood-based estimators is their

Box 5.2 Negative within-group correlation

As described on p. 139, the variance of a random group effect implies a positive *within-group correlation* (intra-class correlation). However, a within-group correlation can also be negative if groups consist of small numbers of individuals who compete for limited resources. A classic example is a litter of pigs who compete for a fixed amount of food. Two members of a group will differ more than two individuals sampled randomly from the entire population of individuals.

Let σ_y^2 be the variance of y in the population. Assume that $r_i = \text{cor}(y_{ij}, y_{ij'})$ is constant for all pairs (ij, ij'). Then, $\text{cov}(y_{ij}, y_{ij'}) = r_i \sigma_y^2$. The lower limit of r_i, i.e. $r_i \geq -\frac{1}{n_i - 1}$, can be derived from the fact that the variance of the group mean needs to be non-negative:

$$\text{var}(\bar{y}_i) = \text{var}\left(\frac{\sum_{j=1}^{n_i} y_{ij}}{n_i}\right) = \frac{n_i \sigma_y^2 + n_i (n_i - 1) r_i \sigma_y^2}{n_i^2} \geq 0 \,. \tag{5.24}$$

In a mixed model $y_{ij} = \boldsymbol{x}'_{ij}\boldsymbol{\beta} + b_i + \epsilon_{ij}$, a negative within-group correlation is allowed if var(b_i)-parameter is allowed to be negative and it is interpreted as an estimate of $\text{cov}(y_{ij}, y_{ij'})$. ANOVA methods (see Section 5.3.1) can produce negative variance components, which are usually converted to zero. In (RE)ML methods, the parameter space of variance parameters is usually restricted to non-negative values. MlWin has an option for allowing negative variances.

A general mixed model $\boldsymbol{y}_i = \boldsymbol{X}_i\boldsymbol{\beta} + \boldsymbol{Z}_i\boldsymbol{b}_i + \boldsymbol{\epsilon}_i$ with $\boldsymbol{b}_i \sim N(\boldsymbol{0}, \boldsymbol{D}_*)$ and $\boldsymbol{\epsilon}_i \sim N(\boldsymbol{0}, \boldsymbol{R}_i)$ can have two possible interpretations. In the conditional, fully hierarchical, interpretation, it is specified as:

$$\boldsymbol{y}_i|\boldsymbol{b}_i \sim N(\boldsymbol{X}_i\boldsymbol{\beta} + \boldsymbol{Z}_i\boldsymbol{b}_i, \boldsymbol{R}_i),\ \boldsymbol{b}_i \sim N(\boldsymbol{0}, \boldsymbol{D}_*)$$

The marginal interpretation is

$$\boldsymbol{y}_i \sim N(\boldsymbol{X}_i\boldsymbol{\beta}, \boldsymbol{Z}_i\boldsymbol{D}_*\boldsymbol{Z}'_i + \boldsymbol{R}_i)$$

In the conditional interpretation, negative diagonal elements of $\boldsymbol{D}_*$ are not allowed, while in the marginal interpretation (see Equation (5.17), p. 149) they are allowed, as long as $\boldsymbol{Z}_i\boldsymbol{D}_*\boldsymbol{Z}'_i + \boldsymbol{R}_i$ is positive semidefinite.

Smith and Murray (1984) discussed negative within-group correlations in the time when ANOVA methods were used to estimate variance components. A more recent discussion is given by Pryseley *et al.* (2011) who also consider negative variance components in the context of generalized linear mixed models (GLMM). We will further discuss negative within-group correlations in Section 6.3. A negative within-group correlation causes *underdispersion* in generalized linear models (see Chapter 8).

generality: they do not need to be tailored for missing data and unbalanced designs. Additional arguments for REML are given in the next subsection.

The method of maximum likelihood is based on the marginal model

$$\boldsymbol{y} \sim N\left(\boldsymbol{X\beta}, \sigma^2\boldsymbol{V}\right)$$

where the structure of the matrices is as defined previously. We rescale the variance-covariance matrices by $1/\sigma^2$ to define $\mathcal{D} = \frac{1}{\sigma^2}\boldsymbol{D}$ and $\mathcal{R} = \frac{1}{\sigma^2}\boldsymbol{R}$, so that $\boldsymbol{b} \sim N(\boldsymbol{0}, \sigma^2\mathcal{D})$ and $\boldsymbol{\epsilon} \sim N(\boldsymbol{0}, \sigma^2\mathcal{R})$. Then also, $\boldsymbol{V} = \boldsymbol{Z}\mathcal{D}\boldsymbol{Z}' + \mathcal{R}$ and $\text{var}(\boldsymbol{e}) = \sigma^2\boldsymbol{V} = \boldsymbol{W}$. The new matrices specify the structure of the random effects and residuals up to a scaling constant σ^2. Under independence and a constant variance of ϵ_{ij}, we have $\mathcal{R} = \boldsymbol{I}_N$.

The structure of $\mathcal{D}$ and $\mathcal{R}$ and, therefore, $\boldsymbol{V}$ also, is specified by parameters $\boldsymbol{\theta}_D$ and $\boldsymbol{\theta}_R$, which are pooled to $\boldsymbol{\theta} = (\boldsymbol{\theta}'_D, \boldsymbol{\theta}'_R)'$. In Example 5.3 $\boldsymbol{\theta}$ includes the two variances and the covariance of random effects. In Example 5.10, $\boldsymbol{\theta}$ additionally includes the parameter δ of the variance function.

Box 5.3 Swamy's GLS estimator

In Equation (5.8) on p. 140, it is assumed that all regressors in the fixed and random parts of the model are the same. Swamy (1970) has shown that the GLS estimator of $\boldsymbol{\beta}$ can then be presented as a "weighted average" of the OLS estimates of $\boldsymbol{c}_i$'s:

$$\widehat{\boldsymbol{\beta}} = \left(\sum_{i=1}^{M} \boldsymbol{W}_i^{-1}\right)^{-1} \sum_{i=1}^{M} \boldsymbol{W}_i^{-1}\widehat{\boldsymbol{c}}_i$$

where

$$\boldsymbol{W}_i = \boldsymbol{D}_* + \sigma_i^2 \left(\boldsymbol{X}_i'\boldsymbol{X}_i\right)^{-1}$$

and $\widehat{\boldsymbol{c}}_i$ is the OLS estimator in group i:

$$\widehat{\boldsymbol{c}}_i = \left(\boldsymbol{X}_i'\boldsymbol{X}_i\right)^{-1} \boldsymbol{X}_i'\boldsymbol{y}_i$$

Notice that the residual variance is allowed to vary from group to group.

The method of maximum likelihood is the application of the method described in Section 4.4.3 for a situation where the variance-covariance matrix of the data has the block-diagonal structure of model (5.17). The log-likelihood becomes

$$\begin{aligned} l(\boldsymbol{\beta}, \sigma^2, \boldsymbol{\theta}) &= -\frac{N}{2}\log(2\pi) - \frac{1}{2}\log\left|\sigma^2 \boldsymbol{V}(\boldsymbol{\theta})\right| - \frac{1}{2\sigma^2}(\boldsymbol{y} - \boldsymbol{X\beta})'\boldsymbol{V}^{-1}(\boldsymbol{\theta})(\boldsymbol{y} - \boldsymbol{X\beta}) \\ &= -\frac{N}{2}\log(2\pi) - \frac{N}{2}\log(\sigma^2) - \frac{1}{2}\sum_{i=1}^{M}\log|\boldsymbol{V}_i(\boldsymbol{\theta})| \\ &\quad - \frac{1}{2\sigma^2}\sum_{i=1}^{M}(\boldsymbol{y}_i - \boldsymbol{X}_i\boldsymbol{\beta})'\boldsymbol{V}_i^{-1}(\boldsymbol{\theta})(\boldsymbol{y}_i - \boldsymbol{X}_i\boldsymbol{\beta}) \end{aligned}$$

Maximization becomes simpler if *profiling* is used to eliminate parameters from the likelihood. In particular, σ^2 as a function of $\boldsymbol{\beta}$ and $\boldsymbol{\theta}$ can be derived by differentiating the log-likelihood with respect to σ^2, setting it to zero and solving for σ^2 to get

$$\widehat{\sigma}^2(\boldsymbol{\beta}, \boldsymbol{\theta}) = \frac{1}{N}(\boldsymbol{y} - \boldsymbol{X\beta})'\boldsymbol{V}^{-1}(\boldsymbol{\theta})(\boldsymbol{y} - \boldsymbol{X\beta})\,. \tag{5.25}$$

Plugging the resulting estimator into the log-likelihood gives $l(\boldsymbol{\beta}, \widehat{\sigma}^2(\boldsymbol{\beta}, \boldsymbol{\theta}), \boldsymbol{\theta})$, which is no longer a function of σ^2. Thereafter, $\boldsymbol{\beta}$ can be eliminated by plugging the GLS (or ML) estimator

$$\widehat{\boldsymbol{\beta}}(\boldsymbol{\theta}) = \left(\boldsymbol{X}'\boldsymbol{V}^{-1}(\boldsymbol{\theta})\boldsymbol{X}\right)^{-1}\boldsymbol{X}'\boldsymbol{V}^{-1}(\boldsymbol{\theta})\boldsymbol{y} \tag{5.26}$$

to the likelihood in place of $\boldsymbol{\beta}$ to get the profile log-likelihood

$$l_p(\boldsymbol{\theta}) = l(\widehat{\boldsymbol{\beta}}(\boldsymbol{\theta}), \widehat{\sigma}^2(\widehat{\boldsymbol{\beta}}(\boldsymbol{\theta}), \boldsymbol{\theta}), \boldsymbol{\theta}),$$

as a function of $\boldsymbol{\theta}$ only. Differentiating the profile log-likelihood with respect to $\boldsymbol{\theta}$, setting to zero and solving for $\boldsymbol{\theta}$ gives the solution. Evaluating the estimator in the data set gives a candidate estimate $\widehat{\boldsymbol{\theta}}$. It is accepted as final ML estimate if the resulting $\mathcal{D}$ and $\mathcal{R}$ are positive definite (recall Section 3.4.4). Note that when the parameters are estimated using ANOVA, such a restriction is not always imposed (see Box 5.2, p. 153). Evaluating Equation (5.26) using the estimate gives the estimate of $\boldsymbol{\beta}$. Using these estimates in Equation (5.25) finally provides the estimate of σ^2.

Example 5.14. The maximum likelihood fit using function `lme` is carried out by using argument `method="ML"`. It is illustrated here using the model with a random intercept in the data set of Table 5.2 (p. 149).

```
> dat<-data.frame(group=c(1,1,2,2,2,3),
+                 y=c(2,4,2,4,5,4),
+                 x=c(10,13,11,14,15,12))
> lme(y~x,random=~1|group,data=dat,method="ML")
Linear mixed-effects model fit by maximum likelihood
  Data: dat
  Log-likelihood: -1.324436
  Fixed: y ~ x
(Intercept)           x
 -5.1064662   0.7047759

Random effects:
 Formula: ~1 | group
        (Intercept)  Residual
StdDev:   0.5076801 0.1304433

Number of Observations: 6
Number of Groups: 3
```

Web Example 5.15. Illustration of manual profiling of the likelihood without function `lme` for the model described in Example 5.14.

5.3.3 Restricted maximum likelihood

The method of restricted maximum likelihood has become the standard fitting method for the following reasons (McCulloch *et al.* 2008):

1. The REML solutions are optimal (minimum-variance unbiased) estimators for balanced data, as we have already discussed in Section 5.3.1. It is not known whether this is also true for unbalanced data, but it is widely believed that they are, at least, nearly optimal (see Harville 1977, for argumentation). Therefore, both REML and ML under normality provide a justified general means for estimation of the variance components even when the random effects are not normally distributed. Recall that under the linear model ML under normality leads to the GLS estimator, which is BLUE for any distribution that has the assumed variance-covariance structure.

2. As pointed out in Section 4.4.3, the method of maximum likelihood provides downward biased estimators for variance parameters. REML takes the degrees of freedom used in estimating the fixed effects $\boldsymbol{\beta}$ into account in an optimal way. This is important when the length of $\boldsymbol{\beta}$ is large in relation to the number of observations. Presumably this feature occurs for unbalanced data too.

3. REML estimates may be less sensitive to outliers than ML estimates.

The restricted likelihood is the likelihood for a specific linear combination of the original data (see Section 4.4.4):

$$\boldsymbol{K}'\boldsymbol{y} \sim N(\boldsymbol{0}, \sigma^2\boldsymbol{K}'\boldsymbol{V}(\boldsymbol{\theta})\boldsymbol{K}),$$

where $\boldsymbol{K}$ is an $N \times (N-p)$ matrix, which has full column rank and fulfills the condition $\boldsymbol{K}'\boldsymbol{X} = \boldsymbol{0}$. Methods to find matrix $\boldsymbol{K}$ were discussed previously in Section 4.4.4. Essentially, the new vector of responses, $\boldsymbol{K}'\boldsymbol{y}$, includes the same information as the original residuals in regard to the variance-covariance structure of the data but has only $N-p$ elements (Patterson and Thompson 1971). The REML log-likelihood function for the LME becomes

$$l_R(\sigma^2, \boldsymbol{\theta}) = -\frac{N-p}{2}\log(2\pi) - \frac{N-p}{2}\log(\sigma^2) - \frac{1}{2}\log|\boldsymbol{K}'\boldsymbol{V}(\boldsymbol{\theta})\boldsymbol{K}| \\ - \frac{1}{2\sigma^2}\boldsymbol{y}'\boldsymbol{K}\left(\boldsymbol{K}'\boldsymbol{V}(\boldsymbol{\theta})\boldsymbol{K}\right)^{-1}\boldsymbol{K}'\boldsymbol{y}.$$

Because $\boldsymbol{K}$ removes the effects of fixed parameters, the likelihood is a function of $\boldsymbol{\theta}$ and σ^2 only. Therefore, only profiling of σ^2 is needed, using the REML estimator

$$\widehat{\sigma}^2_R(\boldsymbol{\theta}) = \frac{1}{N-p}\boldsymbol{y}'\boldsymbol{K}\left(\boldsymbol{K}'\boldsymbol{V}(\boldsymbol{\theta})\boldsymbol{K}\right)^{-1}\boldsymbol{K}'\boldsymbol{y}\,.$$

Plugging this expression to the likelihood gives the profile likelihood

$$l_{p,R}(\boldsymbol{\theta}) = l_R(\widehat{\sigma}^2_R(\boldsymbol{\theta}), \boldsymbol{\theta})$$

which is a function of $\boldsymbol{\theta}$ only. Computing the estimates is then carried out in a similar way as in the ML procedure. Finally, the estimate of $\boldsymbol{\beta}$ is computed using the GLS estimator (4.25), using the REML estimate of $\boldsymbol{V}$ in place of $\boldsymbol{V}$, and ignoring its estimation error.

Example 5.16. The REML estimation in the data set of Table 5.2 (p. 149) gives

```
> lme(y~x,random=~1|group,data=dat,method="REML")
Linear mixed-effects model fit by REML
  Data: dat
  Log-restricted-likelihood: -3.835087
  Fixed: y ~ x
(Intercept)           x
 -5.1068495   0.7048139

Random effects:
 Formula: ~1 | group
        (Intercept)  Residual
StdDev:   0.6233488 0.1593886

Number of Observations: 6
Number of Groups: 3
```

We see substantial increases in the estimates of σ_b^2 and σ^2 when compared to the ML estimates of Example 5.14. Finding the REML estimation manually would include two steps: Firstly determination of the matrix $\boldsymbol{K}$ using similar procedures as we used in Web Example 4.10 (p. 96). Secondly, estimation would follow similar paths to those described in Web Example 5.15. Implementation is left as an exercise.

5.3.4 More about the estimation of $\boldsymbol{R}$

The REML and ML methods can be used to estimate parameter $\boldsymbol{\theta}_R$, which is used to parameterize the inconstant error variance and autocorrelation in matrix $\boldsymbol{R} = \mathrm{var}(\boldsymbol{\epsilon})$. Such estimation methods have been implemented in package `nlme` but not, e.g., with package `lme4`, which has other good properties, such as a more efficient estimation algorithm than `nlme`, and the capability to treat efficiently crossed grouping structures. To allow the flexible use of different R-packages, we will now discuss further the estimation of error variance-covariance structure.

We presented an iterative procedure to estimate the variance function for a linear model in Section 4.4.2. A similar approach can also be used with mixed-effects models as follows:

1. Fit the model by assuming that $\boldsymbol{R} = \sigma^2\boldsymbol{I}$ and save the residuals $\widehat{\epsilon}_{ij}$.
2. Fit a model for the squared residuals $\widehat{\epsilon}^2_{ij}$ using an appropriate predictor of a variance function that shows a good fit.
3. Fit the model again using the inverse of the estimated variance as a weight. Save the residuals $\widehat{\epsilon}_{ij}$. OLS may be sufficient for fitting.
4. Repeat steps 2 and 3 a few times, or until the estimates converge.

Because of the properties of the predicted random effects and residuals (see Section 5.4.1), it is justified to estimate the models of steps 2 and 3 by treating group effects as fixed, especially if the number of observations per group is low. After the model for error variance has been found, the model of the last iteration can then be estimated by treating the group effects as random. Lappi (1997) used this approach, albeit without iterating steps 2 and 3. An example of the iterative procedure is given in Example 6.5 (p. 193).

Similar ideas can be developed for estimation of the covariance structure as well. The estimation would be based on fitting regressions on cross-products $\epsilon_{ij}\epsilon_{i'j'}$ for all possible pairs of $ij, i'j'$. The estimated covariance structure can be used in estimation, following the procedures presented in Equations (4.28) and (4.29) in Section 4.4.2. For an example in a multivariate context, see Lappi (2006). SAS procedure MIXED can be used to estimate many kinds of $\boldsymbol{R}$ matrices.

5.3.5 Bayesian estimation

The Bayesian computational methods (Markov Chain Monte Carlo, MCMC) are becoming popular in the context of mixed-effects models, especially because they allow more flexible model formulation than the methods based on the likelihood and least squares. For example, they allow the modeling of spatially or temporally dependent and non-normal random effects. They are also widely used alternatives to solve the problems of zero-inflation and censoring in the context of generalized linear mixed-effects models (see Chapter 8). If the variance parameters are estimated using Bayesian methods, their estimation errors will be taken into account in the inference about the fixed effects.

As we have already discussed in Section 3.4.5, the Bayesian approach is based on the assumption that the parameters of the model are also random, with a specific prior distribution. In the context of linear mixed-effects models, this means that the fixed effects $\boldsymbol{\beta}$, the parameters specifying the variance-covariance matrix $\boldsymbol{D}$, and the parameters specifying the residual error variance-covariance structure $\boldsymbol{R}$ are all random, and a prior distribution for them needs to be specified.

If the motivation for the Bayesian approach is to enable MCMC computations but not to include any prior information of them in the analysis, then one can use very flat priors. A common choice is to use a normal distribution with a very high variance for the fixed effects, and inverse Wishart distribution with a very high variance for the variance-covariance parameters; these are also the conjugate priors of the linear mixed-effects model. Let $\boldsymbol{\theta}_D$ include the variances and covariances of the random part, and $\boldsymbol{\theta}_R$ the parameters that specify the error variance-covariance structure. Denote the prior densities for $\boldsymbol{\beta}$, $\boldsymbol{\theta}_D$, and $\boldsymbol{\theta}_R$ by $f_{\boldsymbol{\beta}}(\boldsymbol{\beta})$, $f_D(\boldsymbol{\theta}_D)$, and $f_R(\boldsymbol{\theta}_R)$. Assuming that the random parameters are initially independent, the posterior distribution of the parameters given the observed data $\boldsymbol{y}$, is then

$$f(\boldsymbol{\beta},\boldsymbol{\theta}_D,\boldsymbol{\theta}_R|\boldsymbol{y}) = \frac{f(\boldsymbol{y}|\boldsymbol{\beta},\boldsymbol{\theta}_D,\boldsymbol{\theta}_R)f_{\boldsymbol{\beta}}(\boldsymbol{\beta})f_D(\boldsymbol{\theta}_D)f_R(\boldsymbol{\theta}_R)}{\int f(\boldsymbol{y}|\boldsymbol{\beta},\boldsymbol{\theta}_D,\boldsymbol{\theta}_R)f_{\boldsymbol{\beta}}(\boldsymbol{\beta})f_D(\boldsymbol{\theta}_D)f_R(\boldsymbol{\theta}_R)\mathrm{d}\,(\boldsymbol{\beta},\boldsymbol{\theta}_D,\boldsymbol{\theta}_R)}\,.$$

The great benefit of the MCMC methods is that the integral in the denominator need not be evaluated. It suffices to evaluate the numerator of the function in the logarithmic form:

$$\log(f(\boldsymbol{y}|\boldsymbol{\beta},\boldsymbol{\theta}_D,\boldsymbol{\theta}_R)) + \log(f_{\boldsymbol{\beta}}(\boldsymbol{\beta})) + \log(f_D(\boldsymbol{\theta}_D)) + \log(f_R(\boldsymbol{\theta}_R))$$

in order to simulate an MCMC sample from the posterior distribution of the parameters.

In `R`, an easy-to-use package for fitting mixed-effects models in the Bayesian context is package `MCMCglmm` (Hadfield 2010). However, there are also other, more general and flexible tools, see e.g. Stan Development Team (2018).

Example 5.17. We fit model `lmm3` of Example 5.3 using `MCMCglmm`. The function uses the normal and Wishart distributions as priors; the parameters of these distributions are specified in object `prior`. Arguments `nitt`, `burnin`, and `thin` specify the number of iterations, burn-in time and thinning interval of the MCMC algorithm. Before interpreting the results, one should check that the MCMC iterations have sampled the true posterior and the parameters do not have an overly strong temporal autocorrelation; recall Example 3.10 (p. 57).

The posterior means for parameters $\text{var}(b_i^{(1)})$, $\text{var}(b_i^{(2)})$, and $\text{cov}(b_i^{(1)}, b_i^{(2)})$ are 0.0884 ($=0.297^2$), 0.00415 ($=0.0644^2$), and -0.00739 (correlation -0.386). The posterior mean of the residual variance is $0.05991 = 0.245^2$. These values are very similar to the REML estimates in Example 5.3 (see p. 141.) Also, the posterior means for fixed effects are very similar to the values obtained earlier using REML. A convenient feature of the Bayesian approach is that the credible intervals can be easily computed for the parameters using the posterior sample. The bounds of credible intervals reported below are obtained as the marginal 0.025^{th} and 0.975^{th} quantiles of the MCMC sample. These intervals also consider the estimation errors of the variance-covariance parameters and are, therefore, wider than one would get using the procedures presented in Section 4.6.10 for the REML fits.

```
> library(MCMCglmm)
> prior=list(R=list(V=1,nu=0.002),G=list(G1=list(V=diag(2),nu=0.02)))
> lmm3B<-MCMCglmm(fixed=y~d,
+                 random=~us(1+d):plot,
+                 prior=prior, family="gaussian",
+                 nitt=60000, burnin=10000,
+                 thin=25, data=spati2)
> summary(lmm3B)

 Iterations = 10001:59976
 Thinning interval  = 25
 Sample size  = 2000

 G-structure:  ~us(1 + d):plot
                              post.mean  l-95% CI  u-95% CI eff.samp
(Intercept):(Intercept).plot  0.088429  0.051077  0.129646     2000
d:(Intercept).plot           -0.007393 -0.013855 -0.001206     2000
(Intercept):d.plot           -0.007393 -0.013855 -0.001206     2000
d:d.plot                      0.004152  0.002687  0.005836     2000

 R-structure:  ~units
        post.mean l-95% CI u-95% CI eff.samp
units    0.05991  0.05614  0.06423     2000

 Location effects: y ~ d
            post.mean l-95% CI u-95% CI eff.samp  pMCMC
(Intercept)    1.0953   1.0093   1.1787     1897 <5e-04 ***
d              0.2332   0.2140   0.2482     2000 <5e-04 ***
```

5.4 Prediction at group level

5.4.1 Prediction of random effects

Estimation of the linear mixed-effects model yields the estimates for the fixed parameters, the variances and covariances of random effects, and parameters specifying the variance of the residual errors. Estimates for the random group effects, $\boldsymbol{b}_i$ for groups $i = 1, \ldots, M$ are not provided. Their prediction is based on the linear prediction theory, which was presented in general form in Section 3.5.2, and is now presented for this special prediction problem.

Consider the model (5.13) for a single group i. Assuming that $\boldsymbol{\beta}$ are known, recall the following means, variances and covariances from Section 5.2.1:

$$\begin{pmatrix} \boldsymbol{b}_i \\ \boldsymbol{y}_i \end{pmatrix} \sim \left[\begin{pmatrix} \boldsymbol{0} \\ \boldsymbol{X}_i\boldsymbol{\beta} \end{pmatrix}, \begin{pmatrix} \boldsymbol{D}_* & \boldsymbol{D}_*\boldsymbol{Z}_i' \\ \boldsymbol{Z}_i\boldsymbol{D}_* & \boldsymbol{Z}_i\boldsymbol{D}_*\boldsymbol{Z}_i' + \boldsymbol{R}_i \end{pmatrix} \right].$$

Box 5.4 Kalman filter

Kalman filter is a method for predicting random state parameters which evolve randomly over time and from which one can make measurements that contain measurement errors. The dynamic system is described with two equations:

$$\boldsymbol{x}_{t+1} = \boldsymbol{A}\boldsymbol{x}_t + \boldsymbol{w}_t\,, \tag{5.27}$$

$$\boldsymbol{y}_t = \boldsymbol{C}\boldsymbol{x}_t + \boldsymbol{v}_t\,, \tag{5.28}$$

where Equation (5.27) is the *state equation*, and Equation (5.28) is the *observation equation*. In Equation (5.27) $\boldsymbol{x}$ is an n-dimensional random unobservable state vector, $\boldsymbol{A}$ is a fixed $n \times n$-matrix, $\boldsymbol{w}$ is n-dimensional random *state noise* or *process noise* with a mean of 0 and a fixed variance-covariance matrix. The initial state $\boldsymbol{x}_0$ has a fixed expectation and variance-covariance matrix. In Equation (5.28) $\boldsymbol{y}$ is random observable p-dimensional vector, $\boldsymbol{C}$ is a fixed $p \times n$ matrix, and $\boldsymbol{v}$ is p dimensional random *measurement noise* with a mean of 0 and fixed variance-covariance matrix. The purpose is to predict $\boldsymbol{x}_t$ and $\boldsymbol{x}_{t+1}$ when $\boldsymbol{y}_0 \ldots \boldsymbol{y}_t$ are measured. The assumed relationship between $\boldsymbol{x}$ and $\boldsymbol{y}$ is linear. Thus, the predictions can be carried out using Best Linear Predictor (BLP, see Section 3.5.2) which also provides conditional expectations when the multivariate normal distribution is assumed. When the system evolves over time, direct application of BLP would be increasingly laborious. A better prediction method can be derived utilizing the *Markov property* of the system:

$$\mathrm{E}\left(\boldsymbol{x}_t \,|\boldsymbol{x}_0 \ldots \boldsymbol{x}_{t-1}\right) = \mathrm{E}\left(\boldsymbol{x}_t \,|\boldsymbol{x}_{t-1}\right) \tag{5.29}$$

Using the Markov property of the system, Kalman derived a recursive prediction method where $\boldsymbol{x}_t$ is predicted from the predictor $\widetilde{\boldsymbol{x}}_{t-1}$ and $\boldsymbol{y}_t$. Thereafter, $\boldsymbol{x}_{t+1}$ can predicted using Equation (5.27), i.e. by $\widetilde{\boldsymbol{x}}_{t+1} = \boldsymbol{A}\widetilde{\boldsymbol{x}}_t$.

The Kalman filter can be also presented as a special version of the linear mixed-effects model (Piepho and Ogutu 2007). In forestry, Kalman filter has been applied in dendrochronology by VanDeusen (1989) and in forest inventories by Dixon and Howitt (1979).

We want to predict the random effects $\boldsymbol{b}_i$ using the observations $\boldsymbol{y}_i$. This is a special case of the general BLP (Equation 3.19), with $\boldsymbol{h}_1 = \boldsymbol{b}_i$ and $\boldsymbol{h}_2 = \boldsymbol{y}_i$. The predictor and the related variance of prediction errors become

$$\begin{aligned}
\mathrm{BLP}(\boldsymbol{b}_i) &= \boldsymbol{D}_*\boldsymbol{Z}_i'\left(\boldsymbol{Z}_i\boldsymbol{D}_*\boldsymbol{Z}_i' + \boldsymbol{R}_i\right)^{-1}\left(\boldsymbol{y}_i - \boldsymbol{X}_i\boldsymbol{\beta}\right) && (5.30)\\
\mathrm{var}(\mathrm{BLP}(\boldsymbol{b}_i) - \boldsymbol{b}_i) &= \boldsymbol{D}_* - \boldsymbol{D}_*\boldsymbol{Z}_i'\left(\boldsymbol{Z}_i\boldsymbol{D}_*\boldsymbol{Z}_i' + \boldsymbol{R}_i\right)^{-1}\boldsymbol{Z}_i\boldsymbol{D}_*\,. && (5.31)
\end{aligned}$$

Interestingly, there is no lower bound for the number of observations per group to enable prediction. In particular, the number of observations in $\boldsymbol{y}_i$ can be smaller than the number of random effects in $\boldsymbol{b}_i$. The predictor for the complete vector of random effect for all groups, $\boldsymbol{b}$, results from the replacement of $\boldsymbol{Z}_i$ by $\boldsymbol{Z}$, $\boldsymbol{X}_i$ by $\boldsymbol{X}$, $\boldsymbol{D}_*$ by $\boldsymbol{D}$, and $\boldsymbol{R}_i$ by $\boldsymbol{R}$, where the matrices and vectors for all data are as presented in Section 5.2.3.

The BLP assumes that $\boldsymbol{\beta}$ is known, but it needs to be replaced with the (GLS) estimate $\widehat{\boldsymbol{\beta}}$. The resulting formula is Best Linear Unbiased Predictor (BLUP) of random effects (for derivation, see Section 13.4 of McCulloch *et al.* 2008). The formula can be derived as a special case of the equations described in Goldberger (1962) (see p. 129). The variance of prediction error is different from (5.31) because of the estimation error of fixed parameter vector $\mathrm{var}(\widehat{\boldsymbol{\beta}})$ and its covariance with the prediction errors of random effects $\mathrm{cov}(\widehat{\boldsymbol{\beta}}, \widetilde{\boldsymbol{b}}' - \boldsymbol{b}')$. A variance formula that takes the estimation errors of fixed effects into account is most conveniently obtained as the lower-right corner $\boldsymbol{C}_2$ of $\boldsymbol{H}^{-1}$ in the Henderson's mixed model equations (see Box 5.5, p. 162). The off-diagonal block $\boldsymbol{C}_{12}$ of $\boldsymbol{H}^{-1}$ gives $\mathrm{cov}(\widehat{\boldsymbol{\beta}}, \widetilde{\boldsymbol{b}}' - \boldsymbol{b}')$, which is needed to compute the prediction error of a mixed-effects model. For a computationally less intensive solution, the formulas that directly give matrices $\boldsymbol{C}_2$ and $\boldsymbol{C}_{12}$ are presented in Subsection 7.4d of Searle *et al.* (1992).

In addition to $\boldsymbol{\beta}$, the variance-covariance parameters that specify matrices $\boldsymbol{D}$ and $\boldsymbol{R}$ are also unknown. They are replaced with their (REML) estimates in applications. The resulting estimators are called Empirical BLUPs (EBLUPs), and they resemble empirical Bayes, ridge regression and James-Stein estimators (Harville 1977) (see Boxes 4.1 and 4.2, p. 90 and 91, respectively). Also Kalman filter (See Box 5.4), which is widely used in engineering, has similarities with the linear mixed-effect models. To avoid unnecessary complexity in notations, we use notation $\widetilde{\boldsymbol{b}}_i$ both for the BLUPs and EBLUPs of the random effects.

5.4.2 Group-level prediction

Often, we are interested in the group level coefficients or group level predictions. They are special cases of the linear combination

$$l = \boldsymbol{x}_0'\boldsymbol{\beta} + \boldsymbol{z}_0'\boldsymbol{b}\,, \tag{5.32}$$

where $\boldsymbol{x}_0$ and $\boldsymbol{z}_0$ are fixed vectors that are specified according to the prediction situation at hand. Such linear combinations are called *predictable functions* in literature. Marginally unbiased predictions are obtained by using GLS estimates of $\boldsymbol{\beta}$ and the BLUP's of $\boldsymbol{b}$ in (5.32). For example, if one wants to compute the prediction of the group-level intercept for group i' in model $y_{ij} = \beta_1 + \beta_2 x_{ij} + b_i$, then $\boldsymbol{x}_0' = (1, 0)$ picks the fixed effect associated with the intercept, and vector $\boldsymbol{z}_0'$, which includes 1 in the i'^{th} element and zeroes elsewhere, picks the random effect associated with the intercept in group i'. The related prediction error is

$$\begin{aligned}
\operatorname{var}\left(\boldsymbol{x}_0'\widehat{\boldsymbol{\beta}} + \boldsymbol{z}_0'\widetilde{\boldsymbol{b}} - (\boldsymbol{x}_0'\boldsymbol{\beta} + \boldsymbol{z}_0'\boldsymbol{b})\right) &= \operatorname{var}\left(\boldsymbol{x}_0'\widehat{\boldsymbol{\beta}}\right) + \operatorname{var}\left(\boldsymbol{z}_0'(\widetilde{\boldsymbol{b}} - \boldsymbol{b})\right) + 2\operatorname{cov}\left(\boldsymbol{x}_0'\widehat{\boldsymbol{\beta}}, \left(\widetilde{\boldsymbol{b}}' - \boldsymbol{b}'\right)\boldsymbol{z}_0\right) \\
&= \boldsymbol{x}_0' \operatorname{var}(\widehat{\boldsymbol{\beta}})\boldsymbol{x}_0 + \boldsymbol{z}_0' \operatorname{var}\left(\widetilde{\boldsymbol{b}} - \boldsymbol{b}\right)\boldsymbol{z}_0 + 2\boldsymbol{x}_0' \operatorname{cov}\left(\widehat{\boldsymbol{\beta}}, \widetilde{\boldsymbol{b}}' - \boldsymbol{b}'\right)\boldsymbol{z}_0 \\
&= \boldsymbol{x}_0'\boldsymbol{C}_1\boldsymbol{x}_0 + \boldsymbol{z}_0'\boldsymbol{C}_2\boldsymbol{z}_0 + 2\boldsymbol{x}_0'\boldsymbol{C}_{12}\boldsymbol{z}_0\,.
\end{aligned} \tag{5.33}$$

Matrices $\boldsymbol{C}_1$, $\boldsymbol{C}_2$ and $\boldsymbol{C}_{12}$ in Equation (5.33) are conveniently obtained from Henderson's mixed model equations.

Formula (5.33) could be extended to produce the covariances of prediction errors for all pairs of observations. For that purpose, one should define matrices $\boldsymbol{X}_0$ and $\boldsymbol{Z}_0$, where the rows are vectors $\boldsymbol{x}_0'$ and $\boldsymbol{z}_0'$ in the same way that we did in Section 4.7. Using these matrices in the place of $\boldsymbol{x}_0'$ and $\boldsymbol{z}_0'$ in Equation (5.33) would give the variance-covariance matrix of prediction errors.

The prediction of the group mean in group i as a function of predictors $\boldsymbol{x}_0$ and $\boldsymbol{z}_0$ is given by

$$\widetilde{\mu}_i = \widetilde{\mu}_i(\boldsymbol{x}_0, \boldsymbol{z}_0) = \boldsymbol{x}_0'\widehat{\boldsymbol{\beta}} + \boldsymbol{z}_0'\widetilde{\boldsymbol{b}}_i\,. \tag{5.34}$$

The prediction variance is

$$\operatorname{var}(\mu_i - \widetilde{\mu}_i) = \boldsymbol{x}_0' \operatorname{var}(\widehat{\boldsymbol{\beta}} - \boldsymbol{\beta})\boldsymbol{x}_0 + \boldsymbol{z}_0' \operatorname{var}(\widetilde{\boldsymbol{b}}_i - \boldsymbol{b}_i)\boldsymbol{z}_0 + 2\boldsymbol{x}_{ij}' \operatorname{cov}(\widehat{\boldsymbol{\beta}} - \boldsymbol{\beta}, \widetilde{\boldsymbol{b}}_i - \boldsymbol{b}_i)\boldsymbol{z}_{ij}\,,$$

In large data sets, $\operatorname{var}(\widehat{\boldsymbol{\beta}})$ is often small and the following approximate can be used:

$$\operatorname{var}(\mu_i - \widetilde{\mu}_i) = \boldsymbol{z}_0' \operatorname{var}(\widetilde{\boldsymbol{b}}_i - \boldsymbol{b}_i)\boldsymbol{z}_0\,.$$

Notice that with a large data set we mean a data set with both a large number of groups and a large number of individuals per group. However, if all fixed predictors are group-level predictors, it is sufficient to have just a large number of groups.

The *individual level prediction* for a non-measured individual j of group i is given by

$$\widetilde{y}_{ij} = \boldsymbol{x}'_{ij}\widehat{\boldsymbol{\beta}} + \boldsymbol{z}'_{ij}\widetilde{\boldsymbol{b}}_i \,. \tag{5.35}$$

The prediction variance is

$$\begin{aligned}\operatorname{var}(y_{ij} - \widetilde{y}_{ij}) &= \boldsymbol{x}'_{ij} \operatorname{var}(\widehat{\boldsymbol{\beta}} - \boldsymbol{\beta})\boldsymbol{x}_{ij} + \boldsymbol{z}'_{ij} \operatorname{var}(\widetilde{\boldsymbol{b}}_i - \boldsymbol{b}_i)\boldsymbol{z}_{ij} \\ &\quad + 2\boldsymbol{x}'_{ij} \operatorname{cov}(\widehat{\boldsymbol{\beta}} - \boldsymbol{\beta}, \widetilde{\boldsymbol{b}}'_i - \boldsymbol{b}'_i)\boldsymbol{z}_{ij} + \operatorname{var}(\epsilon_{ij}) \,,\end{aligned}$$

Again, the contribution of the estimation errors of $\widehat{\boldsymbol{\beta}}$ is small if the data set is large. One might also consider including the term $\operatorname{cov}(\widetilde{\boldsymbol{b}}_i - \boldsymbol{b}_i, \epsilon_{ij})$; however, it is zero.

If the $y-$variable is known for some individuals whose residual errors are correlated with the residual errors of the individual in question, the individual-level prediction can be further improved by predicting the residual error. Such a case generalizes the discussion of Section 4.7.2 to mixed-effects models and is discussed in Section 6.5.

If no observations of the response y are available from group i, then the BLUP of the random effect is the mean: $\widetilde{\boldsymbol{b}}_i = \boldsymbol{0}$. Its variance is $\operatorname{var}(\widetilde{\boldsymbol{b}}_i - \boldsymbol{b}_i) = \boldsymbol{D}_*$. In that case, the random part of the model cancels out and the group-level prediction includes only the fixed part of the model. This prediction can be considered the prediction in an average group of modeling data, which is based on the mean (which is also the mode for normally distributed random effects) of the distribution of random effects. If information on the group effect is not available, it is justified to assume that the group is an average group. The prediction of the fixed part of the model also has the interpretation as the *marginal* or *population-averaged* prediction (but see Section 6.4 for a critical discussion).

Example 5.18. In Example 5.16 (p. 156), the model

$$y_{ij} = \beta_1 + \beta_2 x_{ij} + b_i + \epsilon_{ij}$$

was fitted to the data set of Table 5.2 using REML and GLS. The parameter estimates were $\widehat{\boldsymbol{\beta}} = \begin{pmatrix} -5.107 \\ 0.705 \end{pmatrix}$, $\widehat{\sigma}_b^2 = 0.623^2$, and $\widehat{\sigma}^2 = 0.159^2$. Consider the prediction of random effect b_1 for group 1, using the two observations of y from the group. The objects `X` and `Z` include the model matrices of all data for the fixed and random parts, respectively. Object `R` includes matrix $0.159^2\boldsymbol{I}_6$ and `D` matrix $0.623^2\boldsymbol{I}_3$.

```
> mod<-lme(y~x,random=~1|group,data=dat,method="REML")
> y<-dat$y
> X<-cbind(1,dat$x)
> Z<-cbind(dat$group==1,dat$group==2,dat$group==3)*1
> R<-diag(rep(mod$sigma^2,dim(dat)[1]))
> D<-diag(rep(getVarCov(mod),3))
```

The random effect for group 1 is predicted below. In addition, the prediction variance is estimated using Equation (5.31).

```
> beta<-c(as.numeric(fixef(mod)))
> y1<-c(2,4)
> Z1<-matrix(Z[1:2,1],ncol=1)
> X1<-X[1:2,]
> D1<-D[1,1]
> R1<-R[1:2,1:2]
> D1%*%t(Z1)%*%solve(Z1%*%D1%*%t(Z1)+R1)%*%(y1-X1%*%beta) # Eq 5.29
            [,1]
[1,] 0.001442943
> D1-D1%*%t(Z1)%*%solve(Z1%*%D1%*%t(Z1)+R1)%*%Z1%*%D1 # Eq. 5.30
           [,1]
[1,] 0.01230026
```

Box 5.5 Henderson's mixed model equations

Henderson (1963) showed that the fixed effects $\widehat{\boldsymbol{\beta}}$ and random effects $\widetilde{\boldsymbol{b}}$ can be simultaneously estimated and predicted as a solution to the following system of equations:

$$\begin{pmatrix} \boldsymbol{X}'\boldsymbol{R}^{-1}\boldsymbol{X} & \boldsymbol{X}'\boldsymbol{R}^{-1}\boldsymbol{Z} \\ \boldsymbol{Z}'\boldsymbol{R}^{-1}\boldsymbol{X} & \boldsymbol{Z}'\boldsymbol{R}^{-1}\boldsymbol{Z}+\boldsymbol{D}^{-1} \end{pmatrix} \begin{pmatrix} \widehat{\boldsymbol{\beta}} \\ \widetilde{\boldsymbol{b}} \end{pmatrix} = \begin{pmatrix} \boldsymbol{X}'\boldsymbol{R}^{-1}\boldsymbol{y} \\ \boldsymbol{Z}'\boldsymbol{R}^{-1}\boldsymbol{y} \end{pmatrix}$$

$$\boldsymbol{H} \begin{pmatrix} \widehat{\boldsymbol{\beta}} \\ \widetilde{\boldsymbol{b}} \end{pmatrix} = \boldsymbol{r}\,.$$

The solution $\begin{pmatrix} \widehat{\boldsymbol{\beta}} \\ \widetilde{\boldsymbol{b}} \end{pmatrix} = \boldsymbol{H}^{-1}\boldsymbol{r}$ gives the GLS estimates (BLUE) of $\boldsymbol{\beta}$ and BLUP's of $\boldsymbol{b}$ (e.g. Searle *et al.* 1992, Section 7.6). The resulting formula for $\boldsymbol{b}$ looks different to Equation (5.30):

$$\widetilde{\boldsymbol{b}} = \left(\boldsymbol{Z}'\boldsymbol{R}^{-1}\boldsymbol{Z}+\boldsymbol{D}^{-1}\right)^{-1}\boldsymbol{Z}'\boldsymbol{R}^{-1}\left(\boldsymbol{y}-\boldsymbol{X}\widehat{\boldsymbol{\beta}}\right), \tag{5.36}$$

but it is equivalent. Henderson's solution has a computational advantage over Equation (5.30): Equation (5.36) requires inversion of a non-diagonal matrix of dimension $q \times q$, where q is the number of random effects per group, whereas Equations (5.30) and (5.31) require inversion of a general $n_i \times n_i$ matrix. In addition, the inverse of the square matrix on the left-hand-side of the system is directly the variance-covariance matrix of the estimation errors (Henderson 1975, Lappi 1986, Appendix A2)

$$\boldsymbol{H}^{-1} = \begin{pmatrix} \boldsymbol{C}_1 & \boldsymbol{C}_{12} \\ \boldsymbol{C}_{12}' & \boldsymbol{C}_2 \end{pmatrix} = \text{var}\begin{pmatrix} \widehat{\boldsymbol{\beta}} \\ \widetilde{\boldsymbol{b}}-\boldsymbol{b} \end{pmatrix}.$$

Therefore, the submatrices give $\text{var}\,\widehat{\boldsymbol{\beta}} = \boldsymbol{C}_1$, $\text{cov}(\widehat{\boldsymbol{\beta}}, \widetilde{\boldsymbol{b}}-\boldsymbol{b}) = \boldsymbol{C}_{12}$, and $\text{var}(\widetilde{\boldsymbol{b}}-\boldsymbol{b}) = \boldsymbol{C}_2$. $\boldsymbol{C}_2$ is the variance of BLUP, which differs from the variance of BLP (5.31) by inclusion of a term related to the estimation errors of $\widehat{\boldsymbol{\beta}}$.

We got $\widetilde{b}_1 = 0.0014$ and $\text{var}(\widetilde{b}_1 - b_1) = 0.0123 = 0.111^2$.

The computed prediction variance does not take the estimation error of fixed effects and their correlation with the prediction error of random effects into account. To do so, we use Henderson's mixed-model equations. The square matrix on the left-hand side of Henderson's system is constructed below into matrix $\boldsymbol{H}$, and the vector in the right-hand side into vector $\boldsymbol{r}$. The solution is simply $\begin{pmatrix} \widehat{\boldsymbol{\beta}} \\ \widetilde{\boldsymbol{b}} \end{pmatrix} = \boldsymbol{H}^{-1}\boldsymbol{r}$, which gives the same values for $\widehat{\boldsymbol{\beta}}$ and $\widetilde{\boldsymbol{b}}$ as fitting using `nlme`.

```
> H11<-t(X)%*%solve(R)%*%X
> H12<-t(X)%*%solve(R)%*%Z
> H22<-t(Z)%*%solve(R)%*%Z+solve(D)
> H<-cbind(rbind(H11,t(H12)),rbind(H12,H22))
> r<-rbind(t(X)%*%solve(R)%*%y,t(Z)%*%solve(R)%*%y)
> solve(H)%*%r
             [,1]
[1,] -5.106849498
[2,]  0.704813859
[3,]  0.001442943
[4,] -0.610692709
[5,]  0.609249765

> fixef(mod)
(Intercept)           x
 -5.1068495   0.7048139
> ranef(mod)
   (Intercept)
```

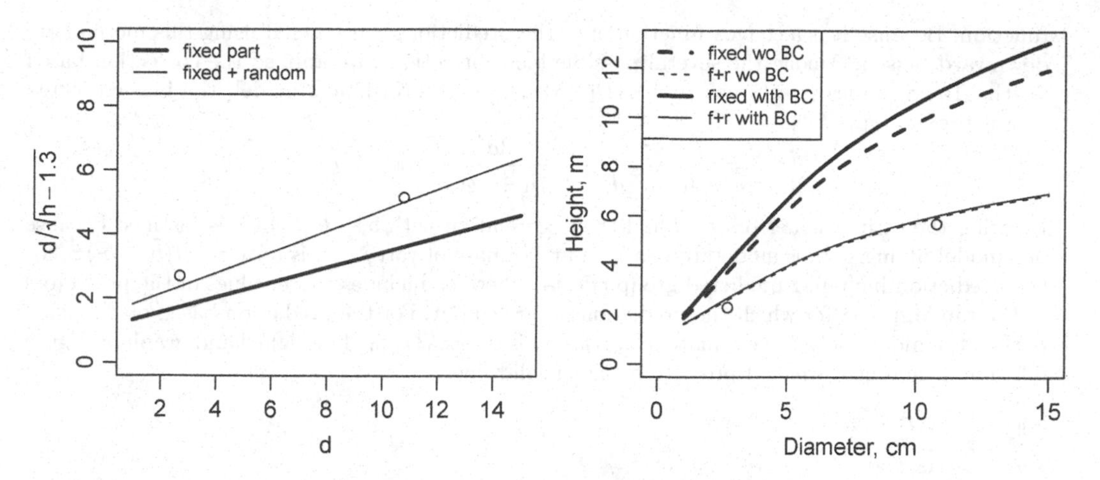

FIGURE 5.5 Prediction based on the fixed part of the model (thick line) and the localized curve (thin line) based on two observations (circles) in the transformed scale (left) and in the original scale (right) in Example 5.19. BC denotes bias correction.

```
1  0.001442943
2 -0.610692709
3  0.609249765
```

Matrix $\boldsymbol{H}^{-1}$ provides an estimate of var $\begin{pmatrix} \widehat{\boldsymbol{\beta}} \\ \widetilde{\boldsymbol{b}} - \boldsymbol{b} \end{pmatrix}$. The estimate of $\text{var}(\widetilde{\boldsymbol{b}} - \boldsymbol{b})$ is in the lower-right corner:

```
> solve(H)[3:5,3:5]
          [,1]      [,2]      [,3]
[1,] 0.1396677 0.1261187 0.1227773
[2,] 0.1261187 0.1392746 0.1231704
[3,] 0.1227773 0.1231704 0.1426160
```

We get $\text{var}(\widetilde{b}_1 - b_1) = 0.140$, which is approximately tenfold greater than the previous estimate, which ignored the uncertainty of fixed effect estimates. The large increase is explained by high uncertainty of $\widehat{\boldsymbol{\beta}}$, which in turn is explained by small size of the data set. Notice also the non-zero off-diagonal elements of $\text{var}(\widetilde{\boldsymbol{b}} - \boldsymbol{b})$, which show that the prediction errors of random effects are correlated across the groups, even though the true random effects are not. This is associated with the estimation error of $\boldsymbol{\beta}$, which is a common term in the prediction variance for all groups.

Example 5.19. The fitted model in Example 5.3 was

$$y_{ij} = 1.098 + 0.233d_{ij} + b_i^{(1)} + b_i^{(2)}d_{ij} + \epsilon_{ij}\,,$$

where

$$\widehat{\text{var}}(\boldsymbol{b}_i) = \begin{pmatrix} 0.0832 & -0.00699 \\ -0.00699 & 0.0832 \end{pmatrix}$$

and

$$\widehat{\text{var}}(\epsilon_{ij}) = 0.245^2\,.$$

Assume that the heights for diameters $(2.7, 10.8)$ cm have been observed to be $(2.3, 5.7)$ m on a sample plot. We use these values to predict the random effects for the plot in question to obtain a local height-diameter model. The R-implementation is shown at the end of the example.

The local model in the transformed scale is shown by the thin line in the left-hand graph of Figure 5.5. The right-hand graph shows the same information following back-transformation to the original Height-Diameter scale. The dashed lines show predictions based on a back-transformation without bias correction, i.e. using the estimates of $\boldsymbol{\beta}$ in the nonlinear Näslund's function. Because it is a convex function of y, the predictions constructed using this method are downward biased. Among the several possible bias corrections, we applied the correction based on the Taylor series (see Equation 10.22 p. 318). For the Näslund's model, the bias-corrected height prediction is

$$\widetilde{h} = \frac{d^2}{(\beta_1 + \beta_2 d)^2} + \frac{3d^2}{(\beta_1 + \beta_2 d)^4} \operatorname{var}(\widehat{y} - y) \,.$$

Ignoring the estimation errors of $\boldsymbol{\beta}$ and their correlation with $\widetilde{\boldsymbol{b}}_i - \boldsymbol{b}_i$, which is justified because our model fitting data is moderately large, our estimate of $\operatorname{var}(\widehat{y} - y)$ is $\widehat{\sigma}^2 + \boldsymbol{z}_0' \widehat{\operatorname{var}}(\widetilde{\boldsymbol{b}}_i - \boldsymbol{b}_i)\boldsymbol{z}_0$ for the prediction including predicted group effects, where $\boldsymbol{z}_0$ includes those values of the predictors of the random part for which the prediction is carried out. For the prediction based on the fixed part of the model only, the estimate of $\operatorname{var}(\widehat{y} - y)$ is $\widehat{\sigma}^2 + \boldsymbol{z}_0' \widehat{\boldsymbol{D}}_* \boldsymbol{z}_0$. The right-hand graph of Figure 5.5 shows the bias-corrected predictions using solid lines.

```
> h<-c(2.3,5.7)
> d<-c(2.7,10.8)
> y<-d/sqrt(h-1.3)
> D<-getVarCov(lmm3)
> X<-Z<-cbind(1,d)
> sigma<-lmm3$sigma
> R<-diag(rep(sigma^2,length(d)))
> beta<-fixef(lmm3)
> D
Random effects variance covariance matrix
             (Intercept)          d
(Intercept)    0.0832410 -0.0069851
d             -0.0069851  0.0035682
  Standard Deviations: 0.28852 0.059735
> Z
         d
[1,] 1  2.7
[2,] 1 10.8
> R
           [,1]       [,2]
[1,] 0.05991194 0.00000000
[2,] 0.00000000 0.05991194
> y
[1] 2.700000 5.148698
> X
         d
[1,] 1  2.7
[2,] 1 10.8
> ranef<-D%*%t(Z)%*%solve(Z%*%D%*%t(Z)+R)%*%(y-X%*%beta)
> fr<-beta+ranef
> fr
                 [,1]
(Intercept) 1.4812003
d           0.3259204
> beta
(Intercept)           d
  1.0978967   0.2332299
> ranef
                  [,1]
(Intercept) 0.38330363
d           0.09269055
> varbb<-D-D%*%t(Z)%*%solve(Z%*%D%*%t(Z)+R)%*%Z%*%D
> d0<-seq(1,15,0.1)
> X0<-Z0<-cbind(1,d0)
> predFixed<-X0%*%beta
> predFixRand<-X0%*%beta+Z0%*%ranef
> varPredFixed<-Z0%*%D%*%t(Z0)+sigma^2       # ignoring var(beta)
> varPredFixrand<-Z0%*%varbb%*%t(Z0)+sigma^2 # ignoring var(beta) and cov(beta,b)
>
> # Predictions of height, no bias correction
> predFixedH<-1.3+d0^2/(beta[1]+beta[2]*d0)^2
> predFixrandH<-1.3+d0^2/(fr[1]+fr[2]*d0)^2
>
> # Bias-corrected predictions of height
> predFixedHCorr<-predFixedH+3*d0^2/(beta[1]+beta[2]*d0)^4*diag(varPredFixed)
> predFixrandHCorr<-predFixrandH+3*d0^2/(fr[1]+fr[2]*d0)^4*diag(varPredFixrand)
```

5.4.3 A closer look at the variance component model

In the variance component model, we have

$$\boldsymbol{W}_i = \sigma^2 \boldsymbol{I}_{n_i} + \sigma_b^2 \boldsymbol{J}_{n_i} ,$$

where $\boldsymbol{J}_{n_i}$ is a square $n_i \times n_i$ matrix filled by ones. The inverse of this residual variance-covariance matrix can be expressed as (McCulloch *et al.* 2008, p. 345)

$$\boldsymbol{W}_i^{-1} = \frac{1}{\sigma^2}\left(\boldsymbol{I}_{n_i} - \frac{\sigma_b^2}{\sigma^2 + n_i\sigma_b^2}\boldsymbol{J}_{n_i}\right) . \tag{5.37}$$

Furthermore

$$\mathrm{cov}(\boldsymbol{b}_i, \boldsymbol{y}_i') = \sigma_b^2 \boldsymbol{1}_{n_i}' , \tag{5.38}$$

The EBLUP of the random effect for group i takes the form:

$$\begin{aligned} \widetilde{b}_i &= \sigma_b^2 \boldsymbol{1}_{n_i}' \left(\frac{1}{\sigma^2}\left(\boldsymbol{I}_{n_i} - \frac{\sigma_b^2}{\sigma^2 + n_i\sigma_b^2}\boldsymbol{J}_{n_i}\right)\right)(\boldsymbol{y}_i - \widehat{\mu}\boldsymbol{1}_{n_i}) \\ &= \frac{n_i\sigma_b^2}{\sigma^2 + n_i\sigma_b^2}(\bar{y}_{i.} - \widehat{\mu}) \\ &= \frac{\sigma_b^2}{\sigma^2/n_i + \sigma_b^2}(\bar{y}_{i.} - \widehat{\mu}) . \end{aligned} \tag{5.39}$$

Specifically, we can summarize the information of $\boldsymbol{y}_i$ to their mean $\bar{y}_{i.}$, which has the well-known variance $\mathrm{var}(\bar{y}_{i.}) = \sigma^2/n_i$. The predicted group mean is then a weighted mean of the marginal mean $\widehat{\mu}$ and the sample mean of the group:

$$\begin{aligned} \widetilde{y}_{i.} &= \widehat{\mu} + \widetilde{b}_i \\ &= w_1\widehat{\mu} + w_2\bar{y}_{i.} \end{aligned}$$

where the weights $w_1 = \frac{\sigma^2/n_i}{\sigma^2/n_i + \sigma_b^2}$ and $w_2 = \frac{\sigma_b^2}{\sigma^2/n_i + \sigma_b^2}$ are proportional to the variance of the random effect and the variance of the sample mean. Reorganizing gives

$$\bar{y}_{i.} - \widehat{\mu} = \widetilde{b}_i + w_2\left(\bar{y}_{i.} - \widehat{\mu}\right) ,$$

which shows that the observed difference between the sample mean and estimated population mean is divided into two components: the predicted random effect and a shrinkage, which are proportional to w_1 and w_2.

Example 5.20. The model `lmm1` described in Example 5.1 had $\widehat{\mu} = 12.30$, $\widehat{\sigma}_b^2 = 4.94^2$ and $\widehat{\sigma}^2 = 2.74^2$. Consider a sample plot where one randomly selected tree had a height of 17 m. Using these values in (5.39) (now $n_i = 1$ because only one tree was measured), the predicted mean height on this particular plot is 15.89 m.

```
12.30+4.94^2/(4.94^2+2.74^2)*(17-12.30)
[1] 15.89
```

A total of 20 trees have been measured on plot 1 of the modeling data. The mean height is 12.56 m, giving $\bar{y}_1 - \widehat{\mu} = 0.256$. Equation (5.39) gives $\widetilde{b}_1 = 0.252$, leading to predicted group mean of 12.557. These same values were extracted in Example 5.1 from the fitted model object using `ranef(lmm1)` and `coef(lmm1)`. The random effect is close to the observed difference $\bar{y}_1 - \widehat{\mu}$ because $n_1 = 20$ is rather high, but it is slightly shrunken towards zero. This implies that the prediction of group mean is shrunken towards the estimated population mean of 12.3 m.

The predicted group mean was 15.89 m in the beginning of the example. The observed mean height of 17 m was shrunken much more towards $\widehat{\mu} = 12.3$ because the mean was based on one observation only.

```
> mu<-12.304          # fixef(lmm1)
> s2b<-4.943^2        # getVarCov(lmm1) or VarCorr(lmm1)
> s2<-2.741^2         # lmm1$sigma
> ybar<-12.560        # mean(spati2$h[spati2$plot==1]) or coef(lm2)[1]
> resid<-ybar-mu
> resid
[1] 0.256
> n<-20               # sum(spati2$plot==1)
> s2b/(1/n*s2+s2b)*resid
[1] 0.252
> ranef(lmm1)[1,1]
[1] 0.252
```

5.4.4 Fixed or random group effect?

As already discussed at the beginning of the current chapter (see Equations (5.2) and (5.3)), the group effects can be treated either as fixed or random. The fundamental difference between these approaches is whether the groups can be naturally regarded as a sample from a population of groups. We now take an additional point of view by exploring the means and prediction/estimation errors of the group effects.

If the group effect is treated as fixed, then only the information of that particular group is used in estimation of the group effect; information related to the average group and variation of groups around it are not used. Under the OLS assumptions, the error variances of the estimated fixed effects are obtained from the diagonal of $\mathrm{var}(\widehat{\boldsymbol{\beta}}) = \sigma^2 \left(\boldsymbol{X}'\boldsymbol{X}\right)^{-1}$. If the group effect is treated as random, then the information of the mean and variability of group effects in the population of groups is utilized, in addition to the information of that particular group. The fitted model can be considered as prior information on group effects, which is updated using the observed data from the group in question. The approach is very similar to the Bayesian approach, and the EBLUP of random effects can be seen as a way to estimate the posterior mean in each group.

The difference can be easily demonstrated using the variance component model and the corresponding fixed-effects model; the one-way ANOVA model. We assume that the number of groups in the data is so large that the variance of $\widehat{\mu}$ can be ignored, and the only important source of uncertainty is the prediction error of the random effect. In the fixed-effects model (5.2), $\mathrm{var}(\widehat{\boldsymbol{\beta}})$ becomes

$$\mathrm{var}(\widehat{\beta}_i) = \frac{\sigma^2}{n_i} .$$

In the variance component model (5.3), the prediction error of random effects results from using (5.37) and (5.38) in (5.31). It gives

$$\mathrm{var}(\widetilde{b}_i - b_i) = \left(\frac{\sigma_b^2}{\sigma^2/n_i + \sigma_b^2}\right)\frac{\sigma^2}{n_i} \tag{5.40}$$

$$= \frac{\sigma_b^2}{1 + \frac{n_i\sigma_b^2}{\sigma^2}} . \tag{5.41}$$

Equation (5.40) shows that multiplication of the estimation variance of the fixed group effect by factor $\left(\frac{\sigma_b^2}{\sigma^2/n_i+\sigma_b^2}\right)$ gives the prediction error variance of a random group effect. This factor is always between 0 and 1, and approaches 1 as n_i gets larger. Therefore, the prediction error of a random group effect always has a smaller variance than the estimation error of a fixed group effect. The difference is large if group size is small, and vanishes as group size increases to ∞. Taking the estimation error of population mean $\widehat{\mu}$ into account lessens the difference between these two variances smaller, but the variance of the random group mean always remains smaller than that of the fixed group mean if the variances are known. Proof is left as an exercise. When $n_i = 0$, the fixed group effect cannot be estimated, and the estimation variance becomes infinity. For a random group effect, BLUP is $\widetilde{b}_i = 0$, and Equation (5.41) shows that $\text{var}(\widetilde{b}_i - b_i) = \sigma_b^2$.

Let us next take a closer look at the bias of BLUP. The predictions of random effects are *marginally* unbiased, i.e. their mean over the population of groups is equal to the mean of true random effects:

$$\text{E}(\widetilde{b}_i) = \text{E}(b_i) = 0\,.$$

However, the expected value of the prediction error conditional on the true random effect is not zero. This *conditional bias* of random effect is equal to the shrinkage of the group mean towards the population mean, which was discussed in Example 5.20.

$$\begin{aligned}
\text{E}\left(\left(\widetilde{b}_i - b_i\right)|b_i\right) &= \text{E}\left(\left(\left(\frac{\sigma_b^2}{\sigma^2/n_i + \sigma_b^2}\right)(\bar{y}_i - \mu) - b_i\right)|b_i\right) \\
&= \left(\frac{\sigma_b^2}{\sigma^2/n_i + \sigma_b^2}\right)\text{E}\left((\bar{y}_i - \mu)\,|b_i\right) - b_i \\
&= \left(\frac{\sigma_b^2}{\sigma^2/n_i + \sigma_b^2}\right) b_i - b_i \\
&= -\left(\frac{1}{1 + \frac{n_i\sigma_b^2}{\sigma^2}}\right) b_i\,.
\end{aligned} \tag{5.42}$$

If we repeatedly sample data from a given group, the mean of predicted random effects is not equal to the true random effect. Instead, positive values of a random effect are underestimated and negative values are overestimated. The absolute bias increases with increasing value of the random effect and decreasing sample size.

If the group effect is treated as fixed, we have

$$\text{E}(\widehat{\beta}_i) = \beta_i\,.$$

Therefore, the estimator is also conditionally unbiased. The price for the decreased variance of a random group mean compared to a fixed group mean is the conditional bias (compare also to Box 4.2, p. 91).

Example 5.21. In example 5.1, mean tree heights were estimated both by treating the group effects as fixed (model `lm2`) and random (model `lmm1`). Figure 5.6 (left) shows the estimated variances of fixed and random group effects using these models as a function of group size. The right-hand plot illustrates the conditional bias of random group effects in model `lmm1` for six different values of true group effect.

It might be interesting to explore the behavior of the MSE of the prediction error for a group effect as a function of b_i; our notation in this discussion does not distinguish between fixed and random group effect. The MSE of a random group effect is the sum of variance and squared bias and increases as a function of b_i because the conditional bias also increases. A fixed group effect is unbiased and the MSE is, therefore, equal to the variance σ^2/n_i, which is not affected by b_i. Therefore, the prediction MSE for a random group effect necessarily exceeds that of the fixed group effect, if b_i is sufficiently large. To compute the conditional prediction MSE of the random group effects, we need the conditional variance of the prediction error of the group effect. Ignoring the estimation error of $\widehat{\mu}$, (5.39) gives (show this!)

$$\text{var}(\widetilde{b}_i - b_i|b_i) = \left(\frac{\sigma_b^2}{\sigma^2/n_i + \sigma_b^2}\right)^2 \frac{\sigma^2}{n_i}\,,$$

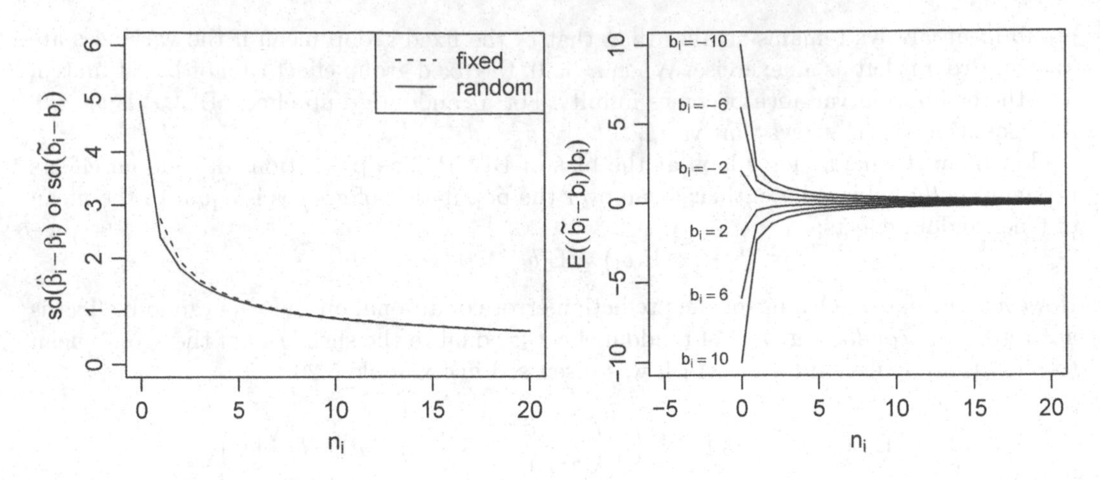

FIGURE 5.6 Standard error of fixed (dashed) and random (solid) group effects as a function of group size (left) and the conditional bias of random group effect (right) using models described in Example 5.1. Notice that when $n_i = 0$ then $\widehat{\beta}$ and sd $\widehat{\beta}_i$ are not defined, but $\widetilde{b}_i = 0$ and $\text{sd}(\widetilde{b}_i - b_i) = \sigma_b$.

which is added to the squared bias from Equation (5.42) to obtain $MSE(\tilde{b}_i - b_i|b_i)$. Exploring this function as a function of b_i shows that it is smallest when $b_i = 0$ and exceeds the variance of the fixed group effect when

$$|b_i| > \sqrt{\sigma^2/n_i + 2\sigma_b^2}\,.$$

The proportion of such groups in the population of groups is, however, so low that the marginal MSE (=variance of Equation 5.41) of random group effects is lower than the MSE (=variance) of a fixed group effect.

This subsection has demonstrated the variance and bias of fixed and random group means using the simple variance component model. The discussion can be generalized to other linear mixed-effects models as well. With continuous predictions, the regression lines/hyperplanes that would be obtained by group-specific OLS or GLS fits, are shrunken towards the population mean line/hyperplane. The degree of shrinkage depends on the variances and covariances of random effects, the observations of that particular group, and the variance-covariance structure of residual errors.

5.5 Evaluation of model fit

Different diagnostic plots are used to evaluate how well the assumptions that were made in model formulation are met in the data set. The diagnostic plots provide information designed to (i) improve model formulation, and (ii) evaluate the validity of the inference and tests. Everything that we discussed in Section 4.5 about diagnostic plots is also valid here. However, the grouped structure of the data leads to additional issues that should be taken into account in the evaluation.

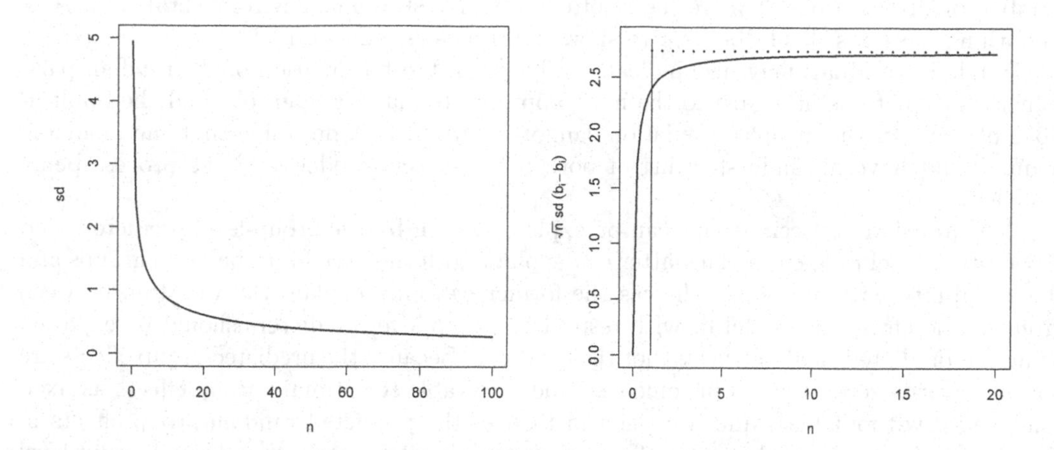

FIGURE 5.7 The graph on the left shows the same curve for the sd of the random effects prediction as shown in Figure 5.6 (left) on p. 168, only the range of n is different. In the graph on the right, the solid line shows $\sqrt{n}\,$sd for random effects prediction, the dashed line is for fixed effect estimation. Vertical line at $n = 1$ is also drawn in order to better show that the fixed effect estimation cannot be carried out for $n = 0$. See also Box 5.6.

Box 5.6 Plotting standard errors

Figure 5.6 (left) showed how standard errors depend on n when the group effect is estimated either as a random effect or a fixed effect. This figure is a common way to visualize the sd with respect to n. The sd seems to stabilize around $n = 5$. This is, however, a visual artefact. In Figure 5.7, the left-hand graph shows the sd associated with the random effect prediction for a different range of n. Now, the sd seems to stabilize around $n = 20$. Thus, the stabilization is an artefact depending on the range of n-values. In fact, sd does not stabilize before n is infinite.

In Figure 5.6 (left), it is difficult to see how the random effect prediction and fixed effect estimation are different. A better way to visualize differences is obtained by noting that in the fixed effect estimation (the reference method) the sd is proportional to $\frac{1}{\sqrt{n}}$. Thus, $\sqrt{n}\,$sd is constant, and plotting $\sqrt{n}\,$sd with respect to n does not encourage to see unwarranted stabilization points. If we plot $\sqrt{n}\,$sd in the same graph for the random effect prediction, the difference between these two methods can be seen better (Figure 5.7 (right)). Lappi (2001) plotted variances multiplied with n and this would be an alternative. Notice that in this example, the within-group variance is quite small in comparison to the between group variance, hence the estimation is already very efficient with a small value of n.

5.5.1 The systematic part

The plot of residuals ($\widetilde{\epsilon}_{ij} = y_{ij} - \boldsymbol{x}'_{ij}\widehat{\boldsymbol{\beta}} - \boldsymbol{z}'_{ij}\widetilde{\boldsymbol{b}}_i$) on all the available predictors and on the predicted value can be used to check for any trends in the mean of residual errors. The model fitting methods ensure that the slope of the average linear trend in these residuals is zero if the predictor is included in the model. Therefore, only nonlinear trends can be found. If such trends are found in a model where the random part has been properly formulated, they may indicate a lack of fit in terms of the predictors *that are defined at the observation level.* Approaches to fix these problems are similar to those used with linear models: nonlinear transformations to the predictors (including the cases where new nonlinear transformations of existing predictors are included) and nonlinear transformations to the response. If the number of observations is large, as it often is in forestry and environmental data sets, smoothing methods should be used to show the trends (see Section 4.5.2).

Trends in residuals may also indicate problems in the formulation of the random part. General instructions in regard to the most appropriate diagnostic graphs and the implications of them in the model formulation cannot be given. It is desirable that the modeling team should have an understanding of both mixed-effects models and the process being modeled.

The variation between groups can be explained both by the group-level predictors and observation-level predictors. The latter can explain both the variability between groups and the variability within groups, whereas the former can only explain the variation between groups. Therefore, the model fit with respect to group-level predictors should be explored using the predicted random effects, not the residuals. Because the predicted group effects are biased towards zero, it could in some cases be preferable to estimate group effects as fixed (and hence without bias) and use them instead of the predicted random group effects in the diagnostic graphs. If the group effects show trends with respect to potential group-level predictors, their inclusion would improve the model. If the trend is nonlinear, an appropriate nonlinear transformation should be applied to the group-level predictor. Inclusion of group-level predictors will usually not cause a large reduction in the residual standard error of the model. Instead, it would reduce the variances of random effects.

Important potential group-level predictors are the group-level aggregates of observation-level predictors, which we denote by $\dot{x}_i$. An intuitive example is the group-level mean of x_{ij}, $\bar{x}_i$. Other alternatives are the group size n_i, which is the sum of the observation-level indicator variable over the group, the group-level variance of x_{ij}, and the true group size N_i. In balanced controlled experiments, these variables do not vary between groups. In unbalanced (observational) data sets, they may show a high variability. Correlation of $\dot{x}_i$ with random effects indicates model misspecification: the expected value of random effects depends on $\dot{x}_i$ and is, therefore, not zero as we had assumed in model formulation. A detailed discussion of the topic is given in Section 6.3; see also a related discussion in Section 6.4. The correlation between covariates and random effects can be modeled by including aggregates $\dot{x}_i$ as predictors in the fixed part, with appropriate nonlinear transformations whenever needed. See Example 5.22 for such an analysis.

5.5.2 The random part

The assumptions about the second-order properties (variance and covariance) of random effect and residuals should also be evaluated. In addition, the shape of distribution may also be of interest, especially if the model is to be used for simulation.

Evaluation of the assumptions of the variance and covariance of residual errors is based on the observed residuals of the model and follows the same principles that were followed in the case of the linear model (see Section 4.5). For random effects, a common variance-covariance structure (through $\boldsymbol{D}_*$) and normality were assumed. The constant variance of random effects can be evaluated by exploring the behavior of each random effect with respect to appropriate group-level predictors and group-level means of predictions. In addition, the random effects can be plotted on each other to explore whether the relationships among them are linear. Normality can be further evaluated using normal q-q plots separately for all predicted random effects. However, even though all marginal distributions show normality and all relationships are linear, the joint distribution is not necessarily multivariate normal. Nevertheless, the constant variance and linear relationship of random effects already indicate that the second-order properties are sufficiently well modeled. Therefore, checking of marginal distributions of random effects and the pairwise joint distributions in most cases

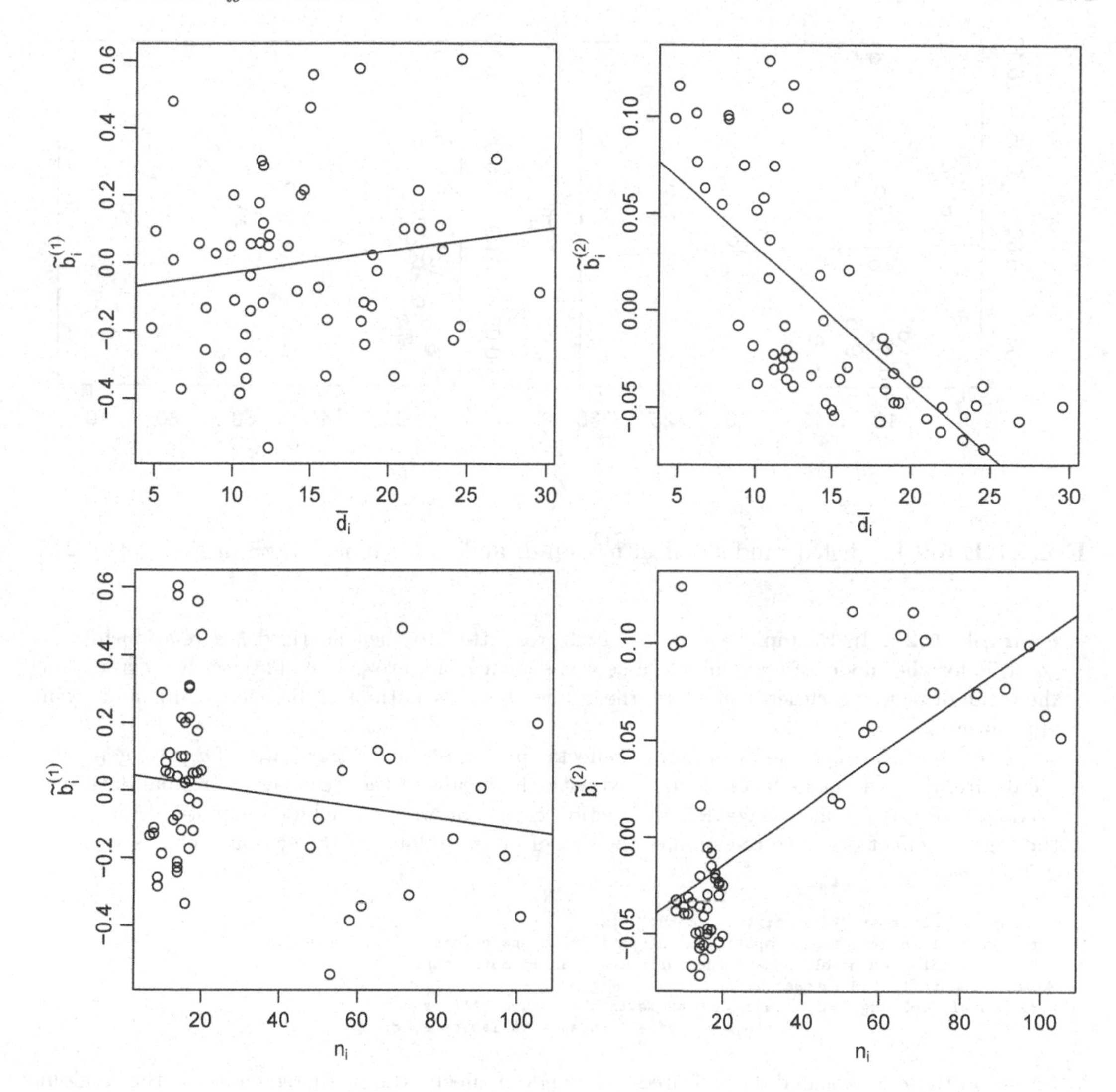

FIGURE 5.8 Predicted random effects of model `lmm3` on $\bar{d}_i$ and n_i in Example 5.22.

gives a sufficiently good picture of the distribution, and the remaining problems on distribution are usually unimportant. In addition, recall that the estimated regression coefficients and (RE)ML are quite robust to violations of normality (see Sections 5.3.1 – 5.3.3). The assumptions about the variance and covariance are more important.

In unbalanced data sets, the effect of shrinkage of the predicted random effects should be recognized. That is, small groups will show lower variability of the predicted random effects than large groups, even if the true variances of the random effects are equal. If group effects are estimated as fixed, small groups show larger variability in estimated group effects than large groups even if the true variances are equal.

In mixed-effects models, all model assumptions are seldom perfectly met. Therefore, there is always some uncertainty as to the quality of the inference based on the model. However, the model is robust to violations of some assumptions and less robust to others. See Chapter 12 in McCulloch *et al.* (2008) for discussion.

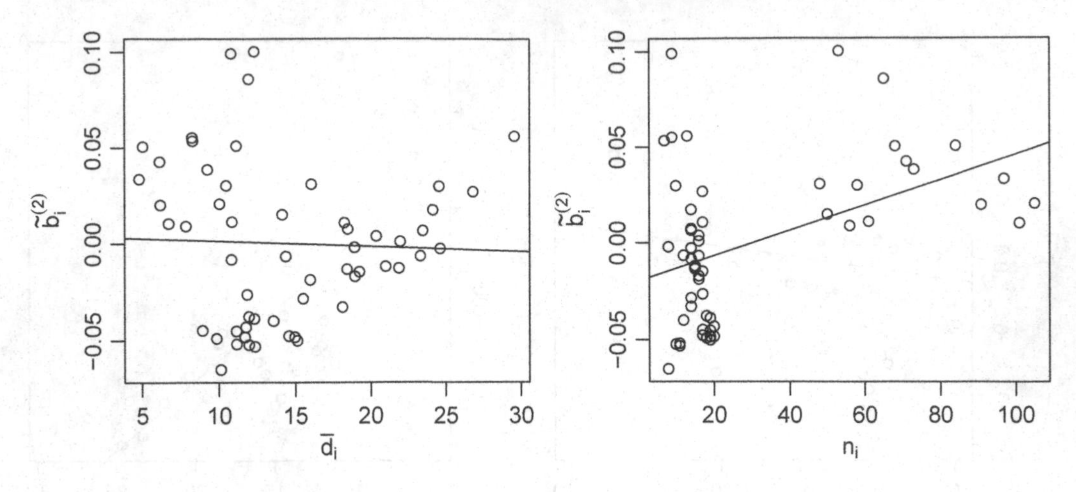

FIGURE 5.9 Predicted random effect $\widetilde{b}_i^{(2)}$ on $\bar{d}_i$ and n_i for model lmm5 in Example 5.22.

Example 5.22. In Example 5.3, model lmm3 was fitted to the linearized h-d relationship. A good fit for the model of residual variance was shown in Figure 5.4 (p. 148), with no trends in the mean of residuals either. Therefore, the applied transformations of diameter and height seem appropriate.

To explore the dependency of random effects on group-specific aggregates of d_{ij}, we generate a data frame of random effects, augmented with the means of the variables of original data set. To do so, we use argument aug=TRUE in function ranef. In order to include the plot-level mean in the random effect data, we re-estimate the model in an updated data set that includes variable $\bar{d}_i$ (dmean):

```
> dmean<-tapply(spati2$d,spati2$plot,mean)
> for (i in 1:length(dmean)) spati2$dmean[spati2$plot==names(dmean)[i]]<-dmean[i]
> # Re-estimate the model to have the variable dmean in data frame
> lmm3<-update(lmm3,data=spati2)
> re<-ranef(lmm3,aug=TRUE) # aug=TRUE augments the random effects with
>                          # the means of all dataset variables by groups.
```

Based on these augmented data of predicted random effects, the predicted values of the random effects on $\bar{d}_i$ and n_i are shown in Figure 5.8. The random intercept does not show a strong correlation with either of these variables, whereas the random effect of the slope does. The systematic part was, therefore, updated so that the slope is written as $\beta_2 = \alpha_1 + \alpha_2 \bar{d}_i$. Term n_i was not yet included because the correlation between $\bar{d}_i$ and n_i is moderately high (0.56), and $\bar{d}_i$ may also model the effect of n_i. The fixed part becomes $\beta_1 + \beta_2 d_{ij} = \beta_1 + \alpha_1 d_{ij} + \alpha_2 \bar{d}_i d_{ij}$.

```
> lmm5<-update(lmm3,fixed=y~d+d:dmean)
```

The diagnostic plots of the new model (Figure 5.9) still show a correlation between $\widetilde{b}_i^{(2)}$ and n_i, even though it is not as strong as it was in the previous model. The model was updated so that n_i is also used to explain the variability in the coefficient of d_{ij}.

```
> lmm6<-update(lmm3,fixed=y~d+d:dmean+d:n)
```

The random effect of slope no longer shows a clear correlation with n_i (Figure 5.10e).

The assumption about the common variance of random effects was explored by plotting the random effects against the group-specific mean of the population-level predictions (Figure 5.10b and 5.10c). No clear evidence of systematic variability in the variance was observed. The relationship between predicted random effects is fairly linear (Figure 5.10d), and the point cloud has a somewhat elliptic shape. Also, both marginal densities of random effects only show a slightly

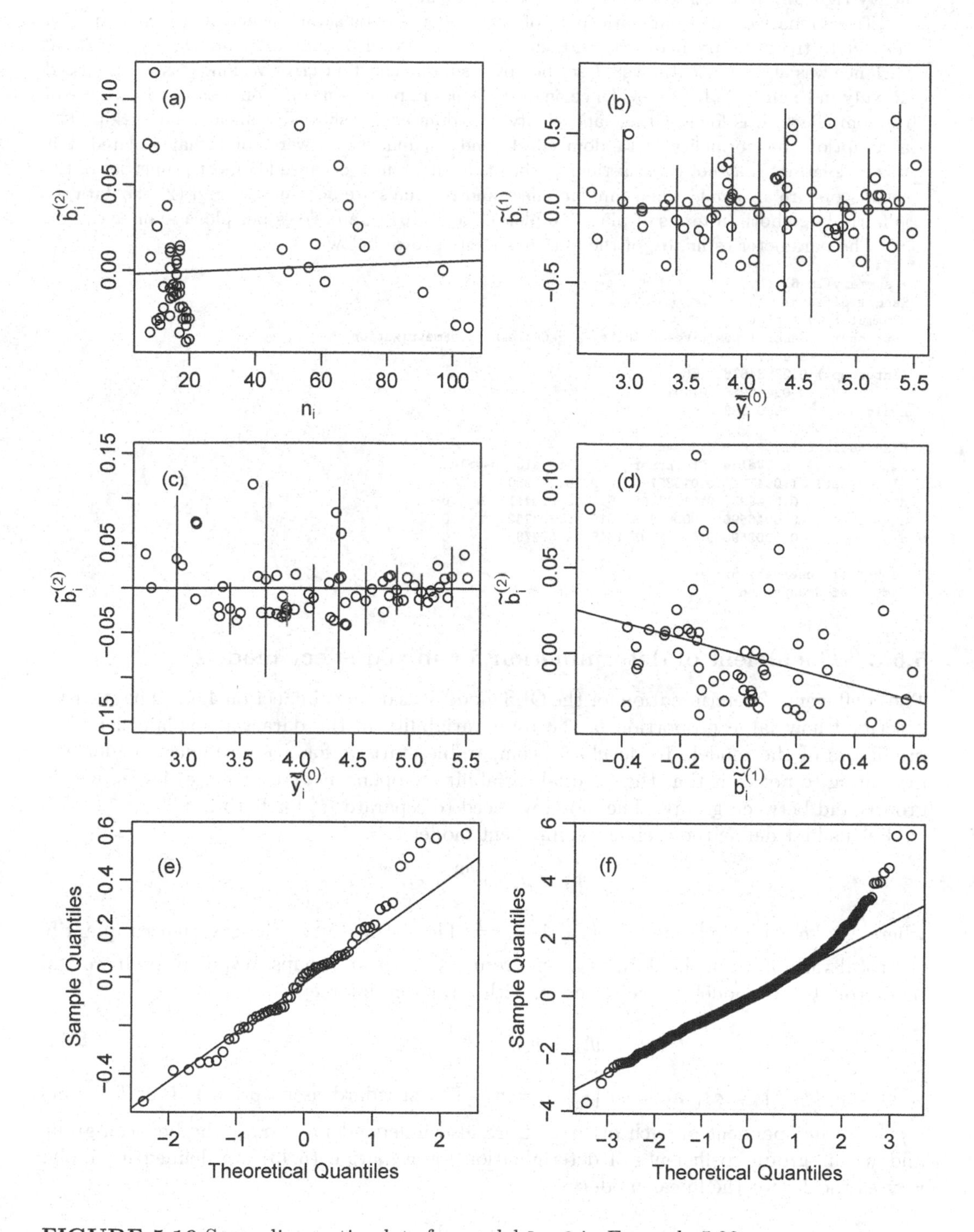

FIGURE 5.10 Some diagnostic plots for model `lmm6` in Example 5.22.

heavy right tail and a light left tail (shown only for the intercept in Figure 5.10e). However, a heavy right tail in the tree-level residuals is evident in Figure 5.10f.

To summarize, the systematic part of the model is satisfactory, when it is explored with respect to tree-level predictor x_{ij} and against the plot-level predictors $\bar{x}_i$ and n_i. The residual variance was also sufficiently well described by assuming constant error variance, as we discussed already in Example 5.10. No evidence against the assumptions on common mean and variance of random effects was found either, and the two random effects showed a linear relationship. The assumptions on normality of random effects and residual errors were somewhat violated, but these violations may not cause serious problems for inference, because (1) most properties of the estimators are also valid when random effect and residuals are not normal, and (2) the data set is fairly large both in terms of plots (56 plots), trees (1678), and trees per plot (30 on average).

The parameter estimates of the final model are shown below.

```
> summary(lmm6)
Random effects:
 Formula: ~d | plot
 Structure: General positive-definite, Log-Cholesky parametrization
            StdDev     Corr
(Intercept) 0.28584676 (Intr)
d           0.03460016 -0.401
Residual    0.24501583

Fixed effects: y ~ d + d:dmean + d:n
                 Value  Std.Error   DF   t-value p-value
(Intercept)  1.0947760 0.04335776 1619 25.249830       0
d            0.2748977 0.01737285 1619 15.823411       0
d:dmean     -0.0046590 0.00087853 1619 -5.303242       0
d:n          0.0008596 0.00018798 1619  4.572782       0

Number of Observations: 1678
Number of Groups: 56
```

5.5.3 Coefficient of determination for mixed-effect models

The coefficient of determination for the OLS model was defined in Section 4.5.3. The statistic described how large proportion of the total variability in the data was explained by the predictors of the model. To develop a comparable statistic for a mixed-effect model, the first thing to notice is that the residual variability happens at two (or more) levels: within groups and between groups. Therefore, we need to separate R^2 for both levels.

Let us first define the variance component model

$$y_{ij} = \mu + b_i{}^{(0)} + \epsilon_{ij}{}^{(0)}$$

where $\text{var}\left(b_i^{(0)}\right) = \sigma^2_{b0}$ and $\text{var}\left(\epsilon_{ij}^{(0)}\right) = \sigma^2_0$. The estimates of these variances quantify the total variability in the data between groups and within groups, respectively. If we add predictors to the model, we get a model with a random intercept:

$$y_{ij} = \boldsymbol{x}'_{ij}\boldsymbol{\beta} + b_i^{(1)} + \epsilon_{ij}{}^{(1)} \tag{5.43}$$

where $\text{var}\left(b_i^{(1)}\right) = \sigma^2_{b1}$ and $\text{var}\left(\epsilon_{ij}^{(1)}\right) = \sigma^2_1$. The standard assumption is that $b_i{}^{(1)}$ and $\epsilon_{ij}{}^{(1)}$ are independent of each other and are also independent of $\boldsymbol{x}_{ij}$. The between-group and within-group coefficients of determination for Equation (5.43) are defined in similar way as the R^2 for the linear model:

$$R^2_b = 1 - \frac{\sigma^2_{b1}}{\sigma^2_{b0}} \tag{5.44}$$

$$R^2_\epsilon = 1 - \frac{\sigma^2_{\epsilon 1}}{\sigma^2_{\epsilon 0}} \tag{5.45}$$

The above definition can also be extended for random slopes. For example, the basic reference model with one regressor with random slope can be:

$$y_{ij} = \beta_1 + \beta_2 x_{ij} + b_i^{(1)} + b_i^{(2)} x_{ij} + \epsilon_{ij} \tag{5.46}$$

When adding individual level predictors, group level predictors or group level predictors multiplied by x_{ij} (which explain $b_i^{(2)}$), we can either monitor the development of $\text{var}\left(b_i^{(1)}\right)$, $\text{var}\left(b_i^{(2)}\right)$ and $\text{var}\left(\epsilon_{ij}\right)$ directly, or we can monitor the development of the R-squares. In Section 6.3, we will discuss how R-squares can also decrease when predictors are added.

The suggested R-square can be extended to models with several random factors and to multilevel models. In the statistical literature, there are also other R-square indexes (e.g. Nakagawa and Scielzeth (2012)), although we prefer Equations 5.44 and 5.45 because they are easy to interpret.

5.6 Tests and confidence intervals

5.6.1 Hypothesis tests

Inference and tests of linear mixed-effects models are based on the procedures and results that were presented in Section 4.6. The Wald's F-tests or corresponding t-tests are suggested for fixed effects $\boldsymbol{\beta}$. If they are not possible, Wald's χ^2 tests are suggested. Likelihood Ratio (LR) tests of ML fits are also an option, but are not generally recommended because of the reported anti-conservative behavior in some cases. The LR test (of either REML or ML fit) is the only option for hypothesis tests about the variance-covariance structure of random effects and residual errors. A third alternative for testing is parametric bootstrapping. If the number of observations per group is sufficiently large, it is also possible to treat the group effects as fixed and then test whether the means are equal in all groups. For a variance component model, this means the one-way analysis of variance.

Tests on fixed effect terms are based on Equation (4.35). Two approaches were presented in Section 4.6.2 to deal with the estimation error of $\boldsymbol{V}$. These were (1) the use of the naive variance estimate, with adjustments to the denominator degrees of freedom, and (2) Kenward-Roger approximation for $\text{var}(\widehat{\boldsymbol{\beta}})$ and degrees of freedom.

In approach (1), at least two different methods have been used to adjust the denominator df. The approach of Pinheiro and Bates (2000, p. 91–92), which is used in `nlme::lme`, defines the degrees of freedom based on the level of predictor (i.e. group-level predictors have a different denominator df than observation-level predictors). The *Satterthwaite approximation* (Satterthwaite 1946, Casella and Berger 2001, p. 314) approximates the distribution of a weighted sum $\sum a_i X_i$ of χ^2-distributed random variables X_i with r_i degrees of freedom, which is tricky to derive exactly, by a χ^2-distribution where the degrees of freedom are $v_2 = \left(\sum a_i X_i\right)^2 / \sum \frac{a_i}{r_i} X_i^2$. A generalization of the Satterthwaite formula was presented in Giesbrecht and Burns (1985), see Stroup (2013, p. 167–168). This approximation has been commonly used to protect against narrow confidence intervals and the anti-conservative test results of the naive approach.

These three methods (Pinheiro and Bates, Satterthwaite or Kenward-Roger) may lead to significantly different results in small data sets. Which one should then be used? Verbeke and Molenberghs (2000) point out that in longitudinal data analysis, the data are usually so large that the differences between approaches are small. Stroup (2013, p. 196) summarizes the rather extensive simulation results of Guerin and Stroup (2000) as follows: Satterthwaite approximation does not necessarily provide an improvement over the naive approach; both produce up to 15% type I error rates using nominal $\alpha = 0.05$. In contrast, the Kenward-Roger approach led to observed p-values that ranged between 0.04–0.06. The Kenward-Roger approximation might, therefore, be the best justified approach for routine use but is not necessary in large data sets. The Kenward-Roger method is the only method presented by McCulloch *et al.* (2008, see p. 167–171) for tests and confidence intervals of fixed effects.

The R-function `nlme::lme` does not include the Satterthwaite and Kenward-Rogers method. The R-package `lme4` does not implement hypothesis tests for fixed effects at all. However, a separate R-package `lmerTest` (Kuznetsova *et al.* 2017) implements both the methods of Satterthwaite and Kenward-Roger for fits based on `lme4`. The differences between these methods are illustrated in Example 5.24. Functions `car::Anova` and `car::lht` provide Wald's χ^2 tests (4.39) (p. 115) of fixed effects for models fitted using both packages `nlme` and `lme4`. However, they ignore the estimation error of σ^2 and is suggested only for large data sets (both a large number of groups and a large number of observations per group).

The LR test can be used for all parameters of the model, not only for fixed effects. However, as already discussed in Section 4.6.8 it is generally not suggested for fixed effects. LR tests based either on ML or REML likelihood can be also used for the variance-covariance parameters, and they are commonly used for that purpose (e.g. Pinheiro and Bates (2000)). As we discussed already in Section 4.6.8, LR test about zero variance violates the precondition for the Wilk's theorem on the χ^2-distribution of the LR statistic. Therefore, the likelihood tests cannot necessarily be trusted for tests of form H_0 : variance $= 0$ (Verbeke and Molenberghs 2000). This problem occurs whenever the restriction is made through removal of random effects from the model. In practice, most meaningful tests on the random effects are of this form. One approach for testing whether the variances of random effects are zero is to estimate the random effects as fixed and then test whether the effects are equal using the F-test (Hui *et al.* 2019). See (Stroup 2013) for more discussion about testing of random effects.

We question the need for testing variance components. The main interest is usually related to the fixed effects. In that case, tests on the random effect structure are not necessary: one can just use the random effect structure supported by the experiment and do the tests on fixed effects. Slight overparameterization of the random part does not usually affect the inference on fixed effects: an estimated zero behaves in the same way as an assumed zero. Overparameterization of the random part, in particular, is often a much less serious problem than an overly simple random effect structure, which often implies an assumption of independence for dependent data. It is also possible to base the inference about the variance-covariance parameters on confidence intervals instead of tests.

Parametric bootstrapping is based on the simulation of a large number of data sets that follow the null model and are of the same size as the original data. Both the null model and the alternative model are fitted to each simulated data set and the test statistics (e.g. logarithmic likelihood ratio) are saved. They provide the reference distribution under the null hypothesis. The proportion of simulated data sets with a test statistic higher than the value in the observed data gives the p-value. Implementation of such a test is available for the most simple models in package `nlme`. Case-specific implementation is needed for more complex models. Function `bootMer` for parametric bootstrapping is available for models fitted using package `lme4`.

Example 5.23. In Example 5.22 a slight relationship was observed between $\bar{d}_i$ and the random intercept (Figure 5.8). Therefore, we consider the following systematic part for the model:

$$\mathrm{E}(y_{ij}) = \beta_1 + \beta_2 \bar{d}_i + \beta_3 d_{ij} + \beta_4 \bar{d}_i d_{ij} + \beta_5 n_i d_{ij}$$

and test hypothesis $\beta_2 = 0$, which can be specified as $H_0 : \boldsymbol{L\beta} = 0$ by defining

$$\boldsymbol{L} = \begin{pmatrix} 0 & 1 & 0 & 0 & 0 \end{pmatrix}$$

in Equation (4.36) (p. 114). The conditional F-tests of `lme`-fits can be carried out, for example, using the following three alternative commands. The fourth command gives the equivalent conditional t-test. Notice the relationship between the test-statistics: $0.7248^2 = 0.525$.

```
> lmm7<-update(lmm6,fixed=y~dmean+d+d:dmean+d:n)
> anova(lmm7,L=c(0,1,0,0,0))
F-test for linear combination(s)
[1] 1
  numDF denDF   F-value p-value
1     1    54 0.5252647  0.4717

> anova(lmm7,Terms=2)
F-test for: dmean
  numDF denDF   F-value p-value
1     1    54 0.5252647  0.4717

> anova(lmm7,type="m")[2,]
      numDF denDF   F-value p-value
dmean     1    54 0.5252647  0.4717

> round(summary(lmm7)$tTable[2,],4)
    Value Std.Error        DF   t-value   p-value
   0.0055    0.0076   54.0000    0.7248    0.4717
```

The p-value of 0.472 confirms that a weak relationship between the mean diameter and random intercept might result from random variability. Note also that because the predictor is group-specific, the denominator degrees of freedom are equal to $M - r - 1 = 54$, where M is the number of groups and $r = 1$ is the number of restrictions (rows) in $\boldsymbol{L}$.

The same hypothesis can also be tested using the LR test. It requires that the constrained and full models are estimated using maximum likelihood.

```
> lmm6ML<-update(lmm6,method="ML")
> lmm7ML<-update(lmm7,method="ML")
> anova(lmm6ML,lmm7ML)
       Model df      AIC      BIC    logLik   Test  L.Ratio p-value
lmm6ML     1  8 392.1806 435.5835 -188.0903
lmm7ML     2  9 393.6480 442.4763 -187.8240 1 vs 2 0.5325956  0.4655
```

The number of groups in our data set is large, and so are the number of observations per group. Therefore, the p-value of the LR test is very similar to the p-value of the conditional F-test.

Based on the marginality principle discussed earlier in Section 4.6.2, one might criticize the decision not to include the main effect of `dmean` in the model, which includes the interaction `dmean:d`. In this case, however, the interpretation of our parameters is not confused by dropping the main effect from the model. This is because the coefficient for `dmean` and `d:dmean` are determined through submodels for the primary parameters of the Näslund's model.

In addition, one might be willing to test whether the `dmean` and `dmean:d`, together, are needed in model `lmm7`, by testing $H_0 : \boldsymbol{L\beta} = \boldsymbol{0}$ where

$$\boldsymbol{L} = \begin{pmatrix} 0 & 1 & 0 & 0 & 0 \\ 0 & 0 & 0 & 1 & 0 \end{pmatrix}.$$

This test cannot be carried out as a conditional F-test, because `dmean` is plot-specific and `dmean:d` is tree-specific. Therefore, they have different denominator degrees of freedom. Such a test can be conducted using the LR test or Wald's χ^2 test. The Wald's χ^2 test is implemented below.

```
> L<-matrix(0,ncol=5,nrow=2)
> L[1,2]<-L[2,4]<-1
> anova(lmm7,L=L) # Wald F-test
Error in anova.lme(lmm7, L = L) :
  L may only involve fixed effects with the same denominator DF

> library(car)
```

```
> lht(lmm7,L)      # Wald chi^2 test
Linear hypothesis test

Hypothesis:
dmean = 0
dmean:d = 0

Model 1: restricted model
Model 2: y ~ dmean + d + dmean:d + d:n

  Df  Chisq Pr(>Chisq)
1
2  2 28.473  6.564e-07 ***
---
Signif. codes:  0 '***' 0.001 '**' 0.01 '*' 0.05 '.' 0.1 ' ' 1
```

The test confirms that at least one of the terms **dmean** or **dmean:d** should be kept in the model. The F-statistic corresponding to the χ^2 test statistic 28.473 would be 28.473/2=14.237; but we do not know its distribution under the null hypothesis.

The graphical evaluation shown in Example 5.10 indicated that the OLS assumption, in regard to the residual errors, might be sufficient. A formal test for this can be done through LR test of the models with and without the variance function. Formally, we test $H_0 : \delta = 0$ in the model that has the variance function (5.15) (p. 147).

```
> anova(lmm3,lmm4)
     Model df      AIC      BIC    logLik   Test  L.Ratio p-value
lmm3     1  6 460.1868 492.7317 -224.0934
lmm4     2  7 459.5534 497.5225 -222.7767 1 vs 2 2.633389  0.1046
```

The high $p-$value from the test confirms that restriction $\delta = 0$ leads to such a small decrease in the (REML-) likelihood that we do not have strong evidence against the null hypothesis of the constant variance.

The need for a random slope could be analyzed using the null hypothesis $H_0 : \mathrm{var}(b_i^{(2)}) = 0$; this assumption implicitly means that also $\mathrm{cov}(b_i^{(1)}, b_i^{(2)}) = 0$. It can be implemented using two-argument **anova** of REML fits:

```
> lmm8<-update(lmm6,random=~1|plot)
> anova(lmm6,lmm8)
     Model df      AIC      BIC    logLik   Test  L.Ratio p-value
lmm6     1  8 433.6877 477.0714 -208.8438
lmm8     2  6 742.3790 774.9168 -365.1895 1 vs 2 312.6913  <.0001
```

The low p-value indicates that the model with a random intercept and slope is significantly better than the model with a random intercept only. As discussed above, this test is questionable because the null hypothesis sets the variance parameter to the boundary of the parameter space.

A better approach for this hypothesis might be parametric bootstrapping, the use of which is demonstrated below. The 1000 simulated values of the LR statistic are shown in Figure 5.11, together with the asymptotic $\chi^2(2)$-density, which was used in the LR test. The observed test statistic (312.69) is much higher than the maximum of simulated statistics. Therefore, the p-value is below 1/1000 which strongly supports the need for random slope in the model. This confirms the previous conclusion, based on graphical evaluation in Example 5.3, about the need for a random slope. The shape of the histogram of simulated statistics is somewhat different to the χ^2 *pdf*, which is used as the null distribution in the LR test.

```
> sim<-simulate.lme(lmm8,m2=list(random=~1+d|plot),nsim=1000) # Takes some time
> LRnul<-2*(sim$alt$REML[,2]-sim$null$REML[,2])
> hist(LRnul,freq=FALSE)
> xapu<-seq(0,20,0.1)
> lines(xapu,dchisq(xapu,2))

> 2*(logLik(lmm6)-logLik(lmm8))
'log Lik.' 312.6913 (df=8)
```

Example 5.24. Consider model **lmm7** in Example 5.23. The tests conducted in Example 5.23 in regard to the fixed effects did not take the bias in the estimation of $\boldsymbol{V}$ into account. To adjust the

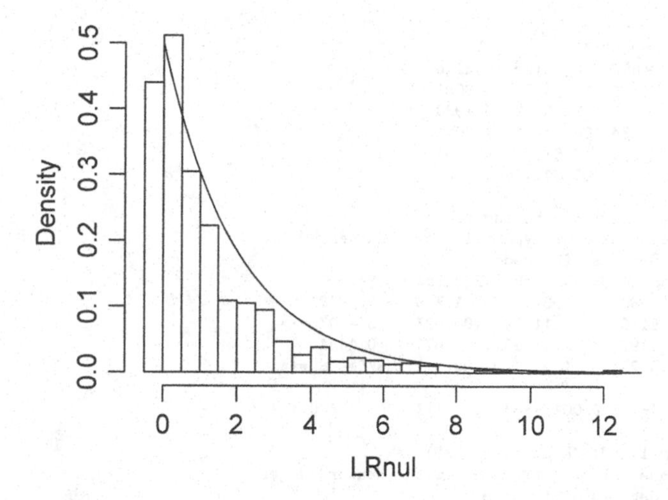

FIGURE 5.11 Simulated LR statistics between model `lmm6` and `lmm8` in Example 5.23. The histogram is based on the REML fit of both models in 1000 simulated data sets that follow model `lmm6`. The solid line shows the density of $\chi^2(2)$-distribution, which was the asymptotic approximate null distribution used in the LR test `anova(lmm6,lmm8)`.

test for the bias using the Satterthwaite or Kenward-Rogers methods, we refit the model using function `lmerTest::lmer` (Kuznetsova *et al.* 2017). Model fitting and the hypothesis test using the Satterthwaite approximation and Kenward-Rogers adjustments are shown below. Note that the tests in `lmer::anova` are based on marginal sums of squares by default, not on sequential sums of squares.

```
> library(lmerTest)
> lmm7lmer<-lmer(y~dmean+d+d:dmean+d:n+(1+d|plot),data=spati2)
> anova(lmm7lmer,ddf="Satterthwaite")
Analysis of Variance Table of type III  with  Satterthwaite
approximation for degrees of freedom
         Sum Sq Mean Sq NumDF  DenDF F.value    Pr(>F)
dmean    0.0315  0.0315     1 56.666   0.525    0.4716
d       14.0329 14.0329     1 53.809 233.724 < 2.2e-16 ***
dmean:d  1.5950  1.5950     1 48.276  26.565  4.70e-06 ***
d:n      1.2452  1.2452     1 49.386  20.740  3.46e-05 ***
---
Signif. codes:  0 '***' 0.001 '**' 0.01 '*' 0.05 '.' 0.1 ' ' 1

> anova(lmm7lmer,ddf="Kenward-Roger")
Analysis of Variance Table of type III  with  Kenward-Roger
approximation for degrees of freedom
         Sum Sq Mean Sq NumDF  DenDF F.value    Pr(>F)
dmean    0.0314  0.0314     1 55.942   0.523    0.4726
d       13.7512 13.7512     1 64.084 229.032 < 2.2e-16 ***
dmean:d  1.5773  1.5773     1 59.280  26.271 3.413e-06 ***
d:n      1.1971  1.1971     1 54.654  19.938 4.051e-05 ***
---
Signif. codes:  0 '***' 0.001 '**' 0.01 '*' 0.05 '.' 0.1 ' ' 1
```

A comparison to Example 5.23 shows that there are practically no meaningful differences in the p-values among the three methods in the test for β_2. However, our data set is quite large with approximately 1678 observations from 56 groups. For curiosity, we repeat the hypothesis tests using the same model fitted in a subset of 70 randomly selected trees from the original data.

```
> set.seed(123) # ensure that the same sample is the same in different runs
> spati2Small<-spati2[sample(1:dim(spati2)[1],size=70),]
> spati2Small$plot<-as.factor(spati2Small$plot)
> lmm7Small<-update(lmm7,data=spati2Small)
> lmm7lmerSmall<-lmer(y~dmean+d+d:dmean+d:n+(1+d|plot),data=spati2Small)
```

```
> anova(lmm7Small,type="m")
            numDF denDF   F-value p-value
(Intercept)     1    34  19.74137  0.0001
dmean           1    31   1.76602  0.1936
d               1    34 106.92681  <.0001
dmean:d         1    34   0.61031  0.4401
d:n             1    34  80.64140  <.0001

> anova(lmm71merSmall,ddf="Satterthwaite")
Analysis of Variance Table of type III  with  Satterthwaite
approximation for degrees of freedom
        Sum Sq Mean Sq NumDF  DenDF F.value    Pr(>F)
dmean   0.0563  0.0563     1 31.809   1.766    0.1933
d       3.4115  3.4115     1 11.142 106.927 4.739e-07 ***
dmean:d 0.0195  0.0195     1 13.252   0.610    0.4484
d:n     2.5729  2.5729     1 20.000  80.641 1.870e-08 ***
---
Signif. codes:  0 '***' 0.001 '**' 0.01 '*' 0.05 '.' 0.1 ' ' 1

> anova(lmm71merSmall,ddf="Kenward-Roger")
Analysis of Variance Table of type III  with  Kenward-Roger
approximation for degrees of freedom
         Sum Sq Mean Sq NumDF  DenDF F.value    Pr(>F)
dmean   0.04968 0.04968     1 29.339   1.557    0.2219
d       2.69966 2.69966     1 10.262  84.615 2.819e-06 ***
dmean:d 0.01681 0.01681     1  9.719   0.527    0.4851
d:n     2.12624 2.12624     1 34.478  66.643 1.442e-09 ***
---
Signif. codes:  0 '***' 0.001 '**' 0.01 '*' 0.05 '.' 0.1 ' ' 1
```

The differences in p-values are now substantially larger between the methods, but they still do not have an effect on our conclusions on any of the parameters, if we use $\alpha = 0.05$ as the significance level in the hypothesis test. However, if the p-values were close to the significance level, they could easily change the conclusions. Notice that in the small data set, the evidence for the dependence of random slope on the mean diameter of the plot is also so weak that the p-values range within 0.440–0.485.

5.6.2 Confidence intervals

The confidence intervals of fixed effects are based on the t-distribution of the regression coefficients, as discussed earlier in Section 4.6.10. The confidence intervals of variance-covariance parameters are based on the asymptotic multivariate normality of the variance-covariance parameters. The confidence intervals are computed in the parameterization (so called natural parameterization, see Pinheiro and Bates (2000)) that is used in estimation, and the endpoints are, thereafter, transformed to the reported scale. For example, in the case of standard deviations of random effects, the two-sided normality-based confidence intervals are first computed for $\theta = \log(\sigma_b^2)$ using

$$\widehat{\theta} \pm z_{\alpha/2}\widehat{\text{sd}}(\widehat{\theta}),$$

which results in a $100\alpha\%$ confidence interval (θ_l, θ_u). The $100\alpha\%$ confidence intervals for σ_b are, thereafter, computed as $(\sqrt{\exp(\theta_l)}, \sqrt{\exp(\theta_u)})$. Notice that this interval will never include 0 because the endpoints are kept positive through the exponential transformations.

Another alternative for computing confidence intervals is bootstrapping. `lme4::bootMer` provides a possibility for such computation. The marginal posterior MCMC samples in the computational Bayes methods can easily be used for computation of confidence intervals as well (see Example 5.17).

Example 5.25. The 95% confidence intervals of parameters of the final model `lmm6` are computed below.

```
> intervals(lmm6)
```

```
Approximate 95% confidence intervals

 Fixed effects:
                   lower          est.         upper
(Intercept)  1.0097327363  1.0947759536  1.179819171
d            0.2408220991  0.2748977320  0.308973365
d:dmean     -0.0063822022 -0.0046590351 -0.002935868
d:n          0.0004908728  0.0008595757  0.001228279

 Random Effects:
  Level: plot
                          lower        est.       upper
sd((Intercept))      0.22629340  0.28584676  0.36107271
sd(d)                0.02742615  0.03460016  0.04365073
cor((Intercept),d) -0.63124783 -0.40060804 -0.10487111

 Within-group standard error:
    lower      est.     upper
0.2365686 0.2450158 0.2537647
```

The confidence intervals based on parametric bootstrapping are computed below using function `lme4::bootMer` (Bates *et al.* 2015). We first refit the model again using `lme4::lmer` to model `lmm6lmer` and define a function `parms` that returns the parameters of interest from the model. Thereafter we compute 1000 parametric bootstrap samples, fit model `lmm6lmer` to each of them and save the parameters returned by function `parms`. Finally, we utilize two functions from the help page of `bootMer` to summarize the bootstrap samples. The resulting 95% confidence intervals based on bootstrapping are very similar to the parametric confidence intervals.

```
> parms<-function(model){
+        p<-c(fixef(model),                                        # fixed effects
+            attributes(VarCorr(model)[[1]])$stddev,               # sd's of random effects
+            attributes(VarCorr(model)[[1]])$correlation[1,2],# correlation
+            summary(model)$sigma)                                 # residual s.e.
+        names(p)[5:8]<-c("sdb1","sdb2","cor","sde")
+        p
+ }

> lmm6lmer<-lmer(y~d+d:dmean+d:n+(1+d|plot),data=spati2)
> library(lme4)
> boot<-bootMer(lmm6lmer,parms,nsim=1000) # Be patient, takes some time

> format.perc<-function(probs,digits){    # see ?bootMer
+             paste(format(100*probs,trim=TRUE,
+                   scientific=FALSE,digits=digits),"%")
+ }

> bCI.tab<-function(b,ind=length(b$t0),type="perc",conf=0.95){     # see ?bootMer
+          btab0<-t(sapply(as.list(seq(ind)),
+                 function(i)boot::boot.ci(b,index=i,conf=conf,type=type)$percent))
+          btab<-btab0[,4:5]
+          rownames(btab)<-names(b$t0)
+          a<-(1-conf)/2
+          a<-c(a,1-a)
+          pct<-format.perc(a,3)
+          colnames(btab)<-pct
+          return(btab)
+ }
> bCI.tab(boot)
                    2.5 %       97.5 %
(Intercept)  1.0059457858  1.185499983
d            0.2384635102  0.310417040
d:dmean     -0.0063122732 -0.002799417
d:n          0.0004812948  0.001254220
sdb1         0.2158508995  0.351248714
sdb2         0.0276219541  0.041796469
cor         -0.6261786343 -0.100798536
sde          0.2368700780  0.254081214
```

	Estimate	se	95% lb	95% ub
Fixed part				
β_1	1.095	0.0434	1.010	1.180
β_2	0.275	0.0174	0.241	0.309
β_3	-0.00466	0.000879	-0.00638	-0.00294
β_4	0.000860	0.000188	0.000491	0.001228
Random part and residual				
$\text{var}\left(b_i^{(1)}\right)$	0.286^2		0.226^2	0.361^2
$\text{var}\left(b_i^{(2)}\right)$	0.035^2		0.027^2	0.044^2
$\text{cor}\left(b_i^{(1)}, b_i^{(2)}\right)$	-0.401		-0.631	-0.105
σ^2	0.245^2		0.237^2	0.254^2

TABLE 5.3 Parameter estimates of model (5.47), which was fitted in Example 5.22.

5.7 Reporting a fitted mixed-effects model

The example below provides one possible way to report a linear mixed-effects model in a scientific journal. The mathematical formula of the model would probably belong in the "Methods" section and the table of parameter estimates in the "Results" section of a scientific article. In addition, the text in "Results" section would interpret the parameters of interest, with a focus on those parameters that are of interest in the particular research question. The table of parameter estimates might also include information about the hypothesis tests of selected fixed effects, if that is justified by the research question.

Example 5.26. A satisfactorily fitting model of Example 5.22 was

$$y_{ij} = \beta_1 + \beta_2 x_{ij} + \beta_3 \bar{x}_i x_{ij} + \beta_4 n_i x_{ij} + b_i^{(1)} + b_i^{(2)} x_{ij} + \epsilon_{ij} \tag{5.47}$$

where $\beta_1, \ldots, \beta_4$ are fixed regression coefficients, $(b_i^{(1)}, b_i^{(2)})$ are normally distributed random plot effects with mean zero and unknown unrestricted variance-covariance matrix, and, ϵ_{ij} is a residual error with mean zero and unknown variance of σ^2. The random effects are independent across plots and residual errors are independent across observations. The model was fitted using REML with package `nlme` (Pinheiro *et al.* 2016) of the `R`-software (R Core Team 2016). Table 5.3 reports the parameter estimates of the final model.

6

More about Linear Mixed-Effects Models

6.1 Nested grouping structure

6.1.1 A nested two-level model

In the nested model, we have observations (k) within subgroups (j) within groups (i). Such a grouping was discussed previously on p. 132. The nested two-level model with random intercept and slope at both levels of grouping is defined as

$$y_{ijk} = \boldsymbol{x}_{ijk}\boldsymbol{\beta} + a_i^{(1)} + a_i^{(2)} x_{ijk}^{(2)} + c_{ij}^{(1)} + c_{ij}^{(2)} x_{ijk}^{(2)} + \epsilon_{ijk} \,,$$

where i ($i = 1, \ldots, M$) refers to a group, j ($j = 1, \ldots, n_i$) to a subgroup and k ($k = 1 \ldots, n_{ij}$) to an observation. We assume that the random effects are normally distributed

$$\begin{pmatrix} a_i^{(1)} \\ a_i^{(2)} \end{pmatrix} = \boldsymbol{a}_i \sim N(\boldsymbol{0}, \boldsymbol{D}_a) \,,$$

$$\begin{pmatrix} c_{ij}^{(1)} \\ c_{ij}^{(2)} \end{pmatrix} = \boldsymbol{c}_{ij} \sim N(\boldsymbol{0}, \boldsymbol{D}_c) \,,$$

$$\epsilon_{ijk} \sim N(0, \sigma^2) \,.$$

The random effects are uncorrelated among groups:

$$\text{cov}(\boldsymbol{a}_i, \boldsymbol{a}'_{i'}) = \boldsymbol{0}$$

$$\text{cov}(\boldsymbol{c}_{ij}, \boldsymbol{c}'_{ij'}) = \boldsymbol{0}$$

and among levels:

$$\text{cov}(\boldsymbol{a}_i, \boldsymbol{c}'_{ij}) = \boldsymbol{0}$$

$$\text{cov}(\boldsymbol{c}_{ij}, \epsilon'_{ijk}) = \boldsymbol{0} \,.$$

A straightforward effect of having a nested grouping structure is that we can specify the expected values of y_{ijk} for subgroup j within group k as $\text{E}(y_{ijk}|\boldsymbol{a}_i, \boldsymbol{c}_{ij})$, for an average subgroup within group i as $\text{E}(y_{ijk}|\boldsymbol{a}_i)$, and for an average subgroup of an average group as $\text{E}(y_{ijk})$. Replacing the regression coefficients and random effects by their estimates and predictions gives the corresponding predictions at different levels of grouping as:

$$\begin{aligned}
\widetilde{y}_{ijk}^{(0)} &= \widehat{\beta}_1 + \widehat{\beta}_2 x_{ijk}^{(2)} + \widehat{\beta}_3 x_{ijk}^{(3)} + \ldots + \widehat{\beta}_p x_{ijk}^{(p)} \,, \\
\widetilde{y}_{ijk}^{(1)} &= \widehat{\beta}_1 + \widetilde{a}_i^{(1)} + \left(\widehat{\beta}_2 + \widetilde{a}_i^{(2)}\right) x_{ijk}^{(2)} + \widehat{\beta}_3 x_{ijk}^{(3)} + \ldots + \widehat{\beta}_p x_{ijk}^{(p)} \,, \\
\widetilde{y}_{ijk}^{(2)} &= \widehat{\beta}_1 + \widetilde{a}_i^{(1)} + \widetilde{c}_{ij}^{(1)} + \left(\widehat{\beta}_2 + \widetilde{a}_i^{(2)} + \widetilde{c}_{ij}^{(2)}\right) x_{ijk}^{(2)} + \widehat{\beta}_3 x_{ijk}^{(3)} + \ldots + \widehat{\beta}_p x_{ijk}^{(p)} \,,
\end{aligned}$$

Group (i)	Subgroup (j)	x	y	Group (i)	Subgroup (j)	x	y
1	1	13	19.4	2	2	28	36.3
1	1	26	32.1	2	3	18	25.3
1	1	18	21.4	3	1	16	31.0
1	2	22	31.8	3	1	15	30.3
1	2	12	20.1	3	2	19	29.9
2	1	27	33.9	3	2	24	34.4
2	2	12	22.6	3	3	25	35.8

TABLE 6.1 A small grouped data set with two nested levels of grouping.

where the hats and tildes are used on the estimates of parameters and predictions of random effects, respectively. The corresponding residuals can be specified at different levels as

$$\begin{aligned}\widetilde{e}_{ijk}^{(0)} &= y_{ijk} - \widetilde{y}_{ijk}^{(0)}, \\ \widetilde{e}_{ijk}^{(1)} &= y_{ijk} - \widetilde{y}_{ijk}^{(1)}, \\ \widetilde{e}_{ijk}^{(2)} &= y_{ijk} - \widetilde{y}_{ijk}^{(2)}.\end{aligned}$$

The term residual usually refers to the highest-level residual $\widetilde{e}_{ijk}^{(2)} = \widetilde{\epsilon}_{ijk}$, which is also the default in `R` method `residuals` for objects fitted using packages `nlme` and `lme4`. Residuals at other levels of grouping are obtained using argument `level`, which is assigned values from 0 (population level) to 2 (the subgroup-level residuals). Note that some authors call the subgroups (j) lower level groups compared to the groups (i), whereas others call the subgroups higher-level groups. We use the latter terminology.

6.1.2 Matrix formulation

The linear mixed-effects model for subgroup j within group i is written as

$$\boldsymbol{y}_{ij} = \boldsymbol{X}_{ij}\boldsymbol{\beta} + \boldsymbol{Z}_{i,j}\boldsymbol{a}_i + \boldsymbol{Z}_{ij}\boldsymbol{c}_{ij} + \boldsymbol{\epsilon}_{ij} \tag{6.1}$$

where $\boldsymbol{a}_i$ are the first-level random effects for group i, and $\boldsymbol{c}_{ij}$ are the second-level random effects for subgroup j within group i. Matrices $\boldsymbol{Z}_{i,j}$ and $\boldsymbol{Z}_{ij}$ are the corresponding model matrices. The structure of $\boldsymbol{Z}_{ij}$ is similar to the group-level model matrix $\boldsymbol{Z}_i$ in model (5.13) but $\boldsymbol{Z}_{i,j}$ is different. From the complete group-level model matrix of group i, $\boldsymbol{Z}_i$, matrix $\boldsymbol{Z}_{i,j}$ includes only those rows that are related to subgroup j. These matrices are best illustrated through an example.

Example 6.1. Consider the data set shown in Table 6.1. Model (6.1) with a random intercept and slope at both levels of grouping is specified for subgroup 1 of group 1 by defining

$$\boldsymbol{y}_{11} = \begin{pmatrix} 19.4 \\ 32.1 \\ 21.4 \end{pmatrix}, \boldsymbol{Z}_{1,1} = \begin{pmatrix} 1 & 13 \\ 1 & 26 \\ 1 & 18 \end{pmatrix}, \boldsymbol{Z}_{11} = \begin{pmatrix} 1 & 13 \\ 1 & 26 \\ 1 & 18 \end{pmatrix},$$

$$\boldsymbol{a}_1 = \begin{pmatrix} a_1^{(1)} \\ a_1^{(2)} \end{pmatrix}, \boldsymbol{c}_{11} = \begin{pmatrix} c_{11}^{(1)} \\ c_{11}^{(2)} \end{pmatrix}, \boldsymbol{\epsilon}_{11} = \begin{pmatrix} \epsilon_{111} \\ \epsilon_{112} \\ \epsilon_{113} \end{pmatrix},$$

$$\boldsymbol{D}_a = \begin{pmatrix} \mathrm{var}\left(a_i^{(1)}\right) & \mathrm{cov}\left(a_i^{(1)}, a_i^{(2)}\right) \\ \mathrm{cov}\left(a_i^{(1)}, a_i^{(2)}\right) & \mathrm{var}\left(a_i^{(2)}\right) \end{pmatrix}, \boldsymbol{D}_c = \begin{pmatrix} \mathrm{var}\left(c_{ij}^{(1)}\right) & \mathrm{cov}\left(c_{ij}^{(1)}, c_{ij}^{(2)}\right) \\ \mathrm{cov}\left(c_{ij}^{(1)}, c_{ij}^{(2)}\right) & \mathrm{var}\left(c_{ij}^{(2)}\right) \end{pmatrix}$$

$$\boldsymbol{R}_1 = \sigma^2 \boldsymbol{I}_{3\times3}.$$

The model for the entire group i is

$$\boldsymbol{y}_i = \boldsymbol{X}_i\boldsymbol{\beta} + \boldsymbol{Z}_i^{(a)}\boldsymbol{a}_i + \boldsymbol{Z}_i^{(c)}\boldsymbol{c}_i + \boldsymbol{\epsilon}_i \tag{6.2}$$

where

$$\boldsymbol{y}_i = \begin{pmatrix} y_{i1} \\ y_{i2} \\ \vdots \\ y_{in_i} \end{pmatrix}, \boldsymbol{X}_i = \begin{pmatrix} \boldsymbol{X}_{i1} \\ \boldsymbol{X}_{i2} \\ \vdots \\ \boldsymbol{X}_{in_i} \end{pmatrix}, \boldsymbol{a}_i = \begin{pmatrix} a_i^{(1)} \\ a_i^{(2)} \end{pmatrix}, \boldsymbol{c}_i = \begin{pmatrix} c_{i1} \\ c_{i2} \\ \vdots \\ c_{in_i} \end{pmatrix},$$

$$\boldsymbol{Z}_i^{(a)} = \begin{pmatrix} \boldsymbol{Z}_{i,1} \\ \boldsymbol{Z}_{i,2} \\ \vdots \\ \boldsymbol{Z}_{i,n_i} \end{pmatrix}, \boldsymbol{Z}_i^{(c)} = \begin{pmatrix} \boldsymbol{Z}_{i1} & \boldsymbol{0} & \cdots & \boldsymbol{0} \\ \boldsymbol{0} & \boldsymbol{Z}_{i2} & \cdots & \boldsymbol{0} \\ \vdots & \vdots & \ddots & \vdots \\ \boldsymbol{0} & \boldsymbol{0} & \cdots & \boldsymbol{Z}_{in_i} \end{pmatrix}, \boldsymbol{\epsilon}_i = \begin{pmatrix} \epsilon_{i1} \\ \epsilon_{i2} \\ \vdots \\ \epsilon_{in_i} \end{pmatrix}.$$

The vector of fixed parameters is as previously defined.

The two separate terms of random effects can be pooled by defining

$$\boldsymbol{b}_i = \begin{pmatrix} \boldsymbol{a}_i \\ \boldsymbol{c}_i \end{pmatrix}, \boldsymbol{Z}_i = \begin{pmatrix} \boldsymbol{Z}_i^{(a)} & \boldsymbol{Z}_i^{(c)} \end{pmatrix}.$$

The model for group i gets the familiar form

$$\boldsymbol{y}_i = \boldsymbol{X}_i\boldsymbol{\beta} + \boldsymbol{Z}_i\boldsymbol{b}_i + \boldsymbol{\epsilon}_i\,.$$

Now

$$\boldsymbol{D}_i = \mathrm{var}(\boldsymbol{b}_i) = \begin{pmatrix} \boldsymbol{D}_a & \boldsymbol{0} & \ldots & \boldsymbol{0} \\ \boldsymbol{0} & \boldsymbol{D}_c & \ldots & \boldsymbol{0} \\ \vdots & \vdots & & \vdots \\ \boldsymbol{0} & \boldsymbol{0} & \ldots & \boldsymbol{D}_c \end{pmatrix},$$

where $\boldsymbol{D}_c$ is replicated on the diagonal n_i times, i.e. once for each subgroup j of group i. If the number of subgroups per group varies, then $\boldsymbol{D}_i$ also varies between groups. The variance-covariance matrix of residual errors in group i becomes

$$\boldsymbol{R}_i = \mathrm{var}(\boldsymbol{\epsilon}_i) = \begin{pmatrix} \boldsymbol{R}_{i1} & \boldsymbol{0} & \ldots & \boldsymbol{0} \\ \boldsymbol{0} & \boldsymbol{R}_{i2} & \ldots & \boldsymbol{0} \\ \vdots & \vdots & \ddots & \vdots \\ \boldsymbol{0} & \boldsymbol{0} & \ldots & \boldsymbol{R}_{in_i} \end{pmatrix}.$$

As before (see page 146), we also have

$$\mathrm{var}(\boldsymbol{y}_i) = \boldsymbol{Z}_i\boldsymbol{D}_i\boldsymbol{Z}_i' + \boldsymbol{R}_i$$

and

$$\mathrm{cov}(\boldsymbol{b}_i, \boldsymbol{y}_i') = \boldsymbol{D}_i\boldsymbol{Z}_i'\,.$$

We have now defined the model for group i. Model (6.2) is of the same form as the model (5.13) for a single group in the context of single level of grouping. Definition of the model for the entire data set is done in the same way as we did earlier in the context of model (5.16), and it will not be repeated here. The model of the entire data set can now be written in the marginal form (5.17) (p. 149), and the estimation of the model parameters and prediction of random effects follows the procedures described earlier in Sections 4.4 and

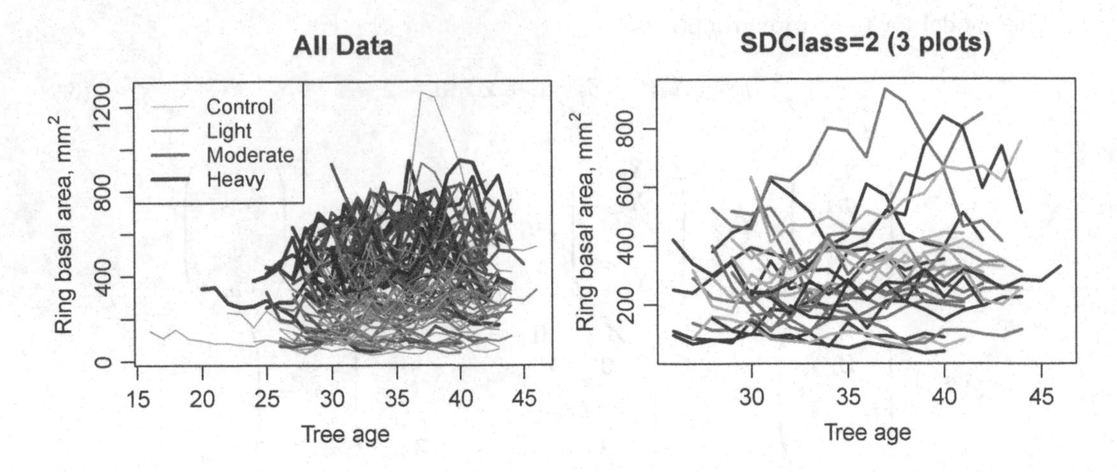

FIGURE 6.1 The basal area of increment cores as a function of tree age in all data (left) and in treatment 2 (light thinning, right). The lines connect observations of a tree.

5.3. The prediction of random effects needs to be carried out for a complete independent block of the data. Therefore, we need to predict the complete vector $\boldsymbol{b}_i$, including the effect for group i ($\boldsymbol{a}_i$) and for all subgroups j ($\boldsymbol{c}_{i1}, \ldots, \boldsymbol{c}_{in_i}$) within group i; see Example 7.12 for an implementation in the context of a nonlinear model. Of course, we can also predict the group effects simultaneously for all data using the matrices and vectors defined for the entire data set.

The discussion in Section 5.5 in regard to model diagnostics and model building extends to multiple levels as well. The assumptions about the random effects should be explored at every level of grouping. Certain random effects are usually justified at a given level only if the corresponding effects are included at the lower levels of grouping and in the fixed part of the model. The fixed effects, which are group-specific at a certain level of grouping, mostly explain the variability of the random effects of that level. The potential extensions of the model include the same possibilities outlined before (see Sections 4.3 and 5.2.2).

Web Example 6.2. Formulating the matrices for group 1 and for the entire data set shown in Table 6.1.

Example 6.3. Data set `lmfor::afterthin` includes repeated measurements of tree growth (measured as the cross-sectional area of the increment core at a height of 1.3 m above the ground) over a 17-year follow-up period after thinning. The data set includes observations from 88 trees (subgroups) from 10 sample plots (groups). The measurements are based on felled trees in a thinning experiment, where one of four thinning treatments (control, light, moderate and heavy (H) thinning) has been randomly assigned to the sample plots. The aim is to explain the variation in the tree-level annual growth using thinning treatment and tree age.

The tree-level growth is measured by the basal area growth, denoted as Ring Basal Area (RBA_{ijk}). It is better related to biomass and volume growth than the diameter growth because it takes into account the geometry related to the initial tree size: a certain diameter growth corresponds to greater basal area growth for trees with large initial diameter than for a tree with small initial diameter.

Visually, the age trend in the data seems quite weak, but the thinning intensity shows a clear effect on the basal area growth (Figure 6.1). We also see that the growth level varies among trees. In addition, there might be a common plot effect for all trees within a given plot. Start

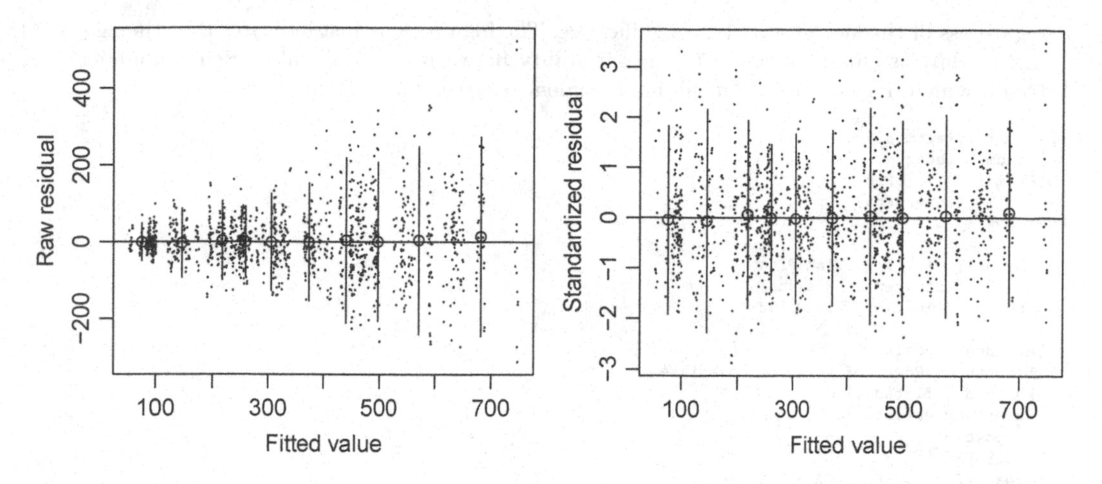

FIGURE 6.2 The raw and Pearson residuals of model `aftmod2` on fitted values at tree level.

with model

$$RBA_{ijk} = \beta_1 + \beta_2 AGE_{ijk} + \beta_3 LIGHT_i + \beta_4 MOD_i + \beta_5 HEAVY_i + a_i + c_{ij} + \epsilon_{ijk} ,$$

where $\beta_1, \ldots, \beta_5$ are fixed parameters; AGE_{ijk} is the age of tree j of plot i in year k; a_i is the random effect for plot i and c_{ij} is the random effects for tree j of plot i. Based on initial analyses (not shown), we already know that the residuals have increasing variance. Therefore, we assume that residuals are independent with variance $\text{var}(\epsilon_{ijk}) = \sigma^2 \widehat{y}_{ijk}^{2\delta}$. The following script fits the model and produces a residual plot:

```
> aftmod2<-lme(RBA~CA+SDClass,
+             random=~1|Plot/Tree,
+             data=afterthin,
+             weights=varPower())

> plot(predict(aftmod2),resid(aftmod2),ylab="Raw residual",xlab="Fitted value",cex=0.1)
> mywhiskers(predict(aftmod2),resid(aftmod2),add=TRUE,se=FALSE)
> abline(0,0)

> plot(predict(aftmod2),resid(aftmod2,type="p"),
+      ylab="Standardized residual",xlab="Fitted value",cex=0.1)
> mywhiskers(predict(aftmod2),resid(aftmod2,type="p"),add=TRUE,se=FALSE)
> abline(0,0)
```

The raw residuals (Figure 6.2, left) are heteroscedastic, which indicates the need for a variance function. The Pearson residuals do not show heteroscedasticity (Figure 6.2, right), indicating that the inconstant error variance was satisfactorily modeled by the power function. Also other plots (not shown) have indicated a good fit of the model. However, the model has problems that will be discussed later in this example and Example 6.5. Ignoring them for now, we illustrate hypothesis testing about the fixed effects.

```
> anova(aftmod2,type="m")
            numDF denDF  F-value p-value
(Intercept)     1  1230 70.26091  <.0001
CA              1  1230  1.45630  0.2278
SDClass         3     6  8.13339  0.0155
```

The tests indicate a significant effect of thinning on the basal area growth. Tree age is not a significant predictor. However, based on the general knowledge, we know that tree growth is affected by tree age: the tree basal area cannot increase constantly without an upper limit, as stated by our null hypothesis $H_0 : \beta_2 = 0$ states. For that reason, we retain tree age in the model regardless of the lack of statistical significance. The high p-value just indicates that the age trend in this data is small compared to the variability between years. We also tried a nonlinear age trend, which did not provide much improvement over the linear trend.

```
> summary(aftmod2)
Random effects:
 Formula: ~1 | Plot
        (Intercept)
StdDev:   0.1071784

 Formula: ~1 | Tree %in% Plot
        (Intercept)  Residual
StdDev:    161.1598 0.7492602

Variance function:
 Structure: Power of variance covariate
 Formula: ~fitted(.)
 Parameter estimates:
    power
0.8015073
Fixed effects: RBA ~ CA + SDClass
                Value Std.Error   DF   t-value p-value
(Intercept) 285.31798  34.03864 1230  8.382178  0.0000
CA           -0.41779   0.34621 1230 -1.206773  0.2278
SDClass2     29.27130  45.46107    6  0.643876  0.5435
SDClass3    145.44541  49.12042    6  2.960997  0.0253
SDClass4    216.97987  49.99053    6  4.340419  0.0049
```

The above output shows the estimates of model parameters. We have $\widehat{\text{var}}(a_i) = 0.11^2$, $\widehat{\text{var}}(c_{ij}) = 161^2$, $\widehat{\text{var}}(\epsilon_{ijk}) = 0.75^2 \widehat{y}_{ijk}^{2\times 0.80}$, $\widehat{\beta}_1 = 285$, $\widehat{\beta}_2 = -0.42$, $\widehat{\beta}_3 = 29.3$, $\widehat{\beta}_4 = 145$ and $\widehat{\beta}_5 = 217$. Because the estimated tree-level variance is much higher than the plot-level variance, the majority of the variation between trees is random variation between trees and is not explained by the plot effects. The difference in mean RBA between the light thinning treatment and the control is 29.3 mm^2. The difference is small compared to the unexplained variability between trees and plots, and is, therefore, not statistically significant (p=0.54). RBA in the moderate thinning treatment is, on average, 145 mm^2 higher than in the control, and the difference between the heavy thinning treatment and the control is 217 mm^2. These effects are statistically significant.

Even though the data includes several trees per plot and several measurements per tree, the number of sample plots is very limited; we have only 10 sample plots (i.e. 2–3 independent replicates per treatment). However, our analysis showed that the variability among plots is not that high, and that much higher variability occurs among trees of the plot. Therefore, the dependence caused by the plots is not that strong. The number of trees per plot is rather high, which means that it is quite unlikely that all trees of a plot grow in a similar way just by chance. Therefore, we are quite confident that the variability between plots is really associated with the thinning treatment, not with random variability among the plot means.

The model could be further improved by modeling the possible temporal autocorrelation among the residual errors. However, before doing such modeling, the possible effects of calendar years should also be modeled. This leads to a model with crossed calendar year effects, which will be discussed in Section 6.2.

The plot-level random effects for the first five plots are

```
> ranef(aftmod2,level=1)[1:5,]
[1]  5.634024e-05 -4.831965e-05  1.398525e-04 -5.917305e-05 -1.137422e-04
```

The tree-level random effects for the first five trees are much larger:

```
> ranef(aftmod2,level=2)[1:5,]
[1] -180.57920  178.25568  198.03843 -172.11372   96.68526
```

The population level, plot-level, and tree level coefficients for the first five trees are:

```
> fixef(aftmod2)
(Intercept)          CA    SDClass2    SDClass3    SDClass4
285.3179783  -0.4177929  29.2712966 145.4454051 216.9798721
```

```
> coef(aftmod2,level=1)
  (Intercept)         CA SDClass2 SDClass3 SDClass4
1    285.3180 -0.4177929  29.2713 145.4454 216.9799
2    285.3179 -0.4177929  29.2713 145.4454 216.9799
3    285.3181 -0.4177929  29.2713 145.4454 216.9799
 < part of the output was removed >

> coef(aftmod2,level=2)[1:5,]
      (Intercept)         CA SDClass2 SDClass3 SDClass4
1/79     104.7388 -0.4177929  29.2713 145.4454 216.9799
1/111    463.5737 -0.4177929  29.2713 145.4454 216.9799
1/181    483.3565 -0.4177929  29.2713 145.4454 216.9799
1/218    113.2043 -0.4177929  29.2713 145.4454 216.9799
1/243    382.0033 -0.4177929  29.2713 145.4454 216.9799
```

To illustrate the problems that result from ignoring the grouped structure of the data, we fit three models that do not properly model the grouping and error variance-covariance structure of the data. The p-values would be very different from those using `aftmod2`, and the conclusions will also differ. These results are presented as a warning and are not to be trusted at all; recall the discussion on page 64.

```
> aftmod2NoReNoVF<-gls(RBA~CA+SDClass,data=afterthin)
> aftmod2NoReVF<-gls(RBA~CA+SDClass,data=afterthin,weights=varPower())
> aftmod2PlotOnlyVF<-lme(RBA~CA+SDClass,random=~1|Plot,
+                        data=afterthin,weights=varPower())

> anova(aftmod2NoReNoVF,type="m")
Denom. DF: 1314
            numDF   F-value p-value
(Intercept)     1   0.32360  0.5695
CA              1  71.49302  <.0001
SDClass         3 108.29755  <.0001

> anova(aftmod2NoReVF,type="m")
Denom. DF: 1314
            numDF   F-value p-value
(Intercept)     1   0.54994  0.4585
CA              1  75.97326  <.0001
SDClass         3 105.26550  <.0001

> anova(aftmod2PlotOnlyVF,type="m")
            numDF denDF  F-value p-value
(Intercept)     1  1308  0.46076  0.4974
CA              1  1308 75.80077  <.0001
SDClass         3     6 28.98446  0.0006
```

6.1.3 More than two nested levels

Model formulation for a nested two-level model started with the highest (innermost) level of grouping. The model for one subgroup was initially formulated in a similar way as the group-level model in the context of single level of grouping. After defining the subgroup-level model, we pooled the subgroup-level models of each group to obtain the group-level model. Thereafter, formulation of the marginal model for the entire data set followed the principles outlined in Section 5.2.3.

Definition of a nested multilevel model follows the same principles. We first define the model for the highest level of grouping. These models are pooled for each second-highest level group, which are further pooled for each third-highest level group etc., until we finally pool the lowest-level groups, resulting in the model for the complete data set. Interpreting all random effects as a part of the residual produces the marginal model, which provides a way to define the likelihood for the data. All random parameters are now included in the matrix $\boldsymbol{V}$, the structure of which is specified by the grouping of the data set, the random effect structure, and the structure of $\boldsymbol{R}$.

6.2 Crossed grouping structure

In the model with crossed random effects, the multiple levels cannot be hierarchically ordered. A typical example of a crossed grouping structure is a data set of repeated measurements with a common year effect. For example, the increment of different trees in a given calendar year may be similar because of the weather conditions in that year within the (restricted) study area. On the other hand, repeated observations of a tree are also similar due to the properties of the tree and its environment. A data set with two crossed levels of grouping can be presented as a two-dimensional table where the columns and rows represent the different groups at the two levels of grouping (recall Table 5.1 and discussion therein). In particular, observations from all cells in Table 5.1 (p. 133) are possible, which was not the case with nested levels of grouping.

As an example, consider the model with two crossed levels of grouping. Assuming random intercepts at both levels, the model is

$$y_{ij} = \boldsymbol{x}_{ij}\boldsymbol{\beta} + a_i + c_j + \epsilon_{ij} \,, \tag{6.3}$$

where $a_i \sim N(0, \sigma_a^2)$, $c_j \sim N(0, \sigma_c^2)$, and $\epsilon_{ij} \sim N(0, \sigma^2)$. The random effects and residuals for the different levels and for the different groups at a certain level are independent. The difference compared to the nested model is quite small in the equations: only the index of the random effect of the second level has changed from c_{ij} to c_j and index k has been dropped. However, there is a big difference in the assumed variance-covariance structure in the data, as we will see shortly.

Model (6.3) implicitly assumes that the data includes only one observation for each combination of the two groups. This is justified in many data sets. For example, we can have only one basal area growth measurement per tree within one calendar year, or a certain tree can be observed only once on a certain aerial image. However, some data sets may include replicates for each combination. We may have, for example, several plants of a certain genotype on a sample plot, or we may move several eggs of the same biological mother into the same bird nest. For such data sets, the grouping structure is fully modeled by a three-level model, where the third level is nested within both crossed levels of grouping:

$$y_{ijk} = \boldsymbol{x}_{ijk}\boldsymbol{\beta} + a_i + c_j + d_{ij} + \epsilon_{ijk} \,, \tag{6.4}$$

where $d_{ij} \sim N(0, \sigma_d^2)$ and $\epsilon_{ijk} \sim N(0, \sigma^2)$; the assumptions in regard to other random effects are as before. The additional level can also be seen as an interaction between the two crossed levels in the same way as the interaction, in two-way ANOVA.

The matrix formulation of the model with crossed effects cannot commence with group-level models, because the observations of the different groups are dependent through the other levels of grouping. Therefore, independent blocks of observations do not exist, and the vectors and matrices need to be directly defined for the entire data set. To illustrate model formulation for data with crossed levels of grouping, we consider an extension of the two-level model (6.3) in more detail. After the reader understands that model, formulation for (6.4) and the more complex grouping structures is rather straightforward.

Consider a data set with two crossed levels of grouping, having n groups at the first level of grouping and m groups at the second level of grouping. The associated group-specific random effects are $\boldsymbol{a}_i$ and $\boldsymbol{c}_j$, respectively. Let us pool the random effects into vectors $\boldsymbol{a} = (\boldsymbol{a}_1', \ldots, \boldsymbol{a}_n')'$ and $\boldsymbol{c} = (\boldsymbol{c}_1', \ldots, \boldsymbol{c}_m')'$. The model for the entire data set is

$$\boldsymbol{y} = \boldsymbol{X}\boldsymbol{\beta} + \boldsymbol{Z}^{(a)}\boldsymbol{a} + \boldsymbol{Z}^{(c)}\boldsymbol{c} + \boldsymbol{\epsilon} \,, \tag{6.5}$$

where $\boldsymbol{Z}^{(a)}$ and $\boldsymbol{Z}^{(c)}$ are the model matrices of the random part for the first and second levels of grouping, respectively. The model matrices are constructed in a similar manner to the model matrices for the entire data set with a single level of grouping. That is, $\boldsymbol{Z}^{(a)}$ is similar to the model matrix for a single-level model that ignores the second level of grouping. Correspondingly, $\boldsymbol{Z}^{(c)}$ is similar to the model matrix of a model that ignores the first level of grouping. With a single level of grouping, the observations are usually ordered according to the groups so that the model matrices become block-diagonal. In the current case, this is not possible for both levels of grouping simultaneously: if the data are ordered according to the first level of grouping (associated with $\boldsymbol{a}$), then $\boldsymbol{Z}^{(c)}$ cannot have the block-diagonal structure and vice versa. Of course, the ordering of observations has no effect on the model itself.

We define the group-level variance-covariance matrix of random effects for the first level of grouping as $\boldsymbol{D}_*^{(a)}$ and for the entire data set as

$$\boldsymbol{D}^{(a)} = \begin{pmatrix} \boldsymbol{D}_*^{(a)} & \boldsymbol{0} & \dots & \boldsymbol{0} \\ \boldsymbol{0} & \boldsymbol{D}_*^{(a)} & \dots & \boldsymbol{0} \\ \vdots & \vdots & \ddots & \vdots \\ \boldsymbol{0} & \boldsymbol{0} & \dots & \boldsymbol{D}_*^{(a)} \end{pmatrix},$$

where the diagonal repeats $\boldsymbol{D}_*^{(a)}$ as many times as there are groups in the first level of grouping. Correspondingly, the block-diagonal matrix $\boldsymbol{D}^{(c)}$ repeats $\boldsymbol{D}_*^{(c)}$ on its diagonal as many times as there are groups in the second level of grouping.

Model (6.5) can now be specified as

$$\boldsymbol{y} = \boldsymbol{X\beta} + \boldsymbol{Zb} + \boldsymbol{\epsilon},$$

where

$$\boldsymbol{Z} = \begin{pmatrix} \boldsymbol{Z}^{(a)} & \boldsymbol{Z}^{(c)} \end{pmatrix}, \boldsymbol{b} = \begin{pmatrix} \boldsymbol{a} \\ \boldsymbol{c} \end{pmatrix}, \boldsymbol{D} = \begin{pmatrix} \boldsymbol{D}^{(a)} & \boldsymbol{0} \\ \boldsymbol{0} & \boldsymbol{D}^{(c)} \end{pmatrix}.$$

Therefore, we were able to define the model as the marginal model (5.17) (p. 149), where the grouped structure of the data defines the variance-covariance structure. These formulations are the basis of (RE)ML estimation of the model and prediction of random effects, which follows the same principles that were presented in Sections 4.4 and 5.3. However, the log-likelihood cannot be simplified to include the sums over the first level groups, because such mutually independent groups of observations do not exist. In addition, the random effects need to be predicted simultaneously for the entire data set. Therefore, fitting a model with crossed effects is computationally more intensive than fitting a nested model, involving computation with a variance-covariance matrix with dimensions equal to the total number of observations.

Example 6.4. Consider the data set in Table 6.2, where the grouping is crossed. All observations that belong to the same group at the second level of grouping are assumed to be correlated (e.g. observations 1, 4, 6 and 7 all belong to group 1). The same holds for members of the groups at the first level of grouping (e.g. observations 1, 2 and 3 are correlated).

Assuming a random intercept at both levels of grouping, the model for the entire data set is

$$\boldsymbol{y} = \boldsymbol{X\beta} + \boldsymbol{Z}^{(a)}\boldsymbol{a} + \boldsymbol{Z}^{(c)}\boldsymbol{c} + \boldsymbol{\epsilon},$$

Observation	Grouping level 1 (i)	Grouping level 2 (j)	x	y
1	1	1	13	19.4
2	1	2	22	31.8
3	1	2	12	20.1
4	2	1	27	33.9
5	2	2	12	22.6
6	3	1	16	31.0
7	3	1	15	30.3
8	3	2	19	29.9

TABLE 6.2 A small grouped data with two crossed levels of grouping

where

$$\boldsymbol{y} = \begin{pmatrix} 19.4 \\ 31.8 \\ 20.1 \\ 33.9 \\ 22.6 \\ 31.0 \\ 30.3 \\ 29.9 \end{pmatrix}, \boldsymbol{Z}^{(a)} = \begin{pmatrix} 1 & 0 & 0 \\ 1 & 0 & 0 \\ 1 & 0 & 0 \\ 0 & 1 & 0 \\ 0 & 1 & 0 \\ 0 & 0 & 1 \\ 0 & 0 & 1 \\ 0 & 0 & 1 \end{pmatrix}, \boldsymbol{Z}^{(c)} = \begin{pmatrix} 1 & 0 \\ 0 & 1 \\ 0 & 1 \\ 1 & 0 \\ 0 & 1 \\ 1 & 0 \\ 1 & 0 \\ 0 & 1 \end{pmatrix}, \boldsymbol{a} = \begin{pmatrix} a_1 \\ a_2 \\ a_3 \end{pmatrix}, \boldsymbol{c} = \begin{pmatrix} c_1 \\ c_2 \end{pmatrix}.$$

The two random effect terms can be pooled by defining

$$\boldsymbol{Z} = \begin{pmatrix} \boldsymbol{Z}_a & \boldsymbol{Z}_c \end{pmatrix} = \begin{pmatrix} 1 & 0 & 0 & 1 & 0 \\ 1 & 0 & 0 & 0 & 1 \\ 1 & 0 & 0 & 0 & 1 \\ 0 & 1 & 0 & 1 & 0 \\ 0 & 1 & 0 & 0 & 1 \\ 0 & 0 & 1 & 1 & 0 \\ 0 & 0 & 1 & 1 & 0 \\ 0 & 0 & 1 & 0 & 1 \end{pmatrix}, \boldsymbol{b} = \begin{pmatrix} \boldsymbol{a} \\ \boldsymbol{c} \end{pmatrix} = \begin{pmatrix} a_1 \\ a_2 \\ a_3 \\ c_1 \\ c_2 \end{pmatrix}.$$

Therefore,

$$\boldsymbol{D} = \begin{pmatrix} \boldsymbol{D}^{(a)} & \boldsymbol{0} \\ \boldsymbol{0} & \boldsymbol{D}^{(c)} \end{pmatrix} = \begin{bmatrix} \sigma_a^2 & 0 & 0 & 0 & 0 \\ 0 & \sigma_a^2 & 0 & 0 & 0 \\ 0 & 0 & \sigma_a^2 & 0 & 0 \\ 0 & 0 & 0 & \sigma_c^2 & 0 \\ 0 & 0 & 0 & 0 & \sigma_c^2 \end{bmatrix}, \boldsymbol{R} = \sigma^2 \boldsymbol{I}_8 ,$$

and

$$\boldsymbol{W} = \boldsymbol{Z}\boldsymbol{D}\boldsymbol{Z}' + \boldsymbol{R} =$$

$$\begin{bmatrix} \mathrm{var}(e_{ij}) & \sigma_a^2 & \sigma_a^2 & \sigma_c^2 & 0 & \sigma_c^2 & \sigma_c^2 & 0 \\ \sigma_a^2 & \mathrm{var}(e_{ij}) & \sigma_a^2+\sigma_c^2 & 0 & \sigma_c^2 & 0 & 0 & \sigma_c^2 \\ \sigma_a^2 & \sigma_a^2+\sigma_c^2 & \mathrm{var}(e_{ij}) & 0 & \sigma_c^2 & 0 & 0 & \sigma_c^2 \\ \sigma_c^2 & 0 & 0 & \mathrm{var}(e_{ij}) & \sigma_a^2 & \sigma_c^2 & \sigma_c^2 & 0 \\ 0 & \sigma_c^2 & \sigma_c^2 & \sigma_a^2 & \mathrm{var}(e_{ij}) & 0 & 0 & \sigma_c^2 \\ \sigma_c^2 & 0 & 0 & \sigma_c^2 & 0 & \mathrm{var}(e_{ij}) & \sigma_a^2+\sigma_c^2 & \sigma_a^2 \\ \sigma_c^2 & 0 & 0 & \sigma_c^2 & 0 & \sigma_a^2+\sigma_c^2 & \mathrm{var}(e_{ij}) & \sigma_a^2 \\ 0 & \sigma_c^2 & \sigma_c^2 & 0 & \sigma_c^2 & \sigma_a^2 & \sigma_a^2 & \mathrm{var}(e_{ij}) \end{bmatrix},$$

where the diagonal elements are $\mathrm{var}(e_{ij}) = \sigma_a^2 + \sigma_c^2 + \sigma^2$.

The multilevel model with nested random effects can be seen as a special case of a model that allows crossed effects, where the groups in the second level of grouping are coded in

Group (i)	Subgroup (j)	x	y	Group (i)	Subgroup (j)	x	y
1	1	13	19.4	2	4	28	36.3
1	1	26	32.1	2	5	18	25.3
1	1	18	21.4	3	6	16	31.0
1	2	22	31.8	3	6	15	30.3
1	2	12	20.1	3	7	19	29.9
2	3	27	33.9	3	7	24	34.4
2	4	12	22.6	3	8	25	35.8

TABLE 6.3 A version of the data set in Table 6.1, where the second-level groups have been recoded so that a model with crossed levels is equivalent to the model with nested levels of grouping.

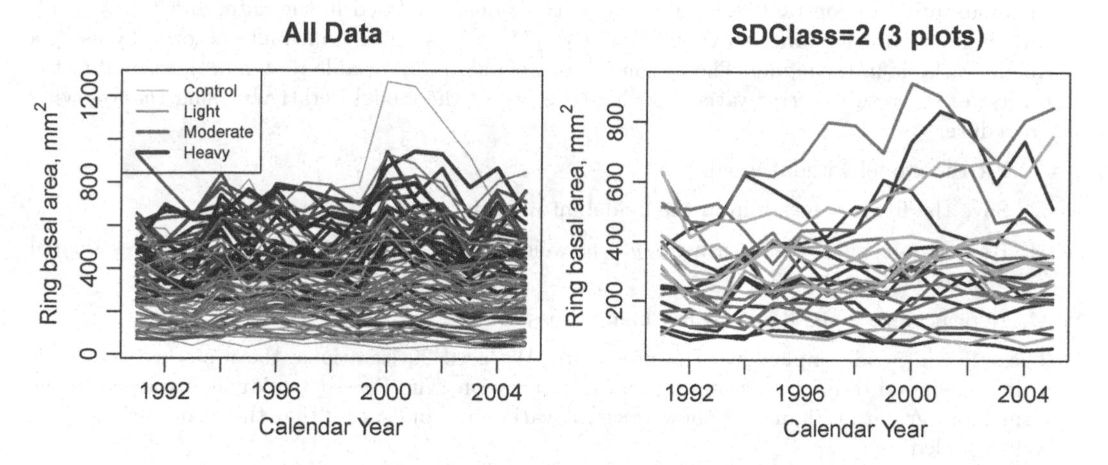

FIGURE 6.3 Observed Ring Basal Area (RBA) after four thinning treatments vs. calendar year for the entire `afterthin` data set (left), and the light thinning treatment only (right) in Example 6.5.

such a way that the same group id never occurs in two different groups in the lower level of grouping (see Table 6.3 for an example). Then matrix $\boldsymbol{Z} = (\boldsymbol{Z}^{(a)}, \boldsymbol{Z}^{(c)})$ has the same columns as the matrix $\boldsymbol{Z}$ for the entire data set for two nested levels (see Example 6.2), and vector $\boldsymbol{b} = (\boldsymbol{a}', \boldsymbol{c}')'$ has the same elements as $\boldsymbol{b}$. The columns of $\boldsymbol{Z}$ and the elements of $\boldsymbol{b}$ are just in a different order.

Handling of crossed random effect structures in the `nlme` package is not very convenient; the model definition is complicated, and the estimation is time-consuming. However, package `lme4` (Bates *et al.* 2015) can handle such structures much more efficiently. Unfortunately, `lme4` does not provide similar flexible tools for modeling inconstant error variance and dependence among observations as `nlme` does. If such modeling is needed, one should use the iterative procedures described in Section 5.3.4, as illustrated in the example below.

Example 6.5. Figure 6.3 shows the same data set as illustrated in Figure 6.1. However, *RBA* has now been plotted against calendar year, instead of tree age. The graph indicates that the relative level of growth has been similar for different trees in a given calendar year. For example, we can see that most trees have had quite low growth level in year 1993, and high growth level in years 1994 and 2000. However, the tree effects are much more prominent: trees that grow well or poorly do so every year. The data structure suggests a model with crossed tree and year effects, and these observations indicate that there is non-zero variability, both at the year- and tree-levels. In addition, we have the plot effect. The full model becomes

$$RBA_{ijk} = \beta_1 + \beta_2 AGE_{ijk} + \beta_3 LIGHT_i + \beta_4 MOD_i + \beta_5 HEAVY_i + a_i + c_{ij} + d_k + \epsilon_{ijk} \,,$$

where a_i is the plot effect, c_{ij} is the tree effect nested within the plot, and d_k is the crossed year effect.

We use package `lme4` to fit this model. Nesting is taken into account when encoding the factors, as illustrated in Example 6.3. Therefore, the interactions with plot and tree need to be specified as a new grouping variable (`Tree2` below).

```
> # Recode tree so that numbering does not start from 1 for each plot.
> # Each tree will now get a unique id in variable "Tree2".
> afterthin$Tree2<-with(afterthin,factor(Plot:Tree))
```

From the previous analysis using `nlme::lme` in Example 6.3 (p. 186), we already know that the assumption of constant residual variance is seriously violated in the data, and the variance can be modeled using function $\mathrm{var}(\epsilon_{ijk}) = \sigma^2 \tilde{y}_{ijk}^{2\times 0.802}$. This can be taken into account by using a weight variable in the model. The weight in `lmer` should be a variable that is proportional to the inverse of the residual error variance. Therefore, we fit the model iteratively using the following procedure:

1. Fit the model without weighting.
2. Save the fitted values $\tilde{y}$ from the model into the data set.
3. Re-estimate the model by using $1/\tilde{y}^{1.6}$ as weight. The exponent 1.6 is based on the estimated δ from model `aftmod2`.
4. Repeat steps 2 and 3 sufficiently many times.

The following code implements the procedure. We used 20 iterations; the first three decimals of log-likelihood did not change after the 15th iteration. The Pearson residuals displayed in the right-hand graph in Figure 6.4 show constant variability, indicating that the error variance was well modeled.

```
> aftmod3<-lmer(RBA~SDClass+CA+(1|Plot)+(1|Tree2)+(1|Year),data=afterthin)
> # We could use RBA~SDClass+CA+(1|Plot/Tree)+(1|Year) as well.
> for (i in 1:20){
+       cat(i,logLik(aftmod3),"\n")
+       afterthin$ytilde<-fitted(aftmod3)
+       aftmod3<-lmer(RBA~SDClass+CA+(1|Plot)+(1|Tree2)+(1|Year),
+                     weights=1/ytilde^{1.6},
+                     data=afterthin)
+ }
1 -7878.233
2 -7708.199
.
.
18 -7677.962
19 -7677.962
20 -7677.962

> plot(fitted(aftmod3),resid(aftmod3,type="pearson"),cex=0.1)        # The plot on the right hand side
> mywhiskers(fitted(aftmod3),resid(aftmod3,type="pearson"),add=TRUE,se=FALSE)
```

The script below shows the estimated variance components. The plot-level variance is very low, and one might wish to simplify the model by assuming that the variance between plots is zero. This is done in model `aftmod4`. This restriction is also supported by the LR test (but recall the warnings about its use in Section 5.6) and AIC, which are also computed below. However, using model `aftmod3` as the final model (e.g. to do inference on fixed effects) would not be problematic here; the models are practically identical: The between-plot variance in `aftmod3` was just estimated to be zero, whereas in `aftmod4` it was assumed to be zero.

```
> summary(aftmod3)
Random effects:
 Groups   Name        Variance  Std.Dev.
 Tree2    (Intercept) 2.575e+04 1.605e+02
```

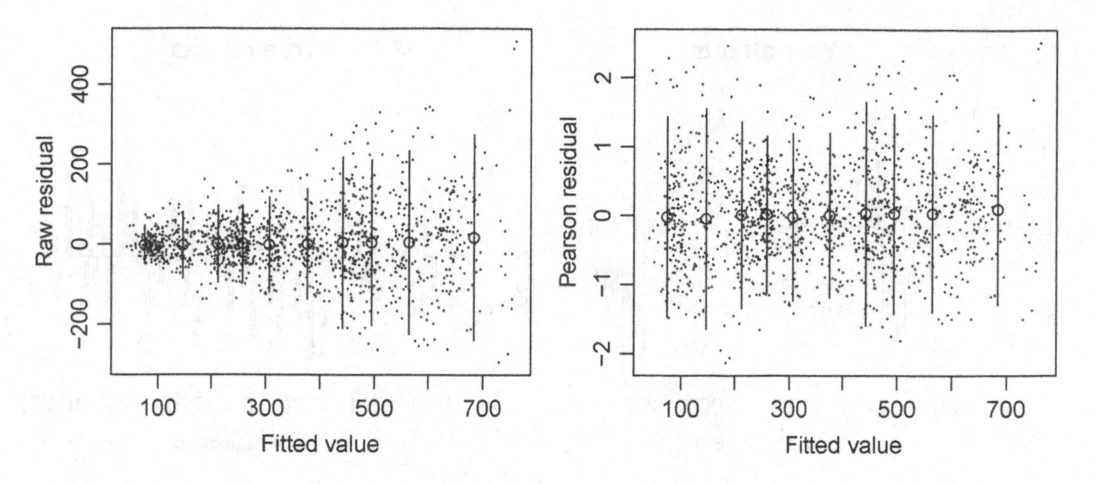

FIGURE 6.4 Plots of raw (left) and Pearson (right) residuals of model `aftmod3` in Example 6.5.

```
 Year      (Intercept) 1.507e+02 1.227e+01
 Plot      (Intercept) 4.537e-11 6.736e-06
 Residual              5.467e-01 7.394e-01
Number of obs: 1319, groups:  Tree2, 88; Year, 15; Plot, 10

> aftmod4<-update(aftmod3,formula=RBA~SDClass+CA+(1|Tree2)+(1|Year))
> for (i in 1:20){
+      cat(i,logLik(aftmod4),"\n")
+      afterthin$ytilde<-fitted(aftmod4)
+      aftmod4<-lmer(RBA~SDClass+CA+(1|Tree2)+(1|Year),
+                    weights=1/ytilde^{1.6},
+                    data=afterthin)
+ }

> anova(aftmod3,aftmod4)
refitting model(s) with ML (instead of REML)
Data: afterthin
Models:
aftmod4: RBA ~ SDClass + CA + (1 | Tree2) + (1 | Year)
aftmod3: RBA ~ SDClass + CA + (1 | Plot) + (1 | Tree2) + (1 | Year)
        Df   AIC   BIC  logLik deviance Chisq Chi Df Pr(>Chisq)
aftmod4  8 15409 15450 -7696.5    15393
aftmod3  9 15411 15458 -7696.5    15393     0      1     0.9984
```

Conditional F-tests on fixed-effects can be carried out using library `lmerTest` (the model needs to be re-estimated to obtain the test results).

```
> library(lmerTest)
> aftmod4<-update(aftmod4)
> anova(aftmod4)
Type III Analysis of Variance Table with Satterthwaite's method
         Sum Sq Mean Sq NumDF  DenDF F value    Pr(>F)
SDClass 13.4083  4.4694     3 83.036  8.1755 7.842e-05 ***
CA       0.0165  0.0165     1 12.878  0.0301    0.8649
---
Signif. codes:  0 '***' 0.001 '**' 0.01 '*' 0.05 '.' 0.1 ' ' 1
```

The conclusions regarding fixed effects did not change compared to model `aftmod2` where the random year effects were ignored (Example 6.3). The most remarkable differences are that now the coefficient of age is even smaller in absolute value than it was in model `aftmod2`, and the estimated between-tree variance has decreased slightly to 160.46^2. The estimated between-year variance is $\widehat{\text{var}}(d_k) = 12.27^2$.

```
> summary(aftmod4)
```

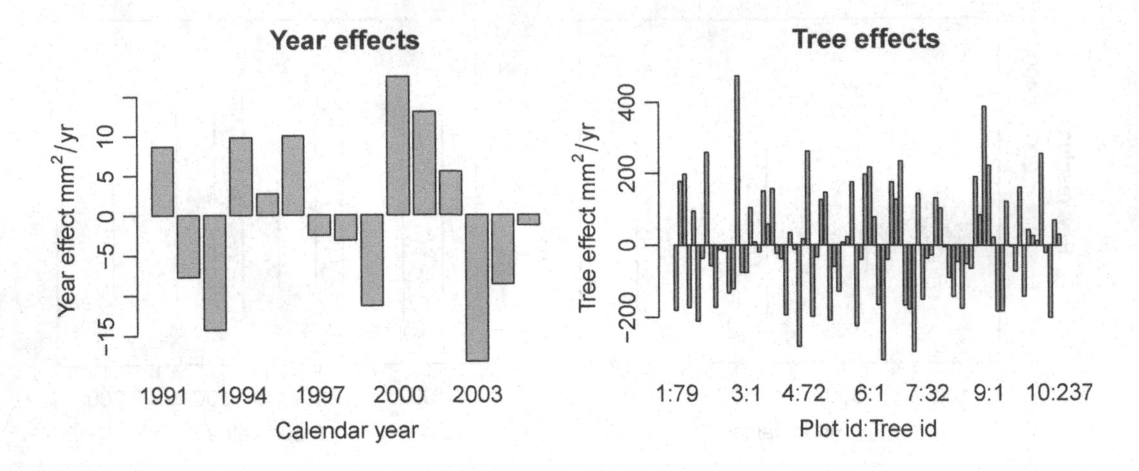

FIGURE 6.5 BLUPs of the year and tree effects of model `aftmod4` of Example 6.5.

```
Random effects:
 Groups   Name        Variance  Std.Dev.
 Tree2    (Intercept) 2.575e+04 160.4612
 Year     (Intercept) 1.507e+02  12.2741
 Residual             5.467e-01   0.7394
Number of obs: 1319, groups:  Tree2, 88; Year, 15

Fixed effects:
            Estimate Std. Error       df t value Pr(>|t|)
(Intercept) 275.2197    42.3908  52.8683   6.492 2.98e-08 ***
SDClass2     29.4953    45.2531  81.9334   0.652  0.51636
SDClass3    145.1423    48.8910  82.8121   2.969  0.00391 **
SDClass4    216.6719    49.7543  83.3870   4.355 3.76e-05 ***
CA           -0.1393     0.8026  12.8783  -0.174  0.86491
```

Parametric bootstrapping (recall Section 5.6) can be used to compute the 95% confidence intervals for all parameters using function `confint`. The first three rows of the output show the confidence intervals for the standard deviations of random effects and the rows 4-8 show the confidence intervals for the fixed effects. As bootstrapping is based on simulated samples, slightly different results are obtained after each run.

```
> confint(aftmod4,method="boot")
Computing bootstrap confidence intervals ...
                   2.5 %       97.5 %
.sig01      136.238509922 187.25711407
.sig02        7.708291248  17.22250820
.sigma        0.009472293   0.01099822
(Intercept) 194.783305722 359.32793868
SDClass2    -63.826952684 121.26541686
SDClass3     49.951236345 239.94299197
SDClass4    120.652743268 323.57752524
CA           -1.503842604   1.30965092
```

Figure 6.5 shows the predicted tree- and year-level random effects of model `aftmod4`. We notice, for example, that the trees have grown on average 18 mm^2 more in 2000 than in an average year. On the other hand, the growth in 2003 was very low compared to the other years. A natural explanation for these growth trends is the weather conditions of the calendar years in the region.

```
> barplot(ranef(aftmod4)$Year[,1],names=rownames(ranef(aftmod4)$Year),main="Year effects") # left graph
```

The random effects allow prediction at the following four levels:

- Population level (fixed part only)

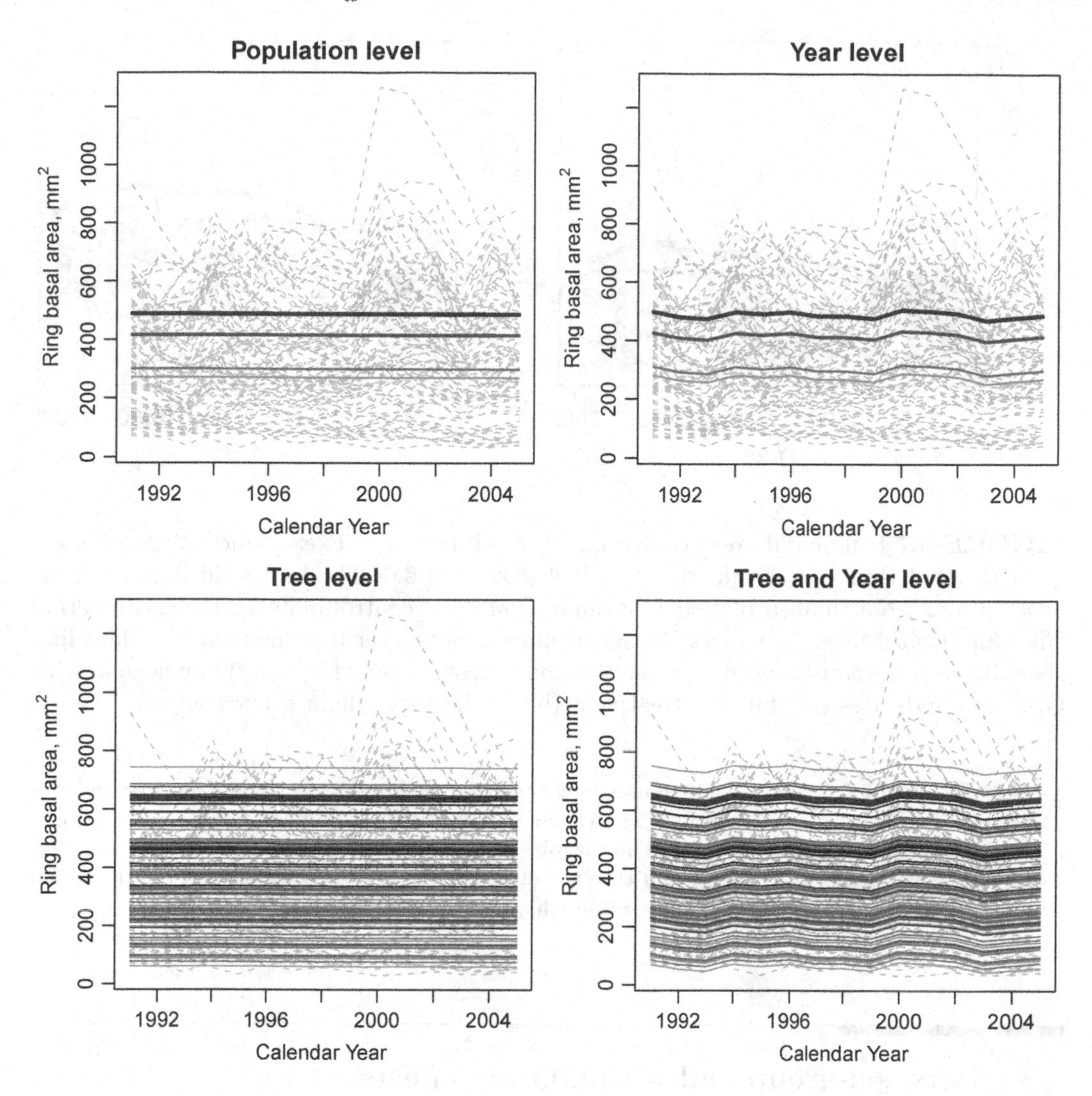

FIGURE 6.6 Predicted values of ring basal area (RBA, mm^2) at different levels of grouping for model `aftmod4` in Example 6.5. Grayscale illustrates the thinning treatment (black = heavy ... light gray = control). The observations are also shown in the background.

- Tree level (fixed part and tree effect)
- Year level (fixed part and year effect)
- Year and tree level (fixed part, tree and year effects)

These predictions can be computed using function `predict` and utilizing argument `re.form`.

```
> predTree<-predict(aftmod4,newdata=afterthin,re.form=~(1|Tree2))
> predPop<-predict(aftmod4,newdata=afterthin,re.form=~0)
> predYear<-predict(aftmod4,newdata=afterthin,re.form=~(1|Year))
> predAll<-predict(aftmod4,newdata=afterthin,re.form=~(1|Tree2)+(1|Year))
```

Figure 6.6 shows the predictions at different levels, as well as the original data. Such predictions may sometimes be highly useful, as illustrated in the following web example.

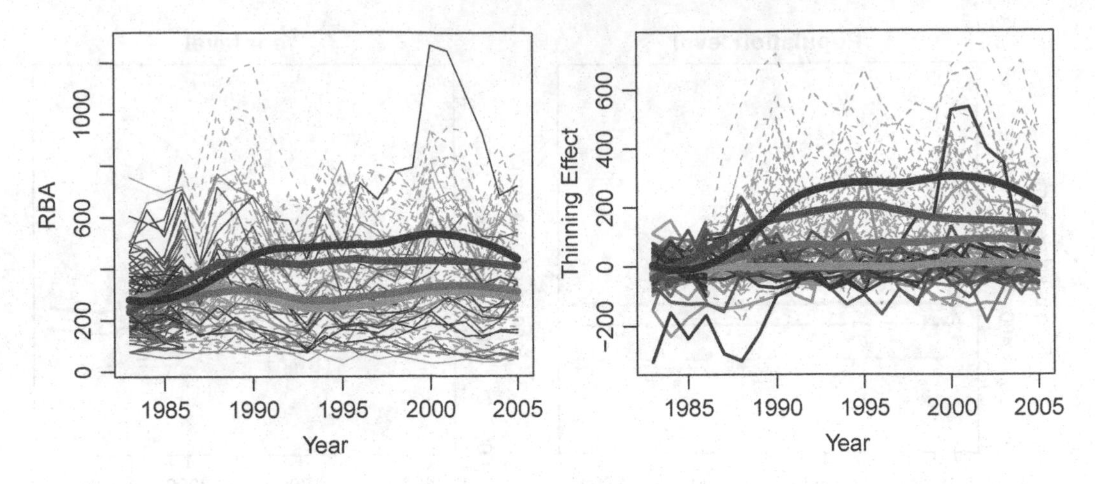

FIGURE 6.7 Ring basal areas (RBA, mm^2) for all trees in all years since 1982 (left), and the extracted thinning effects (right) in Web Example 6.6. The thin solid lines show the observations from thinned plots before thinning and the control plots for the entire period. The thin dashed lines show observations of thinned plots after the thinning. The thick lines show the average trend, based on a lowess smoother, separately for each thinning treatment. Grayscale indicates the thinning treatment (black=heavy ... light gray=control).

Web Example 6.6. This example uses linear mixed-effects models with crossed tree and year level random effects to estimate the pure effects of silvicultural thinning on tree-specific growth time series (Mehtätalo *et al.* 2014). The raw observations of the basal area growth are shown in the left-hand graph in Figure 6.7 and the extracted thinning effects are shown in the right-hand graph. These effects are further analyzed in Chapter 7.

6.3 Between-group and within-group effects

The difference between group-level and individual-level predictors is important in mixed-effect models. A special group-level predictor is the group-level mean of an individual-level predictor. Its coefficient is called a *between-group effect.* Similarly, deviation from the group mean is an important individual-level predictor leading to a *within-group* effect. The need to include them in the model was discussed previously in Example 5.22 (p. 172). This section includes a more detailed and formal discussion.

Let us now more closely analyze between-group and within-group effects when adding observation level variables into a mixed model. Assume that we start analyzing data with the simple variance component model

$$y_{ij} = \mu + b_i^{(0)} + \epsilon_{ij}^{(0)} . \tag{6.6}$$

The standard assumptions are that $\mathrm{E}\left(b_i^{(0)}\right) = \mathrm{E}\left(\epsilon_{ij}^{(0)}\right) = 0$, $\mathrm{var}\left(b_i^{(0)}\right) = \mathrm{cov}\left(y_{ij}, y_{ij'}\right) = \sigma_{b0}^2$ and $\mathrm{var}\left(\epsilon_{ij}^{(0)}\right) = \sigma_0^2$. If we add a new predictor variable x_{ij} to the model, we get the model:

$$y_{ij} = a_0 + a_1 x_{ij} + b_i^{(1)} + \epsilon_{ij}^{(1)} , \tag{6.7}$$

where a_0 and a_1 are fixed coefficients. It is assumed that x_{ij} is a random variable (see Section 10.1). This assumption makes the analysis more straightforward. The standard assumption is that x_{ij}, $b_i^{(1)}$ and $\epsilon_{ij}^{(1)}$ are independent. Often, this assumption is not made explicitly, but we will emphasize here the importance of this assumption. The progress of modeling can be evaluated by monitoring $\widehat{\text{var}}\left(b_i^{(1)}\right)$ and $\widehat{\text{var}}\left(\epsilon_{ij}^{(1)}\right)$ when adding predictors. Equivalently, the between-group coefficient of determination (5.44) on p. 174 and within-group coefficient of determination (5.45) can be monitored.

It can happen that $\widehat{\text{var}}\left(b_i^{(1)}\right)$ is larger than $\widehat{\text{var}}\left(b_i^{(0)}\right)$, i.e. R_b^2 is negative. Why? Adding x_{ij} into the model implicitly assumes that:

1. $b_i^{(0)}$ is explained with $\bar{x}_i$.

2. $\epsilon_{ij}^{(0)}$ is explained with $x_{ij} - \bar{x}_i$.

3. Both $\bar{x}_i$ and $x_{ij} - \bar{x}_i$ have the same coefficient: $y_{ij} = a_0 + a_1\bar{x}_i + a_1(x_{ij} - \bar{x}_i) + b_i^{(1)} + \epsilon_{ij}^{(1)}$.

Often, however:

$$y_{ij} = a_0 + \beta_1\bar{x}_i + \beta_2(x_{ij} - \bar{x}_i) + b_i^{(*)} + \epsilon_{ij}^{(*)}, \tag{6.8}$$

where $\beta_1 \neq \beta_2$. If Equation (6.7) is estimated in the data set where the true model is Equation (6.8), then the expected value of $\widehat{a}_1$ is a weighted average of β_1 and β_2, usually it is close to β_2 because the precision of $\widehat{\beta}_2$ is generally better than the precision of $\widehat{\beta}_1$. If model $y_{ij} = a_0 + a_1 x_{ij} + \epsilon_{ij}{}^{(1)}$ is estimated with OLS, then $\widehat{a}_1$ again is a weighted average of β_1 and β_2, but often it is close to β_1 because the within-group variance of x is often small compared to the between-group variance (see Bell *et al.* (2019) for an explanation for both cases, and see Example 6.7).

Equation (6.8) is equivalent to the following models:

$$y_{ij} = a_0 + (\beta_1 - \beta_2)\bar{x}_i + \beta_2 x_{ij} + b_i^{(*)} + \epsilon_{ij}^{(*)}, \tag{6.9}$$

$$y_{ij} = a_0 + \beta_1 x_{ij} + (\beta_2 - \beta_1)(x_{ij} - \bar{x}_i) + b_i^{(*)} + \epsilon_{ij}^{(*)}. \tag{6.10}$$

When estimating Equation (6.9), $\beta_{12} = \beta_1 - \beta_2$ is a separate parameter, and when estimating Equation (6.10) $\beta_{21} = \beta_2 - \beta_1$ is a separate parameter.

With random regressors, several models can usually be valid simultaneously. If Equation (6.9) is valid and $\beta_1 \neq \beta_2$, then Equation (6.7) is not valid because $b_i^{(1)}$ contains $(\beta_1 - \beta_1)\bar{x}_i$ and is, thus, correlated with $\bar{x}_i$ and, hence, with x_{ij}. If Equation (6.9) is valid, then Equation (6.7) is valid, if $b_i^{(1)}$ is assumed to be fixed.

The easiest way to test whether $\beta_1 = \beta_2$, is to determine whether parameter $\beta_1 - \beta_2 = 0$ in Equation (6.9) or $\beta_2 - \beta_1 = 0$ in Equation (6.10). One indication that Equation (6.7) is invalid is that $\widehat{a}_1$ is very different to the OLS slope obtained by regressing y on x without random effects. One could also check whether the predicted random effects of Equation (6.7) are correlated with group means $\bar{x}_i$, as we did in Example 5.22 (p. 172). Note that if Equation (6.7) is valid and $b_i^{(1)}$ is independent of x_{ij}, then the OLS estimate of a_1 is unbiased. In Example 5.22, the final model was fitted using Equation (6.9).

Predictor $x_{ij} - \bar{x}_i$ in model (6.10) has a negative within-group correlation. To justify that, we note first that (show it!): $\text{cov}(x_{ij} - \bar{x}_i, x_{ij'} - \bar{x}_i) = (\text{cov}(x_{ij}, x_{ij'}) - \text{var}(x_{ij}))/n_i$, if x_{ij} have common variance and $\text{cov}(x_{ij}, x_{ij'})$ is the same for all pairs $(x_{ij}, x_{ij'})$. Because $\text{var}(x_{ij}) \geq \text{cov}(x_{ij}, x_{ij'})$, $\text{cor}(x_{ij} - \bar{x}_i, x_{ij'} - \bar{x}_i)$ is always nonpositive. The lower bound for the correlation is $-1/(n_i - 1)$ (see Equation 5.24, Box 5.2, p. 153). The lower bound would imply that $\bar{x}_i$ would be equal for all i, and it would not be possible to use $\bar{x}_i$ as a predictor. The upper bound 0 would mean that within-group correlation of x_{ij} is 1 and thus $x_{ij} - \bar{x}_i$ is always 0, and it could not describe within-group variation. Thus, $-1/(n_i - 1) < \text{cor}(x_{ij} - \bar{x}_i, x_{ij'} - \bar{x}_i) < 0$. If $\text{cor}(x_{ij}, x_{ij'}) = 0$, then $\text{cor}(x_{ij} - \bar{x}_i, x_{ij'} - \bar{x}_i) = -1/n_i$. The negative within-group correlation of $x_{ij} - \bar{x}_i$ is related to the fact that the mean of $x_{ij} - \bar{x}_i$ is always exactly zero.

Because y_{ij} is linearly related to $x_{ij} - \bar{x}$, the negative within-group correlation of $x_{ij} - \bar{x}$ also implies a negatively correlated component to y_{ij}. Hence, adding $x_{ij} - \bar{x}_i$ to Equation (6.6) (or to a model that includes $\bar{x}_i$) makes $\widehat{\text{var}}(b_i)$ larger and R_b^2 smaller. This can also be seen from the ML equations used to estimate the variance of random effects (Snijders and Bosker 1994). Snijders and Bosker (1994) do not like a decreasing R_b^2 value and suggest an alternative. We regard that a decreasing R_b^2 value is not a problem, provided we understand the reasons behind it.

Let $\bar{X}_i$ be the mean of x_{ij} in the entire group, not only in sample data from the group. Instead of Equation (6.9), it may be more natural to assume that

$$y_{ij} = a_0 + (\beta_1 - \beta_2)\bar{X}_i + \beta_2 x_{ij} + b_i^{(**)} + \epsilon_{ij}^{(**)} . \tag{6.11}$$

If all individuals in group i are not present in the sample, $\bar{x}_i$ is a measurement of $\bar{X}_i$ containing measurement error. If $\beta_1 - \beta_2$ is estimated using Equation (6.9), the absolute value of the estimate is downwards biased, and the estimate $\widehat{\text{var}}\left(b_i^{(**)}\right)$ is upwards biased (see Grilli and Rampichini (2011) and Section 13.1.1). However, most authors ignore this finding.

Even if Equations (6.8), (6.9) and (6.10) are equivalent, they are different with respect to their interpretation. In social sciences $(\beta_1 - \beta_2)\bar{x}_i$ is a *contextual effect* and is a *cohort effect* in panel data. In a forest growth model, Equation (6.9) can appear when x_{ij} is age, and old stands come from poor sites in cross-sectional data. In forestry, $(\beta_2 - \beta_1)(x_{ij} - \bar{x}_i)$ can be a *competition effect*, e.g. when x_{ij} is a measurement of size (e.g. diameter at breast height). Then, the implied negative correlation makes sense: competition is a zero-sum game. When making a causal interpretation of Equation (6.9), it often happens that there is a background variable correlated with $\bar{x}_i$ that has caused differences in the intercept of a model. This is the case, for example, when old stands come from poor sites.

When comparing Equations (6.7) and (6.9), we found that if the latter is valid and $\beta_1 \neq \beta_2$, then the former is invalid, if $b_i^{(1)}$ is assumed to be random and independent of x_{ij}. However Equation (6.7) can be assumed and estimated with random $b_i^{(1)}$, and without independence assumption and without specifying the form of the dependency on $\bar{x}_i$ using the approach of Lappi and Malinen (1994):

1. Estimate model $y_{ij} = a_1 x_{ij} + b_i^{(1)} + \epsilon_{ij}^{(1)}$ as if $b_i^{(1)}$ would be fixed.
2. Estimate a_0 using the arithmetic average of $\widehat{b}_i^{(1)}$.
3. Estimate $\widehat{\text{var}}\left(b_i^{(1)}\right)$ by a modified analysis of variance estimator (see Section 5.3.1).

Lappi and Malinen (1994) also employed random slopes and a two-level hierarchy of random effects.

A new insight to the within- and between-group effects is obtained using a multivariate mixed model which will be discussed in Section 9.4.1. This approach also provides a nice solution to the bias problems that occur when the sample mean $\bar{x}_i$ is used instead of the entire group mean $\bar{X}_i$. Let us consider a bivariate model:

$$y_{ij} = \mu^{(y)} + b_i^{(y)} + \epsilon_{ij}^{(y)} , \tag{6.12}$$

$$x_{ij} = \mu^{(x)} + b_i^{(x)} + \epsilon_{ij}^{(x)}. \tag{6.13}$$

According to this formulation, the true group mean of x is $\mu^{(x)} + b_i^{(x)}$, which is a latent variable.

It is assumed that when these models are estimated, the covariances $\sigma_{bxy} = \text{cov}\left(b_i^{(x)}, b_i^{(y)}\right)$ and $\sigma_{\epsilon xy}$ cov $\left(\epsilon_{ij}^{(x)}, \epsilon_{ij}^{(y)}\right)$ are also estimated, in addition to $\mu^{(y)}$, $\mu^{(x)}$, $\sigma_{by}^2 = \text{var}\left(b_i^{(y)}\right)$, $\sigma_{bx}^2 = \text{var}\left(b_i^{(x)}\right)$, $\sigma_{\epsilon y}^2 = \text{var}\left(\epsilon_{ij}^{(y)}\right)$ and $\sigma_{\epsilon x}^2 = \text{var}\left(\epsilon_{ij}^{(x)}\right)$. Let us then consider the Best Linear Predictor (BLP) of y_{ij} when $\boldsymbol{x_i} = (x_{i1}...x_{in_i})'$ are measured. The random effects $b_i^{(y)}$ and $b_i^{(x)}$ can be predicted from the average of x's, i.e. $\bar{x}_i$, as follows (see Section 9.4.3 for general formulas):

$$\widetilde{b}_i^{(y)} = \frac{\text{cov}\left(b_i^{(y)}, \bar{x}_i\right)}{\text{var}\left(\bar{x}_i\right)}\left(\bar{x}_i - \mu^{(x)}\right) = \frac{\sigma_{bxy}}{\sigma_{bx}^2 + \sigma_{\epsilon x}^2/n_i}\left(\bar{x}_i - \mu^{(x)}\right), \tag{6.14}$$

$$\widetilde{b}_i^{(x)} = \frac{\text{cov}\left(b_i^{(x)}, \bar{x}_i\right)}{\text{var}\left(\bar{x}_i\right)}\left(\bar{x}_i - \mu^{(x)}\right) = \frac{\sigma_{bx}^2}{\sigma_{bx}^2 + \sigma_{\epsilon x}^2/n_i}\left(\bar{x}_i - \mu^{(x)}\right). \tag{6.15}$$

Thereafter, $\epsilon_{ij}^{(y)}$ can be predicted as

$$\widetilde{\epsilon}_{ij}^{(y)} = \frac{\sigma_{\epsilon xy}}{\sigma_{\epsilon x}^2}\left(x_{ij} - \mu^{(x)} - \widetilde{b}_i^{(x)}\right).$$

The predictor can be then presented in the form:

$$\widetilde{y}_{ij} = \mu^{(y)} + \frac{\sigma_{bxy}}{\sigma_{bx}^2 + \sigma_{\epsilon x}^2/n_i}\left(\bar{x}_i - \mu^{(x)}\right) + \frac{\sigma_{\epsilon xy}}{\sigma_{\epsilon x}^2}\left(x_{ij} - \mu^{(x)} - \widetilde{b}_i^{(x)}\right). \tag{6.16}$$

It is straightforward to derive the prediction error variance, but the variance equation becomes complicated.

When n_i increases, Equation (6.16) approaches the equation:

$$\widetilde{y}_{ij} = \mu^{(y)} + \frac{\sigma_{bxy}}{\sigma_{bx}^2}\left(\bar{x}_i - \mu^{(x)}\right) + \frac{\sigma_{\epsilon xy}}{\sigma_{\epsilon x}^2}\left(x_{ij} - \bar{x}_i\right).$$

If n_i varies from group to group, then the estimate of β_1 in Equation (6.8) is some kind of average of coefficients of $\bar{x}$ in Equation (6.16). The estimate of β_2 should be close to the coefficient of $x_{ij} - \mu^{(x)} - \widetilde{b}_i^{(x)}$ in Equation (6.16). The above analysis can be easily extended to treat multiple regressors. We have not seen the above-discussed bivariate mixed-model solution to the measurement error problem of the group mean of x in the literature.

Example 6.7. A hypothetical example of different within- and between-group effects is shown in Figure 6.8. The data are similar to the data in Figure 1.1 on p. 3 of Demidenko (2013). The slope of the regression line of $\bar{y}_i$ with respect to $\bar{x}_i$ is clearly different to the slope of the regression line of $y_{ij} - \bar{y}_i$ with respect to $x_{ij} - \bar{x}_i$. The OLS slope is -0.517.

For simplicity, superscripts of b_i and ϵ_i are now dropped. When Equation (6.6) is estimated, the obtained variance component estimates are $\widehat{\sigma}_{b0}^2 = 11.956^2$ and $\widehat{\sigma}_{\epsilon 0}^2 = 4.932^2$. When Equation (6.7) is estimated, $\widehat{a}_1 = 0.919$, $\widehat{\text{var}}\left(b_i\right) = 21.995^2$ and $\widehat{\text{var}}\left(\epsilon_{ij}\right) = 1.823^2$. Thus, the increase of $\widehat{\text{var}}\left(b_i\right)$ also indicates that the Equation (6.7) is not valid. However, Demidenko (2013) assumes that Equation (6.7) is a proper mixed model analysis of similar data.

When estimating model $y_{ij} = a_0 + \beta_1\left(x_{ij} - \bar{x}_i\right) + b_i + \epsilon_{ij}$, we get $\widehat{\text{var}}\left(b_i\right) = 12.279^2$ and $\widehat{\text{var}}\left(\epsilon_{ij}\right) = 1.816^2$. Thus, $\widehat{\text{var}}\left(b_i\right)$ increases, compared to Equation (6.6), due to the implied negative within-group correlation. When model $y_{ij} = a_0 + \beta_1\bar{x}_i + b_i + \epsilon_{ij}$ is estimated, $\widehat{\text{var}}\left(b_i\right) = 8.572^2$ and $\widehat{\text{var}}\left(\epsilon_{ij}\right) = 4.924^2$. Then, adding $x_{ij} - \bar{x}_i$ to the model, i.e. estimating Equation (6.8),

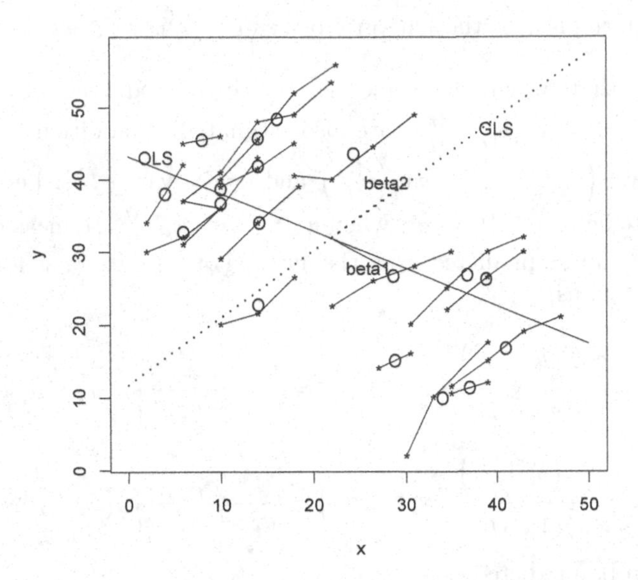

FIGURE 6.8 Illustration of within- and between-group effects. The individual observations are shown with dots. The observations of the same groups are connected. Group means of x and y are shown with circles. The solid line marked with 'OLS' is the OLS regression line. The slashed line marked with 'GLS' is the GLS line, estimated with the assumption that random group intercepts are independent of the x-values (i.e. assuming Equation (6.7)). Both models are mis-specified and biased because the intercepts are clearly correlated with x. This can be seen from the fact that the regression of $\bar{y}_i$ against $\bar{x}_i$ has clearly different slope to the regression of $y_{ij} - \bar{y}_i$ against $x_{ij} - \bar{x}_i$. The estimated slope β_1 in Equation (6.8) is described with the line marked with 'beta1', and the estimated slope β_2 in Equation (6.8) is described with the line marked with 'beta2' (the slope almost coincides with the GLS slope).

we obtain $\widehat{\text{var}}(b_i) = 8.943^2$ and $\widehat{\text{var}}(\epsilon_{ij}) = 1.816^2$, i.e. the group effect variance again increases. The fixed slope parameters are assigned estimates: $\widehat{\beta}_1 = -0.695$ and $\widehat{\beta}_2 = 0.9514$. Thus, both the OLS slope -0.517 and the slope of Equation (6.7) ($\widehat{a}_1 = 0.919$) are between $\widehat{\beta}_1$ and $\widehat{\beta}_2$, the OLS slope is rather close to $\widehat{\beta}_1$ and the slope of Equation (6.7) is very close to $\widehat{\beta}_2$, as usually happens according to Bell *et al.* (2019). The exact relationship between the different parameter estimates are dependent on the data set. If Equation (6.7) is estimated with the assumption that b_i is fixed, then $\widehat{a}_1 = 0.9514$, i.e. practically the same as we get for $\widehat{\beta}_2$ in Equation (6.8).

Summary notes:

1. The above discussion can also be generalized to more complicated models. For instance, we could consider model

$$y_{ij} = a_0 + \beta_1 \bar{x}_i + \beta_2 (x_{ij} - \bar{x}_i) + \beta_3 z_i + b_i^{(1)} + b_i^{(2)} (x_{ij} - \bar{x}_i) + \epsilon_{ij}$$

where z_i is a group level predictor and $b_i^{(2)}$ is a random slope.

2. If the group specific intercept in Equation (6.7) is considered to be fixed, then we would not be worried about the dependence of x and b_i. However, we cannot then have any group-specific predictors in the model because the model matrix would not have full

rank. With the fixed intercepts, we get the same estimate for parameter a_1 as we would get using $x_{ij} - \bar{x}_i$ as the predictor, i.e. a_1 describes the within-group effect of x.

3. Mixed model development can be monitored with ${R_b}^2$ and ${R_e}^2$ as defined in Equations (5.44) and (5.45) on p. 174.

4. If ${R_b}^2$ decreases when an individual level predictor is added, the model probably needs separate between-group and within-group effects. This may also be necessary even if ${R_b}^2$ increases.

5. The addition of deviation variable $(x_{ij} - \bar{x}_i)$ decreases ${R_b}^2$.

6. The addition of x_{ij} after $\bar{x}_i$ implicitly adds $(x_{ij} - \bar{x}_i)$.

7. Random slopes can also be predicted with $\bar{x}_i$. This is done by adding variables $\bar{x}_i x_{ij}$ to the predictors, as we did in Example 5.22.

8. Between-group and within-group effects may be more complicated than linear functions. There may be some bias problems when applying between-group and within-group effects in generalized linear mixed models (GLMM) (Bell *et al.* 2019).

9. The *Hausman test* (Hsiao 2014) can be used to test for the independence of random effects and predictors. However, we agree with Bell *et al.* (2019) that it is more straightforward to start working directly with models that have separate between-group and within-group effects.

Between- and within-group effects have been analyzed e.g. by Neuhaus and McCulloch (2006), and Shen *et al.* (2008). A good introduction to the topic can be found in Fahrmeir *et al.* (2013)(pp. 353-354). The question of different between- and within-group effects is related to *ecological fallacy* and *Simpson's paradox*. Ecological fallacy and Simpson's paradox refer to cases where the correlation between two variables has a different sign when computed at the population-level or separately for different groups. In our analysis this corresponds to the case where β_1 and β_2 have different signs. Having different β_1 and β_2 in a model should also be considered in cases where they do not differ so drastically. Simpson's paradox and ecological fallacy are often considered to be the same phenomenon. According to Kögel (2017), ecological fallacy occurs when $\beta_1 - \beta_2$ in Equation (6.9) is non-zero and one tries to infer the micro-level effect (β_2) from macro-level data consisting only of group-level variables. The Simpson's paradox is the phenomenon when one analyzes individual-level data but does not use predictor $\bar{x}_i$, even if $\beta_1 - \beta_2$ is non-zero.

6.4 Population-averaged prediction

For mixed-effects models, there are several possibilities for prediction, and these predictions have different interpretations. Group level predictions were discussed previously in Section 5.4.2, and individual-level predictions will be discussed in Section 6.5. Such predictions can also be called *conditional predictions* because they are conditional on the predicted group effects and residual errors. *Population-averaged prediction* means a prediction that is averaged over the distribution of group effects. It is sometimes called a *marginal prediction*. However, the terms conditional and marginal are not unambiguous, provided it is not explicitly stated that they are conditional on, or marginal over, the group effects.

Consider again Equation (5.6) on p. 139 with a single level of grouping:

$$y_{ij} = \boldsymbol{x}'_{ij}\boldsymbol{\beta} + \boldsymbol{z}'_{ij}\boldsymbol{b}_i + \epsilon_{ij}\,, \tag{6.17}$$

where $\mathrm{E}(\boldsymbol{b}_i) = \boldsymbol{0}$ and $\mathrm{E}(\epsilon_{ij}) = 0$.

What is the expected value of y_{ij} given the predictors $\boldsymbol{x}_{ij}$? The expected value is evidently $\mathrm{E}(y_{ij}) = \boldsymbol{x}'_{ij}\boldsymbol{\beta}$ (see Equation (5.14) p. 146), which leads to the *population averaged* or *marginal* prediction:

$$\widetilde{y}_{ij} = \boldsymbol{x}'_{ij}\widehat{\boldsymbol{\beta}}\,.$$

This is the standard notion of the population averaged prediction reported in the literature. However, we see inherent problems in this view. What kind of variable is y_{ij}? It is a hierarchical random variable: we first randomly pick a group and then randomly pick an individual. The expectation in this hierarchical selection is $\boldsymbol{x}'_{ij}\boldsymbol{\beta}$. We call the corresponding population a *hierarchical population.*

Let us then consider the *population of individuals*. Let the individuals be numbered by j and let $i(j)$ denote the group of individual j. Equation (6.17) can then be written:

$$y_j = \boldsymbol{x}'_j\boldsymbol{\beta} + \boldsymbol{z}'_j\boldsymbol{b}_{i(j)} + \epsilon_j\,. \tag{6.18}$$

The distribution of $\boldsymbol{b}_{i(j)}$ is a weighted distribution of $\boldsymbol{b}_i$, the weights being equal to the group size N_i. Note that N_i is the total number of individuals in group i, not the number of sampled individuals, which is denoted by n_i. We can assume that the group size N_i is also a random variable. This is a natural assumption in empirical populations.

Let us continue the analysis with a simplified model:

$$y_j = \mu + b_{i(j)} + \epsilon_j\,. \tag{6.19}$$

The results of this model can be generalized to the model above. Let f_w denote the density function of the weighted distribution (see Equation (2.53) on p. 34, Patil and Rao (1978), Lappi and Bailey (1987)). Thus, f_w is :

$$f_w\left(b_{i(j)}\right) = \sum_{i=1}^{\infty} N_i f\left(N_i, b_i\right)/\mathrm{E}\left(N_i\right), \tag{6.20}$$

where $f\left(N_i, b_i\right)$ is the joint density of N_i and b_i. Notice that it is a special two-dimensional density: one argument is discrete and the other argument is continuous. Thus, we obtain:

$$\begin{aligned} E_w\left(b_{i(j)}\right) &= \int_{-\infty}^{\infty} \sum_{i=1}^{\infty} N_i b_i f\left(N_i, b_i\right) db/\,\mathrm{E}\left(N_i\right) \\ &= \frac{\mathrm{cov}\left(N_i, b_i\right)}{\mathrm{E}\left(N_i\right)} = \frac{r\ \mathrm{sd}\left(N_i\right)\mathrm{sd}\left(b_i\right)}{\mathrm{E}\left(N_i\right)} = r\ \mathrm{vc}\left(N_i\right)\mathrm{sd}\left(b_i\right) \end{aligned} \tag{6.21}$$

where r is the correlation between N_i and b_i, vc is the coefficient of variation and sd is the standard deviation. Thus, the expected value of a randomly picked individual is μ only if r, $\mathrm{vc}\left(N_i\right)$ or $\mathrm{sd}\left(b_i\right)$ is zero. Note that if $\mathrm{vc}\left(N_i\right)$ or $\mathrm{sd}\left(b_i\right)$ is zero, then r is not defined but the covariance $\mathrm{cov}\left(N_i, b_i\right)$ is defined and it is zero, which implies that $E_w\left(b_{i(j)}\right)$ is zero. Note also that in the above derivation $\mathrm{cov}\left(N_i, b_i\right) = \mathrm{E}\left(N_i b_i\right) - \mathrm{E}\left(N_i\right)\mathrm{E}\left(b_i\right) = \mathrm{E}\left(N_i b_i\right)$ because $\mathrm{E}(b_i) = 0$.

The result (6.21) can also be understood by considering what is the expected value of $y_{i.} = \sum_{j=1}^{N_i} y_{ij}$:

$$\begin{aligned} \mathrm{E}(y_{i.}) &= \mathrm{E}\left(N_i\mu + N_i b_i + \sum_{j=1}^{N_i} \epsilon_{ij}\right) \\ &= \mathrm{E}(N_i)\mu + \mathrm{E}(N_i b_i) + \mathrm{E}(N_i)\,\mathrm{E}(\epsilon_{ij}) = \mathrm{E}(N_i)\mu + \mathrm{cov}(N_i, b_i) \end{aligned} \quad (6.22)$$

This is compatible with Equation (6.21). To derive $\mathrm{E}\left(\sum_{j=1}^{N_i} \epsilon_{ij}\right) = 0$, we utilized the fact that if $Y = \sum_{j=1}^{m} X_j$, where m is a random variable and X's are iid variables, then $\mathrm{E}(Y) = \mathrm{E}(m)\,\mathrm{E}(X)$.

Thus, the standard predictions from a mixed model are biased at the individual-level, provided the group size and random effects are correlated. If the true group size is known when estimating a mixed model, then the group size could be considered as a predictor in the model. Then, the group size would no longer be correlated with the random effects of the model. However, for example, when developing mixed models with random stand and tree effects the number of trees in a given stand is not generally known. We regard that this question may be important when applying mixed models in inventories. The design-based sampling methods are protected against such bias problems.

It may also happen that n_i, the number of individuals in group i in the estimation data, is correlated with the group-effect of Equation (6.19). This situation is described in the literature under the concept *informative cluster size* (see McCulloch *et al.* (2008), Neuhaus and McCulloch (2011), Seaman *et al.* (2014)). In that case, the standard GLS estimate of μ is biased for both the mean of the group means and for the mean of individuals. The expected value of the estimated μ is between these two means. If n_i is constant in the different groups, i.e. the data are balanced, then the estimated μ is unbiased for μ in Equation (6.19), where μ is the mean of group means. If the entire group i is in the estimation data, i.e. $n_i = N_i$ for all i, then the arithmetic average of y_j's is unbiased for the mean of individuals.

When the group size in the data is informative and is not taken into account, the intercept will thus be biased for the hierarchical distribution assumed in mixed models. But the estimated slope coefficients will, in general, be almost unbiased (Neuhaus and McCulloch 2011). If the group size is informative, we can use n_i as a predictor in the model, as we did in Example 5.22. Then, the model may describe well the estimation data. But the interpretation of the model may be problematic when the model is applied outside the estimation data. If sample trees are selected in the estimation data using fixed-area or relascope plots, then the group size n_i is a variable that describes the stand density or the stand basal area, respectively. Then the interpretation of the model is also clear outside the estimation data in terms of these variables. There could be among predictors both n_i and some measure of the total stand size, e.g. N_i, area of the stand or basal area of the stand, if these are known.

6.5 Prediction of correlated observation using a mixed model

If we want to predict a new observation 0 in group i:

$$y_{i0} = \boldsymbol{x}_{i0}'\boldsymbol{\beta} + \boldsymbol{z}_{i0}'\boldsymbol{b}_i + \epsilon_{i0}\,,$$

and the residual error ϵ_{i0} is correlated with the residual errors of group i, we can utilize these correlations in a similar way to that presented in Section 4.7.2. For the data in group

i the model can be presented as

$$\boldsymbol{y}_i = \boldsymbol{X}_i\boldsymbol{\beta} + \boldsymbol{Z}_i\boldsymbol{b}_i + \boldsymbol{\epsilon}_i \,.$$

Let $\boldsymbol{R}_i = \text{var}(\boldsymbol{\epsilon}_i)$ and $\boldsymbol{w}_i = \text{cov}(\boldsymbol{\epsilon}_i, \epsilon_{i0})$. If we directly apply Equation (4.47) p. 129, the random effects are part of the error term, i.e. $\boldsymbol{e}_i = \boldsymbol{Z}_i\boldsymbol{b}_i + \boldsymbol{\epsilon}_i$. If we take into account that residual errors in group i are uncorrelated with the residual errors of other groups, we get the predictor:

$$\widetilde{y}_{i0} = \boldsymbol{x}_{i0}'\widehat{\boldsymbol{\beta}} + \boldsymbol{c}_i'\boldsymbol{V}_i^{-1}\left(\boldsymbol{y}_i - \boldsymbol{X}_i\widehat{\boldsymbol{\beta}}\right), \tag{6.23}$$

where

$$\boldsymbol{c}_i = \text{cov}\left(\boldsymbol{Z}_i\boldsymbol{b}_i + \boldsymbol{\epsilon}_i, \boldsymbol{z}_{i0}'\boldsymbol{b}_i + \epsilon_{i0}\right) = \boldsymbol{Z}_i\,\text{var}(\boldsymbol{b}_i)\,\boldsymbol{z}_{i0} + \boldsymbol{w}_i\,,$$

and

$$\boldsymbol{V}_i = \text{var}(\boldsymbol{y}_i) = \boldsymbol{Z}\,\text{var}(\boldsymbol{b}_i)\,\boldsymbol{Z}' + \boldsymbol{R}_i\,.$$

Lappi (1986) has shown that the predictor can be presented as:

$$\widetilde{y}_{i0} = \boldsymbol{x}_{i0}'\widehat{\boldsymbol{\beta}} + \boldsymbol{z}_{i0}'\widetilde{\boldsymbol{b}}_i + \boldsymbol{w}_i'\boldsymbol{R}_i^{-1}\left(\boldsymbol{y}_i - \boldsymbol{X}_i\widehat{\boldsymbol{\beta}} - \boldsymbol{Z}_i\widetilde{\boldsymbol{b}}_i\right), \tag{6.24}$$

where $\widetilde{\boldsymbol{b}}_i$ is the BLUP of $\boldsymbol{b}_i$ discussed in Section 5.4.1. Thus, the predictor formula looks the same as in the case where $\boldsymbol{b}_i$ also is fixed, only $\widehat{\boldsymbol{b}}_i$ is replaced with $\widetilde{\boldsymbol{b}}_i$. The prediction variance is (Lappi 1986):

$$\text{var}(\widetilde{y}_{i0} - y_{i0}) = \left(\boldsymbol{x}_*' - \boldsymbol{w}_i'\boldsymbol{R}_i^{-1}\boldsymbol{X}_*\right)\boldsymbol{G}\left(\boldsymbol{x}_* - \boldsymbol{X}_*'\boldsymbol{R}_i^{-1}\boldsymbol{w}_i\right) + \text{var}(e_{i0}) - \boldsymbol{w}_i'\boldsymbol{R}_i^{-1}\boldsymbol{w}_i \tag{6.25}$$

where $\boldsymbol{x}_* = \begin{pmatrix}\boldsymbol{x}_{i0}\\ \boldsymbol{z}_{i0}\end{pmatrix}$, $\boldsymbol{X}_* = (\boldsymbol{X}_i \quad \boldsymbol{Z}_i)$, and $\boldsymbol{G} = \text{var}\begin{pmatrix}\widehat{\boldsymbol{\beta}}\\ \widetilde{\boldsymbol{b}}_i - \boldsymbol{b}_i\end{pmatrix}$.

The prediction variance also looks the same as in the case of fixed $\boldsymbol{b}_i$. With a fixed $\boldsymbol{b}_i$, $\boldsymbol{G}$ would just be equal to $\boldsymbol{G} = \text{var}\begin{pmatrix}\widehat{\boldsymbol{\beta}}\\ \widehat{\boldsymbol{b}}_i\end{pmatrix}$.

6.6 Applying the mixed model BLUP in a new group without re-estimating the model

It is a common procedure in forestry, for example, that a mixed model is formulated using estimation data and the estimated model parameters are then applied for prediction in a new group (usually a stand) without re-estimating the fixed model parameters with the updated data. In this section, we analyze what the prediction variance is with such a procedure. It is assumed that all random parameters are group-specific, i.e. the model for group i is (Equation (5.13) p. 145)

$$\boldsymbol{y}_i = \boldsymbol{X}_i\boldsymbol{\beta} + \boldsymbol{Z}_i\boldsymbol{b}_i + \boldsymbol{\epsilon}_i\,, \tag{6.26}$$

where $\text{var}(\boldsymbol{\epsilon}_i) = \boldsymbol{R}_i$ and $\text{var}(\boldsymbol{b}_i) = \boldsymbol{D}_*$. It is possible that there are also subgroup parameters in $\boldsymbol{b}_i$. Let M be the number of groups in the estimation data. Denote that:

$$\boldsymbol{b} = \begin{pmatrix}\boldsymbol{b}_1\\ \vdots\\ \boldsymbol{b}_M\end{pmatrix}, \quad \boldsymbol{D} = \begin{pmatrix}\boldsymbol{D}_* & & \\ & \ddots & \\ & & \boldsymbol{D}_*\end{pmatrix}$$

It is assumed that the estimated variance and covariance parameters are correct, i.e. $\boldsymbol{R}_i$ and $\boldsymbol{D}_*$ are known. When $\boldsymbol{\beta}$ is estimated with the GLS estimator and $\boldsymbol{b}$ is predicted with BLUP, then (see Box 5.5 on p. 162)

$$\begin{pmatrix} \widehat{\boldsymbol{\beta}} \\ \widetilde{\boldsymbol{b}} \end{pmatrix} = \begin{pmatrix} \sum_{i=1}^{M} \boldsymbol{X}_i'\boldsymbol{R}_i^{-1}\boldsymbol{X}_i & \sum_{i=1}^{M} \boldsymbol{X}_i'\boldsymbol{R}_i^{-1}\boldsymbol{Z}_i \\ \sum_{i=1}^{M} \boldsymbol{Z}_i'\boldsymbol{R}_i^{-1}\boldsymbol{X}_i & \sum_{i=1}^{M} \boldsymbol{Z}_i'\boldsymbol{R}_i^{-1}\boldsymbol{Z}_i + \boldsymbol{D}^{-1} \end{pmatrix}^{-1} \begin{pmatrix} \sum_{i=1}^{M} \boldsymbol{X}_i'\boldsymbol{R}_i^{-1}\boldsymbol{y}_i \\ \sum_{i=1}^{M} \boldsymbol{Z}_i'\boldsymbol{R}_i^{-1}\boldsymbol{y}_i \end{pmatrix} \tag{6.27}$$

$$\text{var}\begin{pmatrix} \widehat{\boldsymbol{\beta}} \\ \widetilde{\boldsymbol{b}} - \boldsymbol{b} \end{pmatrix} = \begin{pmatrix} \sum_{i=1}^{M} \boldsymbol{X}_i'\boldsymbol{R}_i^{-1}\boldsymbol{X}_i & \sum_{i=1}^{M} \boldsymbol{X}_i'\boldsymbol{R}_i^{-1}\boldsymbol{Z}_i \\ \sum_{i=1}^{M} \boldsymbol{Z}_i'\boldsymbol{R}_i^{-1}\boldsymbol{X}_i & \sum_{i=1}^{M} \boldsymbol{Z}_i'\boldsymbol{R}_i^{-1}\boldsymbol{Z}_i + \boldsymbol{D}^{-1} \end{pmatrix}^{-1} . \tag{6.28}$$

If $\boldsymbol{\beta}$ is known, then BLP of $\boldsymbol{b}_i$ is

$$\widetilde{\boldsymbol{b}}_i = \left(\boldsymbol{Z}_i'\boldsymbol{R}_i^{-1}\boldsymbol{Z}_i + \boldsymbol{D}_*^{-1}\right)^{-1}\boldsymbol{Z}_i'\boldsymbol{R}_i^{-1}\left(\boldsymbol{y}_i - \boldsymbol{X}_i\boldsymbol{\beta}\right) \tag{6.29}$$

and

$$\text{var}\left(\widetilde{\boldsymbol{b}}_i - \boldsymbol{b}_i\right) = \left(\boldsymbol{Z}_i'\boldsymbol{R}_i^{-1}\boldsymbol{Z}_i + \boldsymbol{D}_*^{-1}\right)^{-1} . \tag{6.30}$$

If we obtain new data from a new group 0, then we should, in fact, re-estimate the model and obtain new estimates for the fixed parameters and new predictors of all the random effects. However, if we predict $\boldsymbol{b}_0$ assuming that $\widehat{\boldsymbol{\beta}}$ is correct, then BLP, called hereafter BLUP*, (which is Estimated BLP or EBLP) is according to Equation (6.29):

$$\widetilde{\boldsymbol{b}}_0 = \left(\boldsymbol{Z}_0'\boldsymbol{R}_0^{-1}\boldsymbol{Z}_0 + \boldsymbol{D}_*^{-1}\right)^{-1}\boldsymbol{Z}_0'\boldsymbol{R}_0^{-1}\left(\boldsymbol{y}_0 - \boldsymbol{X}_0\widehat{\boldsymbol{\beta}}\right) . \tag{6.31}$$

Notice that Equation (6.31) is the BLUP of $\boldsymbol{b}_0$, if $\widehat{\boldsymbol{\beta}}$ is the GLS estimate based on the entire data set. Let us then derive the prediction variance of Equation (6.31). Note first that Equation (6.31) can be presented as:

$$\begin{aligned} \widetilde{\boldsymbol{b}}_0 = &\left(\boldsymbol{Z}_0'\boldsymbol{R}_0^{-1}\boldsymbol{Z}_0 + \boldsymbol{D}_*^{-1}\right)^{-1}\boldsymbol{Z}_0'\boldsymbol{R}_0^{-1}\left(\boldsymbol{y}_0 - \boldsymbol{X}_0\boldsymbol{\beta}\right) + \\ &\left(\boldsymbol{Z}_0'\boldsymbol{R}_0^{-1}\boldsymbol{Z}_0 + \boldsymbol{D}_*^{-1}\right)^{-1}\boldsymbol{Z}_0'\boldsymbol{R}_0^{-1}\boldsymbol{X}_0\left(\boldsymbol{\beta} - \widehat{\boldsymbol{\beta}}\right) . \end{aligned} \tag{6.32}$$

The first term of Equation (6.32) is the BLP of $\boldsymbol{b}$, based on known $\boldsymbol{\beta}$. As the new group is not used in the estimation of $\boldsymbol{\beta}$, $(\boldsymbol{y}_0 - \boldsymbol{X}_0\boldsymbol{\beta})$ and $\left(\boldsymbol{\beta} - \widehat{\boldsymbol{\beta}}\right)$ are independent. Then the prediction error is

$$\begin{aligned} \widetilde{\boldsymbol{b}}_0 - \boldsymbol{b}_0 = &\left(\boldsymbol{Z}_0'\boldsymbol{R}_0^{-1}\boldsymbol{Z}_0 + \boldsymbol{D}_*^{-1}\right)^{-1}\boldsymbol{Z}_0'\boldsymbol{R}_0^{-1}\left(\boldsymbol{y}_0 - \boldsymbol{X}_0\boldsymbol{\beta}\right) - \boldsymbol{b}_0 + \\ &\left(\boldsymbol{Z}_0'\boldsymbol{R}_0^{-1}\boldsymbol{Z}_0 + \boldsymbol{D}_*^{-1}\right)^{-1}\boldsymbol{Z}_0'\boldsymbol{R}_0^{-1}\boldsymbol{X}_0\left(\boldsymbol{\beta} - \widehat{\boldsymbol{\beta}}\right) \end{aligned}$$

and using Equations (6.29) and (6.30) we obtain

$$\begin{aligned} \text{var}\left(\widetilde{\boldsymbol{b}}_0 - \boldsymbol{b}_0\right) = &\left(\boldsymbol{Z}_0'\boldsymbol{R}_0^{-1}\boldsymbol{Z}_0 + \boldsymbol{D}_*^{-1}\right)^{-1} + \\ &\left(\boldsymbol{Z}_0'\boldsymbol{R}_0^{-1}\boldsymbol{Z}_0 + \boldsymbol{D}_*^{-1}\right)^{-1}\boldsymbol{Z}_0'\boldsymbol{R}_0^{-1}\boldsymbol{X}_0\,\text{var}\left(\widehat{\boldsymbol{\beta}}\right)\boldsymbol{X}_0{}'\boldsymbol{R}_0^{-1}\boldsymbol{Z}_0\left(\boldsymbol{Z}_0'\boldsymbol{R}_0^{-1}\boldsymbol{Z}_0 + \boldsymbol{D}_*^{-1}\right)^{-1} . \end{aligned} \tag{6.33}$$

The predictor of an individual observation y_{0j} is

$$\widetilde{y}_{0j} = \boldsymbol{x}_{0j}'\widehat{\boldsymbol{\beta}} + \boldsymbol{z}_{0j}'\widetilde{\boldsymbol{b}}_0 .$$

Its prediction variance can be derived as follows:

$$\begin{aligned}\mathrm{var}\left(\widetilde{y}_{0j}-y_{0j}\right) &= \mathrm{var}\left(\boldsymbol{x}_{0j}'\widehat{\boldsymbol{\beta}}+\boldsymbol{z}_{0j}'\widetilde{\boldsymbol{b}}_0-\boldsymbol{x}_{0j}'\boldsymbol{\beta}-\boldsymbol{z}_{0j}'\boldsymbol{b}_0-\epsilon_{0j}\right)\\ &= \mathrm{var}\left(\boldsymbol{x}_{0j}'\left(\widehat{\boldsymbol{\beta}}-\boldsymbol{\beta}\right)+\boldsymbol{z}_{0j}'\left(\widetilde{\boldsymbol{b}}_0-\boldsymbol{b}_0\right)-\epsilon_{0j}\right)\\ &= \boldsymbol{x}_{0j}'\,\mathrm{var}\left(\widehat{\boldsymbol{\beta}}\right)\boldsymbol{x}_{0j}+\boldsymbol{z}_{0j}'\,\mathrm{var}\left(\widetilde{\boldsymbol{b}}_0-\boldsymbol{b}_0\right)\boldsymbol{z}_{0j}+\\ &\quad 2\boldsymbol{x}_{0j}'\,\mathrm{cov}\left(\widehat{\boldsymbol{\beta}},\widetilde{\boldsymbol{b}}_0'-\boldsymbol{b}_0'\right)\boldsymbol{z}_{0j}+\mathrm{var}\left(\epsilon_{0j}\right),\end{aligned} \tag{6.34}$$

where $\mathrm{var}\left(\widehat{\boldsymbol{\beta}}\right)$ is obtained from Equation (6.28), $\mathrm{var}\left(\widetilde{\boldsymbol{b}}_0-\boldsymbol{b}_0\right)$ from Equation (6.33), and

$$\begin{aligned}\mathrm{cov}\left(\widehat{\boldsymbol{\beta}},\widetilde{\boldsymbol{b}}_0'-\boldsymbol{b}_0'\right) &= \mathrm{cov}\left(\widehat{\boldsymbol{\beta}},-\left(\boldsymbol{Z}_0'\boldsymbol{R}_0^{-1}\boldsymbol{Z}_0+\boldsymbol{D}^{-1}\right)^{-1}\boldsymbol{Z}_0'\boldsymbol{R}_0^{-1}\boldsymbol{X}_0\widehat{\boldsymbol{\beta}}\right)\\ &= -\,\mathrm{var}\left(\widehat{\boldsymbol{\beta}}\right)\boldsymbol{X}_0'\boldsymbol{R}_0^{-1}\boldsymbol{Z}_0\left(\boldsymbol{Z}_0'\boldsymbol{R}_0^{-1}\boldsymbol{Z}_0+\boldsymbol{D}_*^{-1}\right)^{-1}.\end{aligned} \tag{6.35}$$

As discussed in Section 4.7, the average prediction error does not converge to zero when an ordinary linear regression model is used for prediction, owing to the estimation errors of the fixed parameters. However, if the fixed parameters $\boldsymbol{\beta}$ in a mixed model (6.26) contain only the expected values of the random parameters, then the effect of the estimation errors of the fixed parameters $\boldsymbol{\beta}$ decreases as n_0, the number of observations in the new group, increases. This is demonstrated in the following simple example.

Example 6.8. Let us assume that $y_{ij}=\mu+b_i+\epsilon_{ij}$. Then $\widehat{\mu}_{GLS}=\sum_{i=1}^{M}\bar{y}_i/\sum_{i=1}^{M}w_i$, where $w_i=1/\left(\sigma_b^2+\frac{\sigma^2}{n_i}\right)$. In addition

$$\mathrm{var}\left(\widehat{\mu}\right)=1/\sum_{i=1}^{M}w_i\,. \tag{6.36}$$

Let us then predict b_0 using BLUP* (6.31):

$$\widetilde{b}_0=\left(\frac{\sigma_b^2}{\sigma_b^2+\frac{\sigma^2}{n_0}}\right)\left(\bar{y}_0-\widehat{\mu}\right).$$

According to Equations (6.30),(6.35) and (6.34):

$$\mathrm{var}\left(\widetilde{b}_0-b_0\right)=\left(\frac{\sigma_b^2}{\sigma_b^2+\frac{\sigma^2}{n_0}}\right)\frac{\sigma^2}{n_0}+\left(\frac{\sigma_b^2}{\sigma_b^2+\frac{\sigma^2}{n_0}}\right)^2\mathrm{var}\left(\widehat{\mu}\right),$$

$$\mathrm{cov}\left(\widehat{\mu},\widetilde{b}_0-b_0\right)=-\left(\frac{\sigma_b^2}{\sigma_b^2+\frac{\sigma^2}{n_0}}\right)\mathrm{var}\left(\widehat{\mu}\right),$$

$$\mathrm{var}\left(\widetilde{y}_{0j}-y_{0j}\right)=\left(\frac{\sigma_b^2}{\sigma_b^2+\frac{\sigma^2}{n_0}}\right)\frac{\sigma^2}{n_0}+\left(\frac{\frac{\sigma^2}{n_0}}{\sigma_b^2+\frac{\sigma^2}{n_0}}\right)^2\mathrm{var}\left(\widehat{\mu}\right)+\sigma^2\,.$$

If $\mathrm{var}\left(\widehat{\mu}\right)>\sigma_b^2+\frac{\sigma^2}{n_0}$, then BLUP* has a larger prediction variance than if y_{0j} is predicted with the arithmetic group mean $\bar{y}_0$, i.e. the prediction variance is larger than $\frac{\sigma^2}{n_0}$ (show this!). If there are n observations in each of the M groups in the estimation data, then according to Equation (6.36):

$$\mathrm{var}\left(\widehat{\mu}\right)=\left(\sigma_b^2+\frac{\sigma^2}{n}\right)/M\,. \tag{6.37}$$

Thus, in practice the BLUP* always has a smaller prediction variance than $\bar{y}_0$. However, the above analysis is too optimistic, because it was assumed that the variances of random effects and residual errors are known, when in practice they are not.

7

Nonlinear (Mixed-Effects) Models

7.1 Nonlinear model and nonlinear relationship

If the relationship between the response and predictors is not linear, then the model should describe the nonlinearity. The alternatives are:

1. nonlinear transformations of the predictor,
2. nonlinear transformations of the response, and
3. nonlinear regression.

Models where transformations are made to the predictors, i.e. *the curvilinear models* (see Section 4.2.2) include, among others, regression splines (see Section 10.4.1), polynomial models (recall Section 4.2.2) and response surfaces (see Section 10.4.4). If transformations are made to the response, then all good properties (such as minimum-variance unbiased prediction) are only valid for the transformed y. Of course, the nonlinear transformations can also be done both to the response and the predictors. Using nonlinear transformations in either x, y or both retains the model as linear in terms of its parameters. See Sections 4.2, 10.2 and 10.4 for more discussion about modeling nonlinear relationships with linear models.

This chapter discusses nonlinear models. We start with the fixed-effects models and continue in Section 7.3 with the nonlinear mixed-effects models. Recommended textbooks on the topic are Ratkowsky (1990), Davidian and Giltinan (1995), Bates and Watts (2007), Pinheiro and Bates (2000), Seber and Wild (2003) and Demidenko (2013). In addition, many regression textbooks include chapters devoted to nonlinear regression.

A nonlinear regression model uses such a function for the relationship between $y-$ and $x-$variables that is nonlinear *with respect to its parameters.* If the function is selected wisely, it may lead to theoretically better justified models, with more meaningful parameters than a linear model. Moreover, it is often more robust in regard to extrapolations and may provide a good fit with a smaller number of parameters than a linear model. The nonlinear models are especially beneficial in situations where the subject-matter theory provides a nonlinear function for the relationship, and the parameters of the applied function are themselves of interest. An example is the hyperbolic model for the response of net photosynthesis by a plant to the sunlight. However, nonlinear models are also useful in situations where a well-fitting nonlinear model with a small number of parameters can be found empirically, even though no subject-matter theory is available to support the selected model. Examples of such models in forest sciences are the logistic model for the effect of thinning on tree growth (see examples of this chapter), the Näslund's model for tree height-diameter allometry, the Chapman-Richards model for tree height-age dynamics, and Kozak's model for tree taper.

7.2 Nonlinear fixed-effects model

7.2.1 Model formulation

As we have outlined in Chapter 4, the linear model can be written as

$$g_y(y_i) = \beta_1 + \beta_2 g_2(\boldsymbol{x}_i) + \beta_3 g_3(\boldsymbol{x}_i) + \ldots + \beta_p g_p(\boldsymbol{x}_i) + e_i\,, \tag{7.1}$$

where $\beta_1, \ldots, \beta_p$ are parameters to be estimated, and $g_k(\boldsymbol{x}_i)$, $k = 1, \ldots, p$ and $g_y(y_i)$ are known transformations of predictors $\boldsymbol{x}_i$ and response y_i, respectively. Denoting $u_i = g_y(y_i)$ and $v_i^{(k)} = g_k(\boldsymbol{x}_i)$, the model can be written as

$$u_i = \boldsymbol{v}_i \boldsymbol{\beta} + e_i\,.$$

In the linear model, the number of parameters equals the number of predictors in $\boldsymbol{v}_i$. However, because some predictors may be derived from other predictors (e.g. in polynomial and spline regression or in a model with interaction terms), the number of original predictors is often smaller than the number of parameters. In principle, the number of original predictors can also be larger when terms $g(x_i)$ are defined as functions of several original predictor variables, but that is seldom the case.

The nonlinear model is defined as

$$y_i = f(x_i^{(1)}, \ldots, x_i^{(p)}, \phi_1, \ldots, \phi_q) + e_i = f(\boldsymbol{x}_i, \boldsymbol{\phi}) + e_i\,, \tag{7.2}$$

where $f(\boldsymbol{x}_i, \boldsymbol{\phi})$ is a *response function* that depends on fixed predictors $\boldsymbol{x}_i$ and parameters $\boldsymbol{\phi}$, and e_i is the residual error. We assume that $\mathrm{E}(y_i) = f(\boldsymbol{x}_i, \boldsymbol{\phi})$ and therefore $\mathrm{E}(e_i) = 0$. The number of predictors and parameters can be different in the nonlinear model, and $\boldsymbol{x}_i$ only includes the original predictors. From a mathematical point of view, the linear model is a special case of the nonlinear model. Therefore, all properties of the nonlinear model are valid for the linear model.

The starting assumptions with the residual errors are the standard OLS assumptions: $\mathrm{var}(e_i) = \sigma^2$ for all i and $\mathrm{cov}(e_i, e_j) = 0$ for all pairs $i, j, i \neq j$. These assumptions can be relaxed, however, in a similar approach to that used in the case of linear models (see Section 4.3).

Example 7.1. The following functions are nonlinear with respect to ϕ:

- *Power function*: $y = \phi_1 x^{\phi_2}$, $x \geq 0$,
- *Chapman-Richards*, a flexible growth curve used, e.g. as a site index model: $y = \phi_1 \left(1 - e^{-\phi_2 x}\right)^{\phi_3}$, $\phi_1, \phi_2, \phi_3 > 0; x > 0$,
- *Näslund's model* used for tree height-diameter relationship: $y = \frac{x^2}{(\phi_1 x + \phi_2)^2}$, $\phi_1, \phi_2 > 0; x > 0$,
- *Hyperbolic light saturation curve/Michaelis-Menten curve/rectangular hyperbola* used for net photosynthesis in biology and enzyme reactions in biochemistry: $P = \frac{\phi_1 x}{\phi_2 + x}$, $\phi_1, \phi_2 > 0; x > 0$.

In some cases, a linear function can be found by reorganizing the nonlinear function. The new *linearized* function is based on the same relationship between x and y as the nonlinear functions, as illustrated in the following example.

Example 7.2. Taking logarithms from both sides of the Power function and defining $\phi_1^* = \log(\phi_1)$ yields

$$\log(y) = \phi_1^* + \phi_2 \log(x)\,, \tag{7.3}$$

which is linear with respect to ϕ_1^* and ϕ_2.

Box 7.1 Taylor approximation

Let $f(x)$ be a function that is infinitely many times differentiable with respect to x. The function can be expressed as an infinite-degree polynomial called the *Taylor polynomial*:

$$\sum_{n=0}^{\infty} \frac{f^{(n)}(a)}{n!}(x-a)^n ,$$

where $f^{(n)}$ is the nth derivative of $f(x)$. Using the first k terms of the Taylor polynomial provides k^{th} degree Taylor approximation of function f around point a:

$$f(x) \approx \sum_{n=0}^{k} \frac{f^{(n)}(a)}{n!}(x-a)^n .$$

Any smooth nonlinear function around a fixed value a can be satisfactorily approximated by a polynomial, if the degree of the selected polynomial is sufficiently high.

The idea generalizes to function $f(\boldsymbol{x})$, a smooth scalar function of $\boldsymbol{x} = (x_1, x_2, \ldots, x_p)$. For example, the second-order Taylor approximation around $\boldsymbol{a} = (a_1, a_2, \ldots, a_p)'$ is

$$f(\boldsymbol{x}) \approx f(\boldsymbol{a}) + \sum_{i=1}^{p} (x_i - a_i) \frac{\partial f}{\partial x_i}\bigg|_{\boldsymbol{x}=\boldsymbol{a}} + \frac{1}{2}\sum_{i=1}^{p}\sum_{j=1}^{p} (x_i - a_i)(x_j - a_j) \frac{\partial^2 f}{\partial x_i \partial x_j}\bigg|_{\boldsymbol{x}=\boldsymbol{a}} .$$

The first-order approximation results from only using the first two terms of the approximation.

Taking the square root of both sides of Näslund's model and, thereafter, multiplying both sides by $\frac{\phi_1 x + \phi_2}{\sqrt{y}}$ gives

$$\frac{x}{\sqrt{h}} = \phi_1 x + \phi_2 ,$$

which is linear with respect to ϕ_1 and ϕ_2. We could further divide both sides of the equation by x to get

$$\frac{1}{\sqrt{h}} = \phi_1 + \phi_2 \frac{1}{x} .$$

Therefore, there are (at least) two alternative ways to linearize Näslund's model. The former linearization was used in Example 5.2. For more discussion in regard to transformations in y, see Section 10.2.

Note that the linearization in the above example was carried out without taking the error term of the model into account. If we also include the error term, the logarithm of model

$$y = \phi_1 x^{\phi_2} + e ,$$

where e is the residual error, does not lead to model

$$\log(y) = \phi_1^* + \phi_2 x + e^* . \tag{7.4}$$

Model 7.4 results from the logarithmic transformation of the *multiplicative error model*

$$y = \phi_1 x^{\phi_2} e ,$$

where $\phi_1^* = \log(\phi_1)$ and $e^* = \log(e)$.

If vector $\boldsymbol{\phi}$ in $f(\boldsymbol{x}_i, \boldsymbol{\phi})$ includes too many parameters, the model is not necessarily *identifiable*. This means that the values of the parameters cannot be uniquely determined even if the available data were infinitely large and the residual error would be zero for every single observation. Identifiability in the context of linear models was discussed in Section 4.2. The nonlinear model (7.2) is not identifiable if for every value $\boldsymbol{\phi}_1$ of the parameter vector $\boldsymbol{\phi}$, there exists value $\boldsymbol{\phi}_2 \neq \boldsymbol{\phi}_1$ such that $f(\boldsymbol{x}, \boldsymbol{\phi}_1) = f(\boldsymbol{x}, \boldsymbol{\phi}_2)$ for all $\boldsymbol{x}$. For example, in the hyperbolic function

$$f(x) = \frac{ax}{bx + c}$$

all parameters a, b, and c cannot be estimated. If the parameters have values a_1, b_1, and c_1, the exactly same function with respect to x is obtained using parameter values ka_1, kb_1, and kc_1, where k is an arbitrary, non-zero constant. An identifiable hyperbolic function has only two parameters. There are several ways to parameterize the hyperbolic function using two parameters; one of them was given in Example 7.1.

To understand the behavior of the applied nonlinear function, one should use mathematical analysis to explore the function. Also, graphical analysis by plotting the function using different values of the parameters may be beneficial; see de Miguel *et al.* (2012, Fig. 4), for an example. In particular, maximums, minimums, and inflection points of the model are of special interest. It is also useful to explore the behavior of the function at theoretically interesting points, such as when $x = 0$ and $x \to \infty$. The extremes, inflection points, limits and values at points of interest depend on the function parameters. Therefore, mathematical analysis of the function also provides information about how the parameters affect the function behavior. Understanding the behavior of the function is essential for interpretation of the parameters, in finding reasonable initial guesses for the parameters, and in deciding which parameterization of the selected function should be chosen. A linear model of form (7.1) should also be analyzed in the same way, especially if several nonlinear transformations of the same predictor are included in the model as predictors (recall the polynomial model in Example 4.4).

A specific response function can be parameterized in many (in fact, infinitely many) different ways. Different parameterizations of a function $y = f(\boldsymbol{x}, \boldsymbol{\phi})$ assume the same relationship between $\boldsymbol{x}$ and y, but the parameters $\boldsymbol{\phi}$ are defined in a different way, and have, therefore, different interpretations. Mathematically, two functions $f_1(\boldsymbol{x}, \boldsymbol{\phi}_1)$ and $f_2(\boldsymbol{x}, \boldsymbol{\phi}_2)$ are parameterizations of the same basic function if $\boldsymbol{\phi}_1$ can be expressed in a unique way using the parameters $\boldsymbol{\phi}_2$. In linear modeling, reparameterizations that retained the model as linear were based on linear transformations of the predictors (see Section 4.2.3). If the requirement for linearity of the model is relaxed, the possibility for reparameterizations increases dramatically. For example, Fox and Weisberg (2010, p. 12) present seven different parameterizations of the same mathematical function, which arise from different disciplines and scientific traditions. The most simple reparameterization of a function is based on a monotonic nonlinear transformation of the original parameter.

If the parameters of the function are estimated using least squares, the value of the response function is not affected by the applied parameterization. However, the properties of the parameter estimates (bias, variance) are affected by the parameterization used. Often, a useful parameterization from the viewpoint of interpretation is one where the parameters directly specify the extremes, locations of inflection points, or limits or values of the function at points of interest. Such a parameterization is, however, not necessarily the best from the viewpoint of parameter estimation. For example, the Chapman-Richards function, which is popular in forestry has very poor estimation properties, and Ratkowsky (1990) suggests that it should not be used.

One way to find a parameterization with good estimation and interpretation properties is the use of *expected-value parameters*. An expected value parameter is the value of the regression function for a given value of the predictor. It is also easy to find a reasonable initial value for an expected value parameter by merely examining the plot of y with respect to x. Lappi (1997) applied expected-value parameters in height/diameter curves. However,

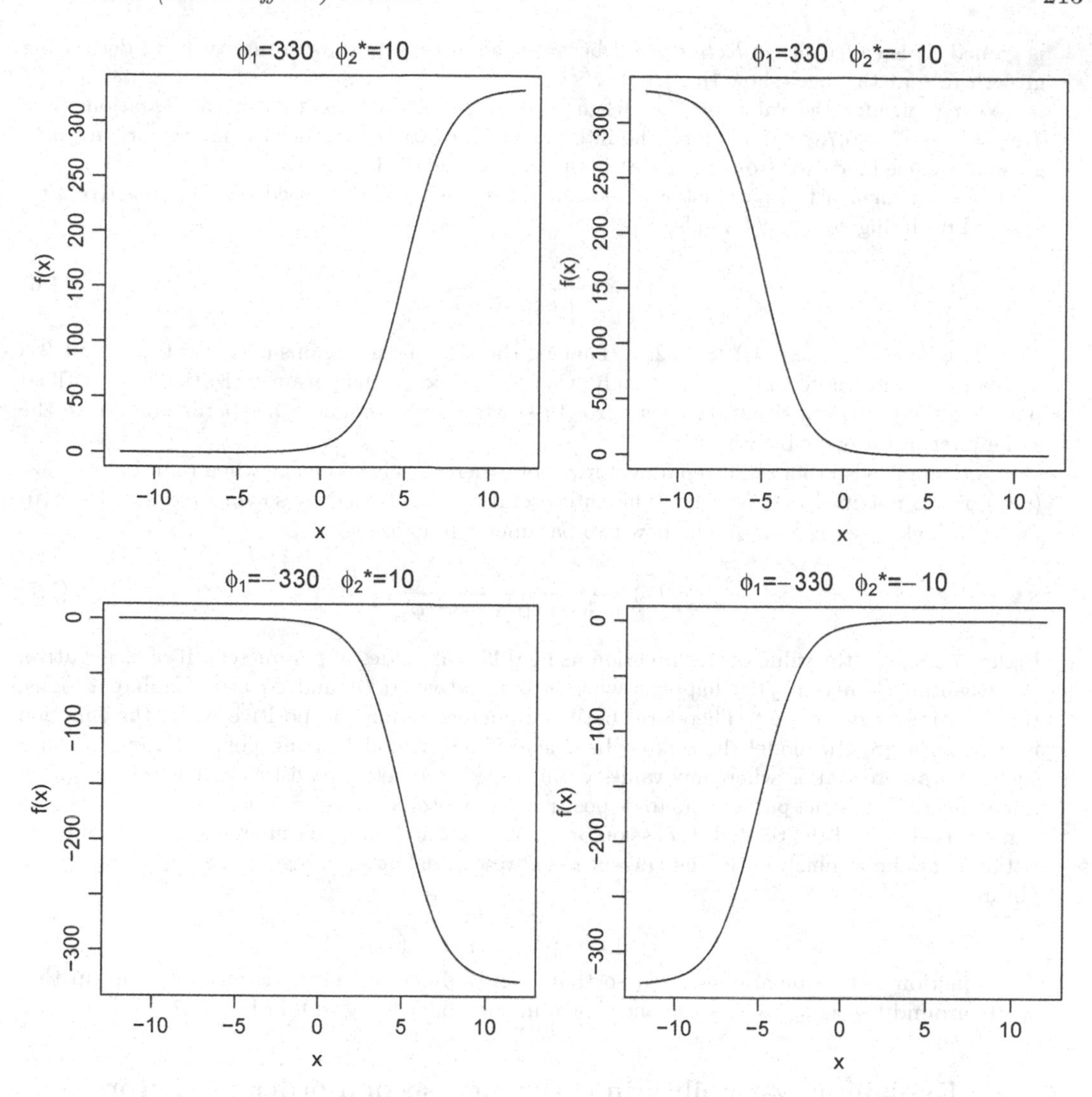

FIGURE 7.1 The value of function (7.7) for different combinations of negative and positive values of parameters ϕ_1 and ϕ_2^*.

the functions based on the expected-value parameters often look cumbersome. For more details, see Ratkowsky (1990, Section 2.3).

Example 7.3. Consider a three-parameter logistic function, which might be used to model the size of an organism that has symmetric growth as a function of age x. A commonly used parameterization of the model is given by Pinheiro and Bates (2000, Section C.7)

$$f(x) = \frac{\phi_1}{1 + \exp\left[\left(\phi_2 - x\right)/\phi_3\right]}. \tag{7.5}$$

For simplicity, we only consider cases where all three parameters are positive. As $x \to \infty$, $\exp\left[\left(\phi_2 - x\right)/\phi_3\right] \to 0$ and $f(x) \to \phi_1$. Therefore, parameter ϕ_1 is interpreted as the *upper asymptote* of the curve, i.e. the maximum size of the organism.

The denominator includes the term $\phi_2 - x$. This term becomes zero when $x = \phi_2$, and then also $f(x) = \frac{\phi_1}{2}$. Therefore, ϕ_2 can be interpreted as the age at which half of the maximum size is gained. It is called the *inflection point* because the increasing growth rate turns to decreasing growth rate at this age (show this!).

Next, consider the value of $f(x)$ when $x = \phi_2 + \phi_3$. Writing this in the equation gives $f(x) = \frac{\phi_1}{1+e^{-1}} \approx 0.73\phi_1$. Therefore, the interpretation of ϕ_3 is time needed for the organism to grow from size $0.5\phi_1$ (or from the inflection point) to size $0.73\phi_1$.

Consider then a two-parameter version of the model, which is used only for positive ages $x > 0$. Specifying $\phi_3 = \phi_2/4$ yields

$$f(x) = \frac{\phi_1}{1 + \exp\left[4 - 4x/\phi_2\right]} . \tag{7.6}$$

Now $f(0) = \phi_1/(1 + \exp(4)) \approx 0.02\phi_1$. That is, the size of the organism at $x = 0$ is about 2% of the maximum (note that $f(x) = 0$ only when $x \to -\infty$). This parameterization is a justified parameterization for a situation where growth starts at $x = 0$. The other parameters have the same interpretation as before.

One might wish to make a reparameterization where ϕ_2 gives the age when (almost) full size (98% of the maximum) is reached. This interpretation is obtained by solving $f(\phi_2^*) = 1 - f(0)$ for ϕ_2^*, which gives $\phi_2^* = 2\phi_2$. The new two-parameter function is

$$f(x) = \frac{\phi_1}{1 + \exp\left[4 - 8x/\phi_2^*\right]} . \tag{7.7}$$

Figure 7.1 shows the value of this function using different values of parameters. If ϕ_2^* is negative, the essential change in $f(x)$ happens when x is negative. If ϕ_1 and ϕ_2^* are of different signs, the function is decreasing. Therefore, both parameters should be positive when the function is used as a growth model. It is often beneficial, from a model fitting point of view, to have such a parameterization where any values within $(-\infty, \infty)$ yield a realistic curve for the modeled phenomena. To restrict parameters to be positive, a monotonic transformation $[0, \infty) \to (-\infty, \infty)$ can be used (recall the related discussion on p. 52). A widely applied function for such purpose is the logarithmic function. In the current case, we can define $\theta_1 = \log(\phi_1)$ and $\theta_2 = \log(\phi_2^*)$ to obtain

$$f(x) = \frac{\exp(\theta_1)}{1 + \exp\left[4 - 8x/\exp(\theta_2)\right]} . \tag{7.8}$$

The function is now reparameterized so that it only describes an increasing relationship that starts around the zero, such as the one shown in the upper-left graph in Figure 7.1.

7.2.2 Explaining variability in ϕ through second-order predictors

It is often justified to formulate the model such that the model parameters are specific for each observation:

$$y_i = f(\boldsymbol{x}_i, \boldsymbol{\phi}_i) + e_i . \tag{7.9}$$

The model is clearly overparameterized. However, we can further assume that each parameter is written as a linear combination $\phi_i^{(k)} = \boldsymbol{a}_i^{(k)}\boldsymbol{\beta}^{(k)}$ of fixed predictors $\boldsymbol{a}_i^{(k)}$ and parameters $\boldsymbol{\beta}^{(k)}$, $(k = 1, \ldots, q)$. That is, we assume that each parameter of the selected response function is a linear function of some predictors of the data. In this context, it is natural to call $\boldsymbol{x}_i$ and $\boldsymbol{\phi}$ first-order (primary) predictors and parameters. Correspondingly, $\boldsymbol{a}_i^{(k)}$ and $\boldsymbol{\beta}^{(k)}$ are second-order predictors and parameters.

Often, the primary predictor $\boldsymbol{x}_i$, is scalar, such as a measure of the age or size of an organism in a growth curve model, intensity of light or carbon dioxide content in a model of photosynthetic activity, or time in longitudinal data. For example, y_i may be the height of tree i and x_i the diameter (in an allometric model) or age (in a growth model).

We can pool the vectors $\boldsymbol{a}_i^{(k)}$ and $\boldsymbol{\beta}^{(k)}$ as follows:

$$\boldsymbol{A}_i = \begin{pmatrix} \boldsymbol{a}_i^{(1)\prime} & \boldsymbol{0} & \dots & \boldsymbol{0} \\ \boldsymbol{0} & \boldsymbol{a}_i^{(2)\prime} & \dots & \boldsymbol{0} \\ \vdots & \vdots & \ddots & \vdots \\ \boldsymbol{0} & \boldsymbol{0} & \dots & \boldsymbol{a}_i^{(q)\prime} \end{pmatrix}, \boldsymbol{\beta} = \begin{pmatrix} \boldsymbol{\beta}^{(1)} \\ \boldsymbol{\beta}^{(2)} \\ \vdots \\ \boldsymbol{\beta}^{(q)} \end{pmatrix}.$$

Then $\boldsymbol{\phi}_i = \boldsymbol{A}_i\boldsymbol{\beta}$.

Model (7.9) reduces to the nonlinear model (7.2) if $\boldsymbol{a}_i^{(k)} = 1$ for all values of k; in that case $\boldsymbol{\phi} = \boldsymbol{\beta}$. Therefore, model (7.2) can be seen as a special case of this model. Alternatively, one can ignore the linear relationship between the secondary predictors and model parameters, and parameterize the nonlinear model directly in terms of $\boldsymbol{\beta}$, which also leads to (7.2). This argumentation sees model (7.9) as a special case of (7.2). Therefore, the two-stage definition is not necessary. However, it is convenient from the following points of view:

1. The model is often based on a nonlinear response function that arises from the subject-matter theory. In this context, interesting research questions can often be formulated in terms of hypotheses about the first-order parameters of the nonlinear function.

2. The model building process is more easily understandable, and parameter interpretation is clearer in the two-stage formulation.

3. The problem of parameter estimation is simplified by this approach.

4. The formulation can be nicely extended to grouped data sets through the introduction of random effects to the linear predictors.

The model for the entire data set of n observations can be written as

$$\boldsymbol{y} = \boldsymbol{f}(\boldsymbol{x}, \boldsymbol{\phi}) + \boldsymbol{e}, \tag{7.10}$$

where

$$\boldsymbol{\phi} = \boldsymbol{A}\boldsymbol{\beta}$$

by defining

$$\boldsymbol{y} = \begin{pmatrix} y_1 \\ y_2 \\ \vdots \\ y_n \end{pmatrix}, \boldsymbol{\phi} = \begin{pmatrix} \boldsymbol{\phi}_1 \\ \boldsymbol{\phi}_2 \\ \vdots \\ \boldsymbol{\phi}_n \end{pmatrix}, \boldsymbol{e} = \begin{pmatrix} e_1 \\ e_2 \\ \vdots \\ e_n \end{pmatrix},$$

$$\boldsymbol{f}(\boldsymbol{x}, \boldsymbol{\phi}) = \begin{pmatrix} f(\boldsymbol{x}_1, \boldsymbol{\phi}_1) \\ f(\boldsymbol{x}_2, \boldsymbol{\phi}_2) \\ \vdots \\ f(\boldsymbol{x}_n, \boldsymbol{\phi}_n) \end{pmatrix}, \boldsymbol{x} = \begin{pmatrix} \boldsymbol{x}_1 \\ \boldsymbol{x}_2 \\ \vdots \\ \boldsymbol{x}_n \end{pmatrix}, \boldsymbol{A} = \begin{pmatrix} \boldsymbol{A}_1 \\ \boldsymbol{A}_2 \\ \vdots \\ \boldsymbol{A}_n \end{pmatrix}.$$

Function $\boldsymbol{f}$ is in bold because its value is an n-length vector. It is assumed that $\mathrm{E}(\boldsymbol{e}) = \boldsymbol{0}$ and $\mathrm{var}(\boldsymbol{e}) = \mathrm{var}(\boldsymbol{y}) = \sigma^2\boldsymbol{V}$. The structure of $\boldsymbol{V}$ is specified according to the data, and it allows the modeling of inconstant residual error variance and dependence among errors. Usually $\boldsymbol{V} = \boldsymbol{I}$, implying OLS assumptions in regard to the residual errors. Furthermore, normality of $\boldsymbol{e}$ is often assumed so as to allow estimation using maximum likelihood, and to justify hypothesis tests in small samples.

Box 7.2 Finney's second-level parameters

Nonlinear regression models often contain so many parameters that it is not possible to estimate them all reliably (i.e. a model may be *overparameterized*). However, it may not be feasible to just drop parameters, because the entire behavior of the function may change. In one data set, the estimated curve may be a reasonable description of the data even if separate parameter estimates have large variances. However, the parameter estimates in different data sets and studies are not comparable, and thus it is difficult to accumulate knowledge. Finney (1979) offers a solution to such a problem in terms of *second-level parameters.*

Second-level parameters are such parameters that are given values using a heuristic evaluation based on graphical analysis, rough quantitative analysis or comparison of different related data sets. After fixing the second-level parameters, the primary parameters are then estimated rigorously. The primary parameters can then be better estimated and compared between different data sets and studies. Convergence problems are common in nonlinear models, especially in nonlinear mixed-effects models (see Section 7.3.3, p. 233). The use of reasonable second-level parameters may be a solution to such problems.

As Finney points out, the statistical praxis is full of implicit second-level parameters. For instance, when we estimate the simplest regression function $y = \beta_1 + \beta_2 x + e$ instead of $y = \beta_1 + \beta_2 x^{\beta_3} + e$, the second-level parameter β_3 is implicitly given the value 1. Similarly, if 1 or some other small constant is added to y before taking logarithm in case y can be zero, then that small constant is a second-level parameter in Finney's sense. In nonlinear models, second-level parameters may need values other than zeroes or ones.

7.2.3 Parameter estimation when $\mathrm{var}(\boldsymbol{e}) = \sigma^2 \boldsymbol{I}$

We consider first the parameter estimation of the model (7.2). Thereafter, we will extend the discussion to model (7.9).

Estimation for model (7.2)

Parameter $\boldsymbol{\phi}$ of model (7.2) under OLS assumptions in regard to residual errors can be estimated using the Ordinary Nonlinear Least Squares (ONLS). It minimizes the residual sum of squares

$$RSS(\boldsymbol{\phi}) = \sum_{i=1}^{n} (y_i - f(\boldsymbol{x}_i, \boldsymbol{\phi}))^2$$

with respect to $\boldsymbol{\phi}$. The resulting least squares estimate is $\widehat{\boldsymbol{\phi}}$.

If the model is nonlinear with respect to only one parameter, it can often be estimated using a trial and error approach. The value of the parameter is fixed at a certain value and the other parameters are estimated using OLS or GLS. Different values of the nonlinear parameter are tested and the value that provides the best fit is selected as the estimate. One can use, for example, a *grid search* algorithm where regularly spaced values over a realistic range are tested systematically. Of course, this approach does not allow for estimation of the standard error of the nonlinear parameter and its effects on the standard errors of the other parameters that are estimated using OLS or GLS.

In some cases, all the parameters of a nonlinear function, which is well justified for the process being modeled cannot be estimated, for example, because of limitations in the available data. In such instances, one can fix some of the less interesting parameters to a justified value, e.g. based on the subject-matter knowledge, and estimate the other parameters rigorously. The less interesting parameters are sometimes called *second-level parameters* (See Box 7.2, p. 216).

If the response function f is differentiable with respect to $\boldsymbol{\phi}$, the ONLS estimates can be computed using the modified Gauss-Newton method. The method is based on the first order Taylor approximation of the function *with respect to parameter* $\boldsymbol{\phi}$ around the current estimates of the parameters. Let $\widehat{\boldsymbol{\phi}}$ be the current estimate of $\boldsymbol{\phi}$. Now (see Box 7.1, p. 211)

$$\begin{aligned} y_i &= f(\boldsymbol{x}_i, \boldsymbol{\phi}) + e_i \\ &\approx f\left(\boldsymbol{x}_i, \widehat{\boldsymbol{\phi}}\right) + \left.\frac{\partial f}{\partial \phi_1}\right|_{\boldsymbol{\phi}=\widehat{\boldsymbol{\phi}}} \left(\phi_1 - \widehat{\phi}_1\right) + \ldots + \left.\frac{\partial f}{\partial \phi_q}\right|_{\boldsymbol{\phi}=\widehat{\boldsymbol{\phi}}} \left(\phi_q - \widehat{\phi}_q\right) + e_i \,. \end{aligned}$$

Therefore

$$y_i - f\left(\boldsymbol{x}_i, \widehat{\boldsymbol{\phi}}\right) \approx \left(\phi_1 - \widehat{\phi}_1\right) \left.\frac{\partial f}{\partial \phi_1}\right|_{\boldsymbol{\phi}=\widehat{\boldsymbol{\phi}}} + \ldots + \left(\phi_q - \widehat{\phi}_q\right) \left.\frac{\partial f}{\partial \phi_q}\right|_{\boldsymbol{\phi}=\widehat{\boldsymbol{\phi}}} + e_i \,. \tag{7.11}$$

The equation is linear and of the form

$$u_i = \psi_1 z_1 + \ldots + \psi_q z_q + e_i \,, \tag{7.12}$$

and the unknown parameters $\psi_k = \phi_k - \widehat{\phi}_k$ $(k = 1, \ldots, q)$, can be estimated using linear least squares. Using matrix formulation, we can write

$$\boldsymbol{u} = \widehat{\boldsymbol{X}} \boldsymbol{\psi} + \boldsymbol{e} \,,$$

where matrix $\widehat{\boldsymbol{X}}_{n \times q}$ includes the partial derivatives $\left.\frac{\partial f}{\partial \phi_k}\right|_{\boldsymbol{\phi}=\widehat{\boldsymbol{\phi}}}$ of parameters $k = 1, \ldots, q$ for all observations $1, \ldots, n$, evaluated at the current estimates of $\boldsymbol{\phi}$. The matrix $\widehat{\boldsymbol{X}}$ is commonly called the *Jacobian.*

The parameter estimates $\widehat{\psi}_k$ are estimates of $\phi_k - \widehat{\phi}_k$. Solving $\widehat{\psi}_k = \phi_k - \widehat{\phi}_k$ for ϕ_k gives

$$\phi_k = \widehat{\phi}_k + \widehat{\psi}_k \,, \tag{7.13}$$

which becomes the new estimate $\widehat{\phi}_k$. In the linear model, the above procedure would lead to OLS estimates after the first step. For the nonlinear model, this leads to an iterative procedure that hopefully reduces $RSS(\widehat{\boldsymbol{\phi}})$ at every step. However, reduction does not necessarily happen because the Taylor series is only an approximation in the nonlinear model. The method can be modified to reduce RSS at every step by multiplying $\widehat{\psi}_1, \ldots, \widehat{\psi}_q$ by the constant λ $(0 < \lambda \leq 1)$, called *the step length,* and using the product in (7.13) in the place of $\widehat{\psi}_k$. The step length is initially set to a value close to 1 and then reduced towards zero until a value is found that leads to a reduction in $RSS(\boldsymbol{\phi})$. The resulting algorithm is called the *modified Gauss-Newton method.* The current estimate at the first iteration needs to be provided by the user, and is called the *initial guess* of parameter $\boldsymbol{\phi}$.

The modified Gauss-Newton method converges to a local minimum of $RSS(\boldsymbol{\phi})$. To ensure that convergence happens and is rapid, sufficiently good initial guesses are required (see below). More discussion on the estimation of a nonlinear model can be found in Seber and Wild (2003).

Some software, such as `nlme::gnls` in `R`, can automatically determine good initial guesses for some commonly used functions, using a linearized version of the response function (see Pinheiro and Bates (2000), Appendix C). By default, `R` function `nls` uses the Gauss-Newton algorithm with numerically evaluated derivatives. Analytical derivatives are used if the response function is specified as an `R`-function, with an attribute `gradient` that evaluates analytically evaluated gradients (see Pinheiro and Bates (2000, Section 8.1) and Example 7.12 on p. 237 for details). There are also functions that use the Levenberg-Marquardt method, which is considered to be more robust against the convergence problems associated with the singularity of the Jacobian.

Estimation for model (7.10)

In model (7.10), we assumed that the first-order parameters are linear functions of second-order predictors and parameters. For such a model, we formulate the Gauss-Newton method for parameters $\boldsymbol{\beta}$. The derivatives corresponding to the kth first-order parameter are

$$\frac{\partial f(\boldsymbol{x}_i, \boldsymbol{\beta})}{\partial \boldsymbol{\beta}^{(k)}} = \frac{\partial f(\boldsymbol{x}_i, \boldsymbol{\phi})}{\partial \phi_k} \boldsymbol{a}_i' .$$

That is, the derivatives of f in terms of all elements $\boldsymbol{\beta}^{(k)}$ can be written by multiplying the derivative in terms of ϕ^k by vector $\boldsymbol{a}$. Therefore, a function giving derivatives with respect to first-order parameters is sufficient for analytic computation of the derivatives with respect to $\boldsymbol{\beta}$.

7.2.4 Initial guesses of parameters

To determine sufficiently good initial guesses for parameters, one should understand the behavior of the applied function and the interpretations of its parameters (recall Example 7.3). The following strategies can be used

- Trust your luck and make a rough guess.
- Plot the function to be estimated onto the data and explore which parameter values make the function roughly follow the data. Often it is sufficient to "find the data", i.e. to give such parameter values that the function passes through the data point cloud and behaves in a somewhat realistic manner within the range of predictors in the data (see Example 7.4).
- Use a grid search: define an anticipated range for every parameter and take values regularly within these intervals. Systematically try all combinations from these vectors.
- Linearize the model using appropriate transformations, fit the linearized model to the data, and recover the estimates of the original parameters that correspond to the estimates of the linearized model. Often, the model can be partially linearized, i.e. with respect to some parameters only. In such cases, approaches based on Finney's second-level parameters can be used (see p. 216 and Box 7.2).
- In model (7.9), it is usually sufficient to use zero as an initial guess for coefficients $\beta_2^{(k)}, \beta_3^{(k)} \ldots$ of vectors $\boldsymbol{\beta}^{(k)}$, $k = 1, \ldots, q$. Therefore, non-zero initial guesses are only needed for the intercept terms of each submodel. If predictors are added to a (null) model that has already been fitted, the parameter estimates of the null model can be used as initial guesses for those terms that have been included in the null model, and zeroes for the other terms of the extended model.

7.2.5 Inference and model diagnostics

The ONLS estimates are biased but consistent (i.e. the bias vanishes as sample size increases). If errors are normally distributed, the least squares estimates are equal to the maximum likelihood estimates. Once the parameters have been estimated, the residual standard error under the OLS assumptions in regard to the residuals can be estimated in the same way as in linear regression:

$$\widehat{\sigma}^2 = \frac{RSS(\widehat{\boldsymbol{\phi}})}{n - q} .$$

A benefit of the Gauss-Newton method is that it also provides an estimate of $\text{var}(\widehat{\boldsymbol{\phi}})$ as a by-product using the applied linear approximation. Let matrix $\widehat{\boldsymbol{X}}$ be the Jacobian at the last iteration. Now

$$\widehat{\text{var}}\left(\widehat{\boldsymbol{\phi}}\right) = \widehat{\sigma}^2 \left(\widehat{\boldsymbol{X}}'\widehat{\boldsymbol{X}}\right)^{-1} . \tag{7.14}$$

Confidence intervals and hypothesis tests about $\boldsymbol{\beta}$ are based on this approximation, and the procedures are similar to the ones used in the context of a linear model. Also tests based on the general F-test statistic (4.35) of the linearized model can be used. The only difference is that the tests are only approximate even for normally distributed data, due to the errors inherent in the Taylor approximation.

The coefficient of determination is computed using the same formulas as used for the linear model. The R^2 given in Box 10.1 (p. 311) should not be used. We prefer the adjusted R-square (4.34) over (4.33). The evaluation of model fit is based on similar graphs of residuals as used in the context of linear models.

Interpretation of the default outputs of hypothesis tests in the context of nonlinear models requires even more care than in the context of linear models. Especially, the default outputs of statistical software report tests against the null hypotheses of the form H_0 : parameter = 0. Care needs to be taken in the interpretation of such results, and understanding of the behavior of the applied response function is, again, necessary. For example, consider a model based on the power function

$$y_i = \phi_1 x_i^{\phi_2} + e_i \, .$$

If $\phi_1 = 0$, we have $\text{E}(y_i) = 0$ for all i. Therefore $H_0 : \phi_1 = 0$ tests the hypothesis that $\text{E}(y_i)$ is zero for all values of x_i, which is seldom of interest. Specifically, it does not test the hypothesis that $\text{E}(y_i)$ is zero only for $x_i = 0$, or that the $\text{E}(y_i)$ is not related to x_i, as one might expect by naïvely following the interpretation of the linear model. If $\phi_2 = 0$, then $x_i^{\phi_2} = 1$ for all x_i, and $y_i = \phi_1 + e_i$. Setting $\phi_2 = 1$ leads to model $y_i = \phi_1 x_i + e_i$. Therefore, the null hypothesis $H_0 : \phi_2 = 1$ tests whether $\text{E}(y_i)$ is linearly related to x_i.

Often, confidence intervals of the first-order parameters are more informative than tests of hypotheses. If the model is based on an existing theory of the underlying process, tests on first-order parameters are not of interest. In contrast, tests on the second-order parameters might be much more interesting.

Example 7.4. Data set `thinning.txt` includes estimates of the effect of silvicultural thinning of the neighboring trees on the basal area growth (mm^2/yr) of an individual Scots Pine tree in Eastern Finland for 23 calendar years (1983 – 2005). The thinning was conducted after the growing season in 1986, so its effect should be visible for the first time in 1987. The annual thinning effects were previously estimated in Example 6.6 for several trees, but this example uses only one of the trees. We ignore the possible prediction errors of the thinning effect and treat the data as if it were a true measured thinning effect.

Figure 7.2 shows a clear thinning effect as a function of time since thinning. However, the tree needs some time to recover from competition and grow a larger crown. Therefore, the full thinning effect is reached gradually after some "switch-on period" after thinning, which is called the reaction time. To model this recovery, we model the thinning effect at year i using the model

$$ThEf_i = f(T_i, \boldsymbol{\phi}) + e_i \, . \tag{7.15}$$

The response function is the reparameterized logistic function from Example 7.3

$$f(T_i, \boldsymbol{\phi}) = \frac{\exp \phi_1}{1 + \exp\left(4 - 8T_i / \exp \phi_2\right)} ,$$

where $ThEf_i$ is the thinning effect and T_i is time since thinning. Note that ϕs are logarithmic parameters, so $\exp(\phi)$ corresponds to ϕ in the original function parameterization. Recall that

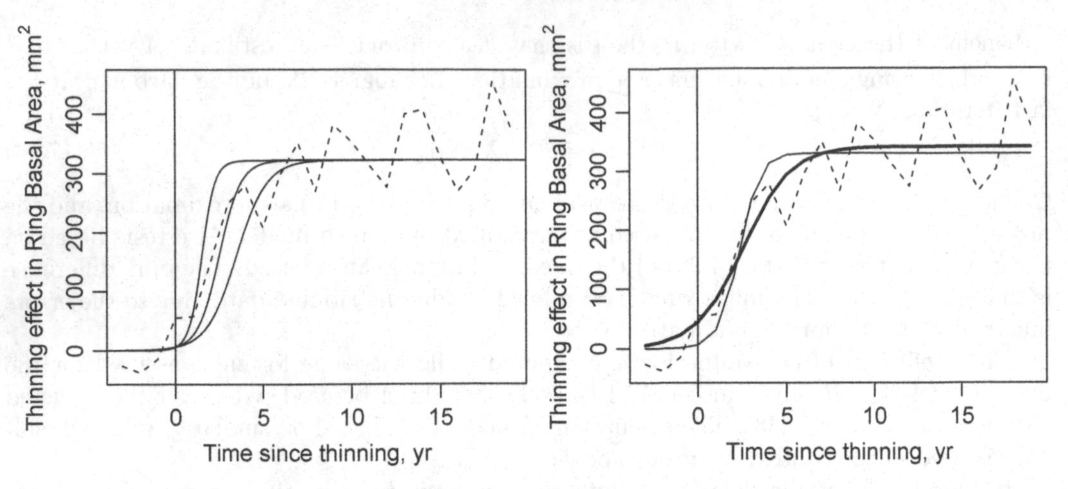

FIGURE 7.2 Observed thinning effect (dashed) after heavy thinning in year 0 (vertical line) for an individual Scots pine tree in North Karelia, Finland. The solid lines on the left-hand graph show logistic curves using a thinning effect of 320 mm^2/yr and reaction times of 3, 5, and 7 years. The solid lines on the right-hand graph show the fitted logistic curves using models `thinnls1` (thin solid) and `thinnls2` (thick solid) of Example 7.4.

the left asymptote of the curve is zero and the right asymptote is $\exp(\phi_1)$. The growth starts approximately at $T_i = 0$ (more specifically, the value of the function at $T_i = 0$ is $0.02\exp(\phi_1)$) and takes its maximum (or more specifically, 98% of the maximum) at $T_i = \exp(\phi_2)$. Therefore, parameter ϕ_1 can be interpreted as the logarithmic maximum thinning effect, and ϕ_2 as the logarithmic reaction time. Based on Figure 7.2, the full thinning effect is approximately 320 mm^2/yr. The left-hand graph shows the curve for reaction times 3, 5, and 7 years. The best-fit seems to occur when the reaction time is 5 years, which suggests the following initial guesses $\widehat{\phi}_1 = \log(320) = 5.76$ and $\widehat{\phi}_2 = \log(5) = 1.61$.

```
> mylogis<-function(x,phi1,phi2){
+          ephi1<-exp(phi1)
+          ephi2<-exp(phi2)
+          ephi1/(1+exp(4-8*x/ephi2))
+          }
>
> thinning$Ythin<-thinning$Year-1986
> thinnls1<-nls(ThEff~mylogis(Ythin,phi1,phi2),
+               data=thinning,
+               start=list(phi1=log(320),phi2=log(5)))
> summary(thinnls1)

Formula: ThEff ~ mylogis(Ythin, phi1, phi2)

Parameters:
     Estimate Std. Error t value Pr(>|t|)
phi1  5.79934    0.04372   132.7  < 2e-16 ***
phi2  1.53505    0.13835    11.1 3.05e-10 ***

Residual standard error: 56.5 on 21 degrees of freedom

Number of iterations to convergence: 10
Achieved convergence tolerance: 8.915e-06
```

We obtain estimates $\widehat{\phi}_1 = 5.80$, $\widehat{\phi}_2 = 1.54$, $\widehat{\sigma}^2 = 56.5^2$, $\widehat{\text{var}}(\widehat{\phi}_1) = 0.0437^2$ and $\widehat{\text{var}}(\widehat{\phi}_2) = 0.138^2$ after 11 iterations. Therefore, our estimate of reaction time is 4.64 years and the full thinning effect is 330 mm^2/yr.

A self-starting three-parameter logistic function called `SSlogis` is available in library `nlme`. It allows fitting of the following model:

$$ThEf_i = \frac{\phi_1}{1 + \exp\left[(\phi_2 - T_i)/\phi_3\right]} + e_i \,.$$

The fitting of this model can be carried out without finding initial parameter estimates. In this function, ϕ_1 is the maximum thinning effect, ϕ_2 is the year when half of the thinning effect has been achieved, and ϕ_3 is a scale parameter. Multiplying ϕ_3 by 8 gives the reaction time. The model differs from the previous model in that the start of the thinning reaction is not restricted to the year of thinning. In addition, the asymptote and scale parameters are parameterized in a different way.

```
> thinnls2<-nls(ThEff~SSlogis(Ythin,phi1,phi2,phi3),data=thinning)
> summary(thinnls2)

Formula: ThEff ~ SSlogis(Ythin, phi1, phi2, phi3)

Parameters:
     Estimate Std. Error t value Pr(>|t|)
phi1 341.4070    15.5300  21.984 1.76e-15 ***
phi2   2.5600     0.4840   5.289 3.55e-05 ***
phi3   1.3455     0.4207   3.198  0.00451 **

Residual standard error: 53.81 on 20 degrees of freedom

Number of iterations to convergence: 0
Achieved convergence tolerance: 3.335e-06
```

The estimated maximum thinning effect is 341 mm^2/yr, and half of the thinning effect is reached 2.6 years after thinning. The reaction starts (i.e. 2% of the effect is reached) almost 3 years before the thinning ($2.56 - 4 \times 1.34$), which is unrealistic. The smooth curve in Figure 7.2 shows a rather good fit to the data. By restricting the ratio of the scale and rate parameters in `thinnls1`, we forced the thinning effect to commence in the thinning year. However, a closer look of the data shows some asymmetry in the reaction (i.e. the first half of the thinning effect happens in a shorter time period than the second half), which indicates that an asymmetric growth function might fit the data better than the logistic model. This lack of fit explains the unrealistic estimate of the starting time of the effect.

Web Example 7.5. To demonstrate the Gauss-newton algorithm, this example fits the model in Example 7.4 without using the function `nls`.

7.2.6 Parameter estimation and inference when $\text{var}(\boldsymbol{e}) = \sigma^2 \boldsymbol{V}$

If the residual errors are correlated and/or heteroscedastic, the ONLS estimates are still consistent. However, the variance-covariance structure of the data needs to be taken into account in hypothesis tests and confidence intervals in same way as with linear models. Furthermore, estimators with lower variance do exist. As with linear models (recall Section 4.3), the variance heterogeneity can be modeled either by making transformations to the response, or by explicitly modeling the error variance. Also, the autocorrelation of residuals can be modeled explicitly using the models that were used with linear models. Function `nlme::gnls` allows for the fitting of such models.

To model the dependence, we use formulation (7.10) where now $\text{var}(\boldsymbol{e}) = \sigma^2 \boldsymbol{V}$. If $\boldsymbol{V}$ is known, the model can be written equivalently (recall Section 4.4.2)

$$\boldsymbol{y}^* = \boldsymbol{f}^*(\boldsymbol{x}, \boldsymbol{\phi}) + \boldsymbol{e}^* \,, \tag{7.16}$$

where $\boldsymbol{y}^* = \left(\boldsymbol{V}^{-1/2}\right)' \boldsymbol{y}$, $\boldsymbol{f}^* = \left(\boldsymbol{V}^{-1/2}\right)' \boldsymbol{f}$ and $\boldsymbol{e}^* = \left(\boldsymbol{V}^{-1/2}\right)' \boldsymbol{e}$; $\left(\boldsymbol{V}^{-1/2}\right)'$ is as defined previously in Section 4.4.2. Similar to (7.10), we assume that $\boldsymbol{\phi} = \boldsymbol{A}\boldsymbol{\beta}$. Now, $\text{var}(\boldsymbol{e}^*) = \sigma^2 \boldsymbol{I}$,

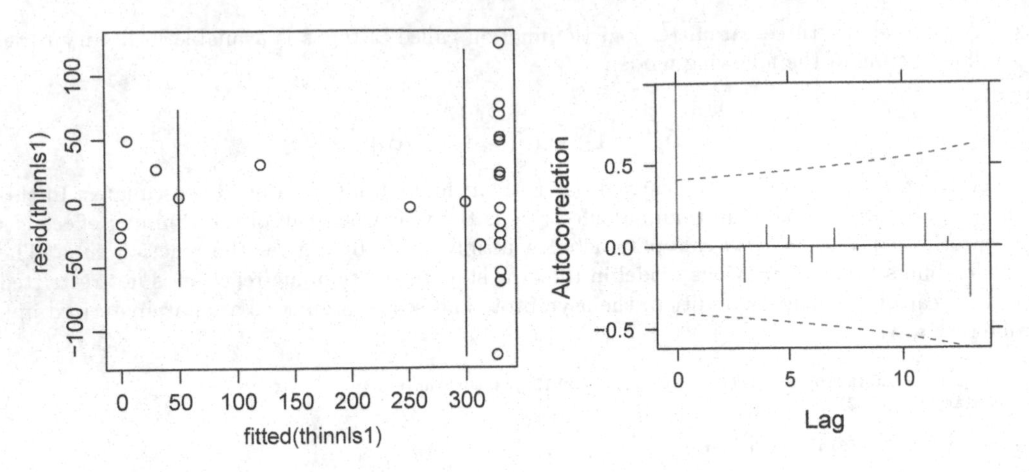

FIGURE 7.3 The residuals of thinning effect model **thinnls1** on fitted values and autocorrelation plot of the residuals with 95% confidence bounds.

so the parameters of this model can be estimated using the standard modified Gauss-Newton procedure. However, because $\boldsymbol{V}$ is not known, an iterative procedure and profiling are needed to estimate the parameters of both $\boldsymbol{V}$ and $\boldsymbol{\beta}$. Details of the procedures can be found in Seber and Wild (2003), Davidian and Giltinan (1995), and Pinheiro and Bates (2000, Section 7.5).

Example 7.6. Figure 7.3 (left) shows the residuals of model `thinnls1` against the fitted values, together with a whiskers plot. As the data set is small, and most fitted values are either 0 or 320, it is not easy to evaluate whether the residual variance is constant or not. The data set is a time series, so it might be realistic to assume temporal autocorrelation in the residual errors. Even though the autocorrelation plot of the residuals does not indicate the need for an autocorrelation model (right-hand graph), its inclusion might still be justified because of the data structure.

Therefore, we consider the power-type variance function $\text{var}(e_i) = \sigma^2 \widehat{y}^{2\delta}$ and AR(1) covariance structure $\text{cor}(e_i, e_j) = \rho^{|i-j|}$. The updated model is fitted below using function `nlme::gnls`. To allow hypothesis tests of the two models using two-argument form `anova`, we also refit model `thinnls1` using `gnls`.

```
> # Refit thinnls1 using gnls
> thinnls1<-gnls(ThEff~mylogis(Ythin,phi1,phi2),
+                data=thinning,
+                start=list(phi1=log(320),phi2=log(5)))

> thinnls3<-gnls(ThEff~mylogis(Ythin,phi1,phi2),
+                data=thinning,
+                start=list(phi1=log(320),phi2=log(5)),
+                cor=corAR1(),
+                weights=varPower())

> anova(thinnls1,thinnls3)
         Model df      AIC      BIC    logLik   Test  L.Ratio p-value
thinnls1     1  3 254.7567 258.1632 -124.3784
thinnls3     2  5 255.2537 260.9312 -122.6269 1 vs 2 3.502994  0.1735
```

The model with heteroscedastic variances and autocorrelated residual errors does not show a significantly better fit than the OLS-model. The parameter estimates of the new model are $\widehat{\phi}_1 = \log(329.5)$, $\widehat{\phi}_2 = \log(4.470)$, $\widehat{\delta} = 0.103$, $\widehat{\rho} = 0.262$, and $\widehat{\sigma}^2 = 32.66^2$. The large difference in the value of parameter σ^2 is explained by its different interpretation in the two models (recall the discussion in Section 4.3.1). The estimates of the mean function parameters $\boldsymbol{\phi}$ are very similar for both models, but the confidence intervals for model `thinnls3` are wider. Our prior knowledge of the process supports the assumption of autocorrelated errors and inconstant error variance.

It is quite likely that the data are just too small to reject the null hypothesis of constant error variance and zero autocorrelation. Therefore, our conservative choice would be to use model **thinnls3** for inference about the fixed effects. The confidence intervals based on an ML fit are shown below (REML is not available in **gnls**):

```
> exp(intervals(thinnls1,which="coef")[[1]])
          lower       est.       upper
phi1 301.396472 330.081294 361.496138
phi2   3.481037   4.641512   6.188856
attr(,"label")
[1] "Coefficients:"
> exp(intervals(thinnls3,which="coef")[[1]])
          lower       est.       upper
phi1 290.127511 329.511809 374.242456
phi2   3.150828   4.470152   6.341908
attr(,"label")
[1] "Coefficients:"
```

7.3 Nonlinear mixed-effects models

7.3.1 Model formulation

Nonlinear mixed-effects models generalize linear mixed-effects models in a similar way that nonlinear models generalize the linear models, and nonlinear model in a similar way as linear mixed-effect model generalize the linear model. Our notations follow those of Pinheiro and Bates (2000), and the model is based on an extension of model (7.10) with the inclusion of random effects to the linear predictors. Denote the response variable y of individual j of group i by y_{ij} and the corresponding first-order predictor variable by $\boldsymbol{x}_{ij}$. Similar to the fixed-effects model (recall Section 7.2.2), $\boldsymbol{x}_{ij}$ is often scalar. The nonlinear mixed-effects model for individual j of group i can be expressed as

$$y_{ij} = f(\boldsymbol{x}_{ij}, \boldsymbol{\phi}_{ij}) + \epsilon_{ij} \quad i = 1, \ldots, M; j = 1, \ldots, n_i . \tag{7.17}$$

Function f is a response function that can be nonlinear with respect to $\boldsymbol{\phi}_{ij}$. Term ϵ_{ij} is the unexplained residual error. It is commonly assumed that the residual errors are independent and normal with a common constant variance, i.e. $\epsilon_{ij} \sim N(0, \sigma^2)$. Extensions of this structure are possible in similar ways as before in Sections 4.3, 5.2.2, and 7.2.6.

The model parameters are expressed as a function of fixed predictors and group-specific random effects. Therefore, we specify a separate linear sub-model for each element $\phi_{ij}^{(k)}$ of vector $\boldsymbol{\phi}_{ij}$, $k = 1, \ldots, q$:

$$\phi_{ij}^{(k)} = \boldsymbol{a}_{ij}^{(k)\prime} \boldsymbol{\beta}^{(k)} + \boldsymbol{z}_{ij}^{(k)\prime} \boldsymbol{b}_i^{(k)} .$$

Term $\boldsymbol{a}_{ij}^{(k)\prime} \boldsymbol{\beta}^{(k)}$ is the same as used in model (7.9). Vector $\boldsymbol{b}_i^{(k)}$ of length s_k includes the random effects and $\boldsymbol{z}_{ij}^{(k)}$ the corresponding predictors associated with the random effects of the first-order parameter $\phi_{ij}^{(k)}$. If only the random intercept is assumed for the first-order parameter $\phi_{ij}^{(k)}$, then $\boldsymbol{b}_i^{(k)}$ is scalar and $\boldsymbol{z}_{ij}^{(k)} = 1$.

Let us pool the second-order parameters and the random effects associated with each parameter $\phi_{ij}^{(1)}, \ldots, \phi_{ij}^{(q)}$ to vectors $\boldsymbol{\beta}$ and $\boldsymbol{b}_i$. The parameter vector for a single observation is defined in matrix form as

$$\boldsymbol{\phi}_{ij} = \boldsymbol{A}_{ij} \boldsymbol{\beta} + \boldsymbol{B}_{ij} \boldsymbol{b}_i , \tag{7.18}$$

where

$$A_{ij} = \begin{pmatrix} a_{ij}^{(1)\prime} & 0 & \dots & 0 \\ 0 & a_{ij}^{(2)\prime} & \dots & 0 \\ \vdots & \vdots & \ddots & \vdots \\ 0 & 0 & \dots & a_{ij}^{(q)\prime} \end{pmatrix}, \beta = \begin{pmatrix} \beta^{(1)} \\ \beta^{(2)} \\ \vdots \\ \beta^{(q)} \end{pmatrix}.$$

It is commonly assumed that

$$b_i \sim N(0, D_*) ,$$

where D_* is a positive definite variance-covariance matrix of random effects. However, restrictions to this matrix are often imposed, so as to overcome convergence problems in estimation (recall the discussion in Section 5.1.3).

If a random effect vector $b_i^{(k)}$ is associated with every first-order parameter $\phi_{ij}^{(k)}$, then matrices B_{ij} and vector b_i are defined as:

$$B_{ij} = \begin{pmatrix} z_{ij}^{(1)\prime} & 0 & \dots & 0 \\ 0 & z_{ij}^{(2)\prime} & \dots & 0 \\ \vdots & \vdots & \ddots & \vdots \\ 0 & 0 & \dots & z_{ij}^{(q)\prime} \end{pmatrix}, b_i = \begin{pmatrix} b_i^{(1)} \\ b_i^{(2)} \\ \vdots \\ b_i^{(q)} \end{pmatrix},$$

where $b_i^{(k)} = \left(b_i^{(k1)}, b_i^{(k2)}, \dots, b_i^{(ks_k)}\right)'$. Random effects are not necessarily associated with all first-order parameters. If they are assumed for only some of the first-order parameters, then only the corresponding columns and elements are included in B_{ij} and b_i. However, all q rows of B_{ij} are always retained; otherwise the lengths of vectors $A_{ij}\beta$ and $B_{ij}b_i$ in (7.18) would not match.

To understand the ingenuity of this model formulation, let us tentatively assume that $\phi_{ij} = \phi_i$. That is, assume that the second-order predictors are defined only at the group level; such an assumption was done in Lindstrom and Bates (1990). Then, the process behind the data can be considered as a two-stage process: (1) each group has a unique ϕ_i. Information on that parameter is provided by the several observations of the response in that group. If the number of observations per group is sufficiently large, we could fit the model separately for each group in order to find the parameter estimates for the group. Thereafter, (2) the dependence of the parameters on some group-specific predictors could be explored using linear regression and separate models for all parameters could be fitted. The residuals of these models would indicate how much the group-specific parameters vary between groups, and they would correspond to the group-specific random effects. Owing to the estimation errors of the parameters, those residuals would be more variable than the true random effects.

In the nonlinear mixed-effects modeling framework, the two steps outlined above are carried out in a single model fitting run. One advantage from the mixed-effects modeling fit is that the number of observations per group needs not be as high as in group-specific ONLS fits per group. In addition, the linear models for all parameters are estimated jointly, and the joint variation among them is described through the covariances of random effects in D_*. Furthermore, as we defined ϕ_{ij} as an observation-level parameter, the model allows the first-order parameters to vary among observations of the same group.

The matrix formulation for the model in group i is

$$y_i = f_i(x_i, \phi_i) + \epsilon_i , \tag{7.19}$$

where $\boldsymbol{\phi}_i = \boldsymbol{A}_i\boldsymbol{\beta} + \boldsymbol{B}_i\boldsymbol{b}_i$ and

$$\boldsymbol{y}_i = \begin{pmatrix} y_{i1} \\ \vdots \\ y_{in_i} \end{pmatrix}, \boldsymbol{\phi}_i = \begin{pmatrix} \phi_{i1} \\ \vdots \\ \phi_{in_i} \end{pmatrix}, \boldsymbol{x}_i = \begin{pmatrix} \boldsymbol{x}_{i1} \\ \vdots \\ \boldsymbol{x}_{in_i} \end{pmatrix}, \boldsymbol{\epsilon}_i = \begin{pmatrix} \epsilon_{i1} \\ \vdots \\ \epsilon_{in_i} \end{pmatrix},$$

$$\boldsymbol{f}_i(\boldsymbol{x}_i, \boldsymbol{\phi}_i) = \begin{pmatrix} f(\boldsymbol{x}_{i1}, \phi_{i1}) \\ \vdots \\ f(\boldsymbol{x}_{in_i}, \phi_{in_i}) \end{pmatrix}, \boldsymbol{A}_i = \begin{pmatrix} \boldsymbol{A}_{i1} \\ \vdots \\ \boldsymbol{A}_{in_i} \end{pmatrix}, \boldsymbol{B}_i = \begin{pmatrix} \boldsymbol{B}_{i1} \\ \vdots \\ \boldsymbol{B}_{in_i} \end{pmatrix},$$

vectors $\boldsymbol{\beta}$ and $\boldsymbol{b}_i$ are as before. Vectors $\boldsymbol{y}_i$, $\boldsymbol{\epsilon}_i$, and $\boldsymbol{f}_i(\boldsymbol{x}_i, \boldsymbol{\phi}_i)$ are all of length n_i, the number of observations in group i. The parameter vector $\boldsymbol{\phi}_i$ includes all parameters for all observations of the group: firstly q parameters for the first observation of the group, then q parameters for the second observation etc. It is, therefore, of length $q \times n_i$. Vector $\boldsymbol{x}_i$ includes the values of all primary predictors for all observations of the group; if the model includes only one primary predictor, then $\boldsymbol{x}_i$ is of length n_i. Also, matrices $\boldsymbol{A}_i$ and $\boldsymbol{B}_i$ include $q \times n_i$ rows. The numbers of columns are equal to the lengths of $\boldsymbol{\beta}$ and $\boldsymbol{b}_i$, respectively. Web Example 7.7 presents a simple example of this model formulation, while more examples are included in Pinheiro and Bates (2000, Section 7.1).

The model for all data pools the group-level models, combining the approaches used with the extended nonlinear regression model (7.10) and the linear mixed-effects model (5.16). The model for all data is

$$\boldsymbol{y} = \boldsymbol{f}(\boldsymbol{x}, \boldsymbol{\phi}) + \boldsymbol{\epsilon}, \tag{7.20}$$

where $\boldsymbol{\phi} = \boldsymbol{A}\boldsymbol{\beta} + \boldsymbol{B}\boldsymbol{b}$. Vectors $\boldsymbol{y}$, $\boldsymbol{x}$, $\boldsymbol{\epsilon}$ and $\boldsymbol{f}(\boldsymbol{x}, \boldsymbol{\phi})$ stack the group-specific vectors into longer vectors, and matrix $\boldsymbol{A}$ stacks the group-specific matrices $\boldsymbol{A}_i$ on each other. Group-specific vectors $\boldsymbol{b}_i$ are stacked into vector $\boldsymbol{b}$, which has the block-diagonal variance-covariance matrix $\boldsymbol{D} = \boldsymbol{I}_M \otimes \boldsymbol{D}_*$. Matrix $\boldsymbol{B}$ is a block-diagonal matrix having matrices $\boldsymbol{B}_i$, $i = 1, \ldots, M$ on the diagonal.

Notice that the model can always be defined by writing each y_{ij} as a function of known $\boldsymbol{x}_{ij}$, unknown fixed $\boldsymbol{\beta}$, and unknown random group specific $\boldsymbol{b}_i$. The introduction of the primary parameters ϕ in the formulation merely supports the formulation of models where the parameters of the model have a special linear structure, and provides a more transparent model formulation, which results in benefits in parameter estimation and interpretation.

Web Example 7.7. Consider the model

$$y_{ij} = \phi_{ij}^{x_{ij}} + \epsilon_{ij}, \tag{7.21}$$

where $\phi_{ij} = \beta + b_i$, $b_i \sim N(0, \sigma_b^2)$, $\epsilon_{ij} \sim N(0, \sigma^2)$, the random effects b_i are mutually independent, residual errors ϵ_{ij} are mutually independent, and random effects are independent of residual errors. Figure 7.4 shows a data set generated from that model. The web example describes the generation of the data set and shows the matrices and vectors related to model (7.19) in this particular case.

7.3.2 Parameter estimation and inference

Parameter estimation of nonlinear mixed-effects models is commonly based on the likelihood

$$L(\boldsymbol{\beta}, \sigma^2, \boldsymbol{\theta}) = \int L(\boldsymbol{\beta}, \sigma^2 | \boldsymbol{b},) f(\boldsymbol{b}|\boldsymbol{\theta}) d\boldsymbol{b}$$

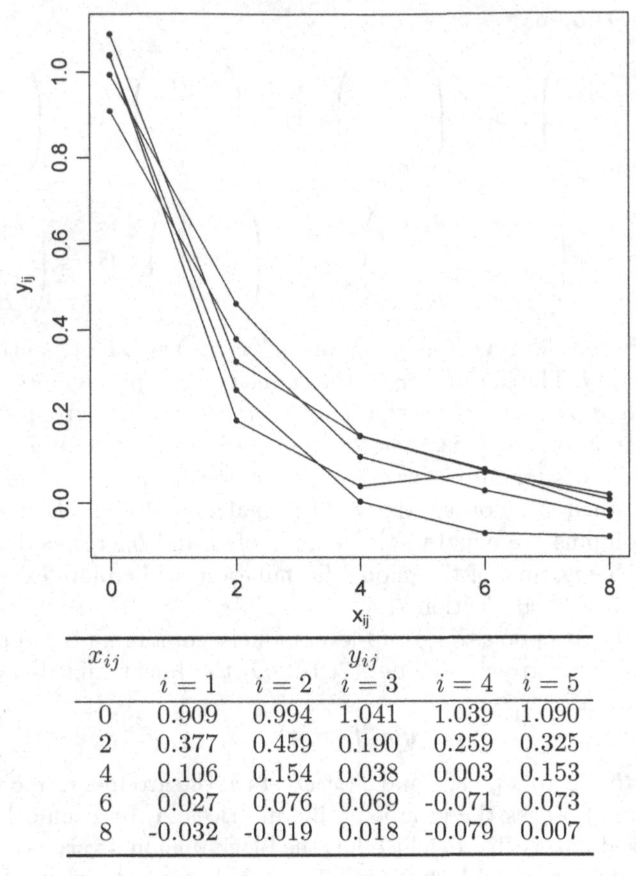

x_{ij}	y_{ij}				
	$i=1$	$i=2$	$i=3$	$i=4$	$i=5$
0	0.909	0.994	1.041	1.039	1.090
2	0.377	0.459	0.190	0.259	0.325
4	0.106	0.154	0.038	0.003	0.153
6	0.027	0.076	0.069	-0.071	0.073
8	-0.032	-0.019	0.018	-0.079	0.007

FIGURE 7.4 A simulated small data set that follows model (7.21) with $\beta = 0.5$, $\sigma_b^2 = 0.15^2$, $\sigma^2 = 0.05^2$, and $\boldsymbol{b}' = (0.0878, 0.1064, -0.0164, -0.0680, 0.0909)$.

where $L(\boldsymbol{\beta}, \sigma^2, \boldsymbol{\theta})$ is the marginal likelihood of the data over the random effects, $L(\boldsymbol{\beta}, \sigma^2|\boldsymbol{b})$ is the likelihood conditional on the random effects, and $f(\boldsymbol{b}|\boldsymbol{\theta})$ is the distribution of the random effects conditional on parameters $\boldsymbol{\theta}$ that parameterize the matrix $\mathcal{D} = \frac{1}{\sigma^2}\boldsymbol{D}$, i.e. they include the scaled and appropriately transformed variances and covariances of the random effects in the similar way to the linear mixed-effects models in Section 5.3. Because the model is nonlinear with respect to the parameters, the integral in the likelihood cannot be expressed in closed form. The integral is multidimensional and difficult to evaluate.

A common solution to this problem is to linearize the model using a Taylor series and then build estimation on the linearized model. In early applications, the derivatives of the model were evaluated at the current estimates of $\boldsymbol{\beta}$ and the expected values $\boldsymbol{b} = \boldsymbol{0}$ of the random effects (Demidenko 2013); an early forestry application for site index modeling using the Chapman-Richard function can be found in Lappi and Bailey (1988). Lindstrom and Bates (1990) proposed an algorithm where the derivatives are evaluated at the conditional modes of $\boldsymbol{b}$. Another alternative is a two-stage estimator, where group-specific curves are first estimated as fixed-effects models and the variability of group-specific parameters is modeled afterwards; this mimics the traditional ANOVA-based methods of LME models (Section 5.3.1). This algorithm can be used if the data includes a sufficient number of observations per group. In addition, one can approximate the likelihood. Alternative approximations are the Laplace approximation, the adaptive Gaussian quadrature approximation, and penalized quasi-likelihood approximation. The Laplace approximation is faster to compute but is less accurate then the adaptive Gaussian quadrature. For discussion of these algorithms, see Demidenko (2013), Pinheiro and Bates (2000), and Section 8.3.2. Other alternatives for estimation are the stochastic approximation expectation maximization (SAEM) (Comets *et al.* 2017) and computational Bayesian methods (see Section 3.4.5).

Lindstrom and Bates algorithm

The Lindstrom-Bates algorithm has been implemented in `nlme::nlme` and is described in more detail here. The algorithm iterates a penalized nonlinear least squares step (PNLS-step) and a linear mixed-effects model step (LME-step) until convergence is reached (Lindstrom and Bates 1990, Pinheiro and Bates 2000, Demidenko 2013). Similar to Section 5.3, we denote $\text{var}(\boldsymbol{b}) = \sigma^2\mathcal{D}$ and $\text{var}(\boldsymbol{\epsilon}) = \sigma^2\mathcal{R}$.

The PNLS-step. This step finds the maximum likelihood estimates of $\boldsymbol{\beta}$ and EBLUP of $\boldsymbol{b}$ when $\mathcal{D}$ and $\mathcal{R}$ are known (compare to Box 5.5, p. 162). Under normality, and conditional on $\mathcal{D}$, $\mathcal{R}$ and σ^2, the estimates can be found by maximizing the function

$$g(\boldsymbol{\beta}, \boldsymbol{b}) = \text{constant} - \frac{1}{2\sigma^2}\left(\boldsymbol{y} - \boldsymbol{f}(\boldsymbol{X}, \boldsymbol{\beta}, \boldsymbol{b})\right)' \mathcal{R}^{-1}\left(\boldsymbol{y} - \boldsymbol{f}(\boldsymbol{X}, \boldsymbol{\beta}, \boldsymbol{b})\right) - \frac{1}{2\sigma^2}\boldsymbol{b}'\mathcal{D}^{-1}\boldsymbol{b},$$

with respect to the second-order parameters $\boldsymbol{\beta}$ and random effects $\boldsymbol{b}$. The solution is found as the least squares solution of the model

$$\boldsymbol{y}^* = \boldsymbol{f}^*(\boldsymbol{X}, \boldsymbol{\beta}, \boldsymbol{b}) + \boldsymbol{e}^*$$

where the augmented data vector $\boldsymbol{y}^*$ of length $n + M \times \sum_{k=1}^{q} s_k$ and the corresponding systematic part are

$$\boldsymbol{y}^* = \begin{pmatrix} \mathcal{R}^{-1/2}\boldsymbol{y} \\ \boldsymbol{0} \end{pmatrix}, \quad \boldsymbol{f}^*(\boldsymbol{X}, \boldsymbol{\beta}, \boldsymbol{b}) = \begin{pmatrix} \mathcal{R}^{-1/2}\boldsymbol{f}(\boldsymbol{X}, \boldsymbol{\phi}) \\ \mathcal{D}^{-1/2}\boldsymbol{b} \end{pmatrix},$$

and $\boldsymbol{e}^* \sim N(\boldsymbol{0}, \sigma^2\boldsymbol{I})$, where $\mathcal{R}^{-1/2}$ and $\mathcal{D}^{-1/2}$ are the Cholesky factors of $\mathcal{R}$ and $\mathcal{D}$ (see Sections 4.4.2 and 4.5.1). Note that $\boldsymbol{X}$ has n rows but the length of $\boldsymbol{y}^*$ is $n + M \times \sum_{k=1}^{q} s_k$. That is, we add $M \times \sum_{k=1}^{q} s_k$ zeroes to the original data, one for each random effect. These pseudo observations pull $\boldsymbol{b}$ towards zero. However, they do not become zero because $\boldsymbol{b}$ is also included in $\boldsymbol{\phi}$ for observations $\boldsymbol{y}$ of $\boldsymbol{y}^*$. These observations, in turn, pull $\boldsymbol{b}$ in such direction that the group-specific curve best fits to the data of that particular group. As a result, such estimates of $\boldsymbol{b}$ are found that are based on the local data of the group but are shrunken towards zero. The degree of shrinkage is determined by $\mathcal{D}$, $\mathcal{R}$ and the number of observations per group, in a similar way to LMEs (recall Section 5.4.1). Applying the above procedure to $\boldsymbol{f}$ that is linear with respect to $\boldsymbol{\beta}$ and $\boldsymbol{b}$ gives the GLS estimates of fixed effects and the BLUP's of random effects as a solution.

LME-step. Matrices $\mathcal{D}$ and $\mathcal{R}$, and scalar σ^2 also need to be estimated. Consider first the nonlinear fixed-effects model, and state it in terms of $\boldsymbol{\beta}$. Model (7.11) can be compactly written as

$$\boldsymbol{y} - \boldsymbol{f}\left(\boldsymbol{X}, \widehat{\boldsymbol{\beta}}\right) \approx \widehat{\boldsymbol{X}}\left(\boldsymbol{\beta} - \widehat{\boldsymbol{\beta}}\right) + \boldsymbol{e},$$

where $\widehat{\boldsymbol{X}}$ is the Jacobian matrix evaluated at the current estimate of $\boldsymbol{\beta}$. Adding $\widehat{\boldsymbol{X}}\widehat{\boldsymbol{\beta}}$ to both sides and simplifying gives

$$\boldsymbol{y} - \boldsymbol{f}\left(\boldsymbol{X}, \widehat{\boldsymbol{\beta}}\right) + \widehat{\boldsymbol{X}}\widehat{\boldsymbol{\beta}} \approx \widehat{\boldsymbol{X}}\boldsymbol{\beta} + \boldsymbol{e}.$$

This shows that an alternative for estimating $\boldsymbol{\beta}$ for the nonlinear fixed-effects model is to compute $\boldsymbol{w} = \boldsymbol{y} - \boldsymbol{f}\left(\boldsymbol{X}, \widehat{\boldsymbol{\beta}}\right) + \widehat{\boldsymbol{X}}\widehat{\boldsymbol{\beta}}$ and fit a linear model $\boldsymbol{w} = \widehat{\boldsymbol{X}}\boldsymbol{\beta} + \boldsymbol{e}$ iteratively until convergence. However, this method does not allow the step halving through parameter λ (recall Section 7.2.3), and therefore will not always converge.

To extend the idea to the nonlinear mixed-effects model, we need to incorporate the random effects into the linearized model. Consider a single group of grouped data. Let us define matrices $\widehat{\boldsymbol{X}}_i$ that includes the partial derivatives $\left.\frac{\partial f}{\partial \beta_k}\right|_{\boldsymbol{\beta}=\widehat{\boldsymbol{\beta}},\boldsymbol{b}_i=\widetilde{\boldsymbol{b}}_i}$ and $\widehat{\boldsymbol{Z}}_i$ that includes the partial derivatives $\left.\frac{\partial f}{\partial b_i^{(k)}}\right|_{\boldsymbol{\beta}=\widehat{\boldsymbol{\beta}},\boldsymbol{b}_i=\widetilde{\boldsymbol{b}}_i}$ for all observations of the group, where $\widehat{\boldsymbol{\beta}}$ and $\widetilde{\boldsymbol{b}}_i$ are the current estimates of fixed parameters and random effects. Now

$$\boldsymbol{y}_i - \boldsymbol{f}_i\left(\boldsymbol{X}_i; \widehat{\boldsymbol{\beta}}, \widetilde{\boldsymbol{b}}_i\right) + \widehat{\boldsymbol{X}}_i\widehat{\boldsymbol{\beta}} + \widehat{\boldsymbol{Z}}_i\widetilde{\boldsymbol{b}}_i \approx \widehat{\boldsymbol{X}}_i\boldsymbol{\beta} + \widehat{\boldsymbol{Z}}_i\boldsymbol{b}_i + \boldsymbol{\epsilon}_i \,. \tag{7.22}$$

The formula can be defined for the entire data set by formulating $\widehat{\boldsymbol{X}}$ and $\widehat{\boldsymbol{Z}}$ in a similar manner to that we used with the linear mixed-effects models in Section 5.2.3. Therefore, $\mathcal{D}$ and $\mathcal{R}$ can be estimated by computing $\boldsymbol{w} = \boldsymbol{y} - \boldsymbol{f}\left(\boldsymbol{X}; \widehat{\boldsymbol{\beta}}, \widetilde{\boldsymbol{b}}\right) + \widehat{\boldsymbol{X}}\widehat{\boldsymbol{\beta}} + \widehat{\boldsymbol{Z}}\widetilde{\boldsymbol{b}}$ and fitting the linear mixed-effects model of form $\boldsymbol{w} = \widehat{\boldsymbol{X}}\boldsymbol{\beta} + \widehat{\boldsymbol{Z}}\boldsymbol{b} + \boldsymbol{\epsilon}$ using the approaches presented in Section 5.3. The Lindstrom-Bates algorithm iterates between the PNLS and LME-steps until convergence.

Web Example 7.9 implements the algorithm for a simple model. After converged iteration, the EBLUPs of the random effects $\boldsymbol{b}$ can be found either as a result from the PNLS step or by applying the methods presented in Section 5.4.1 to the linearized model (see Section 7.3.5).

For the first PNLS step, one needs initial guesses for $\boldsymbol{\beta}$, $\boldsymbol{b}$, $\boldsymbol{\theta}$ and σ^2. A good initial guess for the random effect vector is $\boldsymbol{b} = \boldsymbol{0}$. A separate expectation maximization (EM) algorithm can be used to find initial guesses for the variance-covariance parameters (Lindstrom and Bates 1990). Therefore, user-defined initial guesses are needed only for $\boldsymbol{\beta}$ in computer software, such as `nlme::nlme`; however non-zero initial guesses of $\boldsymbol{b}$ can also be provided.

Inference

As with nonlinear fixed-effects models, approximate inference of the nonlinear model can be based on the linearized model at converged iteration. Therefore, everything that has been discussed in the previous chapters in the context of linear fixed-effects and mixed-effects models remains valid for the linear approximation of the nonlinear model. Because the linearized model is only an approximate of the nonlinear model, the inference is also approximate. The quality of the approximation depends on the shape of the model function f with respect to $\boldsymbol{\phi}$.

Example 7.8. The model specified in Example 7.7 (p. 225) was fitted to the simulated data of Figure 7.4 using `nlme` below. The estimates are $\widehat{\beta} = 0.558$, $\widehat{\sigma}_b^2 = 0.0671^2$, and $\widehat{\sigma}^2 = 0.0506^2$.

```
> fun<-function(x,phi) phi^x
> mod<-nlme(y~fun(x,phi),
+           fixed=phi~1,
+           random=phi~1|group,
+           start=mean(groupmeans))
> summary(mod)
Nonlinear mixed-effects model fit by maximum likelihood
  Model: y ~ fun(x, phi)
 Data: NULL
        AIC       BIC   logLik
  -64.84093 -61.18431 35.42047

Random effects:
 Formula: phi ~ 1 | group
              phi   Residual
StdDev: 0.0671198 0.05063158
```

```
Fixed effects: phi ~ 1
        Value  Std.Error DF  t-value p-value
phi 0.5581786 0.03511328 20 15.89651       0

Number of Observations: 25
Number of Groups: 5
```

Web Example 7.9. Implementations of the Lindstrom-Bates algorithm using the data and model of Example 7.8.

Example 7.10. Data set `thefdata` (see right-hand graph in Figure 6.7, p. 198) includes the predicted effects of thinning on the basal area growth of Scots Pine trees in eastern Finland. The effects of tree age, climatic year effects, and tree-specific random effects were removed using the procedures described in Web Example 6.6. The variable `ThEf` includes the thinning effects. In addition, the trees current age `CA`, tree diameter prior to the thinning `Diam1986` and thinning treatment `SDClass` (1=light, 2=moderate, 3=heavy) are known for each tree. The data are grouped according to plots and trees within a plot, but because the variability between plots is marginally small, we take into account only the tree level in this example.

We model the thinning effect using function (7.8). The thinning effect of tree i at time j is

$$ThEf_{ij} = \frac{\exp\left(\phi_{ij}^{(1)}\right)}{1 + \exp\left[\left(4 - 8T_{ij}/\exp(\phi_{ij}^{(2)})\right)\right]} + \epsilon_{ij}\,, \tag{7.23}$$

where

$$\phi_{ij}^{(1)} = \beta_1^{(1)} + b_i^{(1)}\,,$$
$$\phi_{ij}^{(2)} = \beta_1^{(2)} + b_i^{(2)}\,.$$

The random effects $\boldsymbol{b}_i = (b_i^{(1)}, b_i^{(2)})'$ include the random intercepts of tree i in the logarithmic full thinning effect and logarithmic reaction time. We assume that $\boldsymbol{b}_i \sim N(\boldsymbol{0}, \boldsymbol{D}_*)$ and $\epsilon_{ij} \sim N(0, \sigma^2)$, and that the residual errors and random effects are mutually independent and independent of each other. This model assumes that the thinning effect is not affected by thinning intensity, which is unrealistic based on Figure 6.7. However, this model serves as a starting point for our analysis. The effects of thinning treatments are now included in the random effects. This model allows graphical exploration of how the thinning and other potential tree-level predictors are associated with the thinning effects.

For estimation, the initial guesses of $\beta_1^{(1)}$ and $\beta_1^{(2)}$ are taken from Figure 6.7, which indicates that the thinning effect size is around 200 mm^2 and the reaction time is roughly 5 years. However, we experienced convergence problems even with this simple model. The estimation was interrupted with an error informing us that the maximum number of iterations was reached without convergence. Increasing the maximum number of PNLS iterations was beneficial (see `?nlmeControl`).

```
> thefdata$ythin<-thefdata$Year-1986
> thefdata$diamS<-(thefdata$Diam1986-mean(thefdata$Diam1986))/sd(thefdata$Diam1986)
> ThinNlme0<-nlme(ThEf~mylogis(ythin,phi1,phi2),
+                fixed=list(phi1~1,phi2~1),
+                random=phi1+phi2~1|Tree,
+                data=thefdata,
+                start=c(log(200),log(5)),
+                verbose=TRUE,
+                control=list(pnlsMaxIter=20))

> re<-ranef(ThinNlme0,aug=TRUE)[,c(1,2,3,5,7,10)]
> head(re)
          phi1        phi2 Plot SDClass  Diam1986      diamS
3:1  0.4386422 -0.31644184    3       4  65.00000 -1.2850815
3:2  0.2738323 -0.45329793    3       4 114.60000  1.8995735
3:3  0.9211323 -0.07252985    3       4  68.25000 -1.0764096
3:4 -0.2083211 -0.76014118    3       4 120.80001  2.2976559
3:5  0.9391514  0.14574669    3       4  83.34999 -0.1068882
3:6  1.0282937  0.75676533    3       4  77.20000 -0.5017590
```

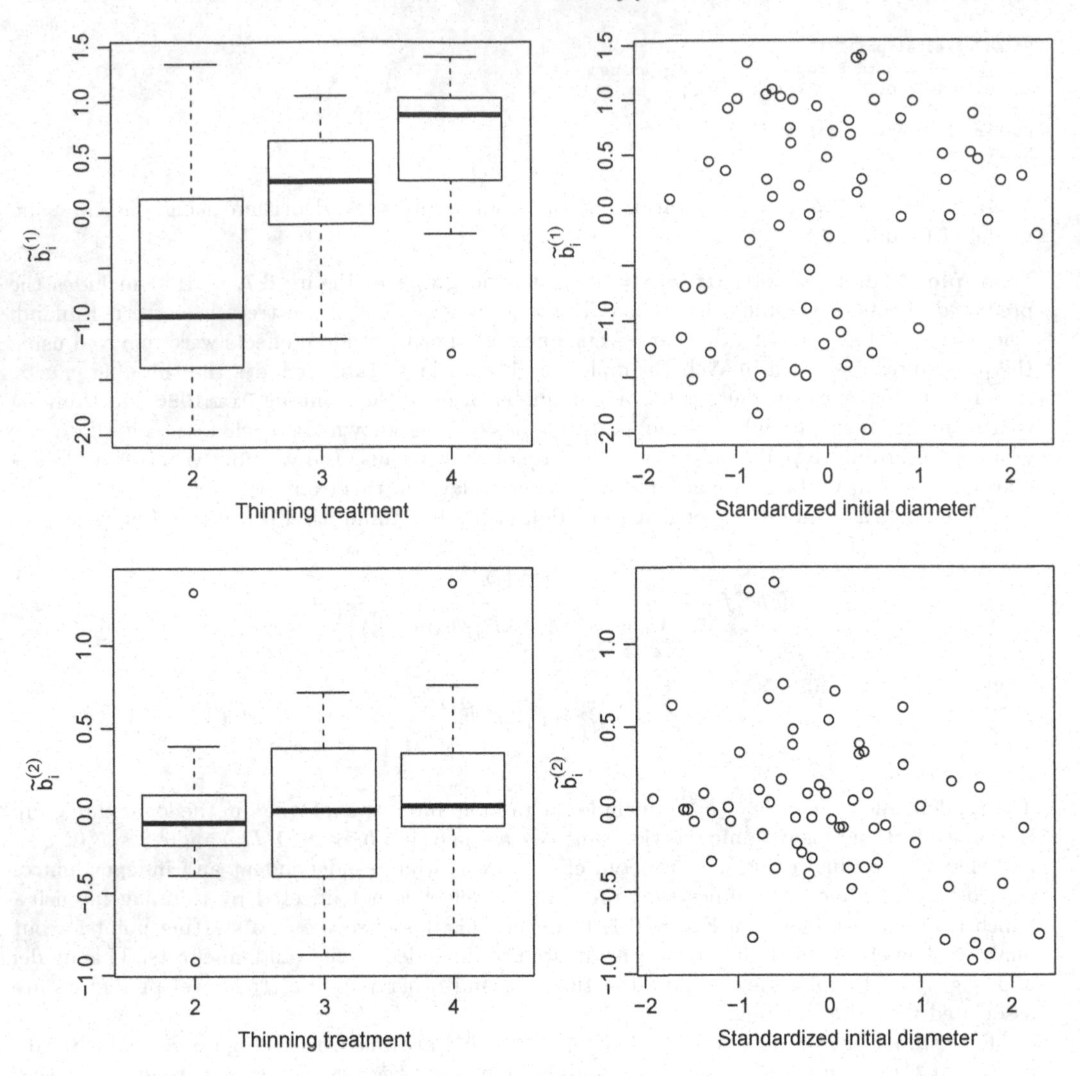

FIGURE 7.5 Random effects of the logarithmic maximum thinning effect (upper) and logarithmic reaction time (lower) on the thinning treatments (1=light, 2=moderate, 3=heavy) (left-hand graphs) and on standardized tree diameter before thinning (right-hand graphs).

We use this model to explore the potential benefit from using thinning treatment and tree diameter prior to thinning as predictors in the model. To do so, we extract the random effects of the model and augment the predictions with the tree-specific mean of standardized diameter before thinning and the tree-specific mode of thinning treatment. Their association with the thinning treatment is shown in Figure 7.5. We notice that both random effects have some correlation with thinning, but the logarithmic maximum thinning effect, especially, is strongly correlated with the treatment. The reaction time shows negative correlation with tree size before thinning. Based on these observations, we update the submodels as follows:

$$\phi_{ij}^{(1)} = \beta_1^{(1)} + \beta_2^{(1)} T2_i + \beta_3^{(1)} T3_i + b_i^{(1)} ,$$
$$\phi_{ij}^{(2)} = \beta_1^{(2)} + \beta_2^{(2)} dS_i + b_i^{(2)} ,$$

where $T2_i$ and $T3_i$ are indicator variables for levels 2 and 3 of the treatment (moderate and heavy thinning) and dS_i is the standardized diameter before thinning. We use $\widehat{\beta}_1^{(1)}$ and $\widehat{\beta}_1^{(2)}$

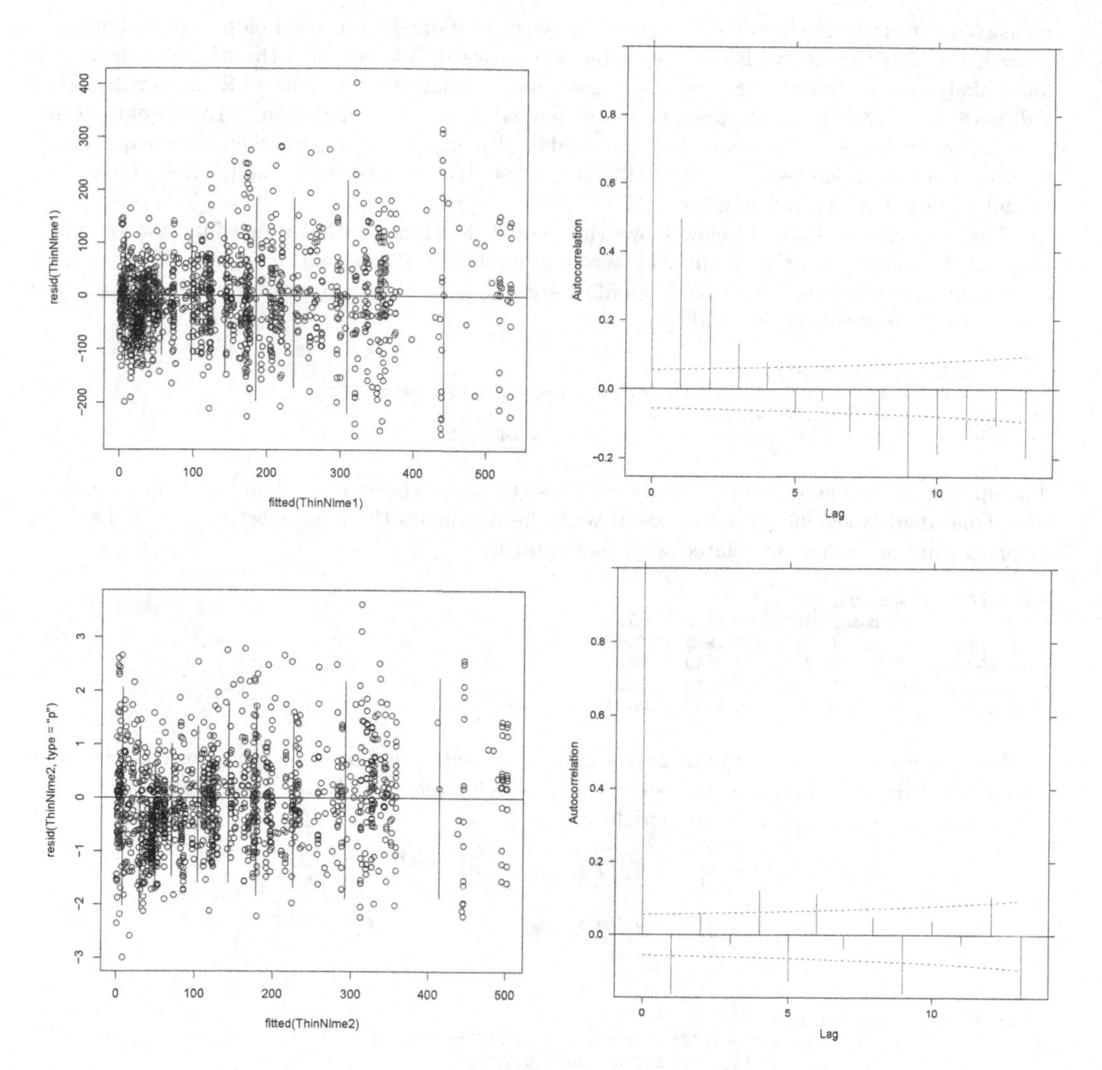

FIGURE 7.6 Residual and autocorrelation plots of model `ThinNlme1` (top) and `ThinNlme2` (bottom).

of model `ThinNlme0` as the initial guesses for parameters $\beta_1^{(1)}$ and $\beta_1^{(2)}$ of the new model. For the other parameters, we use 0. The parameters are included in vector $\boldsymbol{\beta}$ in the following order $\boldsymbol{\beta}' = (\beta_1^{(1)}, \beta_2^{(1)}, \beta_3^{(1)}, \beta_1^{(2)}, \beta_2^{(2)})$, therefore we define **`start=c(beta0[1],0,0,beta0[2],0)`**.

```
> beta0<-fixef(ThinNlme0)
> ThinNlme1<-nlme(ThEf~mylogis(ythin,phi1,phi2),
+        fixed=list(phi1~SDClass,phi2~diamS),
+        random=phi1+phi2~1|Tree,
+        data=thefdata,
+        start= c(beta0[1],0,0,beta0[2],0),
+        control=list(pnlsMaxIter=20))
```

Figure 7.6 (upper) shows that the residuals of the new model exhibit inconstant variance and that there is a strong temporal autocorrelation among the residuals. Therefore, we re-estimate the model by assuming the AR(1) structure for the residuals and a power-type variance function. The behavior of the residuals is better in the updated model (Figure 7.6, lower). Interestingly, the lower right graph indicates a slight negative autocorrelation between Pearson residuals with odd lag and positive correlation with even lags. An obvious explanation for this behavior is measurement error: the increment cores are observed by dividing a wood chip of fixed length to growth rings. If the ring width of a specific year is overestimated, then the neighboring year is most likely underestimated because the rings cannot overlap. The variance of Pearson residuals is still heteroscedastic, being the lowest when fitted values are around 80 mm^2/yr. Therefore, there is still some room for improvement in the model, although those improvements are expected to have only a marginal effect on interpretations of the fixed effects of the model, and the current model is regarded as satisfactory.

The approximate LR test below shows that model `ThinNlme2` with variance function and autocorrelated residuals fits significantly better than model `ThinNlme1` with constant error variance and independent errors. These components were also tested separately (not shown) to confirm that both components are needed.

```
> anova(ThinNlme2,ThinNlme1)
          Model df      AIC      BIC    logLik   Test L.Ratio p-value
ThinNlme2     1 11 14264.57 14320.90 -7121.283
ThinNlme1     2  9 14807.67 14853.76 -7394.835 1 vs 2 547.104  <.0001
```

The approximate conditional $F-$test below shows that, as expected based on Figure 6.7, the thinning treatment is significantly associated with the maximum thinning effect. The tree diameter prior to thinning is also associated with reaction time.

```
> anova(ThinNlme2,type="m")
                  numDF denDF  F-value p-value
phi1.(Intercept)      1  1172 614.2909  <.0001
phi1.SDClass          2  1172  12.2949  <.0001
phi2.(Intercept)      1  1172 467.2379  <.0001
phi2.diamS            1  1172  12.4905   4e-04
```

To test whether the treatment is also associated with reaction time, and the tree size with maximum thinning effect, we include these effects in the model and carry out approximate conditional $F-$tests. The new submodels are

$$\phi_{ij}^{(1)} = \beta_1^{(1)} + \beta_2^{(1)} T2_i + \beta_3^{(1)} T3_i + \beta_4^{(1)} dS_i + b_i^{(1)} ,$$

$$\phi_{ij}^{(2)} = \beta_1^{(2)} + \beta_2^{(2)} T2_i + \beta_3^{(2)} T3_i + \beta_4^{(2)} dS_i + b_i^{(2)} ,$$

```
> ThinNlme3<-update(ThinNlme2,
+                   fixed=list(phi1~SDClass+diamS,phi2~SDClass+diamS),
+                   start=c(beta0[1],0,0,0,beta0[2],0,0,0))

> anova(ThinNlme3,type="m")
                  numDF denDF  F-value p-value
phi1.(Intercept)      1  1169 609.9975  <.0001
phi1.SDClass          2  1169  11.7921  <.0001
phi1.diamS            1  1169   2.4811  0.1155
phi2.(Intercept)      1  1169 117.4030  <.0001
phi2.SDClass          2  1169   0.2593  0.7717
phi2.diamS            1  1169  10.9438  0.0010
```

The p-value for the effect of standardized diameter on maximum thinning effect is 0.116, and that for the effect of treatment on reaction time is 0.772. Therefore, the null hypotheses $H_0 : \beta_4^{(1)} = 0$ and $H_0 : \beta_2^{(2)} = \beta_3^{(2)} = 0$ are not rejected. This provides a justification to retain model `ThinNlme2` as the final model. However, it did not prove that these hypotheses are true, instead we just do not have sufficiently strong evidence against them.

For interpretation of model parameters, the summary of the model is shown below.

```
> summary(ThinNlme2)
Nonlinear mixed-effects model fit by maximum likelihood
  Model: ThEf ~ mylogis(ythin, phi1, phi2)
 Data: thefdata

Random effects:
 Formula: list(phi1 ~ 1, phi2 ~ 1)
```

```
 Level: Tree
 Structure: General positive-definite, Log-Cholesky parametrization
                StdDev    Corr
phi1.(Intercept) 0.6443630 p1.(I)
phi2.(Intercept) 0.3928901 0.155
Residual        48.8149801

Correlation Structure: AR(1)
 Formula: ~1 | Tree
 Parameter estimate(s):
     Phi
0.6546791
Variance function:
 Structure: Power of variance covariate
 Formula: ~fitted(.)
 Parameter estimates:
    power
0.1470081
Fixed effects: list(phi1 ~ SDClass, phi2 ~ diamS)
                     Value Std.Error   DF   t-value p-value
phi1.(Intercept)  4.397048 0.17740841 1172 24.784893  0.0000
phi1.SDClass3     0.759479 0.24161441 1172  3.143351  0.0017
phi1.SDClass4     1.177167 0.23896674 1172  4.926072  0.0000
phi2.(Intercept)  1.797402 0.08315267 1172 21.615686  0.0000
phi2.diamS       -0.301812 0.08539762 1172 -3.534193  0.0004

> exp(fixef(ThinNlme2))
phi1.(Intercept)    phi1.SDClass3    phi1.SDClass4 phi2.(Intercept)       phi2.diamS
      81.2108145        2.1371624        3.2451689        6.0339507        0.7394773
```

The first three fixed effects are related to the effect of thinning treatment on maximum thinning effect. The maximum thinning effect in the light thinning treatment is $\exp(4.40) = 81.2$ mm^2/yr. In the moderate thinning treatment, the maximum effect is approximately $\exp(0.759) = 2.14$ times as high, which gives $2.14 \times 81.2 = \exp(4.40+0.759) = 173.6$mm^2/yr. For the heavy thinning treatment, the maximum effect is $\exp(1.177) = 3.24$ times as high as in the light thinning, giving 263.7 mm^2/yr.

The reaction time of a tree that was of average size in 1986 (i.e. for which $dS_i = 0$) is $\exp(1.800) = 6.03$ years. An increase of one standard deviation in diameter changes the reaction time to $\exp(-0.302) = 0.74$ fold (i.e. decreases it by 26%). Approximately 95% of the trees are within two standard deviations of diameter, therefore the reaction time of the smallest trees, which are most heavily dominated by larger trees, is $\exp(1.80 + 2 \times 0.30) = 11$ years. For the largest, least dominated trees it is $\exp(1.80 - 2 \times 0.30) = 3.3$ years. To conclude, it seems that all the trees in a specific thinning treatment will eventually reach an approximately equal thinning effect. However, the time taken by the trees to recover from past competition varies widely; 3 years for the dominant trees and 11 years for the most heavily dominated trees.

The variances of the random effects indicate that much of the variation among trees is not explained by the fixed effects of the model. Such variability exists both in the reaction time and in the maximum thinning effect. In particular, the trees with the shortest reaction time react 54% faster $(1 - \exp(-2 \times 0.393))$ and the trees with the longest reaction time react 118% slower $(\exp(2 \times 0.393) - 1)$ compared to a typical tree of a given size and treatment. The maximum thinning effect also varies; the minimum is 72% lower $(1 - \exp(-2 \times 0.644))$ and the maximum is 2.6 times higher $(\exp(2 \times 0.644) - 1)$ than for a typical tree of a given size and treatment.

For analysis of the same data using a slightly different model formulation, see Mehtätalo *et al.* (2014).

7.3.3 Convergence problems

A practical difference between nonlinear models and linear models is that convergence problems occur in estimation of the former much more often, especially when random effects are included. Convergence can be used as first indicator of good model formulation.

The first thing to do after a failure to converge is to increase the number of iterations (in different parts of the estimation procedure) and try different starting values. If these do not help, one could monitor the iterations (e.g. using `verbose` argument in `nlme`) and see how the convergence criteria develop over iterations. It is also a good idea to consider another model formulation. In some cases, the convergence criteria can be relaxed, but not too much, so as to ensure that the solution is not too far from the true solution.

In package `nlme`, the model fitting functions use, by default, numerical algorithms to find the gradients of the model function. The gradients can also be computed analytically using user-defined functions; see Pinheiro and Bates (2000, Section 8.1) and Example 7.12 for the definition of the model function in such case. Function `nlme::nlme` also allows user-defined initial guesses for the random effects. Using these options might help in some cases.

With nonlinear models, full models that include several predictors and random effects and a complex variance-covariance structure often fail to converge. Useful ways to simplify the model are, for example, by fixing some covariances of random effects (usually to zero), or by dropping some random effects. If one uses a variance function that depends on fitted values, convergence can sometimes be attained by first fitting the model without the variance function and saving the fitted values to the data, and thereafter fitting a model where the saved fitted value is used as the predictor in the variance function, instead of defining the weighting using `weights=varPower()` (or some other available variance function), which does the iteration inside function `nlme`.

In many cases, one would like to start with a full model to ensure that the model fits well to the data. This is often impossible with nonlinear mixed-effects models because the full model cannot be estimated. Therefore, one needs to start with a simple model in many cases, and add terms sequentially. However, the initial model to start with should not be overly simple, because a model that lacks essential properties of the data may not converge either. A good understanding of the data, the applied model function and the process being modeled are, therefore, essential also here. The initial model should describe the most essential properties of the data. In some cases, the trickiest issue is to find the initial converging model. Thereafter the other models are easily found by utilizing the parameter estimates from that model as initial guesses of the parameters of more advanced models.

If none of the above-specified techniques is useful, one can treat some parameters that are not of primary interest as Finney's second-level parameters, as we have discussed in Box 7.2 (p. 216).

7.3.4 Multiple levels of grouping

Nonlinear mixed-effects models can be generalized to multiple nested and crossed levels of grouping. The model formulation generalizes the ideas presented earlier in (7.17) and Section 6.1.1. Currently, R-package `nlme` only provides easy-to-use tools for nested classifications. Package `nlme4` provides tools for crossed classifications, but the function, at present, is rather restricted.

Example 7.11. A multilevel version of the model `ThinNlme2` can be fitted using

```
> ThinNlme4<-update(ThinNlme2,
+                   random=phi1+phi2~1|Plot/Tree,
+                   start=fixef(ThinNlme2))
> summary(ThinNlme4)
Nonlinear mixed-effects model fit by maximum likelihood
  Model: ThEf ~ mylogis(ythin, phi1, phi2)
 Data: thefdata

Random effects:
 Formula: list(phi1 ~ 1, phi2 ~ 1)
 Level: Plot
 Structure: General positive-definite, Log-Cholesky parametrization
                  StdDev       Corr
phi1.(Intercept) 1.486180e-27 p1.(I)
phi2.(Intercept) 1.796132e-23 0
```

```
 Formula: list(phi1 ~ 1, phi2 ~ 1)
 Level: Tree %in% Plot
 Structure: General positive-definite, Log-Cholesky parametrization
                 StdDev    Corr
phi1.(Intercept)  0.6369234 p1.(I)
phi2.(Intercept)  0.3611982 0.176
Residual         48.9774204

Correlation Structure: AR(1)
 Formula: ~1 | Plot/Tree
 Parameter estimate(s):
      Phi
0.6582365
Variance function:
 Structure: Power of variance covariate
 Formula: ~fitted(.)
 Parameter estimates:
    power
0.1479732
Fixed effects: list(phi1 ~ SDClass, phi2 ~ diamS)
                     Value  Std.Error   DF   t-value p-value
phi1.(Intercept)  4.405380 0.17664885 1172 24.938630  0.0000
phi1.SDClass3     0.763068 0.24001682 1172  3.179228  0.0015
phi1.SDClass4     1.168588 0.23719541 1172  4.926690  0.0000
phi2.(Intercept)  1.824159 0.07980199 1172 22.858567  0.0000
phi2.diamS       -0.390675 0.08155876 1172 -4.790101  0.0000

Number of Observations: 1238
Number of Groups:
          Plot Tree %in% Plot
             7             62
```

The model has two nested levels: plots and trees within plots. However, the variances at the plot-level are very small, indicating that this level of hierarchy is probably not needed. We test it with the corresponding single-level model using an approximate LR test, remembering the previous warnings about this procedure. The test indicates that the two-level model is not significantly better than the one-level model. However, we could still retain both levels so as to make inference for fixed effects (recall the related discussion in Section 5.6).

```
> anova(ThinNlme2,ThinNlme4)
          Model df      AIC      BIC    logLik   Test L.Ratio p-value
ThinNlme2     1 11 14264.57 14320.90 -7121.283
ThinNlme4     2 14 14269.28 14340.98 -7120.639 1 vs 2 1.28916  0.7317
```

7.3.5 Prediction of random effects and y-values

For linear mixed-effects models, the random effects can be predicted using EBLUP after estimation of model parameters using REML or ML, even though, in practice, these two steps are done simultaneously using Henderson's mixed model equations. A benefit of this separate step of random-effect prediction is the possibility of local calibration of a previously fitted mixed-effects model using local measurements of the response variable. This approach, which is widely used in forest sciences, was demonstrated with the linear models in Examples 5.19, 5.20 and in Section 6.6. In the Lindstrom-Bates algorithm, parameter estimation and random-effect prediction were not presented as two separate steps. However, the need for local calibration in forest sciences for nonlinear models, e.g. those for site index, tree taper and tree height, has also motivated the development of a separate calibration step for nonlinear models.

Assume that the fixed parameters $\boldsymbol{\beta}$ in model (7.17) are known and state simply

$$y_{ij} = f(x_{ij}, \boldsymbol{b}_i) + \epsilon_{ij}.$$

Consider a situation where vector $\boldsymbol{y}_i$ includes the observed values of the response variable for some members of group i. Three different local predictions of y for a known $\boldsymbol{x}_0$ and unknown random effect $\boldsymbol{b}_i$ can be defined:

$$\widetilde{y} = f(\boldsymbol{x}_0, \widehat{\boldsymbol{\beta}}, \boldsymbol{0}) + \widehat{\boldsymbol{z}}_0' \operatorname{E}(\boldsymbol{b}_i|\boldsymbol{y}_i) \quad (7.24)$$

$$\widetilde{y} = f(\boldsymbol{x}_0, \operatorname{E}(\boldsymbol{b}_i|\boldsymbol{y}_i)) \quad (7.25)$$

$$\widetilde{y} = \operatorname{E}(f(\boldsymbol{x}_0, \boldsymbol{b}_i)|\boldsymbol{y}_i). \quad (7.26)$$

That is, one can (i) use predicted random effects in the linear approximation of the random part of the model, (ii) use the predicted random effects directly in the nonlinear function, or (iii) directly predict y for $\boldsymbol{x}_0$. For the linear mixed-effects model, these three approaches lead to an identical result because $\widehat{\boldsymbol{z}}_0 = \boldsymbol{z}_0$ and

$$\operatorname{E}(\boldsymbol{X}_i\boldsymbol{\beta} + \boldsymbol{Z}_i\boldsymbol{b}_i + \boldsymbol{\epsilon}_i|\,\boldsymbol{y}_i) = \boldsymbol{X}_i\boldsymbol{\beta} + \boldsymbol{Z}_i \operatorname{E}(\boldsymbol{b}_i|\boldsymbol{y}_i) + \boldsymbol{\epsilon}_i\,,$$

but this is not the case for the nonlinear model. If f is nonlinear with respect to $\boldsymbol{b}_i$, the use of the conditional expected value of $\boldsymbol{b}_i$ in the nonlinear formula does not lead to the conditional expected value of $\boldsymbol{y}_i$. This means that using the unbiased predictor of $\boldsymbol{b}_i$ in (7.25) does not imply unbiased prediction of $\boldsymbol{y}$.

Approach (7.24) was used in an early paper by Lappi and Bailey (1988), and has usually been used instead of (7.25) whenever the Jacobian $\widehat{\boldsymbol{Z}}$ (see below) used in the prediction of random effects is based on the Taylor expansion at $\boldsymbol{b}_i = \boldsymbol{0}$. This approach is called *the zero-expansion method* (Hall and Bailey 2001). In this instance, (7.24) is theoretically better justified than (7.25) because the random effects are also predicted for the linearized model; the use of (7.25) may produce inferior predictions (Meng and Huang 2009). The zero-expansion method is the result after the first step of the iterative procedure presented below.

Equation (7.25) has been widely used but should be used only if the Jacobians are evaluated at the predicted values of the random effects, which requires the iterative procedure described below (Hall and Clutter 2004, Temesgen *et al.* 2008, Meng and Huang 2009, Yang and Huang 2011, Sirkiä *et al.* 2014). Equation (7.25) is consistent with the common practice of "plug-in" predictors, where the parameter estimates and predicted group effects are plugged into the original nonlinear function to obtain group-specific predictions. Equation (7.26) has been used by Hall and Bailey (2001), Hall and Clutter (2004), and Sirkiä *et al.* (2014). It leads to an unbiased prediction of $\boldsymbol{y}$, although the predicted group-specific $y - x$ relationship is not of the form f. Equation (7.25) gives a group-specific $y - x$ relationship that is of form f.

Because $\boldsymbol{b}_i$ is not linear in $\boldsymbol{y}_i$, the formulas for $\operatorname{E}(\boldsymbol{b}_i|\boldsymbol{y}_i)$ for general f do not exist in closed form. However, one may use the Taylor approximation of the nonlinear model in a similar way to that used in model estimation. For such an approach, one can predict the random effect of the approximate model using BLUP in a similar way to that used with the linear model. Temesgen *et al.* (2008) and Meng and Huang (2009) have suggested an iterative procedure to compute the approximation at $\boldsymbol{b}_i = \widetilde{\boldsymbol{b}}_i$, where $\widetilde{\boldsymbol{b}}_i$ is the predicted random effect for group i. Hall and Clutter (2004) had used a similar approach four years earlier, and Sirkiä *et al.* (2014) showed that this procedure converges to the conditional mode of $\boldsymbol{b}_i$ given the observations $\boldsymbol{y}_i$. The procedure, which is a straightforward application of the LME step of the Lindstrom-Bates algorithm, is presented below in detail.

Recall the linear approximation of the nonlinear model from the LME step of the Lindstrom and Bates algorithm (p. 228)

$$\boldsymbol{w}_i \approx \widehat{\boldsymbol{X}}_i\boldsymbol{\beta} + \widehat{\boldsymbol{Z}}_i\boldsymbol{b}_i + \boldsymbol{\epsilon}_i$$

where the response variable is

$$\boldsymbol{w}_i = \boldsymbol{y}_i - \boldsymbol{f}_i(\boldsymbol{x}_i, \widehat{\boldsymbol{\beta}}_i, \widetilde{\boldsymbol{b}}_i^*) + \widehat{\boldsymbol{X}}_i\widehat{\boldsymbol{\beta}} + \widehat{\boldsymbol{Z}}_i\widetilde{\boldsymbol{b}}_i^* \quad (7.27)$$

and $\widehat{\boldsymbol{X}}_i$ and $\widehat{\boldsymbol{Z}}_i$ are Jacobians of $\boldsymbol{f}_i(\boldsymbol{x}_i, \boldsymbol{\beta}, \boldsymbol{b}_i)$ with respect to $\boldsymbol{\beta}$ and $\boldsymbol{b}_i$, respectively, evaluated at GLS estimates of $\boldsymbol{\beta}$ and the current predictions of $\boldsymbol{b}_i$, denoted by $\widehat{\boldsymbol{\beta}}$ and $\widetilde{\boldsymbol{b}}_i^*$.

The approximate model is of form (5.13), for which the random effect can be predicted using the BLUP (recall Section 5.4.1)

$$\widetilde{\boldsymbol{b}}_i = \boldsymbol{D}_* \widehat{\boldsymbol{Z}}_i' \left(\widehat{\boldsymbol{Z}}_i \boldsymbol{D}_* \widehat{\boldsymbol{Z}}_i' + \boldsymbol{R}_i \right)^{-1} \left(\boldsymbol{w}_i - \widehat{\boldsymbol{X}}_i \boldsymbol{\beta}_i \right), \tag{7.28}$$

where the last term simplifies to $\boldsymbol{w}_i - \widehat{\boldsymbol{X}}_i \boldsymbol{\beta}_i = \boldsymbol{y}_i - \boldsymbol{f}_i(\boldsymbol{x}_i, \widehat{\boldsymbol{\beta}}_i, \widetilde{\boldsymbol{b}}_i) + \widehat{\boldsymbol{Z}}_i \widetilde{\boldsymbol{b}}_i$. Therefore, the following iterative procedure is needed:

1. Set $\widetilde{\boldsymbol{b}}_i = \boldsymbol{0}$.
2. Set $\widetilde{\boldsymbol{b}}_i^* = \widetilde{\boldsymbol{b}}_i$. Compute $\widehat{\boldsymbol{Z}}_i$.
3. Update $\widetilde{\boldsymbol{b}}$ using $\widetilde{\boldsymbol{b}}_i = \boldsymbol{D}_* \widehat{\boldsymbol{Z}}_i' \left(\widehat{\boldsymbol{Z}}_i \boldsymbol{D}_* \widehat{\boldsymbol{Z}}_i' + \boldsymbol{R}_i \right)^{-1} \left(\boldsymbol{y}_i - \boldsymbol{f}_i(\boldsymbol{x}_i, \widehat{\boldsymbol{\beta}}_i, \widetilde{\boldsymbol{b}}_i) + \widehat{\boldsymbol{Z}}_i \widetilde{\boldsymbol{b}}_i \right)$
4. Repeat steps 2 and 3 until the difference between $\widetilde{\boldsymbol{b}}_i$ and $\widetilde{\boldsymbol{b}}_i^*$ becomes marginally small.

Notice that variances and covariances of the random effects are not updated as in the full LME step of the Lindstrom-Bates algorithm. An `R` implementation of the algorithm is given in the following example. For a SAS-implementation, see Meng and Huang (2009).

An alternative to predicting $\boldsymbol{b}_i$ or $\boldsymbol{y}_i$ using the Taylor approximation of the model is direct computation of $\mathrm{E}(\boldsymbol{b}|\boldsymbol{y}_i)$ in (7.25) or direct computation of $\mathrm{E}(y|\boldsymbol{y}_i)$ in (7.26). These can be done, for example, by using Laplace or Gauss-Hermite approximations of the integrals in the expectations, or by using Markov Chain Monte Carlo methods (see Sirkiä *et al.* (2014)). They also provide an `R`-implementation of such procedures. However, such an approach does not necessarily provide practically meaningful improvements over approaches based on a linearized model.

Example 7.12. Consider a site index model, which is based on the Chapman-Richards function

$$H_{ijt} = \phi_{ijt}^{(1)} \left(1 - \exp\left(-\phi_{ijt}^{(3)} T_{ijt}\right)\right)^{\phi_{ijt}^{(2)}} + \epsilon_{ijt}$$

where H_{ijt} and T_{ijt} are the height (in feet) and age (yr), respectively, of tree j on plot i at time t. The parameters are

$$\phi_{ijt}^{(1)} = 86.2 + a_i^{(1)} + c_{ij}^{(1)},$$
$$\phi_{ijt}^{(2)} = 1.385 + a_i^{(2)} + c_{ij}^{(2)},$$
$$\phi_{ijt}^{(3)} = 0.0702,$$

where

$$\mathrm{var}(\boldsymbol{a}_i) = \mathrm{var}\begin{pmatrix} a_i^{(1)} \\ a_i^{(2)} \end{pmatrix} = \begin{pmatrix} 82.8 & 0.492 \\ 0.492 & 0.0835 \end{pmatrix}$$

and

$$\mathrm{var}(\boldsymbol{c}_{ij}) = \mathrm{var}\begin{pmatrix} c_{ij}^{(1)} \\ c_{ij}^{(2)} \end{pmatrix} = \begin{pmatrix} 19.4 & 0.178 \\ 0.178 & 0.00706 \end{pmatrix}$$

and $\mathrm{var}(\epsilon_{ijt}) = 2.16^2 = 4.67$. This model was fitted in a linearized form to slash pine data in Lappi and Bailey (1988). They also used the model for the prediction of tree- and plot-level height-age curves. Because the Taylor approximation around $\boldsymbol{b} = \boldsymbol{0}$ was used, the prediction of dominant height was based on (7.24). To the best of our knowledge, this model was the first published nonlinear mixed-effects model in forestry. Furthermore, the paper introduced the idea of model calibration using random effects with a previously fitted model even though the same idea had been used for taper curves earlier in Lappi (1986). This example illustrates the above-specified calibration procedures using the model of Lappi and Bailey (1988).

We assume that the parameter estimates of the linearized model are the known true parameters for the nonlinear Chapman-Richard model. The partial derivatives of the function with respect to the first-order parameters $\phi_{ijt}^{(1)}$ and $\phi_{ijt}^{(2)}$ are

$$\begin{aligned} \frac{\partial f(\boldsymbol{\phi}_{ijt})}{\partial \phi_{ijt}^{(1)}} &= \left(1-\exp\left(-\phi_{ijt}^{(3)}T_{ijt}\right)\right)^{\phi_{ijt}^{(2)}} \\ \frac{\partial f(\boldsymbol{\phi}_{ijt})}{\partial \phi_{ijt}^{(2)}} &= \phi_{ijt}^{(1)}\left(1-\exp\left(-\phi_{ijt}^{(3)}T_{ijt}\right)\right)^{\phi_{ijt}^{(2)}} \log\left(1-\exp\left(-\phi_{ijt}^{(3)}T_{ijt}\right)\right) . \end{aligned}$$

Because the sub-models for parameters $\boldsymbol{\phi}_{ijt}^{(1)}$ and $\boldsymbol{\phi}_{ijt}^{(2)}$ only include a random intercept, these expressions are also the derivatives with respect to $a_i^{(1)}$ and $a_i^{(2)}$, respectively, and with respect to $c_i^{(1)}$ and $c_i^{(2)}$. For completeness, function `f` below also implements the derivative with respect to $\phi_{ijt}^{(3)}$, even though it is not needed in this example.

```
> # This function could be used with nlme and user-supplied gradients.
> f<-function(age,phi1,phi2,phi3) {
+     value<-phi1*(1-exp(-phi3*age))^phi2
+     grad<-cbind((1-exp(-phi3*age))^phi2,
+                 phi1*(1-exp(-phi3*age))^phi2*log(1-exp(-phi3*age)),
+                 phi1*phi2*(1-exp(-phi3*age))^(phi2-1)*exp(-phi3*age)*age)
+     attr(value,"gradient") <- grad
+     value
+     }
```

Consider a sample plot where two dominant trees have been observed: the tree #1 at ages 10 and 15 years and the tree #2 at 15 years. The observed heights are 43, 58, and 53 ft, respectively. The site index is defined as the mean height of dominant trees at the age of 25 years. We want to predict the site index for this particular plot. Therefore, we are interested in the plot-level prediction based on the mixed-effects model and these three observed heights. However, we cannot predict the plot effects without also predicting the tree effects. Therefore, we implement the procedure for predicting both plot and tree effects (recall Example 6.2 and the discussion on p. 186).

The parameter vector for observation t of tree j on plot i is

$$\boldsymbol{\phi}_{ijt} = \boldsymbol{A}_{ijt}\boldsymbol{\beta} + \boldsymbol{B}_{ijt}^{(1)}\boldsymbol{a}_i + \boldsymbol{B}_{ijt}^{(2)}\boldsymbol{c}_{ij} ,$$

where

$$\boldsymbol{A} = \boldsymbol{I}_3 , \boldsymbol{B}_{ijt}^{(1)} = \boldsymbol{B}_{ijt}^{(2)} = \begin{pmatrix} 1 & 0 \\ 0 & 1 \\ 0 & 0 \end{pmatrix} .$$

For tree #1, we have

$$\boldsymbol{\phi}_{i1} = \boldsymbol{A}_{i1}\boldsymbol{\beta} + \boldsymbol{B}_{i1}^{(1)}\boldsymbol{a}_i + \boldsymbol{B}_{i1}^{(2)}\boldsymbol{c}_{i1} ,$$

where

$$\boldsymbol{A}_{i1} = \begin{pmatrix} \boldsymbol{A}_{ijt} \\ \boldsymbol{A}_{ijt} \end{pmatrix} \quad \text{and} \quad \boldsymbol{B}_{i1}^{(1)} = \boldsymbol{B}_{i1}^{(2)} = \begin{pmatrix} \boldsymbol{B}_{ijt}^{(1)} \\ \boldsymbol{B}_{ijt}^{(2)} \end{pmatrix} .$$

The matrices include two blocks because we have two observations of this particular tree. For tree #2, these matrices include only one block. For the entire plot, we have

$$\boldsymbol{\phi}_i = \boldsymbol{A}_i\boldsymbol{\beta} + \boldsymbol{B}_i^{(1)}\boldsymbol{a}_i + \boldsymbol{B}_i^{(2)}\boldsymbol{c}_i ,$$

where

$$\boldsymbol{A}_i = \begin{pmatrix} \boldsymbol{A}_{i1} \\ \boldsymbol{A}_{i2} \end{pmatrix}, \boldsymbol{c}_i = \begin{pmatrix} \boldsymbol{c}_{i1} \\ \boldsymbol{c}_{i2} \end{pmatrix}, \boldsymbol{B}_i^{(1)} = \begin{pmatrix} \boldsymbol{B}_{i1}^{(1)} \\ \boldsymbol{B}_{i2}^{(1)} \end{pmatrix}, \boldsymbol{B}_i^{(2)} = \begin{pmatrix} \boldsymbol{B}_{i1}^{(2)} & \boldsymbol{0} \\ \boldsymbol{0} & \boldsymbol{B}_{i2}^{(2)} \end{pmatrix} .$$

Vector $\boldsymbol{\beta}$ includes the regression coefficients, which are assumed to be known, $\boldsymbol{a}_i$ includes the unknown plot effects for plot i, and $\boldsymbol{c}_{ij}$ the unknown tree effects of tree j of plot i, $(j = 1, 2)$ and $\boldsymbol{c}_i = (\boldsymbol{c}'_{i1}, \boldsymbol{c}'_{i2})'$. The parameter vector $\boldsymbol{\phi}_i$ of length 9 now includes the three parameters for all three measurements. We further pool the matrices and vectors to obtain

$$\boldsymbol{\phi}_i = \boldsymbol{A}_i\boldsymbol{\beta} + \boldsymbol{B}_i\boldsymbol{b}_i\,,$$

where

$$\boldsymbol{B}_i = \begin{pmatrix} \boldsymbol{B}_i^{(1)} & \boldsymbol{B}_i^{(2)} \end{pmatrix} \quad \text{and} \quad \boldsymbol{b}_i = \begin{pmatrix} \boldsymbol{a}'_i & \boldsymbol{c}'_i \end{pmatrix}'.$$

Specifically, $\boldsymbol{b}_i$ is of length 6, and includes the plot effects in the first two elements and, thereafter, the tree effects for tree #1 in elements 3 and 4, and for tree #2 in the last two elements.

The linearized model is based on Taylor approximation of the model at $\boldsymbol{b}_i = \widetilde{\boldsymbol{b}}_i$, where $\widetilde{\boldsymbol{b}}_i$ is the current prediction of the random effect vector. The approximation can be written in the familiar form (recall model (6.2), p. 185)

$$\boldsymbol{w}_i = \widehat{\boldsymbol{X}}_i\boldsymbol{\beta} + \widehat{\boldsymbol{Z}}_i^{(1)}\boldsymbol{a}_i + \widehat{\boldsymbol{Z}}_i^{(2)}\boldsymbol{c}_i + \boldsymbol{\epsilon}_i\,,$$

where $\widehat{\boldsymbol{Z}}_i^{(1)}$ is the Jacobian with respect to plot effects $\boldsymbol{b}_i$, evaluated at the current predictions of random effect for all three trees, and matrix $\widehat{\boldsymbol{Z}}_i^{(2)}$ is a block-diagonal matrix where the first two-row block includes the Jacobian with respect to $\boldsymbol{c}_{i1}$ and the second one-row block includes the Jacobian with respect to $\boldsymbol{c}_{i2}$. For prediction of random effects, we pool the matrices as $\widehat{\boldsymbol{Z}}_i = (\widehat{\boldsymbol{Z}}_i^{(1)}, \widehat{\boldsymbol{Z}}_i^{(2)})$ to obtain

$$\boldsymbol{w}_i = \widehat{\boldsymbol{X}}_i\boldsymbol{\beta} + \widehat{\boldsymbol{Z}}_i\boldsymbol{b}_i + \boldsymbol{\epsilon}_i\,,$$

where $\boldsymbol{b}_i$ now includes the plot effects in the first two elements and, thereafter, tree effects for the two trees.

A single iteration of the prediction procedure can now be carried out using the formulas that were presented in Section 5.4.1. We also implement the iterative procedure, where matrix $\widehat{\boldsymbol{Z}}_i$ is updated after every iteration. This is continued until the change in $\widetilde{\boldsymbol{b}}_i$ becomes marginally small. The final predictions and their standard errors are obtained from the linearized model at the last iteration. Using these predictions, we computed both plot-level and tree-level predictions for the mean height of dominant trees as a function of age. The prediction, utilizing the predicted random effects after the first step, was computed using (7.24), and the prediction using random effect, based on the iterative procedure using (7.25).

The R-script below implements the procedure. A total of six iterations were needed for convergence, $\widetilde{\boldsymbol{b}}_i = (3.59, -0.244, 2.10, -0.0224, -1.46, 0.0104)'$ was the prediction in the last iteration. The plot effects are, therefore, $\widetilde{\boldsymbol{a}}_i = (3.59, -0.244)'$. Using these values in the model and predicting for age 25 yr, the predicted site index is 72.3 feet, and the 95% confidence interval (ignoring the estimation errors of fixed effects) is 65.1 . . . 79.5 feet. The predicted tree effects are $\boldsymbol{c}_{i1} = (2.10, -0.0224)'$ and $\boldsymbol{c}_{i2} = (-1.46, 0.0104)'$ for trees #1 and #2, respectively. Figure 7.7 illustrates the predictions based on these random effects.

Using the same parameter estimates and measurements in the linearized model, Lappi and Bailey (1988) reported a site index of 72.6 feet. The approach is similar to our approach after the first iteration, which provided a site index value of 72.53 feet. The differences between these two values result from rounding errors of the parameter estimates.

```
> library(magic)

> D1<-diag(c(9.1^2,0.289^2))
> D1[1,2]<-D1[2,1]<-9.1*0.289*0.187

> D2<-diag(c(4.4^2,0.084^2))
> D2[1,2]<-D2[2,1]<-4.4*0.084*0.482

> T1<-c(10,15); H1<-c(43,58)
> T2<-15; H2<-c(53)
> T<-c(T1,T2); H<-c(H1,H2)

> D<-adiag(D1,D2,D2)
```

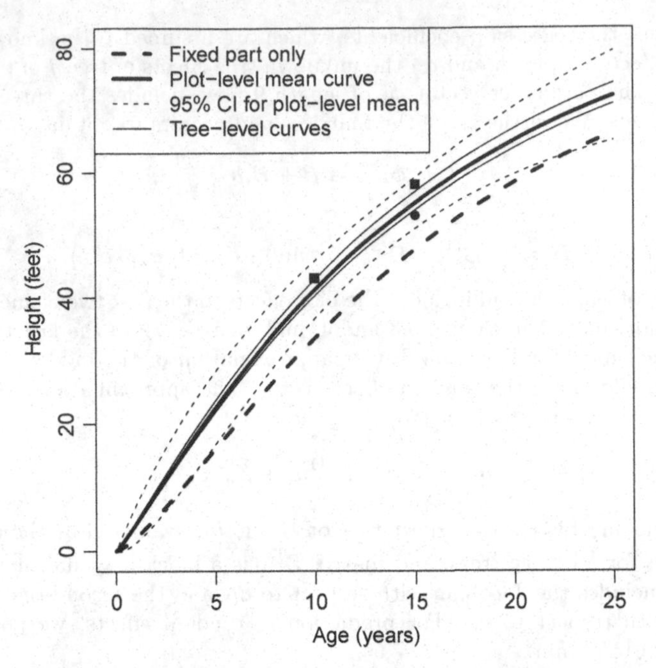

FIGURE 7.7 Predicted height-age curves of dominant trees in Example 7.12. The fixed part shows the population average prediction. The dots show the measurements of age and height of two trees, which were used to produce the plot-level and tree-level predictions. Approximate 95% confidence intervals for the plot-level mean are also shown.

```
> sigma2<-2.16^2
> R<-diag(rep(sigma2,3))

> Aijk<-diag(rep(1,3))
> A<-rbind(Aijk,Aijk,Aijk)
> Bijk<-cbind(c(1,0,0),c(0,1,0))
> zeroblock<-matrix(0,ncol=2,nrow=3)
> B1<-rbind(Bijk,Bijk,Bijk)
> B2<-cbind(rbind(Bijk,Bijk,zeroblock),rbind(zeroblock,zeroblock,Bijk))
> B<-cbind(B1,B2)
> beta<-c(86.2,1.385,0.0702)

> # Initial guesses for random effects
> b0<-rep(0,6)    # pool the random effects to b0
> b0ld<-b0-1      # initialize for the loop
> # Initial guesses for random effects
> b0<-rep(0,6)    # pool the random effects to b0
> b0ld<-b0-1      # initialize for the loop
> history<-matrix(ncol=6,nrow=0)

> while (sum((b0-b0ld))^2>1e-10) {
+         phi<-A%*%beta+B%*%b0
+         phi1<-phi[seq(1,9,by=3)]
+         phi2<-phi[seq(2,9,by=3)]
+         phi3<-phi[seq(3,9,by=3)]
+         pred<-f(T,phi1,phi2,phi3)
+         X<-attributes(pred)$gradient
+         Za<-X[,c(1,2)]
+         Zb<-adiag(Za[1:2,],matrix(Za[3,],ncol=2))
+         Z<-cbind(Za,Zb)
+         b0ld<-b0
+         b0<-D%*%t(Z)%*%solve(Z%*%D%*%t(Z)+R)%*%(H-pred+Z%*%b0)
+         history<-rbind(history,t(b0))
+         cat(b0,"\n")
+         }
3.763176 -0.267426 2.224527 -0.02366303 -1.571858 0.01019456
3.570648 -0.2445138 2.098181 -0.0225388 -1.471668 0.01044628
```

```
3.587479 -0.2438648 2.096076 -0.0223775 -1.465198 0.01036205
3.587418 -0.2438625 2.095589 -0.02238374 -1.464723 0.01036835
3.587477 -0.2438611 2.095599 -0.02238367 -1.464719 0.01036849
3.587475 -0.2438611 2.095597 -0.0223837 -1.464718 0.01036851

> b<-b0
> cat(b,"\n")
3.587475 -0.2438611 2.095597 -0.0223837 -1.464718 0.01036851

> age<-seq(0,25,0.1)
> predFixed<-f(age,beta[1],beta[2],beta[3])
> predPlot<-f(age,beta[1]+b[1],beta[2]+b[2],beta[3])
> predTree1<-f(age,beta[1]+b[1]+b[3],beta[2]+b[2]+b[4],beta[3])
> predTree2<-f(age,beta[1]+b[1]+b[5],beta[2]+b[2]+b[6],beta[3])
> sePlot<-(D-D%*%t(Z)%*%solve(Z%*%D%*%t(Z)+R)%*%Z%*%D)[1:2,1:2]
> Z0<-attributes(predPlot)$gradient[,1:2]
> sePred<-sqrt(diag(Z0%*%sePlot%*%t(Z0)))
> ubPlot<-predPlot+1.96*sePred
> lbPlot<-predPlot-1.96*sePred

> # Site index and confidence bounds
> print(SI<-predPlot[age==25])
[1] 72.29923
> SI+c(-1.96,1.96)*sePred[age==25]
[1] 65.08703 79.51143

> # SI after the first iteration
> # Plug the predicted RE after first iteration
> # to the linearized model, see Meng and Huang (2009) for justification.
> b0<-history[1,] # predicted RE after first iteration
> predf<-f(25,beta[1],beta[2],beta[3])
> predf+attributes(predf)$gradient[,c(1,2)]%*%b0[c(1,2)]
         [,1]
[1,] 72.52796
attr(,"gradient")
          [,1]      [,2]     [,3]
[1,] 0.7687986 -12.58068 479.6988
```

7.3.6 Principal components of random parameters

When a regression function has one independent variable but several random parameters, it is difficult to see how the function varies when the parameters vary. One can plot the regression function when setting each random parameter equal to its estimated expected value, and ± its standard deviation. But as the random parameters are practically always correlated, this does not give a good picture of the variation of the curves. The variation of curves can be nicely illustrated using *principal components* of the random parameters. Principal components of random parameters or any random variables are uncorrelated linear combinations of the random variables. Principal component analysis is a standard method of multivariate statistical analysis (see e.g. Everitt and Hothorn (2011)) which we do not cover in this book, except to briefly demonstrate its use in the application below.

Principal component analysis is based on the properties of *eigenvalues* and *eigenvectors*. If $\boldsymbol{A}$ is a square $p \times p$ matrix and

$$\boldsymbol{A}\boldsymbol{v} = \lambda \boldsymbol{v} \tag{7.29}$$

where $\boldsymbol{v}$ is a $p \times p$ vector and λ is a scalar, then $\boldsymbol{v}$ is an eigenvector (characteristic vector) and λ is the corresponding eigenvalue (characteristic root). Eigenvectors are scaled so that $\boldsymbol{v}'\boldsymbol{v} = 1$. A square $n \times n$ matrix has n eigenvectors. Different eigenvectors are orthogonal, i.e. for any two eigenvectors $\boldsymbol{v}_1$ and $\boldsymbol{v}_2$ it holds that $\boldsymbol{v}_1'\boldsymbol{v}_2 = 0$. Let $\boldsymbol{Q}$ be the matrix having eigenvectors as its columns. Since $\boldsymbol{Q}$ is orthogonal, its transpose is its inverse.

If $\boldsymbol{Q}$ is the eigenmatrix of a variance matrix $\boldsymbol{V}$ of a random $p \times 1$ vector $\boldsymbol{x}$ then

$$\boldsymbol{c} = \boldsymbol{Q}'\boldsymbol{x} \tag{7.30}$$

are the *principal components* of the original variables $\boldsymbol{x}$. The principal components are linear transformations of the original variables. Principal components are uncorrelated. The variance of each principal component is equal to the eigenvalue of the corresponding eigenvector used to multiply $\boldsymbol{x}$. As the eigenmatrix is orthogonal, then the original variables $\boldsymbol{x}$ can be obtained simply as:

$$\boldsymbol{x} = \boldsymbol{Q}\boldsymbol{c} \tag{7.31}$$

The principal components can be used to present correlated variables in terms of uncorrelated components. Lappi (1986) also used principal components in *dimension reduction* when he expressed random stand parameters of a multivariate mixed model in terms of a few principal components.

Let us assume that a nonlinear mixed model can be expressed as:

$$y_{ij} = f\left(x_{ij}, \boldsymbol{\beta}, \boldsymbol{b}_i\right) + \epsilon_{ij} \tag{7.32}$$

where $\boldsymbol{\beta}$ is the vector of fixed parameters and $\boldsymbol{b}$ is the vector of random parameters with expectation zero and with dimension p. For simplicity, we do not allow the dependency of random parameters on z variables as carried out in Equation (7.17). Let $\boldsymbol{D} = \text{var}\left(\boldsymbol{b}_i\right)$, $\boldsymbol{R} = \text{cor}\left(\boldsymbol{b}_i\right)$ and let $\boldsymbol{Q}$ be the $p \times p$ matrix with eigenvectors of $\boldsymbol{R}$ as its columns. The scales of the different parameters are not comparable, so we suggest that the principal components are computed for the standardized parameters. This can be done by computing the eigenmatrix and eigenvalues of the correlation matrix instead of the variance matrix. Note that the correlation matrix is the variance matrix of standardized variables. Let $\boldsymbol{\lambda}$ be the vector of eigenvalues. The following procedure is suggested for visualization of the effect of the variation of the random parameters.

1. Draw the regression function in Equation (7.32) using $\boldsymbol{\beta} = \widehat{\boldsymbol{\beta}}$ and $\boldsymbol{b} = \boldsymbol{0}$.

2. For each value $r = 1, \ldots, p$ do the following:

   (a) Put $b_k = \sqrt{\lambda_r D_{kk}} Q_{kr}$ for $k = 1, \ldots, p$. Recall that $\sqrt{\lambda_r}$ is the standard deviation of the rth principal component. Multiplication with $\sqrt{D_{kk}}$ returns the parameters back to the original scale. Draw the regression function in the same figure.

   (b) Put $b_k = -\sqrt{\lambda_r D_{kk}} Q_{kr}$ for $k = 1, \ldots, p$. Draw the regression function in the same figure.

It is useful to order the eigenvectors so that the first eigenvector is the eigenvector with the largest eigenvalue, etc. This is done automatically in the `R` function `eigen`. Instead of using $\pm\sqrt{\lambda_r}$, one could use $\pm 2\sqrt{\lambda_r}$ to illustrate the 95% variation of the curves. The method can be directly used in linear mixed-effects models, when several predictor variables are nonlinear functions of the one-x variable. The principal component analysis could also be used in general linear mixed-effects models even if the results cannot be shown graphically.

We are not aware of literature references to the suggested illustration method.

Example 7.13. Net photosynthesis as a function of light intensity can often be modeled with a *rectangular hyperbola*:

$$P = \frac{x P_m}{\alpha + x} + R_0 \,, \tag{7.33}$$

where P is net photosynthesis, x is the light intensity, P_m is the maximum gross photosynthesis, R_0 is the dark respiration and parameter α determines how fast the curve approaches the upper limit $P_m + R_0$ (see Example 4.2, p. 73). In data set `lmfor::foto`, the parameters were assigned the estimates: $\widehat{P}_m = 55.61$, $\widehat{\alpha} = 168.99$, and $\widehat{R}_0 = -15.20$. The parameters were assumed to vary randomly from sample to sample. In the following matrix the standard deviations are on the diagonal, and the nondiagonal elements are correlations:

$$\begin{array}{c} \\ P_m \\ \alpha \\ R_0 \end{array} \begin{array}{c} \begin{array}{ccc} P_m & \alpha & R_0 \end{array} \\ \begin{pmatrix} 22.83 & 0.225 & -0.995 \\ 0.225 & 22.63 & -0.297 \\ -0.995 & -0.297 & 2.68 \end{pmatrix} \end{array}$$

Parameters P_m and R_0 are almost perfectly correlated. The eigenvalues of the correlation matrix $\boldsymbol{R}$ were $\boldsymbol{\lambda} = (2.1169, 0.8806, 0.0025)'$. The sum of the eigenvalues is three, as is the sum of the diagonal elements of $\boldsymbol{R}$. The eigenmatrix was

$$\boldsymbol{Q} = \begin{pmatrix} 0.6660 & -0.2608 & 0.6989 \\ 0.3141 & 0.9478 & 0.0544 \\ -0.6766 & 0.1833 & 0.7131 \end{pmatrix} \tag{7.34}$$

Parameters P_m and R_0 have equally large *loadings* in the first principal component. Parameter α has large loading in the second component.

The random variation of the curves is illustrated in Figure 7.8. There is practically no variation in the direction of the third principal component. The R code for carrying out the computations and figures is below.

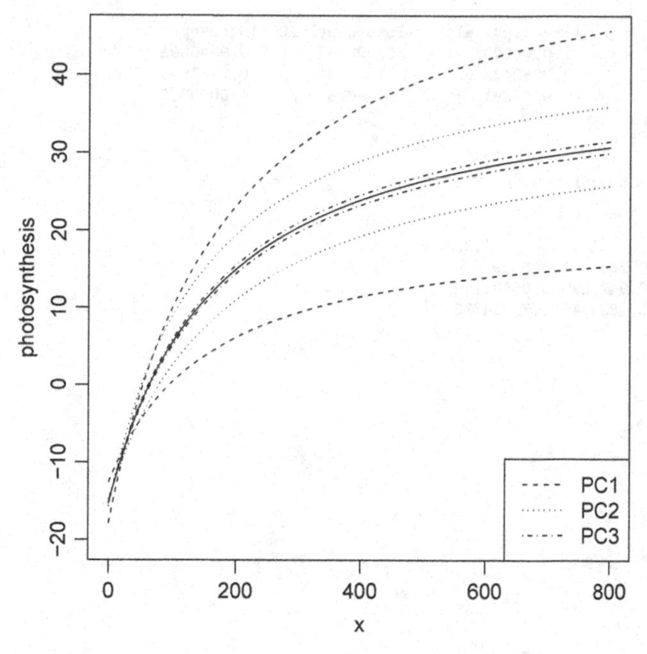

FIGURE 7.8 Principal components of the standardized random parameters. The solid line is computed using the fixed parameters. The dashed lines illustrate the first principal component, the dotted lines the second principal component, and the dash-dotted lines the third component.

```
> data(foto)
>
> LightResp<-function(PPFD,Pmax=10,alpha=0.1,A0=0) {
+             A0+Pmax*PPFD/(alpha+PPFD)
+             }
>
> model5<-nlme(A~LightResp(PARtop,Pmax,alpha,A0),
+         fixed=list(Pmax~Site+Treatment,alpha~Site+Treatment+moisture,A0~Site),
+         random=list(sample=Pmax+alpha+A0~1),
+         data=foto,
+         start=c(c(100,0,0,0,0,0,0),c(80,0,0,0,0,0,0,0),c(-20,0,0,0,0,0)),
+         verbose=TRUE)
```

```
> D<-model5$sigma^2*pdMatrix(model5$modelStruct$reStruct)$sample
> R<-cov2cor(D)
> E<-eigen(R)
> lambda<-E$values
> Q<-E$vectors

> phi0<-fixef(model5)[c(1,8,16)]
> lty<-c("dashed","dotted","dotdash")
> PPFD<-seq(0,800)

> plot(PPFD,LightResp(PPFD,phi0[1],phi0[2],phi0[3]),type="l",ylim=c(-20,45),
         ylab="photosynthesis",xlab="x")
> for (i in 1:3) {
+     phi1<-phi0+sqrt(lambda[i]*diag(D))*Q[,i]
+     phi2<-phi0-sqrt(lambda[i]*diag(D))*Q[,i]
+     lines(PPFD,LightResp(PPFD,phi1[1],phi1[2],phi1[3]),lty=lty[i])
+     lines(PPFD,LightResp(PPFD,phi2[1],phi2[2],phi2[3]),lty=lty[i])
+     }

> D
                   Pmax.(Intercept) alpha.(Intercept) A0.(Intercept)
Pmax.(Intercept)          521.00806         116.34923     -60.801979
alpha.(Intercept)         116.34923         512.13270     -17.985009
A0.(Intercept)            -60.80198         -17.98501       7.171603
> R
                   Pmax.(Intercept) alpha.(Intercept) A0.(Intercept)
Pmax.(Intercept)          1.0000000         0.2252423     -0.9946891
alpha.(Intercept)         0.2252423         1.0000000     -0.2967640
A0.(Intercept)           -0.9946891        -0.2967640      1.0000000
> E
eigen() decomposition
$values
[1] 2.116816013 0.880626115 0.002557873

$vectors
           [,1]       [,2]       [,3]
[1,]  0.6659771 -0.2607814 0.69890454
[2,]  0.3141088  0.9478299 0.05435244
[3,] -0.6766167  0.1833346 0.71314673
```

8

Generalized Linear (Mixed-Effects) Models

With linear and nonlinear models and the corresponding mixed-effects models, it is assumed that the expected value of the response variable is unbounded and that the residual errors are, at least, approximately normally distributed (but recall what we have discussed in Sections 4.1.1, 4.6.3 and 5.3.1 about the importance of the normality assumption). However, a response variable may also have a non-normal distribution, with bounded expected value. Common examples are

- *Binary data.* The response is an outcome of a Bernoulli trial, which can produce two different values. Examples are dead or alive, success or failure, presence or absence of some property etc. The outcome of the experiment is then coded as 0 and 1 and modeled using the Bernoulli distribution.
- *Grouped binary data.* The response is the number of successes *in a fixed number of Bernoulli trials.* For example, the number of dead plants in a plot with a fixed number of plants. Often grouped binary data are called binomial data.
- *Count data.* Examples include the number of trees in a forest sample plot, number of insects or lizards in a trap after one day, and the number of polypore species in a sample plot. The essential difference to grouped binary data is that the data are not an outcome from a fixed known number of Bernoulli trials, and the upper limit for the count cannot be determined.
- *Continuous proportion or percentage data.* Examples include canopy cover on an aerial image, log volume reduction caused by defects in a tree stem, and Vaccinium vitis-idaea-cover in a sample plot.
- *Time-to-event data.* Examples include the number of days until willow bud burst in the spring, and time to the death of a tree after a forest fire.

A model where non-normal distribution is assumed for the response variable is called *a generalized linear model (GLM)*. Furthermore, a GLM with random effects is called *a generalized linear mixed-effects model (GLMM)*. GLMs and GLMMs are widely used in the analysis of binary data, grouped binary data and count data. They can also be used for analysis of continuous proportions. Models to handle time-to-event data are covered in the field of survival analysis and will not be covered in this book. The methods are mostly used in the health sciences (Collett 2014), but time-to-event data also occur within forestry and environmental research (e.g. Hämäläinen *et al.* 2016).

Notice the distinction between the terms GLS and GLM. GLS is an acronym for generalized least squares in the context of a linear model (recall Chapter 4.4.2). Some confusion is caused in that some statistical packages and literature use the acronym GLM for the term general linear model, which refers to the model that we call the GLS-model (see Section 4.1.2). Majority of statistical literature uses the acronym GLM for the generalized linear model, and that is also our convention here.

Generalized linear models are strongly related to the method of maximum likelihood. Fitting a generalized linear model can be seen as an ML fit of independent data where the

Box 8.1 Transformations or GLMM?

Before the availability of generalized linear (mixed) models, non-normal data were usually modeled by transforming the dependent variable. Count data were usually transformed with log or square root transformation and relative frequencies were transformed with arcsin square root or logit transformation. The purpose of transformations is to normalize the distribution, stabilize the variance and linearize the dependency with respect to the explanatory variables. Often these goals are in conflict. In this book we have not emphasized normality.

If y is a count variable distributed according to a Poisson distribution with expected value μ (which may depend on predictor variables), then $y = \mu + \epsilon$ where $\text{var}(\epsilon) = \mu$. If we take the first-order Taylor series approximation of $y^{0.5}$ around μ, we get $y^{0.5} \approx \mu^{0.5} + 0.5\mu^{-0.5}\epsilon$. Thus, $\text{var}(y^{0.5}) \approx 0.25$, i.e. the square root transformation approximately stabilizes the variance of a Poisson variable with mean μ. Hence, the square root transformation can be used in an ANOVA model to test differences between fixed class effects. But there are no special reasons why $y^{0.5}$ would be linear with respect to regressor variables.

Log transformation was also used for count data in many cases. This looks similar to the use of log-link in GLM, but it is not equivalent as $\text{E}(\log(y)) \neq \log(\text{E}(y))$. One drawback in log transformation is that it cannot be applied if there are zero counts in the data. One solution to this problem is to add a small constant (e.g. 1) to y. For a Poisson variable y, log transformation does not stabilize the variance as well as square root transformation, but it may be better described as a linear function of predictors if the predictors have multiplicative effects. In power law models, the explanatory variables are also log-transformed. See St-Pierre *et al.* (2018) for further discussion and references. Nonetheless, GLMM with a log link and Poisson or negative binomial distribution is a theoretically better basis for the analysis of count data.

For frequency data, $\arcsin\left(\sqrt{y}\right)$ has been often used to normalize observed relative frequencies (grouped binary data). The transformed relative frequencies are assigned values between 0 and $\frac{\pi}{2}$. Thus, they cannot be predicted with linear predictor equations, which are assigned values outside of this range. The logit transformation, i.e. $\log \frac{y}{1-y}$ is also used. This also imitates the use of logit link in GLMM where the true probabilities are transformed using the logit function. Logit transformation is not defined when the observed relative frequency is zero or one, thus some *ad hoc* modification is needed as when dealing with zeroes in count data. Nowadays GLMM provides a better basis for the analyses of binary data.

expected value, or a nonlinear transformation of the expected value, is a linear function of predictors. The nonlinear transformation is called *a link function*, and it is used to ensure that the estimated mean is always within its support; e.g. probability of success in binary data is always between 0 and 1, or the expected count in count data is positive.

8.1 Common parent distributions

Before we define the GLM, let us consider models for the binary, grouped binary and count data. We focus especially on the possible values of the parameter that specify the expected value, and the relationship between the expected value and variance.

8.1.1 The normal distribution

The normal distribution is commonly used for interval and ratio scale variables. The central limit theorem declares that the sample means of large samples have a normal distribution.

Otherwise, there is no general result that would guarantee normality of a ratio or interval scale response variable in the regression context. The density of normal distribution $N(\mu, \sigma^2)$ was presented in Example 2.3. The two parameters are the expected value μ and variance σ^2 (or alternatively, standard deviation σ) of the distribution. A specific property of the normal distribution is that the mean and variance are unrelated, which allowed separate modeling of the expected value and variance in the contexts of linear and nonlinear models.

8.1.2 Bernoulli and binomial distribution

A Bernoulli trial has two possible discrete outcomes: success and failure. A random variable Y has *Bernoulli*(p) distribution if it takes the value of 1 ("success") with probability p, and value 0 ("failure") with probability $1-p$, where $0 \leq p \leq 1$ (recall Example 2.2, p. 10). The expected value and variance of Y are

$$\begin{aligned} \mathrm{E}(Y) &= p \times 1 + (1-p) \times 0 = p \\ \mathrm{E}(Y^2) &= p \times 1^2 + (1-p) \times 0^2 = p \\ \mathrm{var}(Y) &= \mathrm{E}(Y^2) - [\mathrm{E}(Y)]^2 = p - p^2 = p(1-p) \,. \end{aligned}$$

The expected value is equal to the probability of success and is, therefore, bounded between 0 and 1. The variance is bounded between 0 and 0.25. In particular, the variance approaches 0 as $p \to 0$ or $p \to 1$ and takes its maximum value when $p = 0.5$ (show this!). The mean and variance have the relationship

$$\mathrm{var}(Y) = \mathrm{E}(Y)(1 - \mathrm{E}(Y)) \,.$$

A mathematical relationship between the mean and variance is an essential difference to the normal distribution where the first two moments are not related.

The *binomial* distribution expresses the number of successes in a series of n independent Bernoulli trials. The probability to have y successes in n independent trials with success probability p is

$$P(Y = y|n, p) = \binom{n}{y} p^y (1-p)^{n-y} \,.$$

The expectation and variance of binomial distribution are related through

$$\begin{aligned} \mathrm{E}(Y) &= np \\ \mathrm{var}(Y) &= np(1-p) \,. \end{aligned}$$

If X describes a single Bernoulli trial, we can further state the binomial mean and variance in terms of $\mathrm{E}(X)$ as

$$\begin{aligned} \mathrm{E}(Y) &= n\,\mathrm{E}(X) \\ \mathrm{var}(Y) &= n\,\mathrm{var}(X) = n\,\mathrm{E}(X)(1 - \mathrm{E}(X)) \,. \end{aligned}$$

8.1.3 Poisson distribution

Recall from Example 2.4 the definition of the Poisson distribution for random counts, such as the number of points on a line segment or within a 2- or 3-dimensional area, if the points occur independently from each other and are uniformly distributed over the space with intensity λ. A Poisson-distributed random variable takes non-negative integer values.

As an example, consider a large forest area with density λ trees per hectare. Assume that the trees are uniformly distributed over the area and are independent so that the locations of the trees do not interact. Consider a sample plot A, which is very small compared to the entire forest area. Now the random variable Y "The number of trees within the sample plot of area $|A|$" has a $Poisson(\lambda|A|)$ distribution (see illustration in the left-hand panel of Figure 8.5, p. 267). Mathematically, a Poisson distribution is the limiting distribution of Y when sampling from an infinitely large forest area. Another similar process is the number of bark beetles caught in a trap during one day (assuming that the bark beetle population is so large that the probability to catch a new individual is constant regardless of the number of beetles already caught). In general, the Poisson distribution is a model for count data without an upper limit. The missing upper limit of the Poisson distribution makes an essential difference to the binomial distribution, which models the number of successes in a fixed, known number of Bernoulli trials.

Let us then consider $Poisson(\mu)$ distribution (see Example 2.4 and Figure 2.1). If we consider uniformly distributed points on a set A (e.g. a sample plot), then we might further have $\mu = \lambda|A|$, where $|A|$ is the area and λ the intensity. The mean and variance are also related in the count data. In particular,

$$\mathrm{E}(Y) = \mu$$

$$\mathrm{var}(Y) = \mu \, .$$

That is, the variance is equal to the expected value.

The $Poisson(\mu)$ distribution can also be seen as a limiting distribution of the binomial distribution where $np = \mu$. If n increases (and p decreases), then the distribution converges towards the $Poisson(\mu)$ distribution.

8.1.4 Link and mean functions

Let us now take a closer look at the expected values of the distributions presented above. For simplicity, we denote the expected value of the distribution by μ for all distributions. The expected value has the range $-\infty < \mu < \infty$ for the normal distribution, $0 < \mu < 1$ for the Bernoulli distribution, and $0 < \mu < \infty$ for the *Poisson* distribution.

In the GLM, we want to state the expected value of the assumed distribution as a function of model parameters and predictors. Specifically, we want to state the expected value using the linear predictor $\boldsymbol{x}_i'\boldsymbol{\beta}$. However, the value of this linear predictor is unrestricted and can take any value between $-\infty$ and ∞. In the linear model, this did not cause a problem because μ of the normal distribution is unrestricted (recall Example 2.3). For binary, grouped binary and count data, the restricted support of the expected value needs to be taken into account. Otherwise, our model could predict success probabilities below 0 or above 1 (recall Example 2.2) or expected counts below zero. This restriction is specified using an appropriately selected link function.

The *link function* is a continuous, monotonic mapping from the support of the expected value to the real axis. Its inverse, the *mean function* (also called the *inverse link function*, *response function* (Fahrmeir *et al.* 2013) and *probability function* (Demidenko 2013)), is a mapping from the real axis to the support of μ. We denote the link function by $g(\mu)$ and the mean function by $h(\gamma)$.

Consider the *Bernoulli* case as an example. The link function is a mapping $g : (0, 1) \rightarrow (-\infty, \infty)$ and the mean function is a mapping $h : (-\infty, \infty) \rightarrow (0, 1)$. We could use the inverse of any continuous *cdf* that is defined in the real axis as the link function. The most

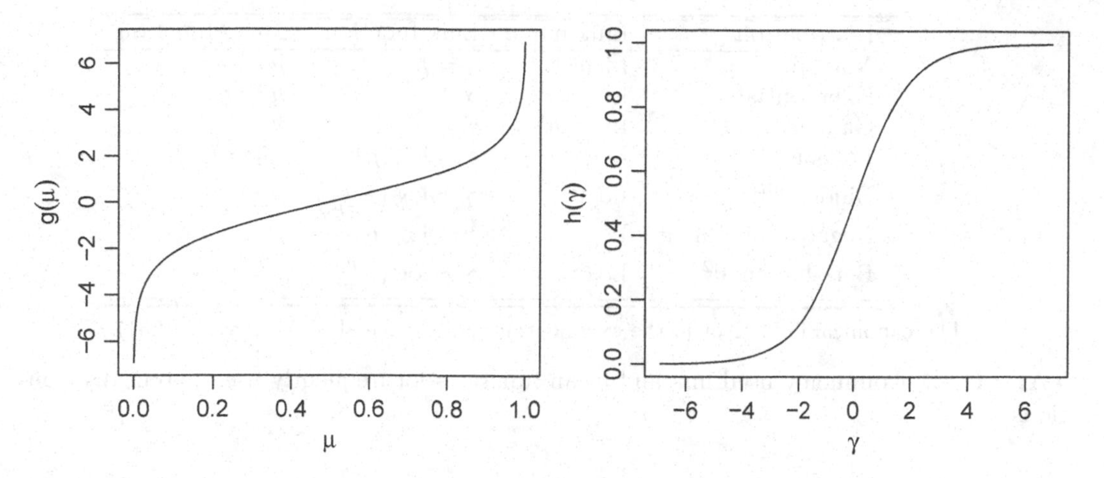

FIGURE 8.1 The logit link (left) and mean (right) functions.

common choice is the *logit link*

$$\gamma = g(\mu) = \log\left(\frac{\mu}{1-\mu}\right). \tag{8.1}$$

The corresponding mean function results from solving the equation for μ:

$$\mu = h(\gamma) = \frac{e^\gamma}{1+e^\gamma} = \frac{1}{1+e^{-\gamma}}.$$

These functions are illustrated in Figure 8.1. The model for binary data using a logit link leads to the model that is commonly called the *logistic regression model.*

For the count data modeled using a Poisson model, the expected value needs to be strictly positive. In that case, the *log link* and the corresponding mean function are commonly used:

$$\gamma = g(\mu) = \log(\mu)$$

$$\mu = h(\gamma) = e^\gamma.$$

In GLMs, the link function ensures that the expected value of the response variable as a function of predictors is always within the support of μ. Note that the same functions can also be used to restrict the range of distribution parameters in estimation, even though term link function is not used in this context (see e.g. Examples 3.7 and 3.8 on p. 55 and Section 11.2).

8.2 Generalized linear model

8.2.1 Formulation of the LM as a GLM

With an LM, the common way of thinking is that there is a linear regression function that is usually of interest to the modeler. In addition, there is an error term (noise) which is added to the regression line to obtain the realized observations. This way of thinking is used in

Distribution	Link name	Link function	Mean function
Normal	Identity[1]	$\gamma = \mu$	$\mu = \gamma$
Exponential	Inverse[1]	$\gamma = \mu^{-1}$	$\mu = \gamma^{-1}$
Gamma	Inverse[1]	$\gamma = \mu^{-1}$	$\mu = \gamma^{-1}$
Poisson	Log[1]	$\gamma = \log(\mu)$	$\mu = e^{\gamma}$
Binomial[3]	Logit[1]	$\gamma = \log\left(\frac{\mu}{1-\mu}\right)$	$\mu = \frac{e^{\gamma}}{1+e^{\gamma}}$
Negative binomial[2]	Log	$\gamma = \log(\mu)$	$\mu = e^{\gamma}$
Beta-binomial[2]	Logit	$\gamma = \log\left(\frac{\mu}{1-\mu}\right)$	$\mu = \frac{e^{\gamma}}{1+e^{\gamma}}$

[1]The canonical link. [2]Not in the exponential family. [3]See also Table 8.2 (p. 251).

TABLE 8.1 Commonly used link and mean functions for frequently used parent distributions.

the development of the least squares estimation methods, so we call it the *LS-formulation of the linear model.* It can be written as

$$\boldsymbol{y} = \boldsymbol{X}\boldsymbol{\beta} + \boldsymbol{e}$$

where $\boldsymbol{e} \sim N(\boldsymbol{0}, \sigma^2\boldsymbol{V})$. We can also employ the formulation that was used in Section 4.4.3 to derive the ML-estimators for the model parameters. This formulation, which we call the *ML-formulation of the linear model,* is specified as

$$\boldsymbol{y} \sim N(\boldsymbol{X}\boldsymbol{\beta}, \sigma^2\boldsymbol{V}) .$$

That is, we specify the assumption about $\boldsymbol{y}$, not about $\boldsymbol{e}$, and state the expected value of $\boldsymbol{y}$ as a linear function of the predictors.

The LS- and ML-formulations are equivalent but arise from two different ways of thinking. The ML-formulation can be generalized to a non-normal response and is used in GLMs. For example, with count data, it makes sense that the counts, which are integers, have a Poisson distribution. However, the residuals (count - mean of counts) of the model are not integers and it is not very clear what is the implicit assumption on their distribution. Correspondingly, the binary response has a Bernoulli distribution (with values 0 or 1), but the distribution of differences (observed response - mean) is not of any commonly used form. Therefore, the LS-formulation is justified only with the linear and nonlinear models, where the mean and variance are unrelated. It should be overlooked in the context of non-normal distributions (but will be recalled when we demonstrate that a GLM can be seen as a nonlinear GLS model).

The GLM applies the ML-formulation of the linear model to non-normal cases. The distribution is always parameterized in terms of its mean μ and probably some other parameters, but usually μ as a function of predictors is the parameter of primary interest. The bounded range of μ is taken into account through appropriately selected link and mean functions:

$$g(\mu_i) = \boldsymbol{x}_i'\boldsymbol{\beta}$$

$$\mu_i = h(\boldsymbol{x}_i'\boldsymbol{\beta}) ,$$

If the link function is not the identity link, the expected value becomes a nonlinear function of the predictors, but the predictors enter the model through the linear predictor $\boldsymbol{x}_i'\boldsymbol{\beta}$. Usually, *the canonical link and mean functions* are used (Table 8.1), but that is not always the case. The starting point for parameter estimation is ML. The following subsections present the GLM for the two most common cases: binary and count data. Thereafter, we generalize the formulation for a general member of a univariate exponential family.

Model	Link function	Mean function
Logistic regression	$g(\mu) = \log \frac{\mu}{1-\mu}$	$h(\boldsymbol{x}_i'\boldsymbol{\beta}) = \frac{\exp(\boldsymbol{x}_i'\boldsymbol{\beta})}{1+\exp(\boldsymbol{x}_i'\boldsymbol{\beta})}$
Probit regression	$g(\mu) = \Phi^{-1}(\mu)$	$h(\boldsymbol{x}_i'\boldsymbol{\beta}) = \Phi(\boldsymbol{x}_i'\boldsymbol{\beta})$
Complementary log-log regression	$g(\mu) = \log(-\log(1-\mu))$	$h(\boldsymbol{x}_i'\boldsymbol{\beta}) = 1 - \exp(-\exp(\boldsymbol{x}_i'\boldsymbol{\beta}))$
log-log regression	$g(\mu) = -\log(-\log(\mu))$	$h(\boldsymbol{x}_i'\boldsymbol{\beta}) = \exp(-\exp(-\boldsymbol{x}_i'\boldsymbol{\beta}))$

TABLE 8.2 Four alternative models for binary data. Φ is the *cdf* of the standard normal distribution and Φ^{-1} its inverse; these functions cannot be written in closed form.

8.2.2 Models for binary data

The GLM for binary data is based on the Bernoulli distribution. Let y_i be the binary response (0=success, 1=failure) for observation i and let $\boldsymbol{x}_i$ include the predictors of that observation and $\boldsymbol{\beta}$ the regression coefficients; everything that was discussed earlier in regard to the predictors in Section 4.2 applies here as well. The model is formulated as

$$y_i \sim Bernoulli(\mu_i) \,.$$

Notice the subscript i in μ_i, which indicates that the mean is allowed to vary among the observations. We restrict μ_i to the range $(0, 1)$ by the link function g. Most commonly, the *logit link* function (8.1) is used. Other frequently used link functions for the Bernoulli/binomial case are the *probit link*, which is the quantile function (i.e. the inverse *cdf*) of the standard normal distribution, and the *complementary log-log link* (Table 8.2). The models that use these link functions are called the *probit model* and *complementary log-log regression*. For more discussion about these link functions and models, see e.g. Demidenko (2013, Section 7.1).

The commonly applied *binary logistic regression* or just *logistic regression* is a GLM for binary data using the canonical logit link function. The applied link function needs to be taken into account when interpreting the regression coefficients. The interpretation is best explained through an example.

Example 8.1. Assume that the response y_i codes for the occurrence of *Sphagnum magellanicum*, a red-colored peatland moss species, on vegetation sample plot i so that $y_i = 1$ denotes occurrence and $y_i = 0$ denotes absence, and x_i is the long-term average water table of plot i in cm, relative to the level 10 cm below the ground so that an increase in water table implies a higher water level. We consider the range $-4 \leq x_i \leq 5$ in this example and assume the binary logistic model where $g(\mu_i) = \beta_1 + \beta_2 x_i$. We assume that the sample plots are small so that the occurrence of the species indicates that the plot is fully covered by *S. magellanicum*, and the probability of occurrence can be directly interpreted as the coverage for a given value of x_i.

Let us define the *odds* of occurrence for the value $x_i = x_0$ as the ratio of the probability of occurrence to the probability of absence. The logistic mean function (Table 8.2) gives

$$O_1 = \frac{P(y_i = 1|x_i = x_0)}{P(y_i = 0|x_i = x_0)} = \frac{\frac{\exp(\beta_1+\beta_2 x_0)}{1+\exp(\beta_1+\beta_2 x_0)}}{1 - \frac{\exp(\beta_1+\beta_2 x_0)}{1+\exp(\beta_1+\beta_2 x_0)}} = \exp(\beta_1 + \beta_2 x_0) \,.$$

If the water table is 1 cm higher, then the odds become:

$$O_2 = \frac{P(y_i = 1|x_i = x_0 + 1)}{P(y_i = 0|x_i = x_0 + 1)} = \exp(\beta_1 + \beta_2(x_0 + 1)) \,.$$

The ratio of these two odds is

$$OR = \frac{O_2}{O_1} = \frac{\exp(\beta_1 + \beta_2(x_0 + 1))}{\exp(\beta_1 + \beta_2 x_0)} = e^{\beta_2} \,.$$

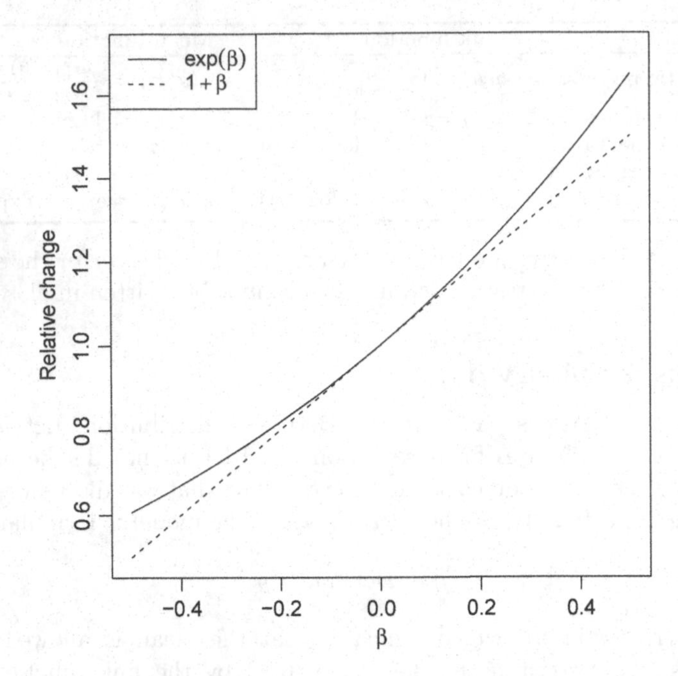

FIGURE 8.2 Illustration of the approximation $\exp(\beta) \approx 1 + \beta$ for small absolute values of β.

The intercept can be used to compute the probability of occurrence when $x_i = 0$ as $\mu_0 = E(y_i|x_i = 0) = \frac{e^{\beta_1}}{1+e^{\beta_1}}$. To interpret coefficient β_2, we make two notes. First, if μ_0 is small, then $\frac{P(success)}{P(failure)} = \frac{\mu_0}{1-\mu_0} \approx \mu_0$, because the probability of failure, $1 - \mu_0$, is close to 1. This means that the odds are an approximate of success probability μ if μ is small. Second, if β_2 is close to zero, then $\exp(\beta_2) \approx (1 + \beta_2)$ (see Figure 8.2). Therefore, coefficient β_2 can be directly interpreted as the relative change in the odds when the predictor is changed by one unit. If both μ_0 and β_2 are close to zero, the value of the regression coefficient is an approximate of the relative change in the probability of success when x_i is increased by one unit.

This can be illustrated using hypothetical peatland sites A, B, C and D, with the following parameter values:

$$
\begin{array}{lll}
\text{Site A} & \beta_1 = -3.89 & \beta_2 = -0.07 \\
\text{Site B} & \beta_1 = -3.89 & \beta_2 = -0.40 \\
\text{Site C} & \beta_1 = 0.405 & \beta_2 = -0.07 \\
\text{Site D} & \beta_1 = 0.405 & \beta_2 = -0.40
\end{array}
$$

When $x_i = 0$, the *S. magellanicum* cover is 2% on sites A and B (Table 8.1) and 60% on sites C and D. On site A, where μ is small for the range $-4 < x_i < 5$, a one-unit increase in water table causes an approximate 7% decrease in species cover regardless of the value of x_i; this decrease is well approximated by $\beta_2 = -0.07$. On site B, a one-unit increase in x_i causes a 45–49% decrease in cover. Because β_2 is large, this is much greater than the approximation (40% decrease regardless of the value of x_i) based on $\beta_2 = -0.4$. On site C, where μ ranges from 51 to 67%, the relative decrease caused by a one-unit increase in x_i ranges from 2.4 to 3.4%, which is more than two times smaller than the approximate decrease based on $\beta_2 = -0.07$. Finally, on site D, where both μ and β_2 are far from zero, the relative decrease in μ caused by a one-unit increase in x_i varies considerably (6–38%), and is mostly far from the value of 40%, the approximation based on direct interpretation of $\beta_2 = -0.4$ as change in probability.

Table 8.3 demonstrates that in all four sites, a one-unit increase in x_i causes a similar relative decrease in the odds for all values of x_i. This decrease is well approximated by β_2 on sites A and

	Site A		Site B		Site C		Site D	
	small μ, small β_2		small μ, large β_2		large μ, small β_2		large μ, large β_2	
x_i	μ	$\frac{\mu_2}{\mu_1}-1$	μ	$\frac{\mu_2}{\mu_1}-1$	μ	$\frac{\mu_2}{\mu_1}-1$	μ	$\frac{\mu_2}{\mu_1}-1$
-4	0.026	-	0.092	-	0.665	-	0.881	-
-3	0.025	-0.071	0.064	-0.447	0.649	-0.024	0.833	-0.058
-2	0.023	-0.071	0.044	-0.461	0.633	-0.025	0.769	-0.082
-1	0.021	-0.071	0.030	-0.470	0.617	-0.027	0.691	-0.113
0	0.020	-0.071	0.020	-0.477	0.600	-0.028	0.600	-0.152
1	0.019	-0.071	0.014	-0.482	0.583	-0.029	0.501	-0.197
2	0.017	-0.071	0.009	-0.485	0.566	-0.030	0.403	-0.245
3	0.016	-0.071	0.006	-0.487	0.549	-0.031	0.311	-0.294
4	0.015	-0.071	0.004	-0.489	0.531	-0.033	0.232	-0.339
5	0.014	-0.071	0.003	-0.490	0.514	-0.034	0.169	-0.378

	Site A		Site B		Site C		Site D	
	small μ, small β_2		small μ, large β_2		large μ, small β_2		large μ, large β_2	
x_i	ODDS	OR − 1	ODDS	OR − 1	ODDS	OR − 1	ODDS	OR − 1
-4	0.027	-	0.101	-	1.984	-	7.426	-
-3	0.025	-0.073	0.068	-0.492	1.850	-0.073	4.978	-0.492
-2	0.024	-0.073	0.046	-0.492	1.725	-0.073	3.337	-0.492
-1	0.022	-0.073	0.031	-0.492	1.608	-0.073	2.237	-0.492
0	0.020	-0.073	0.020	-0.492	1.499	-0.073	1.499	-0.492
1	0.019	-0.073	0.014	-0.492	1.398	-0.073	1.005	-0.492
2	0.018	-0.073	0.009	-0.492	1.303	-0.073	0.674	-0.492
3	0.017	-0.073	0.006	-0.492	1.215	-0.073	0.452	-0.492
4	0.015	-0.073	0.004	-0.492	1.133	-0.073	0.303	-0.492
5	0.014	-0.073	0.003	-0.492	1.057	-0.073	0.203	-0.492

TABLE 8.3 The odds of occurrence of *S. magellanicum* on four hypothetical peatland sites of Example 8.1 (ODDS) and the relative change in the odds when x_i is increased by one unit for different values of x_i (OR − 1).

C where β_2 is small, but not well approximated by β_2 on sites B and D where β_2 is large. The script below demonstrates the computation of Tables 8.1 and 8.3 for site A.

```
> ilogit<-function(l) exp(l)/(1+exp(l))
> betaA<-c(-3.89,-0.07)
> x<-seq(-4,5)
> X<-cbind(1,x)
> predA<-ilogit(X%*%betaA)
> oddsA<-ilogit(X%*%betaA)/(1-inv.logit(X%*%betaA))
> changeA<-(predA[-1]-predA[-10])/predA[-1]
> changeAO<-(oddsA[-1]-oddsA[-10])/oddsA[-1]
> round(cbind(predA,c(NA,changeA)),3)
> round(cbind(oddsA,c(NA,changeAO)),3)
```

If the predictors of a binary model are not specific for each observation but they are specific for a group of observations, a lot of space (on paper or computer disc) can be saved with no loss of information by coding the binary data as grouped binary data. For example, consider the effect of elevated temperature treatment (two levels: control and elevated temperature) on the occurrence of severe rust infection (more than 1000 rust spots per leaf) on leaves of the willow species *Salix myrsinifolia.* Instead of coding one row per leaf with a value of 1 or 0 for the response variable, we could code them in two rows: one row for the control treatment and another row for the elevated temperature treatment. For both treatments, we need to know two values: the total number of leaves and the number of leaves that are infected. Of course, if the data were based on several individual plants, one

	Success (at least one fungal species)			Failure (no fungi)		
Mites	Spring	Summer	Fall	Spring	Summer	Fall
0	59	96	37	33	4	49
1	1	1	5	0	0	3
2	0	0	2	1	0	2
3	2	0	1	0	0	1
4	0	0	0	1	0	0

TABLE 8.4 The number of *Ips typographus* bark beetles carrying at least one ophiostomatoid fungal species (success) and with no fungal species (failure), classified by the season and the number of mites in the `ips` data set.

should only pool the observations from the same individual plant together, and take into account the grouping of the data, e.g. by using random effects for each individual plant (see Section 8.3).

The grouped binary data can be modeled using the $Binomial(n_i, \mu_i)$ distribution, where parameter n_i is the (known) number of trials and μ_i is the probability of success for group i. The expected number of successes in group i is then $n_i\mu_i$. However, the parameter of interest is not the binomial mean $n_i\mu_i$ but the mean of a single Bernoulli trial, μ_i, which is stated as $h(\boldsymbol{x}_i'\boldsymbol{\beta})$, as done with the binary model. The alternative link functions are the same as well. The only difference is how the data are coded in the data frame. A binary model can be considered a special case of the binomial model where $n_i = 1$ for all groups i.

Example 8.2. The ophiostomatoid fungal families *Microascales* and *Ophiostomatales* are common associates of the bark beetle *Ips typographus*, which they use to spread within the wood material. The number of fungal species in these families is large, and an individual beetle can carry several fungal species. The bark beetles may have mites attached to them, and it may be possible that some fungal species are associated with the beetles via mites.

Data set `ips` (Linnakoski *et al.* 2016) includes measurements of 289 bark beetle individuals from a storm-felled Norway spruce (*Picea abies*) forest in eastern Finland. For each beetle individual, the number of attached mites was determined using a microscope. In addition, the number of fungal species per bark beetle was determined genetically. However, it was not possible to determine whether the fungi were associated with the mites or with the bark beetle itself. The observations were collected during three different seasons: spring, summer, and fall of the same year; approximately 100 individuals in each season. The data are used to analyze the effects of season and number of mites on the number of fungal species per bark beetle.

We recoded the data as binary data, where success refers to the presence of at least one fungal species with the bark beetle individual, and failure refers to the absence of fungal species (Table 8.4). We fitted the following logistic regression model:

$$y_i \sim Bernoulli(\mu_i)$$

where

$$\mu_i = \frac{\exp(\beta_1 + \beta_2 S_i + \beta_3 F_i + \beta_4 m_i)}{1 + \exp(\beta_1 + \beta_2 S_i + \beta_3 F_i + \beta_4 m_i)},$$

which can be equivalently specified as

$$\log\left(\frac{\mu_i}{1-\mu_i}\right) = \beta_1 + \beta_2 S_i + \beta_3 F_i + \beta_4 m_i,$$

where y_i is the binary response (with value 1 indicating presence and 0 absence), S_i and F_i are binary indicators for spring and fall, respectively, and m_i is the number of mites on bark beetle individual i. The following script recodes the data, fits the model, and prints the summary.

```
> library(lmfor)
> data(ips)
> ips$FBin<-ifelse(ips$Fungi==0,0,1)
> ipsmodBin<-glm(FBin~Season+Mites,family=binomial,data=ips)
> coef(ipsmodBin)
 (Intercept)  SeasonSpring    SeasonFall         Mites
     3.18796      -2.62234      -3.39900       0.04697

> exp(coef(ipsmodBin))-1
 (Intercept) SeasonSpring   SeasonFall         Mites
  23.2389600   -0.9273676   -0.9665933     0.0480874
```

The parameter estimates show that the probability of presence of a fungi in summer on a bark beetle with no mites is $e^{3.188}/(1+e^{3.188}) = 0.960$. The odds of presence in spring and fall are 93% and 97% lower than in summer, respectively. These changes in odds should not be interpreted as changes in presence, because neither the probabilities nor the coefficients for seasons are close to zero. One mite increases the odds by 4.8%, practically the same conclusion as reached by directly interpreting the regression coefficient $\widehat{\beta}_4 = 0.047$ as OR − 1. However, this value should not be interpreted as a relative change in probability either, because the probabilities are not close to zero. We see, for example, that increasing the number of mites from 0 to 1 in summer, changes the probability of presence from 0.960 to 0.962 ($\exp(3.188+0.047)/(1+\exp(3.188+0.047)) = 0.962$), which is only a 0.2% increase. The probability of presence of fungi in spring on a beetle with no mites is $\exp(3.188 - 2.622)/(1 + \exp(3.188 - 2.622)) = 0.64$. Probabilities of presence for other seasons and the numbers of mites are computed correspondingly.

Data set **ips** includes one row for each bark beetle, and the response was coded as a binary variable. We could alternatively code the data as grouped binary data so that each row corresponds to one combination of predictors. The possible combinations in our data are determined by the three seasons and the number of mites, which varies between 0 and 4 in our data. However, only 11 of the 15 possible combinations of predictors are present in our data. Therefore, we get the following data set of 11 observations (groups):

```
> ips2
    Season Mites Success Failure
1   Spring     0      59      33
76  Spring     1       1       0
75  Spring     2       0       1
74  Spring     3       2       0
73  Spring     4       0       1
98  Summer     0      96       4
165 Summer     1       1       0
199   Fall     0      37      49
205   Fall     1       5       3
220   Fall     2       2       2
276   Fall     3       1       1
```

The script below does the fitting based on binomial distribution. We have a bivariate response, where the first column specifies the number of successes per group and the second column specifies the number of failures. The assumed model is the same as before, and so the parameter estimates are also the same.

```
> glm(cbind(Success,Failure)~Season+Mites,family="binomial",data=ips2)
Coefficients:
 (Intercept)  SeasonSpring    SeasonFall         Mites
     3.18796      -2.62234      -3.39900       0.04697

Degrees of Freedom: 10 Total (i.e. Null);  7 Residual
Null Deviance:       81.41
Residual Deviance: 7.991     AIC: 34.89
```

As demonstrated in the example, the logistic regression is beneficial due to simple calculus and interpretation of the coefficients through OR. It is especially justified in case-control studies (Collett 1991). In addition to the logistic link function, Table 8.2 listed three other models for binary data. Figure 8.3 illustrates the differences between the alternative link functions.

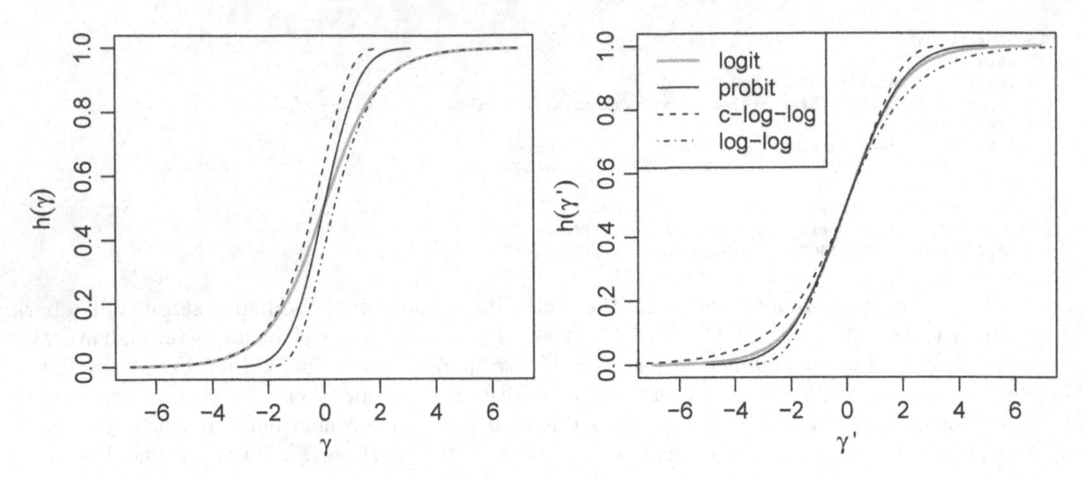

FIGURE 8.3 Illustration of the four mean functions shown in Table 8.2. In the right-hand graph, all functions are adjusted to have $h(0) = 0.5$ and $h'(0) = 1/4$ to better illustrate the differences in shape.

The probit regression uses the inverse of the standard normal *cdf* as the link function. The behavior of the probit link is rather similar to the logit link, except for the tails, where the probit link is much flatter than the logit. That is, the logistic distribution has heavier tails than the normal distribution. Therefore, one could expect differences between these models when the probabilities are close to 0 or 1. Notice that a straightforward interpretation of the coefficients through odds is not valid for the probit model. The probit model was initially developed by Bliss (1934) for dose-response models, and later made popular by Finney (1971). It is widely used in engineering and econometrics. A slight disadvantage of the link is that the calculation needs to be carried out numerically.

Both the probit and logit links are symmetric, which means that the results are not affected by how the two values of the binary response are coded to the binary response. This is not the case with the complementary log-log and log-log links, which are asymmetric: the lower tail of the complementary log-log link is heavy compared to the upper tail. The log-log and complementary log-log links are related: the upper/lower tail of the log-log link has a similar shape to the lower/upper tail of the complementary log-log link. Therefore, log-log models lead to results that are identical to the complementary log-log-model, provided that the coding of the binary response is switched. It can also be seen in Figure 8.3 that the lower tail of the complementary log-log link / upper tail of the log-log link is even heavier than the tail of the logit link.

The development of the models for binary data was originally motivated by biological assay studies. In those studies, different doses of a stressor (e.g. chemicals, temperature) are applied to a batch of subjects (animals, plants) and the binary outcome of each subject to the dose (e.g. dead/alive, sick/healthy) is recorded. It is justified to assume that the tolerance of individuals to the stressor varies in the population. Let random variable $U \sim N(\eta, \sigma^2)$ describe the tolerance of individuals to the stressor, so that the probability of success (e.g. death) at level $U = x_i$ of the dose is $\mu_i = \Phi(\frac{x_i - \eta}{\sigma})$. Denoting $\beta_1 = -\eta/\sigma$ and $\beta_2 = 1/\sigma$, we see that the normal tolerance distribution leads to the following model for success y_i:

$$\begin{aligned} y_i &\sim Bernoulli(\mu_i) \\ \text{probit}(\mu_i) &= \beta_1 + \beta_2 x_i \,, \end{aligned}$$

which is the probit model for binary data. Correspondingly, assuming logistic distribution for tolerance leads to the logistic regression model. Therefore, the link function in biological assays is related to the tolerance distribution in the underlying population. Often, the tolerance distribution is summarized as the median lethal dose (LD_{50}), the value of the stressor that leads to success (i.e. death or injury) for 50% of the subjects at given values of the predictors. In the simple logit and probit models, it is related to the regression coefficients through $LD_{50} = -\beta_1/\beta_2$. For a forestry-related discussion about hypothesis testing of LD_{50}, see Lappi and Luoranen (2018). For more details on biological assays, see Collett (1991). Demidenko (2013, p. 335) discusses the same issue in a more general context, where the binary response is considered as a categorized continuous latent variable.

8.2.3 Models for count data

The Poisson model can be used for count data without upper limit. The link function is a mapping $g : (0, \infty) \to (-\infty, \infty)$. The most commonly applied link function is the log link $g(\mu) = \log(\mu)$ (see Section 8.2.4), which is the canonical link for the Poisson distribution. Another alternative is the square root link $g(\mu) = \sqrt{\mu}$.

The Poisson regression with log link is specified as:

$$\begin{aligned} y_i &\sim Poisson(\mu_i) \\ \log(\mu_i) &= \boldsymbol{x}_i'\boldsymbol{\beta} \end{aligned}$$

so that $\mu_i = e^{\boldsymbol{x}_i'\boldsymbol{\beta}}$. This model is a special case of the *log-linear model.* To interpret the parameters of the Poisson model (or any log-linear model), consider the model where

$$\boldsymbol{x}_i'\boldsymbol{\beta} = \beta_1 + \beta_2 x_i \, .$$

Now

$$\mathrm{E}(y_i|x_i = 0) = e^{\beta_1} \, ,$$

i.e. e^{β_1} gives the expected count μ_i when $x_i = 0$. In addition, notice that

$$\begin{aligned} \mathrm{E}(y_i|x_i = x_0) &= e^{\beta_1 + \beta_2 x_0} \\ &= e^{\beta_1} e^{\beta_2 x_0} \\ \mathrm{E}(y_i|x_i = x_0 + 1) &= e^{\beta_1 + \beta_2 (x_0 + 1)} \\ &= e^{\beta_1} e^{\beta_2 x_0} e^{\beta_2} \end{aligned}$$

and

$$\frac{\mathrm{E}(y_i|x_i = x_0 + 1)}{\mathrm{E}(y_i|x_i = x_0)} = e^{\beta_2} \, .$$

Therefore, e^{β_2} gives the relative change in μ_i caused by a one-unit increase in x_i. In the log-linear models, it is implicitly assumed that this relative change is constant regardless of the value of μ. However, if β_2 is close to zero, one does not need to compute e^{β_2}, because $e^{\beta_2} \approx 1 + \beta_2$ (recall Figure 8.2 and Example 8.1). In that case, β_2 directly gives a good approximate of the relative change in μ caused by one-unit increase in x_i.

The Poisson model is useful provided the observed counts are small, especially if zero counts are present in the data. For count data with large counts, the Poisson model is still the best justified model although the linear model for untransformed counts could be applied as well, provided the variance function $\mathrm{var}(y_i) = \mathrm{E}(y_i)$ is assumed (recall Section 4.3); non-normality of the data is not a problem even with small data sets because the Poisson distribution is well approximated by the normal distribution if μ is large. Such a model formulation would not ensure non-negativity of expected counts through model formulation, but that is only a theoretical problem if the expected counts are far from zero.

Another alternative for count data with non-zero counts is modeling of log- or square root transformed counts in the context of linear model. Such a model would always lead to positive predicted counts, and the transformation could be selected so that it stabilizes the variance so that OLS-fitting is justified (see Box 8.1, p. 246). However, the predicted counts from such a model would be biased due to the back-transformation bias (see Section 10.2 for discussion about bias correction).

Example 8.3. The data set `ips` was analyzed as binary data in Example 8.2, thus only the presence of fungi on an individual bark beetle was analyzed. In this example, we directly analyze the fungal count per bark beetle. The maximum number of fungal species is large, and is actually unknown because new species can always be found in such analysis. The observed number of fungal species varies from 0 to 5, with an average 1.35 species per beetle.

We model the number of fungal species per bark beetle individual, y_i, using the Poisson GLM

$$y_i \sim Poisson(\mu_i)\,,$$

where

$$\mu_i = \exp\left(\beta_1 + \beta_2 S_i + \beta_3 F_i + \beta_4 m_i\right)\,.$$

The model is fitted below using function `glm`.

```
> ipsmod1<-glm(Fungi~Season+Mites,family=poisson,data=ips)
> ipsmod1
Coefficients:
 (Intercept)  SeasonSpring     SeasonFall           Mites
      0.8474       -0.8771        -1.2193          0.1202

Degrees of Freedom: 297 Total (i.e. Null);  294 Residual
Null Deviance:       409
Residual Deviance: 301.1    AIC: 818
> exp(coef(ipsmod1))
  (Intercept) SeasonSpring   SeasonFall         Mites
    2.3336821    0.4160034    0.2954431     1.1277395
```

The estimated regression coefficients show that the expected number of fungal species for a bark beetle with no mites was 2.33 in summer, $0.42 \times 2.33 = 0.97$ in spring, and $0.30 \times 2.33 = 0.69$ in fall. One additional mite was associated with a 13% increase in the expected number of fungal species per beetle. The same general conclusion about the effect of mites is obtained by interpreting the coefficient $\widehat{\beta}_4 = 0.12$ as the relative change associated with a one-unit increase in the number of mites.

8.2.4 GLM for the exponential family

The normal, Bernoulli, binomial and Poisson distributions are members of the *exponential family* of distributions. Other widely used univariate members of the family are the gamma, exponential, lognormal and two-parameter beta distribution. The GLMs (McCullagh and Nelder 1989) refer to a class of models that are developed for the members of the exponential family. We now generalize the model formulation of the previous sections to any member of the univariate exponential family. To do so, let us first formulate the exponential family and discuss its key properties.

The density of a univariate random variable of the exponential family can be stated in the following *canonical form* (McCulloch *et al.* 2008, Fahrmeir *et al.* 2013):

$$f(y|\gamma,\tau) = \exp\left(\frac{y\gamma - b(\gamma)}{\tau^2} - c(y,\tau)\right)\,. \tag{8.2}$$

The parameters of the canonical form are the canonical parameters, of which γ is related to the mean of the distribution and τ is related to the variance. Functions b and c are known fixed functions. Parameter γ is related to the expected value of y through the *canonical mean function*:

$$\mu = b'(\gamma)\,.$$

The inverse of this function is called the *canonical link function*. The variance of y is

$$\sigma^2 = \tau^2 b''(\gamma) .$$

For example, the normal density can be written as

$$f(y|\gamma, \tau) = \exp\left(\frac{y\gamma - \gamma^2/2}{\tau^2} - \frac{1}{2}\left(y^2/\tau^2 + \log(2\pi\tau^2)\right)\right)$$

where $\gamma = \mu$ and $\tau = \sigma$. Now $b(\gamma) = \gamma^2/2$ and therefore $\mu = b'(\gamma) = \gamma$ and $\sigma^2 = \tau^2 b''(\gamma) = \tau^2$.

The Bernoulli *pmf* (see Examples 2.2, p. 10 and 3.7, p. 54) can be written as (show this!)

$$f(y|\mu) = \exp\left(\log\left(\mu^y(1-\mu)^{1-y}\right)\right) = \exp\left(y\log\left(\frac{\mu}{1-\mu}\right) + \log(1-\mu)\right).$$

Defining $\gamma = \log\left(\frac{\mu}{1-\mu}\right)$, implies $\mu = \frac{e^\gamma}{1+e^\gamma}$ and $1 - \mu = (1 + e^\gamma)^{-1}$, giving

$$f(y|\mu) = \exp\left(y\gamma - \log(1 + e^\gamma)\right) .$$

This is of the form (8.2), where $b(\gamma) = \log(1 + e^\gamma)$, $\tau^2 = 1$, and $c(y, \tau) = 0$. The canonical mean function becomes

$$b'(\gamma) = \mu = \frac{e^\gamma}{1 + e^\gamma} .$$

Solving for γ gives the logit function, which is the canonical link function. The variance becomes

$$\sigma^2 = b''(\gamma) = \frac{e^\gamma}{(1 + e^\gamma)^2} = \frac{e^\gamma}{1 + e^\gamma}\frac{1}{1 + e^\gamma} = \mu(1 - \mu) ,$$

the variance of the Bernoulli distribution. Canonical link and mean functions for the most commonly used members of the exponential family are given in Table 8.1.

The models for binary, grouped binary, and count data are all special cases of the GLM of exponential family, which is specified as: (McCulloch *et al.* 2008)

$$\begin{aligned} y_i &\sim \exp\left(\frac{y_i\gamma_i - b(\gamma_i)}{\tau^2} - c(y_i, \tau)\right) \\ g(\mu_i) &= \boldsymbol{x}_i'\boldsymbol{\beta} . \end{aligned} \tag{8.3}$$

Usually, the canonical link function is used as $g(\mu_i)$, which implies that its inverse, the mean function, is $h(\boldsymbol{x}_i'\boldsymbol{\beta}) = b'(\boldsymbol{x}_i'\boldsymbol{\beta})$. The GLM formulations extend the ideas presented earlier for binary and count data to any member of the exponential family (see Table 8.1). For example, gamma distribution could be used to model a non-negative continuous response variable.

Notice that the specific distribution family, such as binomial or Poisson, is assumed conditional on predictors. The same assumption was also made in the context of linear models, as we discussed earlier in Section 4.1. The marginal distribution of the response over the predictors is not usually of the assumed family, because it is also affected by the values of the x-variables. Therefore, the marginal distribution of the observed y-values, which can be visualized, e.g. by the histogram of the response variable, does provide much information about which distribution should be assumed in the GLM.

8.2.5 Estimation

The starting point of parameter estimation of a GLM is based on the ML. The log-likelihood of a random variable that is distributed according to the exponential family is obtained from Equation (8.2) as

$$l = \frac{1}{\tau^2}\sum_{i=1}^{n}\left[y_i\gamma_i - b(\gamma_i)\right] - \sum_{i=1}^{n} c(y_i, \tau) .$$

Utilizing the properties of the exponential family (see, e.g. McCulloch *et al.* (2008, Section 5.4) for details), the first derivative of the log-likelihood (the score function) with respect to $\boldsymbol{\beta}$ becomes

$$\frac{\partial l}{\partial \boldsymbol{\beta}} = \frac{1}{\tau^2}\sum_{i=1}^{n}\frac{(y_i - \mu_i)h'(\gamma_i)}{v(\mu_i)}\boldsymbol{x}_i' ,$$

where $v(\mu_i) = \frac{1}{\tau^2}b''(\mu)$ is the variance function and $h'(\gamma_i)$ is the first derivative of the mean function with respect to γ_i. It can be written in matrix form as

$$\frac{\partial l}{\partial \boldsymbol{\beta}} = \frac{1}{\tau^2}\boldsymbol{X}'\boldsymbol{W}^{-1}\boldsymbol{\Delta}\left(\boldsymbol{y} - \boldsymbol{\mu}\right) , \tag{8.4}$$

where $\boldsymbol{X}_{n\times p}$ is the model matrix that has a similar structure to that used in linear models, $\boldsymbol{y}$ is the vector of the response variable, $\boldsymbol{\mu}$ is the vector of expected values $h(\boldsymbol{x}_i'\boldsymbol{\beta})$, $\boldsymbol{W}_{n\times n}$ is a diagonal matrix of weights $v(\mu_i)$, and $\boldsymbol{\Delta}_{n\times n}$ is a diagonal matrix of derivatives $h'(\mu_i)$. Matrices $\boldsymbol{W}$ and $\boldsymbol{\Delta}$, and vector $\boldsymbol{\mu}$ all include the unknown $\boldsymbol{\beta}$, which makes the solution complicated. Solving $\boldsymbol{\beta}$ is done iteratively. The most used approach is the Fisher scoring algorithm (see Fahrmeir *et al.* 2013, Section 5.4). The algorithm also provides an estimate of asymptotic var$\left(\widehat{\boldsymbol{\beta}}\right)$. The iteratively reweighted least squares (see p. 93) can be used as well (Nelder and Wedderburn 1972, McCullagh and Nelder 1989).

Interestingly, the assumed distribution function of the data only affects the score function through the variance function. Therefore, the estimation problem reduces to a nonlinear GLS problem, where the distribution of the data is taken into account through a justified selection of the variance function. For example, we assume var$(y_i) = \mu_i$ with count data and we assume var$(y_i) = \mu_i(1 - \mu_i)$ with Bernoulli data. It is also possible to formulate the model as a nonlinear regression model and use the algorithm that was described in Section 7.2.6 for a nonlinear regression model with general variance-covariance structure.

Example 8.4. In this example, we illustrate that the estimation of the GLM can be seen either as (1) an ML fit to the data, where μ depends on the predictors through the assumed link function, or (2) a nonlinear model estimated by generalized nonlinear least squares.

Function `nll` below defines the negative log-likelihood of the *Poisson* model that was assumed for variable `Fungi` of data set `ips`, and fitted using function `glm` in Example 8.3:

```
> nll<-function(b0,b1,b2,b3) {
+       cat(b0," ",b1," ",b2," ",b3," ")
+       value<--sum(log(dpois(ips$Fungi,
+             exp(b0+b1*(ips$Season=="Spring")+b2*(ips$Season=="Fall")+b3*ips$Mites))))
+       cat(value,"\n")
+       value
+       }
```

To find the MLE of the regression coefficients, we use function `mle` of package `stats4`. As a starting value for the intercept, we use the logarithmic mean of the response variable. For the other regression coefficients, we use zeroes.

```
> start<-list(log(mean(ips$Fungi)),0,0,0)
> names(start)<-c("b0","b1","b2","b3")
> library(stats4)
```

```
> ipsmodMLE<-mle(minuslogl=nll,start=start)
0.3018431   0   0   0  458.8991
0.3028431   0   0   0  458.8993
.
.
.
0.8474228   -0.8770349   -1.219207   0.120186  404.9787
0.8474228   -0.8770349   -1.219207   0.118186  404.9789
> ipsmodMLE
Coefficients:
        b0         b1          b2          b3
 0.8474228 -0.8770349 -1.2192073  0.1201860

> logLik(ipsmod1)
'log Lik.' -404.9787 (df=4)
> logLik(ipsmodMLE)
'log Lik.' -404.9787 (df=4)
```

The log-likelihoods for both models are equal to four decimal places. The resulting parameter estimates are very close to those of **ipsmod1** in Example 8.3 (p. 258).

Let us then consider the following nonlinear model

$$y_i = e^{\beta_1+\beta_2 S_i+\beta_3 F_i+\beta_4 m_i} + e_i$$

where

$$\text{var}(y_i) = \text{E}(y_i) = e^{\beta_1+\beta_2 S_i+\beta_3 F_i+\beta_4 m_i},$$

and residual errors e_i are uncorrelated and normally distributed. The parameters can be estimated using function **gnls** of package **nlme** (see Example 7.6, p. 222). To implement the assumed variance function $\text{var}(y_i) = E(y_i)$, we use the power-type variance function (4.13) (p. 83) but restrict the parameters to $\delta = 0.5$ and $\sigma^2 = 1$.

```
> ipsmodNLS<-gnls(Fungi~exp(theta),
+                 data=ips,
+                 params=list(theta~Season+Mites),
+                 start=start,
+                 weights=varPower(fixed=list(delta=0.5)),
+                 control=list(sigma=1))

> ipsmodNLS
Generalized nonlinear least squares fit
  Model: Fungi ~ exp(theta)
  Data: ips
  Log-likelihood: -435.4204

Coefficients:
 theta.(Intercept) theta.SeasonSpring   theta.SeasonFall          theta.Mites
         0.8474473         -0.8770684         -1.2192474            0.1201751

Variance function:
 Structure: Power of variance covariate
 Formula: ~fitted(.)
 Parameter estimates:
power
  0.5
Degrees of freedom: 298 total; 294 residual
Residual standard error: 1
```

The only difference between the nonlinear model and GLM is in the assumed distribution of the data. In models **ipsmod1** and **ipsmodMLE**, we assumed a Poisson distribution for the data, whereas in the nonlinear model we assumed normal distribution. That is why the value of likelihood is very different from the likelihoods in models **ipsmod1** and **ipsmodMLE**. However, the parameter estimates are equal up to the fourth decimal place.

The small differences in the parameter estimates between the models **ipsmod1**, **ipsmodMLE** and **ipsmodNLS** result from the differences in the convergence criteria and from the way the algorithm searches for the maximum of the likelihood function. If the estimates from **ipsmod1** were used as the starting value in the other two models, the resulting estimates would not differ at all between the methods. This demonstrates that all three methods lead to the same parameter estimates. The Fisher scoring algorithm implemented in **glm** is just a better algorithm for estimation than the (more general) algorithms used in `mle` and `gnls`. The equality of the `gnls` and `glm` solutions demonstrates that the mean-variance relationship should be based on the assumed random process, whereas the shape of the distribution is unimportant.

8.2.6 Evaluating the fit of a GLM

As with linear models, graphical evaluation of model fit is based on the residuals. The raw residuals of a GLM are defined as

$$\widehat{e}_i = y_i - h(\boldsymbol{x}_i'\widehat{\boldsymbol{\beta}}) .$$

However, these are not very useful. Scaling the raw residuals by the estimated standard deviation of y_i leads to the Pearson residuals

$$\widetilde{e}_i = \frac{\widehat{e}_i}{\sqrt{v(\widehat{\mu}_i)}} = \frac{\widehat{e}_i}{\sqrt{v(h(\boldsymbol{x}_i'\widehat{\boldsymbol{\beta}}))}} .$$

Definitions of raw and Pearson residuals are equivalent to the definitions in the context of linear and nonlinear models (recall Section 4.5.1).

Still another type of residual for non-normal responses are the *deviance residuals*, which have not been defined previously for linear and nonlinear models. Their definition starts from a *saturated model*: a model where μ_i is estimated separately for each observation of the data as $\widehat{\mu}_i = y_i$. Such a model has a maximally good fit to the data. The likelihood of that model for observation i is the assumed exponential family density with $\mu_i = y_i$, i.e. $f(y_i|y_i)$. The deviance residual of observation i, $\ddot{e}_i$, is based on the difference in the log-likelihood between the estimated model and the saturated model

$$\ddot{e}_i = \text{sign}(y_i - \widehat{\mu}_i)\sqrt{2\left(\log f(y_i|y_i) - \log f(y_i|\widehat{\mu}_i)\right)} ,$$

where $\text{sign}(y_i - \widehat{\mu}_i) = -1$ if the raw residual $y_i - \widehat{\mu}_i$ is negative and 1 otherwise. In R, the call of function `residuals` for a GLM model fit provides the deviance residuals by default.

Important numerical criteria to evaluate the goodness of fit are the *Pearson statistic*, which is the sum of squared Pearson residuals:

$$P = \sum_{i=1}^{n} \widetilde{e}_i^2 ,$$

and the *residual deviance*, which is the sum of squared deviance residuals:

$$D = \sum_{i=1}^{n} \ddot{e}_i^2 .$$

Both these statistics have an asymptotic $\chi^2(n-p)$-distribution if the model is correct.

If the model fits, the variance of the Pearson and deviance residuals should be close to $(n-p)/n \approx 1$. Therefore, P and D should be close to their expected value $n-p$. If the values of these statistics are far from $n-p$, then the model does not fit well to the data. Quite often these statistics are much higher than $n-p$, which indicates a lack of fit. Possible reasons are (1) missing predictors or poorly selected transformations of them, (2) a poor fit of the link function, and (3) outliers. If graphical evaluation does not indicate any of these problems, then the observations may be dependent, which is seen as *overdispersion*. Sometimes P and D are too small, which may indicate *underdispersion*. It is also possible that the response variable does not follow the assumed distribution, e.g. because of *zero-inflation*. These topics will be discussed later.

The raw, Pearson and deviance residuals could all be used for evaluation of the fit of systematic part of a fitted GLM. The local *mean of these residuals* should not show trends when plotted against fitted values or predictors. If that is not the case, the systematic part $\boldsymbol{x}_i'\boldsymbol{\beta}$ of the model should be updated. Everything that was discussed in Sections 4.2 and 4.5.2 about the formulation of the systematic part of a linear model also applies here.

The *variance of raw residuals* should behave according to the variance function suggested by the applied parent distribution. To explore whether the variance behaves as expected, one can check whether the Pearson or deviance residuals show constant variance when plotted against predictors and fitted values (either on $\boldsymbol{x}_i'\boldsymbol{\beta}$ or on $h(\boldsymbol{x}_i'\boldsymbol{\beta})$).

To explore whether the assumed distribution family fits to the data, one could explore the shape of the distribution. The deviance residuals can be used for that purpose. If the assumed distribution of y_i is supported by the data, then the deviance residuals should have an approximate normal distribution. However, the raw or Pearson residuals should not show normal distribution. The existence of potential outliers and influential observations can also be explored using similar procedures as we used with linear models, but instead using the deviance residuals.

When the fit of a GLM is evaluated using Pearson and deviance statistics, the null hypothesis is that the model fits adequately. The current model is the null model that is compared to the full saturated model, which fits the data as closely as possible. For a well-fitting model, the difference between it and the saturated model should be small. Therefore, the statistic should be small and the p-value from the hypothesis test should be large. This is opposite to the linear model, where the null model is the simplest possible linear model, the current model presents the alternative hypothesis, and a large value of the test statistic and a small p-value indicate that the estimated model deviates significantly from the null model. Therefore, a high p-value from a test that evaluates the fit of the model is "good" for a GLM, whereas a low p-value from corresponding test is "good" in a linear model. This may sometimes cause confusion in the analysis.

Note that the above-described evaluation of goodness of fit cannot be done with the linear model, because normal distribution has separate parameters for expected value and variance. Therefore, a fully saturated model cannot be defined, and deviance residuals cannot be computed. Furthermore, the Pearson residuals from a linear model always have the variance of 1, regardless of the fit of the model. In the linear model, the tests are made with the assumption that the full model is correctly specified. However, there is generally no way to test whether the estimated model fits well or not. Only when there are several observations with the same x values, can one test whether the lack-of-fit of the assumed regression function is significant (see Section 4.6.1, p. 115). The same problem occurs with GLM if over- or underdispersion is treated with a new parameter (see the next section).

Example 8.5. The fitted values and raw, Pearson, and deviance residuals for model `ipsmod1` are computed below. Of course, a more convenient way to compute them would be to use functions `fitted` and `residuals` with an appropriately selected value for argument `type`, but we compute them manually to illustrate the underlying procedure. The left-hand graphs in Figure 8.4 show the raw, Pearson and deviance residuals against fitted values. As expected, we see that the variability of raw residuals increases as a function of fitted values. However, the variability of Pearson and deviance residuals decreases, which indicates that the variance does not increase as much as a function of fitted values as one would expect based on the Poisson assumption. An explanation for this might be that the possible number of available fungal species is limited, or a bark beetle individual cannot carry too many fungal species.

The deviance residuals plots with respect to season and number of mites do not show trends in the variance. The normal qq-plot of deviance residuals shows that the left tail is slightly too

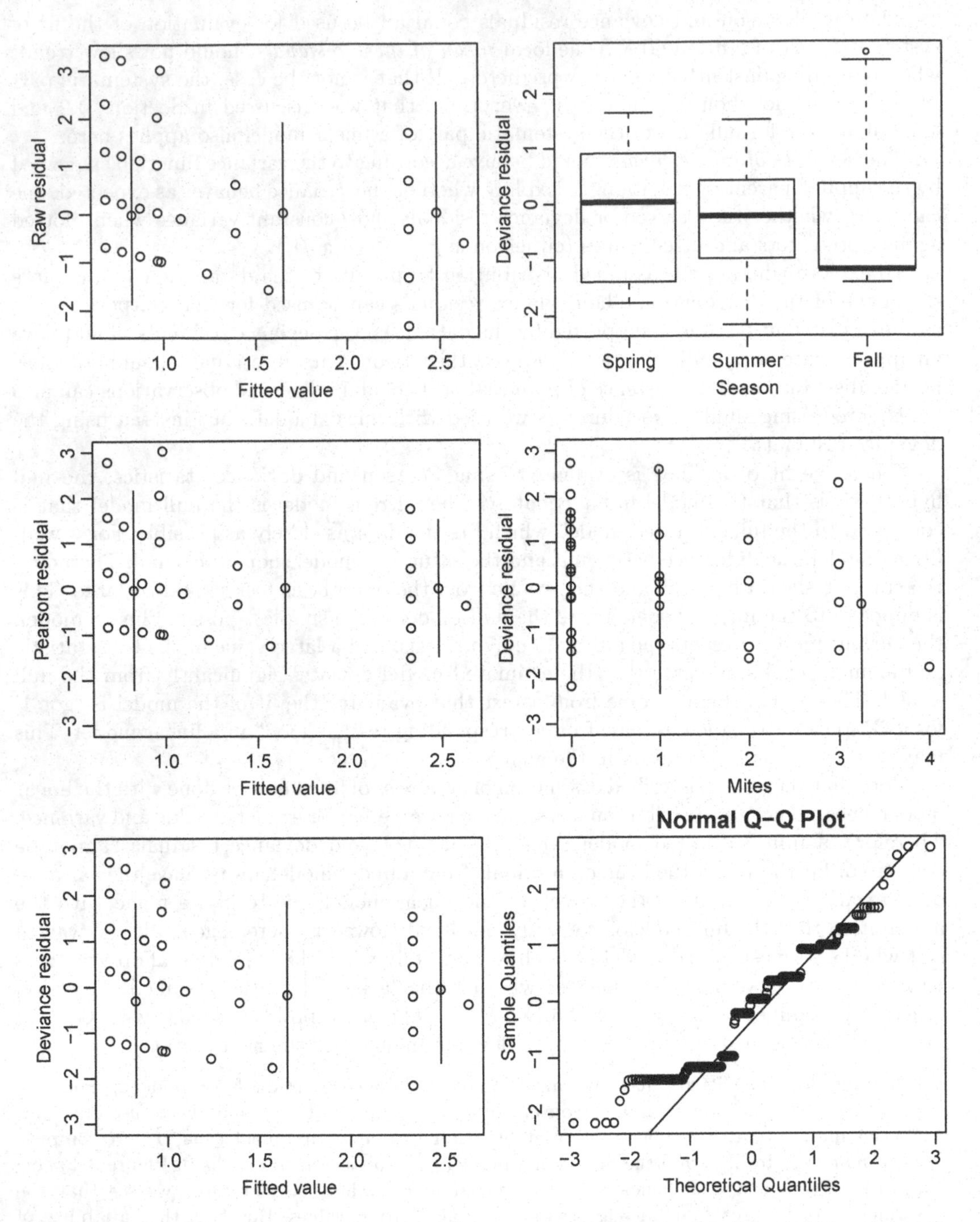

FIGURE 8.4 Diagnostic plots of model `ipsmod1` in Example 8.5. The left-hand graphs show the raw, Pearson and deviance residuals against fitted values. The upper two right-hand graphs show the deviance residuals against the two predictors, and the lower graph shows the normal q-q plot of the deviance residuals.

light, which indicates that the data may have slightly too many zeroes compared to the Poisson assumption.

```
> X<-cbind(1,ips$Season=="Spring",ips$Season=="Fall",ips$Mites)
> muhat<-exp(X%*%coef(ipsmod1))
> # or use ips$muhat<-fitted(ipsmod1)
> eraw<-ips$Fungi-muhat
> ePea<-(ips$Fungi-muhat)/sqrt(muhat)
> d<--2*(log(dpois(ips$Fungi,muhat))-log(dpois(ips$Fungi,ips$Fungi)))
> edev<-sign(ips$Fungi-muhat)*sqrt(d)
```

The Pearson and residual deviance statistics are computed manually below. Note that the residual deviance was already reported in the default output of the model in Example 8.3. Both statistics are fairly close to $n - p = 298 - 4 = 294$, thereby indicating a good fit of the model.

```
> sum(ePea^2)
[1] 273.2903
> sum(edev^2) # or use deviance(ipsmod1)
[1] 301.1184
```

8.2.7 Over- and underdispersion

The mean and variance in the Poisson, binomial and Bernoulli distributions are related, as we have discussed in Section 8.1. Therefore, the model for $\mathrm{E}(y_i|\boldsymbol{x}_i)$ simultaneously expresses the model for $\mathrm{var}(y_i|\boldsymbol{x}_i)$. A very common problem in the context of count and binary data is that these relationships between conditional mean and conditional variance do not hold, even though a graphical evaluation indicates that the systematic part is properly formulated and that the Pearson or deviance residuals show constant variability with respect to fitted values and predictors. This phenomenon is called overdispersion (conditional variance is too high compared to the theoretical value) or underdispersion (conditional variance is too low). This means that the Bernoulli and Poisson assumptions do not hold. Common reasons for such departures of assumptions are

1. Lack of independence. Subjects in count data may enter the line or the area in groups (causing overdispersion) or in a systematic fashion with respect to time or space (causing underdispersion). In general, there are interactions among them (see Example 8.6). If the subjects belong to groups and the grouping factor is known, then the dependence can be modeled explicitly using GLMMs. If the order with respect to time, or spatial locations, are known, then more sophisticated models could be used to explicitly model the spatio-temporal dependence structure. However, if grouping, order, or locations are not known, or if none of these is the factor behind the overdispersion, then the only option is to model the dependence as overdispersion.

2. Some unknown factors affect the mean of y_i, so that the Poisson or Bernoulli assumption for y_i, conditional on the predictors, does not hold. This means that some predictors are missing and the function $h(\boldsymbol{x}_i'\boldsymbol{\beta})$ does not give the common conditional mean for a given value of predictors.

The linear predictor $\boldsymbol{x}_i'\boldsymbol{\beta}$ describes both the mean and variance of y_i. If the assumed relationship between the expected value and the variance does not hold, the model will describe $\mathrm{E}(y_i)$ well but describe $\mathrm{var}(y_i)$ poorly. That is, the estimation method "prioritizes" a good description of the mean at the cost of poorly describing the variance. Therefore, $\mathrm{E}(y_i)$, which is usually of primary interest, is well described even if overdispersion is ignored. However, ignoring overdispersion leads to severe problems in estimation of the variances of regression coefficients and in the inference on the model, resulting in anti-conservative tests and too narrow confidence and prediction intervals. Correspondingly, if underdispersion is ignored, the tests are conservative, and the confidence intervals are too wide. Because bias in an anti-conservative direction is usually regarded as a more serious problem than bias in a conservative direction, overdispersion is regarded as a more severe problem than underdispersion. It is also more commonly found in empirical data.

Example 8.6. Simulated tree locations in three different hypothetical one-hectare stands with a true density of 1000 trees per hectare are shown in Figure 8.5. In the random point pattern (left), x- and y-coordinates of the trees are drawn independently from the $Uniform(0, 100)$ distribution, resulting in a point pattern where tree locations do not interact; such a pattern is called complete spatial randomness. In the regular pattern (middle), potential tree locations were also drawn from the uniform distribution, but the simulation was conducted sequentially, and the potential new location was accepted only if it was located more than 2 m away from an existing (earlier) point. Such a regular (or systematic) point pattern is called a *hard-core pattern*. In the clustered pattern (right), the new potential location was accepted whenever it was located within a 3 m distance from an existing (earlier) point. If it was more than 3 m apart from an existing point, it was accepted only with the probability 0.1. These three patterns are just examples of models that produce regular and clustered patterns. For more details about possible point process models, see Illian *et al.* (2008).

Each simulated plot was divided into 400 non-overlapping 5m×5m subplots, and the number of trees in each of these plots was counted. This resulted in 400 observed identically distributed counts with a mean of 2.5 trees per plot. The observed probability mass functions of these counts are shown in the lower panel of Figure 8.5 together with the $Poisson(2.5)$ *pmf*. In the plot with a random pattern, observed counts nicely follow the Poisson distribution. In the plot with a regular pattern, the distribution of observed counts is clearly narrower than the Poisson distribution, indicating lower variance among the observed counts than the Poisson model would suggest. In the clustered pattern, the situation is the opposite: the distribution of observed counts is clearly wider, and the variance is larger, than in the Poisson case. These illustrate that the regular pattern of events (in time or space) would lead to underdispersion in the count data. Correspondingly, a clustered pattern would lead to overdispersion.

Note that if the total number of trees on the plot was fixed exactly to 1000 trees, the counts in the subplots would not be completely independent Poisson variables. To ensure that the counts in the random point pattern are independent Poisson subplot counts, the total number of trees on the plot was generated using the $Poisson(1000)$ distribution.

There are several ways to take the overdispersion into account in model fitting:

1. Quasi-likelihood fitting.
2. Switching to a generalized linear mixed-effects model where the random effects are allowed in the linear predictor, and
3. Using a model where μ_i is treated as a random variable.

The maximum quasi-likelihood method is probably the most used method to model overdispersion. In this method, the value of τ^2 in the score function (8.4) (p. 260) is not restricted to the value supported by the assumed distribution family. The assumed distribution of y_i is not exactly specified, only the variance-mean relationship is specified up to the scaling constant τ^2 using the assumed distribution family. The resulting point estimates of $\boldsymbol{\beta}$ are the same as with ML, but the estimated standard errors of parameters are adjusted according to the overdispersion parameter τ^2, which is estimated as the following function of Pearson residuals (compare to Equation (4.27))

$$\widehat{\tau}^2 = \frac{1}{n-p} \sum_{i=1}^{n} \widetilde{e}_i^2 .$$

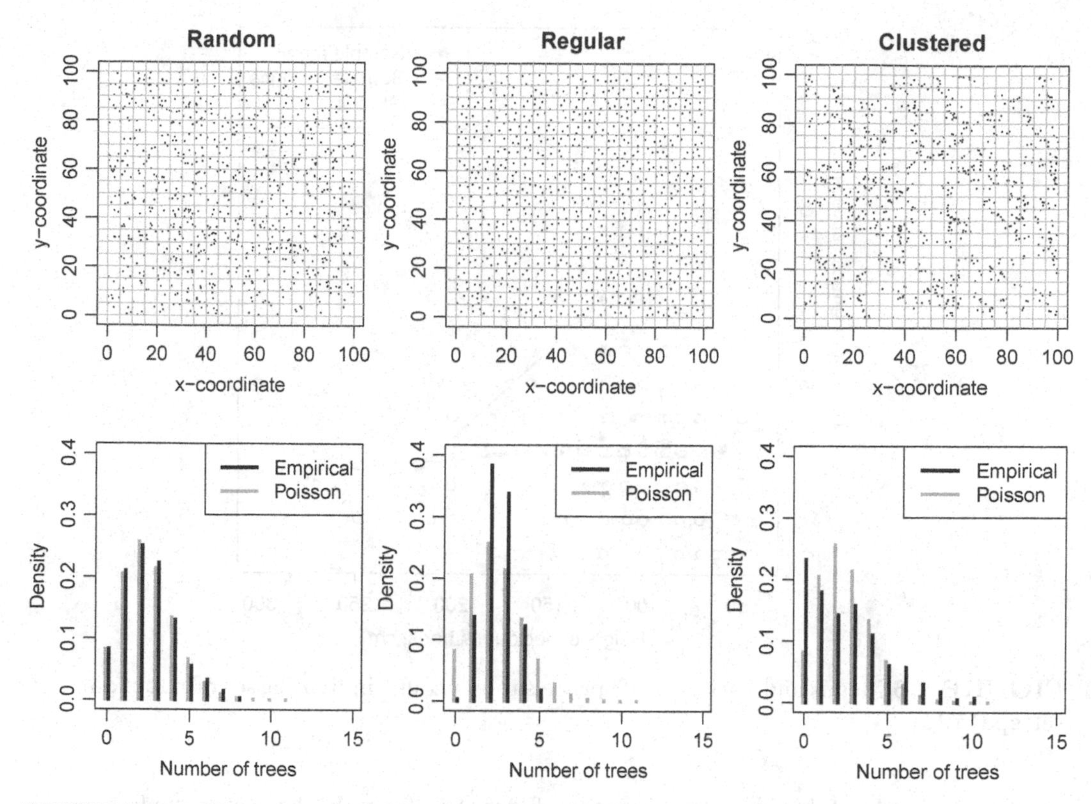

FIGURE 8.5 Illustration of the connection of point pattern to the distribution of count data in Example 8.6.

In the second approach, random effects can be included at the group level or for each observation. The resulting model is a generalized linear mixed-effects model, which is discussed in detail in Section 8.3. If random effects are assumed at the observation level, the linear predictor is specified as $\boldsymbol{x}_i'\boldsymbol{\beta} + e_i$, where $e_i \sim N(0, \sigma^2)$ takes a unique value for each observation. Interestingly, this model is identifiable for some families, including Poisson and Bernoulli.

The third approach uses the idea of compound distributions, which were discussed in Section 2.9. For binary data, we could assume that the success probability of the Bernoulli distribution for observation i is drawn from the beta distribution. Therefore, the probability may differ for two observations with the same values of predictors. The assumption leads to the beta-binomial model. For count data, we could assume that the expected value of the Poisson distribution follows the gamma distribution, which leads to negative binomial distribution for the counts (see Example 8.7). The Bernoulli and Poisson distributions are obtained as the limit when the variances of the beta and gamma distributions approach zero. The beta-binomial distribution does not belong to the exponential family, nor does the negative binomial distribution in its general form.

Example 8.7. Data set `plants2` includes the observed number of spruce saplings in a total of 123 sample plots. Each of the plots has been taken from a different regeneration area, so that they can be taken as independent counts. We analyze the sum of planted and natural spruce saplings per plot. The spruce saplings are competing for resources with faster-growing deciduous trees. Therefore, we explore the association between the total number of spruce saplings (`spruces`) and

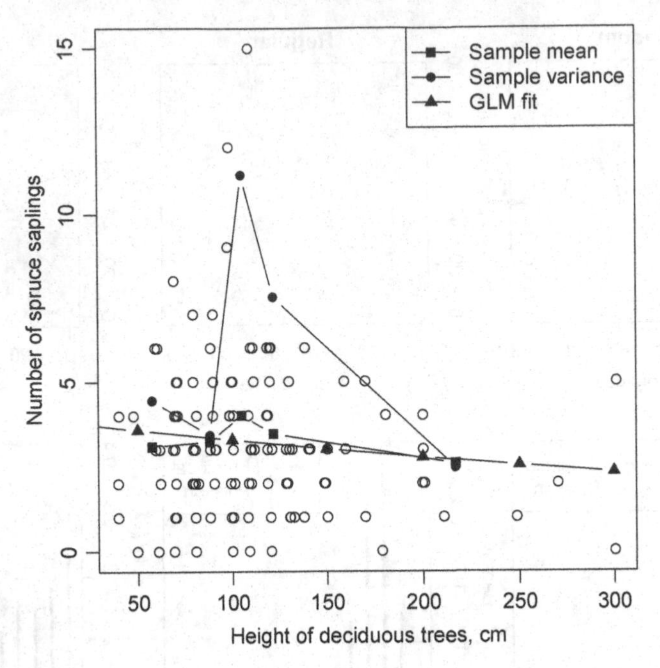

FIGURE 8.6 Means and variances of spruce sapling count in five classes of `hdecid` in data set `spati2`.

the mean height of deciduous trees (`hdecid`). Figure 8.6 shows the data set; a small amount of random noise has been added to deciduous tree heights to indicate whether several plots have exactly the same values for variables `spruces` and `hdecid`. Sample mean and sample variance of spruce sapling count in five classes of deciduous tree height are also shown. The variance is mostly greater than the mean, which indicates overdispersion. In our case, this might be associated with a clustered spatial pattern of the (natural) spruce saplings, or with missing regeneration-area level predictors, such as planting density. Overdispersion is also suggested by the deviance and Pearson statistics, which are approximately 1.67 times and 1.71 times greater than $n - p$ ($D = 187.46/(114 - 2) = 1.67$ and $P = 191.52/(114 - 2) = 1.71$):

```
> plants2$spruces<-plants2$planted+plants2$spruces #include also planted spruces
> glm1 <- glm(spruces ~ hdecid, family=poisson(), data=plants2)
> deviance(glm1)
[1] 187.4611
> sum(resid(glm1,type="pearson")^2)
[1] 191.5158
```

Therefore, we re-estimate the model using maximum quasi-likelihood

```
> glm2 <- glm(spruces ~ hdecid, family=quasipoisson(), data=plants2)
> summary(glm2)
Coefficients:
             Estimate Std. Error t value Pr(>|t|)
(Intercept)  1.360962   0.176684   7.703 5.77e-12 ***
hdecid      -0.001739   0.001503  -1.157     0.25

(Dispersion parameter for quasipoisson family taken to be 1.709962)

    Null deviance: 189.85  on 113  degrees of freedom
Residual deviance: 187.46  on 112  degrees of freedom
AIC: NA

Number of Fisher Scoring iterations: 5
```

If overdispersion had not been taken into account, the parameter estimates would have been the same, but the estimated standard errors would have been much smaller and, therefore, also the p-values would have been much smaller:

```
> summary(glm1)
Coefficients:
              Estimate Std. Error z value Pr(>|z|)
(Intercept)  1.360962   0.135115  10.073   <2e-16 ***
hdecid      -0.001739   0.001149  -1.513     0.13
```

Another approach to model overdispersion in count data is to use the negative binomial model. Assume that there is a latent random variable l_i such that

$$\mu_G l_i \sim Gamma(\mu_G, \theta)\,,$$

where μ_G and θ are the expected value and shape parameter of the gamma distribution, respectively. Now l_i has the expected value 1 and variance $1/\theta$. Assumption

$$y_i|l_i \sim Poisson(\mu_i l_i)$$

leads to the negative binomial marginal distribution (compare with Example 2.24, p. 35), which has the *pdf*

$$f(y_i|\mu_i, \theta) = \frac{\Gamma(\theta + y_i)}{\Gamma(\theta) y_i!} \frac{\mu_i^{y_i} \theta^\theta}{(\mu_i + \theta)^{\theta + y_i}}, \qquad \mu_i, \theta > 0\,.$$

Now, $\mathrm{E}(y_i) = \mu_i$, $\mathrm{var}(y_i) = \mu_i + \mu_i^2/\theta$. The log link function is commonly used, therefore $\mu_i = e^{\boldsymbol{x}'_i \boldsymbol{\beta}}$. As $\theta \to \infty$, the distribution approaches the Poisson distribution. Because any other value of θ (only positive values of the inverted variance parameter θ are allowed) would imply a greater variance, the negative binomial model cannot be used for modeling underdispersion. In R, function **glm** requires parameter θ be specified by the user. However, function **MASS::glm.nb** also allows estimation of that parameter (Venables and Ripley 2002, p. 206-208). Notice that if θ is zero, then the variance is infinite. Thus, the null hypothesis $H_0 : \theta = 0$ is not justified, even though some estimation procedures do such a test. In some cases, the negative binomial distribution is parameterized in terms of $1/\theta$.

```
> library(MASS)
> glm3 <- glm.nb(spruces ~ hdecid, data=plants2)
> summary(glm3)
Coefficients:
             Estimate Std. Error z value Pr(>|z|)
(Intercept)  1.369600   0.168071   8.149 3.67e-16 ***

(Dispersion parameter for Negative Binomial(5.6665) family taken to be 1)

    Null deviance: 125.44  on 113  degrees of freedom
Residual deviance: 123.83  on 112  degrees of freedom
AIC: 490.97

Number of Fisher Scoring iterations: 1

              Theta:  5.67
          Std. Err.:  2.10

 2 x log-likelihood:  -484.973
```

As we assumed a different model to the model used previously, the estimates of $\boldsymbol{\beta}$ were also changed. However, their interpretation is similar to that in the Poisson GLM because the same link function was used. The estimated variance function is the second-order polynomial:

$$\widehat{\mathrm{var}}(y_i) = \widetilde{y}_i + 0.176 \widetilde{y}_i^2\,,$$

where $\widetilde{y}_i = e^{\boldsymbol{x}'_i \widehat{\boldsymbol{\beta}}}$ and the coefficient is based on $1/5.6665 = 0.176$. The residual deviance is even less than the number of observations, which indicates that the overdispersion of the Poisson model was sufficiently modeled. If one, hereafter, would like to use inferential tools that are not available for **glm.nb**, one can refit the model using **glm**:

```
> glm3b <- glm(spruces ~ hdecid, family=negative.binomial(5.6665), data=plants2)
```

Finally, recall from Example 8.4 (p. 260) that we could see the Poisson GLM as a nonlinear regression model:

$$y_i = e^{\beta_1 + \beta_2 hdecid_i} + e_i$$

where $\mathrm{var}(y_i) = \mathrm{E}(y_i)$ and the residual errors e_i are uncorrelated and normally distributed. The quasi-likelihood method for estimation is equivalent to changing our assumption about the error variance from $\mathrm{var}(y_i) = \mathrm{E}(y_i)$ to $\mathrm{var}(y_i) = \sigma^2\,\mathrm{E}(y_i)$, where σ^2 is the overdispersion parameter to be estimated. Such a model is fitted below. When the estimates from **glm2** are used as initial guesses, the estimates of $\boldsymbol{\beta}$, their standard errors and p-values are exactly the same as in the fit using **glm**. However, if other starting values had been used, the fitting algorithm of **gnls** would not have found this exact solution but would have instead found a solution that is slightly worse in terms of likelihood. The residual variance of the **gnls**-model can now be interpreted as the estimated overdispersion parameter, and it is also equal to the estimate of model **glm2**. The only benefit from the use of **glm2** is better numerical accuracy of the estimates.

```
> glm4<-gnls(spruces~exp(theta),
+            data=plants2[,c("spruces","hdecid")],
+            params=list(theta~hdecid),
+            start=coef(glm2),
+ #          start=c(log(mean(plants2$spruces)),0),
+            weights=varPower(fixed=list(delta=0.5)))
> summary(glm4)
Generalized nonlinear least squares fit
  Model: spruces ~ exp(theta)
  Data: plants2[, c("spruces", "hdecid")]
       AIC      BIC    logLik
  521.5543 529.7629 -257.7771

Variance function:
 Structure: Power of variance covariate
 Formula: ~fitted(.)
 Parameter estimates:
power
  0.5

Coefficients:
                        Value  Std.Error   t-value p-value
theta.(Intercept)   1.3609621 0.17668406  7.702801  0.0000
theta.hdecid       -0.0017385 0.00150273 -1.156926  0.2498

> glm4$sigma^2
[1] 1.709962
```

8.2.8 Inference on GLM

The concept of a hypothesis test in GLM's is similar to what has been discussed in the previous sections. One commonly tests the following hypotheses:

$$H_0 : \boldsymbol{L}\boldsymbol{\beta}_F = \boldsymbol{k}$$
$$H_1 : \boldsymbol{L}\boldsymbol{\beta}_F \neq \boldsymbol{k}\,.$$

There are three possible alternatives for tests: the LR test, Wald test, and Rao's score test. In the context of a linear OLS model, all these tests lead to identical results, and the test is exact for normally distributed and asymptotic for non-normal data, as we have discussed in Section 4.6.3. In the GLM context, the situation gets more complicated. First, these three tests are not necessarily identical. Second, they are asymptotic, even when the assumption of the distribution is met.

The LR test is based on comparison of two models: the full model with p parameters and likelihood $l(\widehat{\boldsymbol{\beta}}_F)$, and the constrained model with q parameters and likelihood $l(\widehat{\boldsymbol{\beta}}_C)$ under the $r = p - q$ constraints posed by H_0. The test statistic is

$$LR = 2[l(\widehat{\boldsymbol{\beta}}_F) - l(\widehat{\boldsymbol{\beta}}_C)]$$

which has an asymptotic χ^2-distribution with r degrees of freedom, if H_0 is true.

The Wald test is based on the estimated variance-covariance matrix of the regression coefficients $\widehat{\text{var}}(\widehat{\boldsymbol{\beta}}_F)$, which is obtained as an inverted Fisher's information matrix at $\boldsymbol{\beta} = \widehat{\boldsymbol{\beta}}_F$. The test statistic is

$$w = (\boldsymbol{L}\boldsymbol{\beta}_F - \boldsymbol{k})' \, \boldsymbol{L}\widehat{\text{var}}(\widehat{\boldsymbol{\beta}}_F)\boldsymbol{L}' \, (\boldsymbol{L}\boldsymbol{\beta}_F - \boldsymbol{k}) \, .$$

Also, the Wald's statistic has an asymptotic $\chi^2(r)$-distribution, if H_0 is true. Because the Wald test does not require estimation of a restricted model, it is computationally simpler than the LR-test. However, according to McCulloch *et al.* (2008), it may lead to an inferior approximation of the p-value with moderate or small sample sizes.

The third alternative, Rao's score test, is based on the score function $\boldsymbol{s}(\boldsymbol{\beta}) = \frac{\partial l}{\partial \boldsymbol{\beta}}$ (Equation (8.4), p. 260). The ML estimate of the full model is found as a solution to $\boldsymbol{s}(\boldsymbol{\beta}) = \boldsymbol{0}$, therefore the value of the score function at $\widehat{\boldsymbol{\beta}}_F$ is zero and $\boldsymbol{s}(\boldsymbol{\beta}_C) - \boldsymbol{s}(\boldsymbol{\beta}_F) = \boldsymbol{s}(\boldsymbol{\beta}_C)$. The test statistic is the weighted sum

$$u = \boldsymbol{s}'(\widehat{\boldsymbol{\beta}}_C)\widehat{\text{var}}(\widehat{\boldsymbol{\beta}}_C)\boldsymbol{s}(\widehat{\boldsymbol{\beta}}_C) \, .$$

Also, u has an asymptotic $\chi^2(r)$-distribution, if H_0 is true.

For models where the overdispersion parameter has been estimated using the maximum quasi-likelihood, one can use the Wald F-statistic,

$$F_{obs} = \frac{w}{r} \, ,$$

which also takes the estimation error of the overdispersion parameter into account. The statistic is the same approximation of the test statistic (4.35) (p. 114) that was used with nonlinear models. The test is based on the approximate, asymptotic $F(r, n-p)$-distribution of the test statistic, if H_0 is true.

As with the linear model, testing hypotheses in practice involves carrying out several tests using the marginal and sequential procedures (recall Section 4.6.4). Many tests reported by the automatic procedures are useless, and all the tests that might be of interest are not included there. In R, general linear hypotheses of GLM can be tested, for example, using functions `anova`, `car::Anova`, `car::lht`, and the functions of package `multcomp`. The following example demonstrates the use of the three tests with the models fitted in the previous examples.

Example 8.8. Consider testing whether the number of mites, m_i, is significantly associated with the number of fungi in the model fitted in Example 8.3. The LR test is most conveniently carried out by using the two-argument form `anova`, after first estimating the null model `ipsmod1b`:

```
> ipsmod1b<-update(ipsmod1,formula = Fungi ~ Season)
> anova(ipsmod1b,ipsmod1,test="LRT") # You could use test="Chisq" as well
Analysis of Deviance Table

Model 1: Fungi ~ Season
Model 2: Fungi ~ Season + Mites
  Resid. Df Resid. Dev Df Deviance Pr(>Chi)
1       295     302.28
2       294     301.12  1   1.1578   0.2819
```

The test indicates that the number of mites is not a statistically significant predictor in the model.

The Wald χ^2-test can be carried out using `car::Anova`. We use a type III procedure, i.e. drop each predictor in turn from the full model.

```
> library(car)
> Anova(ipsmod1,test="Wald",type="III")
Analysis of Deviance Table (Type III tests)
Response: Fungi
```

```
            Df    Chisq Pr(>Chisq)
(Intercept)  1 169.4311     <2e-16 ***
Season       2 101.6498     <2e-16 ***
Mites        1   1.2754     0.2588
```

The p-value is slightly different from that based on the LR-test, but the conclusion does not change. The Wald test results based on marginal sums of squares are also produced by the model summary.

```
> summary(ipsmod1)
             Estimate Std. Error z value Pr(>|z|)
(Intercept)   0.84745    0.06511  13.017  < 2e-16 ***
SeasonSpring -0.87706    0.12253  -7.158 8.21e-13 ***
SeasonFall   -1.21928    0.13813  -8.827  < 2e-16 ***
Mites         0.12022    0.10645   1.129    0.259
```

Rao's score test can be carried out using **anova** and gives a result that differs slightly from the previous two:

```
> anova(ipsmod1,ipsmod1b,test="Rao")
Analysis of Deviance Table

Model 1: Fungi ~ Season + Mites
Model 2: Fungi ~ Season
  Resid. Df Resid. Dev Df Deviance     Rao Pr(>Chi)
1       294     301.12
2       295     302.28 -1  -1.1578 -1.2848    0.257
```

More sophisticated tests can be done using function **car::lht**. For example, we might want to test whether the number of fungi differs significantly between spring and fall using the LR test. This is carried out using the following hypothesis matrix

$$\boldsymbol{L} = \begin{pmatrix} 0 & 1 & -1 & 0 \end{pmatrix}$$

in a call to **car::lht**.

```
> lht(ipsmod1,hypothesis.matrix=c(0,1,-1,0),test="Chisq")
Linear hypothesis test
Hypothesis:
SeasonSpring - SeasonFall = 0

Model 1: restricted model
Model 2: Fungi ~ Season + Mites

  Res.Df Df  Chisq Pr(>Chisq)
1    295
2    294  1 4.7659    0.02903 *
```

The difference between the two seasons, $0.877 - 1.219 = -0.342$ is statistically significantly. There are less fungal species carried by the bark beetles in the fall than in the spring. One could also test whether the number of fungi in spring and fall differ significantly from the summer, i.e. whether there are some differences among seasons.

```
> lht(ipsmod1,hypothesis.matrix=rbind(c(0,1,0,0),c(0,0,1,0)),test="Chisq")
Hypothesis:
SeasonSpring = 0
SeasonFall = 0

Model 1: restricted model
Model 2: Fungi ~ Season + Mites

  Res.Df Df  Chisq Pr(>Chisq)
1    296
2    294  2 101.65  < 2.2e-16 ***
```

The same test was already carried out above using **car::Anova**. Also command **anova (ipsmod1,ipsmod1c)**, where **ipsmod1c** includes only mites as the predictor, would do the same test. Command **anova(ipsmod1,test="lrt")** would not do the same test, however, because it is based on a sequential procedure, but **anova** for a model where the order of predictors is changed would do the same test.

If maximum quasi-likelihood is used in fitting, the Wald's F-test can be caried out using **anova** and **car::Anova**. We illustrate the test using models from Example 8.7.

```
> anova(glm2,test="F")
Analysis of Deviance Table
Model: quasipoisson, link: log
Response: spruces
Terms added sequentially (first to last)
       Df Deviance Resid. Df Resid. Dev      F Pr(>F)
NULL                     113     189.85
hdecid  1    2.392       112     187.46 1.3989 0.2394
```

The p-value is very different from the value we would get if overdispersion was not taken into account:

```
> anova(glm1,test="Chisq")
hdecid  1    2.392       112     187.46    0.122
```

This illustrates the importance of taking overdispersion into account.

Example 8.9. One of the very first tests covered in elementary statistical courses is the χ^2-test of independence for cross-tabulated data. In the table, each cell includes counts (the number of observations in the cell). The test statistic compares the observed counts to the expected counts, which are based on the observed marginal counts. As an example, consider the following table of **ips** data that classifies the individual beetles as either those who have (one or more) mites and those who do not have mites, separately for each of the three seasons.

```
> ipstab
        MitesB
Season   FALSE TRUE
  Spring    92    5
  Summer   100    1
  Fall      86   14
```

The χ^2-test of independence for this table would analyze whether the existence of mites and season are independent. The test can be done in **R** as follows:

```
> chisq.test(ipstab, correct=F)
        Pearson's Chi-squared test

data:  ipstab
X-squared = 14.141, df = 2, p-value = 0.0008499
```

We would reject the null hypothesis about independence and conclude that the existence of mites and season are dependent.

The table shown above represents counts and, therefore, one might analyze it as count data using a Poisson GLM. For that purpose, the data set is reorganized as

```
> ips2
  Season Mites Beetles
1 Spring    NO      92
2 Summer    NO     100
3   Fall    NO      86
4 Spring   YES       5
5 Summer   YES       1
6   Fall   YES      14
```

to fit a Poisson GLM of form

$$\mathrm{E}(B_i) = \exp(\beta_1 + \beta_2 M_i + \beta_3 S_i + \beta_4 F_i + \beta_5 M_i S_i + \beta_6 M_i F_i)\,,$$

where B_i is the number of beetles in cell i and M_i, S_i, and F_i are binary predictors for the existence of mites, spring and fall, respectively. The model includes as many parameters as there are observations, so it is almost overparameterized. The test for hypothesis $H_0 : \beta_5 = \beta_6 = 0$ tests the independence of the two predictors. If the test is based on Rao's score test, the test statistics and p-value are identical to the Pearson's χ^2-test of independence.

```
> modPois<-glm(Beetles~Mites+Season+Mites*Season,family=poisson,data=ips2)
> modPoisNul<-update(modPois,formula=Beetles~Mites+Season)
> anova(modPoisNul,modPois,test="Rao")
Analysis of Deviance Table

Model 1: Beetles ~ Mites + Season
Model 2: Beetles ~ Mites + Season + Mites * Season
  Resid. Df Resid. Dev Df Deviance    Rao  Pr(>Chi)
1         2     15.078
2         0      0.000  2   15.078 14.141 0.0008499 ***
```

The formulation of the χ^2-test of independence as a Poisson GLM also allows analysis of the causes for the lack of independence. For example, we could further analyze whether there is a statistically significant difference between spring and summer by testing $H_0 : \beta_5 = 0$, or between spring and fall by testing $H_0 : \beta_5 = \beta_6$.

8.2.9 Zero-inflation

In addition to overdispersion, a problem that commonly occurs with count data is *zero-inflation*. In zero-inflated data, the proportion of zeroes in the data is much greater than one would expect based on the Poisson or negative binomial assumptions.

Zero-inflation commonly exists in data sets that are based on a two-step data-generating process. There might, for example, be a binary process that determines whether a certain willow leaf has rust infection and another for the number of rust spots on an infected leaf. To model such data, one could adjust the Poisson distribution by allowing more zeroes than one would expect based on the Poisson assumption. In practice, one needs to specify two predictor functions: one for occurrence (of the rust infection) and another for the rust spot count. If the expected count for infected leaves is so great that the probability of having zero spots on an infected leaf is practically zero, then a satisfactory solution might be fitting two different models: one for the presence of infection (using all leaves) and another for the number of spots on infected leaves.

A proportion of the zeroes is usually associated with the zero-inflation and a proportion of them are zeroes from the Poisson process. Thus, we need two linked submodels for each observation; a model component for the extra zeroes (e.g. a binary model with logit link) and a (parent) model component for the expected number of counts of the main process (e.g. a Poisson or negative binomial model with a log-link). The overall model is called a *zero-inflated* model.

An alternative to a zero-inflated model is a *hurdle model*. In a hurdle model, the probability of all zeroes is modeled with one model component, and the expected non-zero counts are modeled with another (parent) model component (e.g. with a truncated Poisson or negative binomial model; recall Section 2.8 for the definition of a truncated random variable). Hurdle models can also describe *zero-deflation*, i.e. situations where the number of zeroes is smaller than the parent model implies. Hurdle models are more flexible than zero-inflated models, thus they usually fit better than zero-inflated models. But zero-inflated models may be theoretically more appealing as a bigger part of the variation is explained by the parent model, which may have a theoretical justification and be of primary interest. The role of exposure variables (see next section) is more problematic in hurdle models than in zero-inflated models.

Zero-inflation (and one-inflation) also occurs commonly in modeling continuous proportions, such as vegetation cover (see Section 10.6). However, we will not cover modeling of zero-inflated data further here; readers interested in the topic can read Stroup (2013, Section 11.5), for example.

8.2.10 Exposure variables in count data

When modeling counts in forestry data, counting is usually done over some time period or over some spatial area (or line segment), or both. The counting area and counting time are called *exposure variables*. A basic assumption of the exposure variables can be that the expected number of counts is proportional to the exposure variable. If the exposure variables are constant in a data set, then the modeling can be done without explicitly accounting for the amount of exposure. If the estimated model is extrapolated for other exposures assuming proportionality, then the effect of exposure needs to be explicitly formulated. Assume for simplicity that the exposure variable is the counting area A, time exposure can be treated similarly.

If the counting area A is constant and log-link is applied, then there are two equivalent ways to write A into the linear predictor of $\log(\mu)$. First, $\log(A)$ can be used as an *offset variable* when defining the model. Second, $\log(A)$ can be subtracted from the intercept of the estimated model and added as an explicit constant. In both cases we get model $\log(\mu_i) = \boldsymbol{x}_i'\boldsymbol{\beta} + \log(A)$. If the model is applied to a new value of A, then the new value of $\log(A)$ is used in the equation.

If A_i has several values in the data set, then $\log(A_i)$ can be used as a predictor variable in the model. Then its coefficient can be compared to one to test the hypothesis of proportionality. This solution was used in Vuorio *et al.* (2015) to take into account the varying fence length in a drift fence trap study of newts, which is described below. It may be easier to find an explanation for non-proportionality with respect to exposure time than with respect to counting area. In zero-inflated Poisson or negative binomial models, the counts are proportional to an exposure variable, if the model for the extra zeroes does not depend on the exposure variable and the coefficient of the exposure variable is one in the parent model.

8.3 Generalized linear mixed-effects models

The GLMM is a generalized linear model with random effects. Such a model is appropriate for modeling grouped data sets where groups constitute a random sample from a population of groups, as we discussed in Chapter 5.

For example, Vuorio *et al.* (2015) used a Poisson GLMM to analyze the effect of forest structure on the occurrence of a newt species, that lives in forests around small breeding ponds, surrounded by several forest stands. Eight breeding ponds were randomly selected in a drift fence trap study, which were used to catch the newt individuals. The drift fence trap includes a plastic fence of length 2 – 10 m, which was installed in the forest floor so that it prohibits the movement of an animal. When an animal encounters the fence, it follows the fence until it drops into a plastic bucket, which had been dug at the end of the fence, so that the rim of the bucket was at the level of ground surface. A total of 16 traps were installed around each of the ponds in different stands at varying distances from the edge of the pond. The traps were visited daily to count and release the animals from the traps. The final data included the number of newt individuals caught in each of the traps during the campaign, averaged to a few individuals per trap. The model included nested random effects for ponds and stands within ponds.

As another example, Melin *et al.* (2016b) analyzed the effect of forest structure on grouse brood occurrence, measured by volunteer hunters using a clustered sampling design with 98 clusters. In each cluster, which is called a wildlife monitoring triangle, a team of three people walked along a 12-km triangular path and marked all locations where they observed a grouse brood (157 broods in total). In the analysis stage, the 12-km path was divided into 60-m segments with either presence or absence of a grouse brood. The effect of forest structure, measured with airborne lidar, was analyzed on the presence-absence data using the binary logistic GLMM, with random effects associated for the clusters.

8.3.1 Model formulation

For simplicity, consider only a single level of grouping. Extensions to multiple levels are straightforward, similar to the case of linear and nonlinear mixed-effects models, see Sections 6.1.1, 6.2 and 7.3.4. Let y_{ij} be the response variable for observation j in group i ($i = 1, \ldots, M$ and $j = 1, \ldots, n_i$), $\boldsymbol{\beta}$ the vector of fixed effects corresponding to predictors $\boldsymbol{x}_{ij}$, $\boldsymbol{b}_i$ the vector of random effect(s) corresponding to the predictors of the random part $\boldsymbol{z}_i$, and g is the applied link function. We directly define the model for the exponential family (recall Section 8.2.4); models for binary and count data are special cases of that model. The generalized linear mixed-effects model is specified as:

$$
\begin{aligned}
y_{ij}|\boldsymbol{b}_i &\sim \exp\left(\frac{y_{ij}\gamma_{ij} - b(\gamma_{ij})}{\tau^2} - c(y_{ij}, \tau)\right) && (8.5)\\
g(\mu_{ij}) &= \boldsymbol{x}'_{ij}\boldsymbol{\beta} + \boldsymbol{z}'_{ij}\boldsymbol{b}_i && (8.6)\\
\boldsymbol{b}_i &\sim N(\boldsymbol{0}, \boldsymbol{D}_*)\,. && (8.7)
\end{aligned}
$$

It is assumed that the observations within group are independent, and also that the group effects are independent realizations from the above-specified normal distribution. The assumptions about the random effects are similar to the assumptions in the context of LME models. The model specification is similar to the GLM, but now we have a linear combination of random group effects $\boldsymbol{z}'_{ij}\boldsymbol{b}_i$ added to the linear predictor, in a similar way to the linear and nonlinear mixed-effects models. The GLMM assumes a specific distribution for the observations conditional on the random effects. We might, for example, assume that tree species within a given two-species mixed forest stand is described by a *Bernoulli* distribution, or that the number of trees of a given species from plots in one stand follows a *Poisson* distribution.

In GLM, the marginal distribution of the response variable for a given value of fixed predictors $\boldsymbol{x}_{ij}$ is of the exponential family. In GLMM, this is not the case. Even for two groups with exactly the same $\boldsymbol{x}_{ij}$ values, the marginal distribution is not of the assumed exponential family because of group-specific random effects. This is a difference to the LME models, where for a given value of fixed predictors $\boldsymbol{x}_{ij}$, the marginal distribution of y_{ij} over the distribution of random effects is also normal, and the fixed part can be seen either as the mean for a typical group or as the marginal mean over all groups. The second interpretation was not valid for an NLME model, and it is not valid for GLMM either. With GLMM, the distribution of the response variable is affected by the non-normal distribution of the data, the normal distribution of the random effects, and the nonlinear link function.

The general formula for the marginal mean, which is marginalized over the random effects (but not over the fixed predictors), is of form (see Eq. (2.45), p. 25)

$$
\mathrm{E}(y) = \mathrm{E}_{\boldsymbol{b}}(\mathrm{E}(y|\boldsymbol{b})) = \mathrm{E}_{\boldsymbol{b}}(h(\boldsymbol{x}'\boldsymbol{\beta} + \boldsymbol{z}'\boldsymbol{b}))\,. \tag{8.8}
$$

This leads to integrating the product of the mean function and the (multivariate) normal *pdf* of the random effects over the range of the (multivariate) normal *pdf*. Such an integral can be written in closed form in some special cases only. Correspondingly, the marginal variance of y can be derived using (see Eq. (2.46), p. 25)

$$
\mathrm{var}(y) = \mathrm{E}_{\boldsymbol{b}}(\mathrm{var}(y|\boldsymbol{b})) + \mathrm{var}_{\boldsymbol{b}}(\mathrm{E}(y|\boldsymbol{b}))
$$

and the marginal covariance between two observations that share the group effect $\boldsymbol{b}$ using

$$\text{cov}(y_1, y_2) = \text{cov}_{\boldsymbol{b}}(\text{E}(y_1|\boldsymbol{b}), \text{E}(y_2|\boldsymbol{b}))\,.$$

In particular, the marginal means, variances and covariances all involve an integral over the random effects, and the value of that integral depends on the choice of the random-effect structure and the link function. For a more detailed discussion and examples, see McCulloch *et al.* (2008), Demidenko (2013), and Fahrmeir *et al.* (2013).

Example 8.10. Motivated by the experiment of Vuorio *et al.* (2015), assume that n_i drift fence traps of equal fence length were placed around breeding ponds i, $i = 1, \ldots, M$. Let y_{ij} be the number of newt individuals observed in trap j of pond i during the survey period. A model for the number of animals per trap is specified as

$$\begin{aligned} y_{ij}|b_i &\sim Poisson(\mu_{ij}) \\ \log(\mu_{ij}) &= \boldsymbol{x}_{ij}'\boldsymbol{\beta} + b_i \\ b_i &\sim N(0, \sigma_b^2)\,. \end{aligned}$$

This model assumes only a random intercept for each pond, so that the logarithmic animal count behaves in a similar way with respect to the predictors on all ponds, but the level of the hyperplane may vary among ponds.

For illustration, let us then simulate a data set with $M = 100$ ponds and $n_i = 24$ traps per pond for all ponds i, with distances $d_{ij} = 5, 15, \ldots, 75$ m from the pond, so that each pond includes three replicates for each distance. Assume that $\boldsymbol{x}_i'\boldsymbol{\beta} = 0.5 - 0.03d_{ij}$ and $\sigma_b^2 = 0.5^2$. Such a data set was simulated using the following script.

```
> set.seed(12345)
> M<-100; n<-24
> sigmab<-0.5
> pond<-rep(1:M,each=n)
> dist<-rep(seq(5,75,10),3*M)
> beta<-c(0.5,-0.03)
> X<-cbind(1,dist)
> Z<-sapply(1:M,function(i) as.numeric(pond==i))
> b<-rnorm(M,0,sigmab)
> mu<-exp(X%*%beta+Z%*%b)
> y<-rpois(length(mu),mu)
> newtdat<-data.frame(pond=pond,dist=dist,newts=y)
```

The black lines in the upper left-hand panel in Figure 8.7 illustrate the marginal *pmf* of variable `newt` for traps that are located at a 5 m distance to the pond. The marginal distribution is contributed by all ponds, and so it is wider and of a different shape to the Poisson distribution with the same mean, which is illustrated by the gray lines. The black lines in the upper right-hand graph illustrate the distribution that is marginalized both over the ponds and distance. It is also of a different shape to the Poisson distribution with the same mean (gray lines). Notice that the mean over distance would not be a very useful quantity in practice. Especially, it cannot be interpreted as a mean density within 75 m radius around small ponds because the traps at small distances represent a smaller forest area than the traps at greater distances.

The lower left-hand graph illustrates the linear predictor, which expresses the logarithmic expected value of the response variable in the 100 simulated ponds as a function of distance. The lines are parallel because we have only a random intercept. The lower right-hand graph shows the corresponding expected values using gray lines. In addition, the thick dashed line shows the expected value for a pond with $b_i = 0$. The thick solid line shows the marginal mean based on Equation (8.8). The marginal mean can be computed as

$$\text{E}(y) = \int_{-\infty}^{\infty} \exp(\beta_1 + \beta_2 d + b) f_b(b) db$$

where f_b is the $N(0, \sigma_b^2)$ *pdf* of the random intercept. The integral was evaluated numerically for distances $5, 6, \ldots, 75$m using

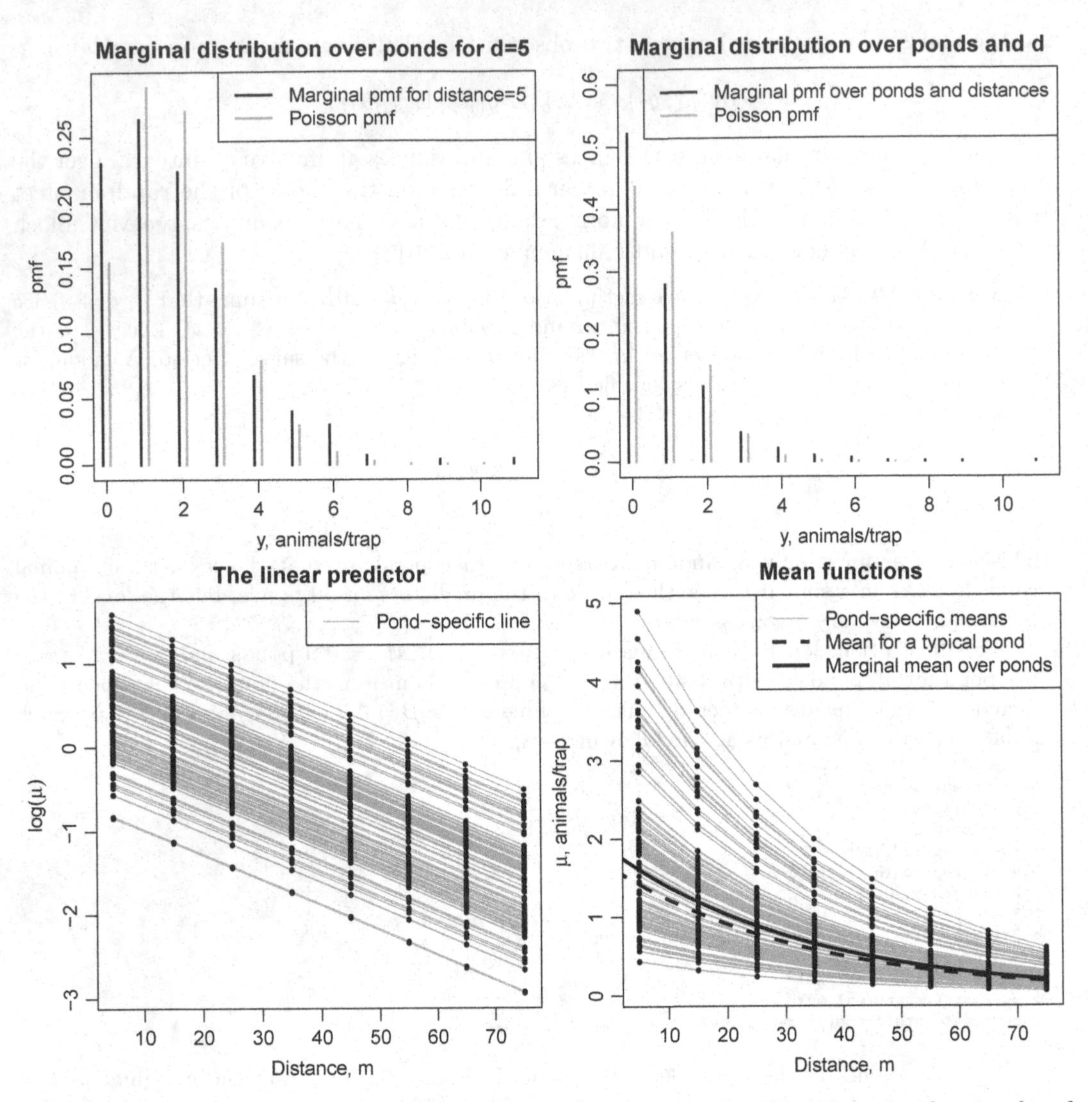

FIGURE 8.7 Illustration of the Poisson GLMM with random intercept in the simulated newt data set described in Example 8.10.

```
> margmean<-rep(NA,75)
> for (i in 1:75) {
+       margmean[i]<-integrate(function(b) exp(beta[1]+beta[2]*i+b)*dnorm(b,mean=0,sd=sigmab),
+                              -100,100)$value
+ }
```

Example 8.10 numerically evaluated the marginal mean for the Poisson model with a log link. However, the Poisson assumption was not needed to compute the marginal mean, and the result is applicable with any model that uses the log link, such as the negative binomial model. Furthermore, a closed-form solution to the integral exists (show this!):

$$\mathrm{E}(y) = \exp(\boldsymbol{x}'\boldsymbol{\beta} + \sigma_b^2/2)\,. \tag{8.9}$$

Notice that this same formula can be used for the correction of transformation bias in logarithmic regression (see Section 10.2.1). With other link functions, the conditional and marginal means are not necessarily of the same mathematical form. For example, using the logit link in a GLMM does not lead to an inverse logit function for the marginal mean. Example 8.14 (p. 285) illustrates the use, and demonstrates the importance, of the bias correction when marginal prediction is done using a GLMM.

In modeling grouped non-normal data using GLMM, all issues that have been discussed in the context of LME and GLM models should be taken into account. These include the specification of random effect structure, the link function, as well as possible overdispersion and zero-inflation. We will not repeat these issues here, but instead refer readers to the previous chapters on these topics.

For one-parameter distribution families, it is possible to also use a random effect ϵ_{ij} at the observation level, giving

$$g(\mu_{ij}) = \boldsymbol{x}_{ij}'\boldsymbol{\beta} + \boldsymbol{z}_{ij}'\boldsymbol{b}_i + \epsilon_{ij}\,, \tag{8.10}$$

where

$$\epsilon_{ij} \sim N(0, \sigma^2)$$

and $\mathrm{cov}(\epsilon_{ij}, \epsilon_{i'j'}) = 0$ for $ij \neq i'j'$ and $\mathrm{cov}(\boldsymbol{b}_i, \epsilon_{i'j}) = \boldsymbol{0}$ for all i, i', j. Inclusion of effect ϵ_{ij} to the linear predictor allows a separate expected value μ_{ij} for every observation of the data. Such a model is identifiable for Poisson and Bernoulli families. The observation-level random effect models the overdispersion, as we have already discussed in Section 8.2.7. If the estimate of σ^2 is close to zero, there is no overdispersion in the data. This approach is also applicable for data that is not grouped. Other approaches for modeling overdispersion in GLMM are discussed in the context of estimation methods in the next section.

8.3.2 Estimation

The likelihood of the GLMM in one group is obtained by averaging the likelihood of the GLM, over the distribution of random effects. Because the random effect vector consists of independent group-specific random effects, $\boldsymbol{b} = (\boldsymbol{b}_1, \boldsymbol{b}_2, \ldots, \boldsymbol{b}_M)'$, the likelihood for all data is the product of group-specific likelihoods:

$$L(\boldsymbol{\beta}, \boldsymbol{\theta}) = \prod_{i=1}^{M} \int_{\boldsymbol{b}_i} \left(\prod_{j=1}^{n_i} f(y_{ij}|\boldsymbol{b}_i, \boldsymbol{\beta}) \right) f_{\boldsymbol{b}_i}(\boldsymbol{b}_i|\boldsymbol{\theta}) d\boldsymbol{b}_i \tag{8.11}$$

where $f(y_{ij}|\boldsymbol{b}_i, \boldsymbol{\beta})$ is the conditional *pdf*/*pmf* of observation y_{ij} (e.g. Poisson or Bernoulli *pmf*) and $f_{\boldsymbol{b}_i}(\boldsymbol{b}_i|\boldsymbol{\theta})$ is the normal distribution of random effects in group i, parameterized through $\boldsymbol{\theta}$ that includes the variances and covariances of the random effects.

Example 8.11. In Example 8.10, the log-likelihood for the *Poisson* model with log link and random intercept becomes

$$\begin{aligned} l(\boldsymbol{\beta}, \sigma_b^2) =& \log \left(\prod_{i=1}^{M} \int_{-\infty}^{\infty} \prod_{j=1}^{n_i} \frac{\mu_{ij}^{y_{ij}} e^{-\mu_{ij}}}{y_{ij}!} \frac{1}{\sqrt{2\pi\sigma_b^2} e^{-\frac{1}{2\sigma_b^2} b_k^2}} db_i \right) \\ =& \boldsymbol{y}'\boldsymbol{X}\boldsymbol{\beta} - \sum_{i,j} (\log y_{ij}!) + \\ & \sum_i \log \int_{-\infty}^{\infty} \exp \left[\left(\sum_j y_{ij} \right) b_i - \sum_j e^{\boldsymbol{x}_{ij}'\boldsymbol{\beta} + b_i} \right] \frac{1}{\sqrt{2\pi\sigma_b^2}} e^{-\frac{1}{2\sigma_b^2} b_k^2} db_i \end{aligned}$$

where $\mu_{ij} = \exp(\boldsymbol{x}_{ij}'\boldsymbol{\beta} + b_i)$. It is important to note that the integral in the last term cannot be written in closed form, and maximization needs to be based on numerical approximation.

In the example above, the log-likelihood includes an integral that cannot be expressed in closed form. A similar problem will also occur if the model for binary data is defined using logit, probit, or complementary log-log links. In our example, we should evaluate the one-dimensional integral separately for each group, which is possible using the standard numerical integration methods implemented, e.g. in function `integrate`. However, the computations become complicated with more levels and more random effects per level. Therefore, estimation of model parameters in GLMM becomes more problematic than in the context of GLM due to the need to approximate that integral.

There are several methods to solve this problem. We consider the following three options:

1. Approximate the likelihood using Laplace or Adaptive Gaussian Quadrature (AGQ) techniques.
2. Develop the idea of quasi-likelihood of GLM, which is based only on the variance-mean relationships, to GLMM. This leads to the Penalized Quasi-Likelihood (PQL).
3. Switch to the Bayesian framework and estimate the posterior distribution of the parameters using Markov Chain Monte Carlo (MCMC) techniques.

A general approximation to the integral in the likelihood is provided by AGQ. The method approximates a smooth integral by a weighted sum of k terms, where the number of terms is specified by the user. Increasing the number of terms increases the accuracy, but also the computational requirements of the approximation. Therefore, if the approximation with two different values of k differs, one should use the approximation based on larger k. In practice, one should try different values and attempt to monitor computing time and the convergence of the parameter estimates. For details and examples in the context of GLMM, see e.g. McCulloch *et al.* (2008, Section 14.3). If the number of points is one, the approximation is equivalent to the so called Laplace approximation, see Demidenko (2013) for details. These integral approximation methods have been implemented in function `lme4::glmer`.

The integral approximations only provide estimates of the fixed effects and the parameters that specify the variance-covariance structure, but not the random effects. Using the Bayes formula, the *pdf* of random effect in group i, given the data of that group, can be stated as

$$f_{\boldsymbol{b}_i|\boldsymbol{y}_i}(\boldsymbol{b}_i|\boldsymbol{y}_i) = \frac{f_{\boldsymbol{y}_i|\boldsymbol{b}_i}(\boldsymbol{y}_i|\boldsymbol{b}_i) f_{\boldsymbol{b}_i}(\boldsymbol{b}_i)}{f_{\boldsymbol{y}_i}(\boldsymbol{y}_i)} .$$

The predicted value of the random effect is the expected value of this distribution. For a scalar random effect, the denominator is $f_{y_i}(y_i) = \int_{-\infty}^{\infty} \prod_{j=1}^{n_i} f_{y_{ij}|b_i}(y_{ij}|b_i) f_{b_i}(b_i) db_i$ and the expected value becomes

$$\widetilde{b}_i = \frac{\int_{-\infty}^{\infty} b_i \prod_{j=1}^{n_i} f_{y_{ij}|b_i}(y_{ij}|b_i) f_{b_i}(b_i) db_i}{\int_{-\infty}^{\infty} \prod_{j=1}^{n_i} f_{y_{ij}|b_i}(y_{ij}|b_i) f_{b_i}(b_i) db_i} . \tag{8.12}$$

Computation of these integrals is computationally intensive for models with several levels and random effects, and can be based on the AGQ approximation.

The PQL is an extension of the quasi-likelihood method of GLM to mixed-effects models. For a formal presentation of the method, see McCulloch *et al.* (2008) and Demidenko (2013). Recall that the quasi likelihood estimation of GLM was equivalent to specifying the GLM as a nonlinear model with normally distributed errors, using a variance function justified by the distribution of the response variable. Correspondingly, PQL is equivalent to formulating GLMM as a nonlinear mixed-effects model with normally distributed errors, using a variance function that is supported by the distribution of the response variable and by estimating the model parameters using the Lindstrom-Bates algorithm. An additional benefit is the implicit modeling of overdispersion. The predictions of the random effect are also obtained as a by-product (recall Section 7.3.2). Therefore, PQL is an attractive approach for practice. However, based on some simulation studies, McCulloch *et al.* (2008) cast doubt on the performance of the method, at least with binary data and small clusters. Implementation of the PQL method is available in function `glmmPQL` of library `MASS`. Implementation using `nlme` is shown in the following example.

A third option for the estimation of a GLMM is a Bayesian approach (recall Sections 3.4.5 and 5.3.5). We illustrate a Bayesian estimation of a Poisson model using package `MCMCglmm` (Hadfield 2010). The model is based on the formulation (8.10) (p. 279).

Example 8.12. Consider the simulated fence trap data of newt individuals that was generated previously in Example 8.10. The model is fitted below with an AGQ approximation of the integral. The number of nodes in the AGQ approximation range from 1 to 10. Exploring the approximates of deviance for different values of k shows that no changes in the first three decimals occur when k is increased from 8 to 9 or 10. Therefore, we believe that the last model, computed using $k = 10$, should be based on a sufficiently good approximation of the likelihood.

```
> library(lme4)
> dev<-rep(NA,10)
> for (i in 1:10) {
+     nmod1<-glmer(newts~dist+(1|pond),data=newtdat,family=poisson,nAGQ=i)
+     dev[i]<-deviance(nmod1)
+ }
> dev
 [1] 2211.980 2212.014 2211.949 2211.806 2211.811 2211.808 2211.805 2211.806 2211.806 2211.806

> summary(nmod1)
Generalized linear mixed model fit by maximum likelihood (Adaptive
  Gauss-Hermite Quadrature, nAGQ = 10) [glmerMod]
 Family: poisson  ( log )
Formula: newts ~ dist + (1 | pond)
   Data: newtdat

Random effects:
 Groups Name        Variance Std.Dev.
 pond   (Intercept) 0.2852   0.534
Number of obs: 2400, groups:  pond, 100

Fixed effects:
             Estimate Std. Error z value Pr(>|z|)
(Intercept)  0.638344   0.066022   9.669   <2e-16 ***
dist        -0.029571   0.001116 -26.497   <2e-16 ***
```

Because the data are simulated, we know the true values of the parameters. The estimates of both fixed effects $\widehat{\beta} = (0.638, -0.029)'$ and the variance of random effect $\widehat{\sigma}_b^2 = 0.534^2$ are fairly close to the true values $\beta = (0.5, -0.03)'$ and $\sigma_b^2 = 0.5^2$. The predicted random effects are also strongly correlated with the known true values of vector $\boldsymbol{b}$ in Example 8.10, and have shrunk towards zero as they should (Figure 8.8, left). To illustrate the calculation of random effects, we apply formula (8.12) to our data, pond by pond, to compute the random effects into vector `myb`. Figure 8.8 (right) shows that they are very similar to the random effects returned by `ranef(nmod1)`; the small differences result from the different integral approximations.

```
> re<-function(i,model=nmod1) {
+     y1<-newtdat$newts[newtdat$pond==i]
+     X1<-X[newtdat$pond==i,]
+     beta<-fixef(model)
+     # multiply dpois by 5 to get better accuracy for the integral approximations
+     fnum<-function(b) sapply(b, function(bi) bi*prod(5*dpois(y1,exp(X1%*%beta+bi)))*
+                            dnorm(bi,mean=0,sd=attributes(model)$theta))
+     fden<-function(b) sapply(b, function(bi) prod(5*dpois(y1,exp(X1%*%beta+bi)))*
+                            dnorm(bi,mean=0,sd=attributes(model)$theta))
+     num<-integrate(fnum,-20,20)
+     den<-integrate(fden,-20,20)
+     num$value/den$value
+ }
> myb<-sapply(1:100,re)
```

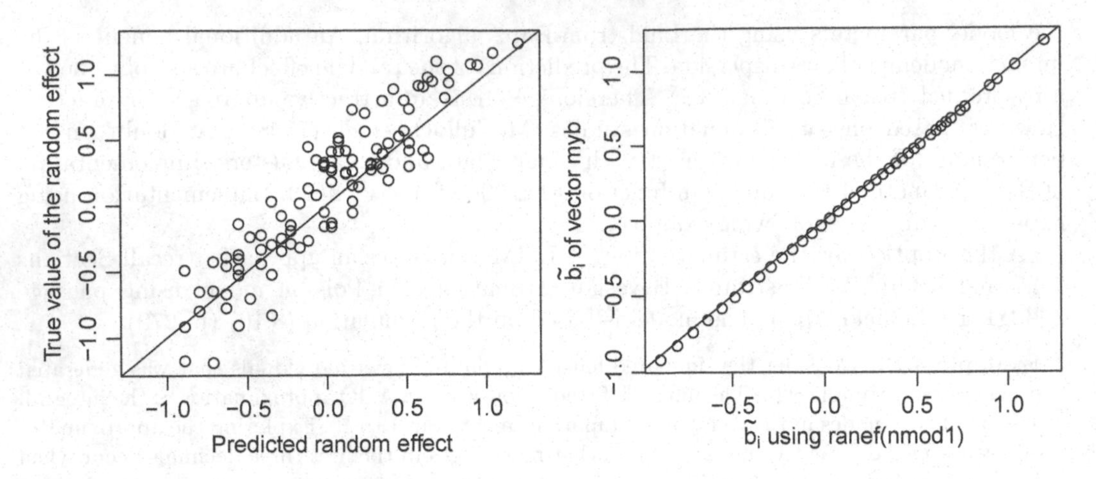

FIGURE 8.8 Predicted random effects of the model `nmod1` plotted against the true values in the simulated data set `newtdat` (left). The predicted random effects computed using function `integrate` on the predicted values returned by `glmer` (right).

Another option for estimation is PQL, which is applied below using function `MASS::glmmPQL`. Practically the same parameter estimates (the differences result from different optimization algorithms and convergence criteria) are obtained by formulating the model as the following nonlinear mixed-effects model

$$y_{ij} = e^{\beta_1+\beta_2 d_{ij}+b_i} + \epsilon_{ij} ,$$

where

$$\mathrm{var}(\epsilon_{ij}) = \sigma^2 |\widehat{y}_{ij}| . \tag{8.13}$$

The fit of this model is also shown below. Compared to the model fitted using `glmer`, the PQL model has an additional parameter, σ^2, which models overdispersion. If that parameter is close to 1, no overdispersion exists in the data. If that is greater than 1, then overdispersion exists, but it has been taken into account in the parameter estimates and their standard errors. Correspondingly, a value below 1 indicates underdispersion. With our data, the estimate of σ^2 is close to 1, which is logical because our simulated data is known to follow the Poisson distribution. The overdispersion parameter could also be restricted to one, as we did in Example 8.4.

```
> library(MASS)
> glmmPQL(newts~dist,random=~1|pond,data=newtdat,family="poisson")
Linear mixed-effects model fit by maximum likelihood
  Data: newtdat
  Log-likelihood: NA
  Fixed: newts ~ dist
(Intercept)        dist
 0.66021719 -0.02957114

Random effects:
 Formula: ~1 | pond
        (Intercept)  Residual
StdDev:   0.5284436 0.9592882

Variance function:
 Structure: fixed weights
 Formula: ~invwt
Number of Observations: 2400
Number of Groups: 100

> nlme(newts~exp(theta),fixed=list(theta~dist),
+      random=list(pond=theta~1),data=newtdat,
+      start=c(0,-0.01),weights=varPower(fixed=list(delta=0.5)))
Nonlinear mixed-effects model fit by maximum likelihood
```

```
  Model: newts ~ exp(theta)
  Data: newtdat
  Log-likelihood: -2771.966
  Fixed: list(theta ~ dist)
theta.(Intercept)          theta.dist
      0.66022255          -0.02957114

Random effects:
 Formula: theta ~ 1 | pond
        theta.(Intercept)  Residual
StdDev:         0.5284429 0.9592854

Variance function:
 Structure: Power of variance covariate
 Formula: ~fitted(.)
 Parameter estimates:
power
  0.5
Number of Observations: 2400
Number of Groups: 100
```

Let us then fit the model using `MCMCglmm`. After some initial analysis (not shown), we decided to use a thinning interval of 1000, which produced acceptable-looking trace plots of the iteration. The posterior means of parameters are $\hat{\beta} = (0.625, -0.0293)'$, which are close to the true values used in the simulation. The posterior mean for the variance of random effect is $\hat{\sigma}_b^2 = 0.2957 = 0.544^2$, which is also close to the true value of 0.5^2. The posterior mean of the unit-to-unit variance is small, $\hat{\sigma}^2 = 0.0935^2$, which indicates that there is no severe overdispersion in the data.

```
> library(MCMCglmm)
> nmodBay<-MCMCglmm(newts~dist,random=~pond,data=newtdat,
+                   family="poisson",nitt=530000,thin=1000)
> plot(nmodBay)
> summary(nmodBay)

 Iterations = 3001:529001
 Thinning interval  = 1000
 Sample size  = 527

 G-structure:  ~pond
     post.mean l-95% CI u-95% CI eff.samp
pond    0.2957   0.1929    0.401    248.7

 R-structure:  ~units
      post.mean l-95% CI u-95% CI eff.samp
units  0.008744  5.7e-05  0.03296     63.9

 Location effects: newts ~ dist
            post.mean l-95% CI u-95% CI eff.samp  pMCMC
(Intercept)   0.62545  0.49593  0.74556    249.5 <0.002 **
dist         -0.02933 -0.03115 -0.02740     78.2 <0.002 **
```

There are also other estimation methods to the three approaches demonstrated above. The conditional likelihood method is based on conditional estimates and tests by treating group effects as fixed. The Generalized Estimating Equations (GEE) are based on the estimation of the marginal mean. They specify the marginal mean (8.8) as a function of predictors, instead of the conditional means specified by the GLMM. The mean function can be estimated by assuming independence of the data, even though that is not true in clustered data. The parameter estimates based on the assumption of independence are consistent and are usually only slightly less efficient than the parameter estimates based on the mixed-effects model. The variance estimators based on the independence assumption are biased, but are adjusted for the lack of independence with a sandwich-type formula, which takes into account the dependence associated with the grouping of the data. For more detail on these approaches, see McCulloch *et al.* (2008) and Demidenko (2013).

8.3.3 Inference

The deviance is a useful statistic in evaluating a GLMM model fit, with a similar interpretation to that in the case of GLM. The inference in models using the likelihood approximations can be based on the LR and Wald tests, which are now based on the approximate likelihoods. In PQL, the inference is based on the same procedures that we used earlier with nonlinear mixed-effects models. With models fitted using MCMC, the inference can be based on the simulated posterior samples of the parameter estimates.

Even if the likelihood could be calculated exactly, the inferential tools available would be based on asymptotic results and therefore require sufficiently large sample sizes. The various approximations used in estimation affect the estimates of standard errors and inference even more than the point estimates of the parameters. Therefore, much care is needed in the inference on GLMM. Usually, it is wise to fit the model with several different methods and even different software to ensure that the results are not affected by potential artifacts caused by the algorithms.

Example 8.13. An extensive spruce sapling data set can be found in `lmfor::plants`. The data includes measurements from 1926 plots from 123 regeneration areas. We explore whether the number of spruce saplings per plot is associated with deciduous tree height, soil preparation method, stoniness, and wetness of the plot. We start by fitting the Poisson model with a random intercept to the data, using AGQ approximation with a sufficiently large number of terms.

```
> data(plants)
> pmod1<-glmer(spruces ~ (1|stand)+hdecid+prepar+stones+wet,
+              family=poisson(),
+              data=plants,nAGQ=10)
> summary(pmod1)
     AIC      BIC   logLik deviance df.resid
  3247.7   3292.2  -1615.9   3231.7     1918

Fixed effects:
              Estimate Std. Error z value Pr(>|z|)
(Intercept)  0.1323067  0.7157282   0.185   0.8533
hdecid       0.0006986  0.0007347   0.951   0.3417
prepar1     -1.3464147  0.8180921  -1.646   0.0998 .
prepar2     -1.1853791  0.7078811  -1.675   0.0940 .
prepar3     -1.1791047  0.7325921  -1.609   0.1075
stones1     -1.0323701  0.4293658  -2.404   0.0162 *
wet1        -0.6198296  0.3890692  -1.593   0.1111
```

The model summary indicates that spruce count might be associated with stoniness. However, the deviance is much higher than the number of plots, which indicates overdispersion. To account for overdispersion, we re-estimate the model using (1) `glmer` with observation-level random effects, (2) package `glmmTMB` with observation-level random effects, (3) penalized quasi likelihood (`MASS::glmmPQL`), (4) a Bayesian model estimated with MCMC (`MCMCglmm`), and (5) the negative binomial model implemented in package `glmmTMB`. Models (1) and (2) should be the same: both are based on the Laplace approximation of the Poisson likelihood.

```
> library(MCMCglmm)
> library(glmmTMB)
> plants$plot<-1:dim(plants)[1]
>
> pmod2glmer<-glmer(spruces ~ (1|stand/plot)+hdecid+prepar+stones+wet,
+                   family=poisson(), data=plants,nAGQ=1)
>
> pmod2TMB<-glmmTMB(spruces ~ (1|stand/plot)+hdecid+prepar+stones+wet,
+                   family=poisson, data=plants)
>
> pmodPQL<-glmmPQL(spruces ~ hdecid+prepar+stones+wet,
+                  random=~1|stand, family=poisson(), data=plants)
>
> pmodBay<-MCMCglmm(spruces ~ hdecid+prepar+stones+wet,
+                   random=~stand, data=plants,
+                   family="poisson", nitt=53000,thin=100)
>
```

	glmer	glmmTMB	glmmPQL	MCMC	negbin
			Fixed part		
Intercept	-0.434 (0.64)	-0.822 (0.39)	0.312 (0.69)	-0.596 (0.54)	0.420 (0.69)
hdecid	0.000 (0.26)	0.001 (0.46)	0.001 (0.35)	0.000 (0.74)	0.001 (0.55)
prepar1	-1.662 (0.10)	-1.331 (0.18)	-1.269 (0.14)	-1.489 (0.15)	-1.543 (0.16)
prepar2	-1.672 (0.08)	-1.314 (0.16)	-1.165 (0.13)	-1.409 (0.16)	-1.379 (0.18)
prepar3	-2.085 (0.03)	-1.646 (0.09)	-1.302 (0.11)	-1.637 (0.12)	-1.615 (0.13)
stones1	-1.117 (0.12)	-0.892 (0.19)	-1.007 (0.12)	-0.967 (0.16)	-1.047 (0.10)
wet1	-0.146 (0.85)	-0.094 (0.90)	-0.536 (0.37)	-0.309 (0.73)	-0.205 (0.78)
			Random part and overdispersion		
var(b_i)	0.713^2	0.686^2	0.823^2	0.998^2	0.890^2
Overdisp	1.638^2	1.608^2	1.531^2	1.510^2	0.337

TABLE 8.5 Parameter estimates of the five models for the spruce sapling counts in Example 8.13. They are based on GLMMs that allow overdispersion. The p-values of tests $H_0 : \beta_j = 0$, based on the standard output of each procedure, are given in parentheses.

```
> pmodNB<-glmmTMB(spruces ~ (1|stand)+hdecid+prepar+stones+wet,
+                 family=nbinom2,data=plants)
```

The parameter estimates are summarized in Table 8.5. All three models indicate overdispersion. To interpret the overdispersion parameters, recall that the models of overdispersion in pmod2 and pmodBay are based on Equation (8.10), whereas that in pmodPQL is modeled by σ^2 in (8.13). For interpretation of the overdispersion parameter of model pmodNB, recall from Example 8.7 (p. 267) that for the negative binomial distribution $\text{var}(Y) = \text{E}(Y) + \frac{1}{\theta}\,\text{E}(Y)^2$.

The estimates of within-plot variances are somewhat similar in pmod2 and pmodBay ($1.638^2 \approx 1.608^2 \approx 1.510^2$) and are much greater than the variance between plots. The PQL fit indicates that the variance of the counts is $1.531^2 = 2.34$ times greater than expected by the Poisson assumption. The negative binomial model indicates that $\widehat{\text{var}}(y_{ij}|b_i) = \widetilde{y}_{ij} + \widetilde{y}_{ij}^2/0.337$, where $\widetilde{y}_{ij} = \widehat{\text{E}}(y_{ij}|b_i) = \exp\left(\boldsymbol{x}_{ij}\widehat{\boldsymbol{\beta}} + \widetilde{b}_i\right)$. The overdispersion may indicate, for example, that the trees have a clustered pattern within the stands (recall Example 8.6), or there are some missing covariates that affect the stand density.

The fixed-effects of the five models differ considerably but are still somewhat similar in terms of sign and order. They even differ between models pmod2glmer and pmod2TMB, which are both based on the same Laplace approximation of the Poisson likelihood. We had to use nAGQ=1 in glmer because values above 1 are not available for two-level models. It seems that none of the predictors shows a statistically significant effect on the mean spruce count. The large variability in parameter estimates is partially explained by the fact that all effects are close to zero when compared to their estimation accuracy.

Example 8.14. If prediction is done using a GLMM, the difference between the conditional and marginal means (recall Example 8.10), needs to be taken into account to get (approximately) unbiased predictions. Here, we illustrate the bias corrections using models pmod2glmer, pmodPQL and pmodNB, which are all based on a different model formulation.

Consider first the group-level prediction, which is the prediction of $\text{E}(y_{ij}|b_i)$ for a new observation, which belongs to group i and has value $\boldsymbol{x}_0$ of the predictor vector. For models pmodPQL and pmodNB, this prediction is obtained directly as $\exp\left(\boldsymbol{x}_0'\widehat{\boldsymbol{\beta}} + \widetilde{b}_i\right)$. However, for model pmod2glmer that prediction would be an underestimate, because the observation-level random effect was included *within* the link function. An approximately unbiased prediction is obtained as $\exp\left(\boldsymbol{x}_0'\widehat{\boldsymbol{\beta}} + \widetilde{b}_i + \widehat{\sigma}^2/2\right)$, where $\widehat{\sigma}^2$ is the estimated variance of the observation-level random effects.

For marginal prediction, we need an unbiased prediction of the marginal mean $\text{E}(y_{ij})$. To find it, one should also take into account the variability of random effects. An approximately unbiased marginal prediction for models pmodPQL and pmodNB is $\exp\left(\boldsymbol{x}_0'\widehat{\boldsymbol{\beta}} + \widehat{\sigma}_b^2/2\right)$. For pmod2glmer,

	Bias-corrected		*Naïve*	
	plot-level	marginal	plot-level	marginal
glmer	1.20	0.61	0.31	0.12
glmmPQL	1.53	0.68	–	0.45
negbin	1.32	0.60	–	0.40

TABLE 8.6 Bias-corrected and naive predictions of mean tree count per plot using three different generalized linear mixed-effects models in Example 8.14.

an approximately unbiased prediction includes also the overdispersion parameter, and becomes $\exp\left(\boldsymbol{x}_0'\widehat{\boldsymbol{\beta}} + \frac{\widehat{\sigma}_b^2+\widehat{\sigma}^2}{2}\right)$. The above-specified bias corrections are approximate, because they are based on the estimates of the model parameters, and on the normality of the random effects. The other bias correction methods discussed in Section 10.2 could also be used.

For illustration, consider a new sample plot in stand 230, where the height of deciduous trees is 50 dm, site preparation method 2 has been applied, and the binary predictors `stones` and `wet` both are zero. The script below extracts the required parameter estimates and random-effect predictions from the models fitted in Example 8.13, and computes the predictions. The results are summarized in Table 8.6. The bias-corrected plot-specific predictions of the spruce sapling counts on this plot are 1.20, 1.53 and 1.32 trees per plot using models `pmod2glmer`, `pmodPQL` and `pmodNB`, respectively. If the prediction of `pmod2glmer` were not corrected for the prediction bias caused by the overdispersion model, the prediction would have been much lower, only 0.31 trees per plot. The marginal, bias-corrected predictions are 0.61, 0.68 and 0.60 trees per plot, respectively. Naïve predictions without bias correction would be 0.12, 0.45 and 0.40 trees per plot. The effect of bias correction is large in this example because there is much variability between stands and overdispersion in the data.

```
> x0<-c(1,50,0,1,0,0,0)
> beta_glmer<-fixef(pmod2glmer)
> b_glmer<-ranef(pmod2glmer)$stand[1,1]
> sigma2b_glmer<-summary(pmod2glmer)$varcor$stand[1]
> sigma2_glmer<-summary(pmod2glmer)$varcor[[1]][1]
>
> beta_PQL<-fixef(pmodPQL)
> b_PQL<-ranef(pmodPQL)[1,1]
> sigma2b_PQL<-exp(attributes(summary(pmodPQL)$apVar)$Pars[1])
>
> beta_NB<-fixef(pmodNB)$cond
> b_NB<-ranef(pmodNB)$cond$stand[1,1]
> sigma2b_NB<-summary(pmodNB)$varcor$cond$stand[1]
>
> c(exp(x0%*%beta_glmer+b_glmer),
+  exp(x0%*%beta_glmer+b_glmer+sigma2_glmer/2),
+  exp(x0%*%beta_PQL+b_PQL),
+  exp(x0%*%beta_NB+b_NB))
[1] 0.3129334 1.1977052 1.5291874 1.3167536
>
> c(exp(x0%*%beta_glmer+(sigma2b_glmer+sigma2_glmer)/2),
+    exp(x0%*%beta_PQL+sigma2b_PQL/2),
+    exp(x0%*%beta_NB+sigma2b_NB/2))
[1] 0.6119006 0.6761869 0.5959948
>
> c(exp(x0%*%beta_glmer),
+  exp(x0%*%beta_PQL),
+  exp(x0%*%beta_NB))
[1] 0.1240070 0.4481770 0.4010131
```

9

Multivariate (Mixed-Effects) Models

The models discussed in the previous chapters include only one response variable. Quite often several response variables are measured from the same sampling units. In many cases, it is sufficient to formulate univariate models separately for each response variable. However, it is also possible to formulate the model jointly for all the response variables. The resulting model is called a *model system*, a *multivariate model*, or a *joint model*. Whether a multivariate model has some benefits over separate univariate models depends largely on the research question at hand.

The individual equations of a model system may be *directly related*, *seemingly unrelated*, or *unrelated*. In directly related models, the response of one model may be the predictor in another model. In a seemingly unrelated model (SUR), direct relationships do not exist between the individual models. However, the models may be related through the dependence of residual errors and random effects. In unrelated models, no direct relationship exists between the models, and the residual errors and random effects are independent across models.

9.1 Why model systems?

In the next sections, we will specifically discuss the seemingly unrelated linear models and linear mixed-effects models. In those models, the most important difference to separately fitted models is the estimation of the cross-model correlation of residual errors and random effects. The estimated cross-model correlation may improve the analysis for the following reasons.

1. The fixed parameters of the model may be better estimated.
2. Predictions using the models will be more efficient, if the response of one of the component models has been observed for the sampling unit in question.
3. A more realistic simulation could be performed.
4. Hypotheses on the effects of a predictor on several response variables can be jointly tested.

The first item seldom causes any major improvement in efficiency, as we will discuss later in Sections 9.2.2 and 9.4.2. The benefits related to the other items are now discussed in more detail.

The second item may produce substantial benefits. Even though information on between-model correlation may not change the estimates of fixed effects, utilization of the estimated cross-model correlation may lead to substantially better and more realistic results in prediction, provided the observed response of one component model can be used in prediction for *cross-model calibration*.

One of the earliest applications of the cross-model calibration can be found in Lappi (1991a), who developed a bivariate mixed-effects model system for tree volume and tree height as a function of diameter. He proposed that the volume model could be calibrated using measured heights for improved prediction of standing tree volume. An earlier but rather poorly known forest application (in the context of taper curves) was proposed by Lappi (1986), and the method was further developed in Lappi (2006). The latter model system included a total of 13 equations: 12 equations for the stem diameter at fixed absolute or relative heights and an additional equation for tree height. The model expressed the expected value for each model conditional on tree diameter. Furthermore, the variance-covariance structure among the model residuals was modeled as a function of tree diameter. In applications, the cross-model correlations of the models are utilized to predict the residuals of all models, typically using one or two measurements of a single tree, such as DBH, height and upper stem diameter. Interpolation is used to generalize the expected values, variances and covariances from fixed relative heights to the relative heights that correspond to the actual measurements. A similar idea was also utilized for diameter distributions in Mehtätalo (2005) (see Section 11.5.3).

Siipilehto (2011) developed a system of models for 20 stand-level variables, some of which were important stand characteristics such as basal area, mean diameter and dominant height, whereas others were parameters of models that describe tree allometry and diameter distribution. Measurements of some stand variables, typically the basal area and mean diameter, can then be used to improve the prediction of all variables of the system. The benefit of the system over a set of predictive regression models is flexibility in the use of the measurements: any set of measurements (including the empty set of no measurements) can be used to produce the prediction. The more variables that have been measured, the more accurate are the predictions. The approach illustrates that the estimates of mean and the variance-covariance matrix of the variables includes all the information that is needed to produce OLS-estimates of the regression coefficients of any regression model based on the available variables. Indeed, regression coefficients were usually estimated this way in the old days when computing power was a limitation (e.g. Yule and Kendall 1958). Once the means and the variance-covariance matrix of the variables have been computed, all possible regression models could be estimated using minimal computing effort (see Section 10.1.1).

Other prediction applications with similar concept include Eerikäinen (2009) for tree height-diameter relationship (the different responses were tree species in mixture stands) and de Souza Vismara *et al.* (2016), who developed a bivariate model for individual tree volumes in eucalyptus plantations (two responses were associated with two rotations of the plantation). We will discuss and demonstrate the cross-model calibration in Sections 9.2.4 and 9.4.3.

Item 3 of the list above is relevant, e.g. in simulators of forest development, which commonly comprise several models for different natural processes (such as ingrowth, diameter growth, and mortality) or different structural components (such as stemwood, branch, needle, and root biomass). In such a situation, it is extremely important to take the dependencies of model predictions into account to avoid unrealistic simulation results and to be able to estimate the prediction errors of the output (Leskinen *et al.* 2009). In general, a realistic model for the dependency structure may be far more important than the effective estimation of regression coefficients. Furthermore, it is not only the variance-covariance structure that matters; the shape of the distribution may also have a strong effect.

The fourth benefit from joint estimation is the possibility to test hypothesis of the form "Does predictor x have an effect in some of the responses" against the null hypothesis "Predictor x does not have an effect on any of the response variables", see Section 9.2.3 for discussion and example of a system of fixed-effect models. Similar benefits occur in the case of mixed-effects models.

9.2 Seemingly unrelated regression models

9.2.1 Model formulation

The seemingly unrelated models are related to each other because the residuals are correlated. Let us consider an example where tree height and volume are both modeled on tree diameter:

$$\log h_i = \beta_1^{(h)} + \beta_2^{(h)} \log d_i + e_i^{(h)} \tag{9.1}$$

$$\log v_i = \beta_1^{(v)} + \beta_2^{(v)} \log d_i + e_i^{(v)}, \tag{9.2}$$

where d_i is tree diameter, $\beta_1^{(h)}$, $\beta_2^{(h)}$, $\beta_1^{(v)}$, $\beta_2^{(v)}$ are parameters to be estimated, and $e_i^{(h)}$ and $e_i^{(v)}$ are the residuals of the models. It is assumed that the residuals have zero mean and variances $\mathrm{var}(e_i^{(h)}) = \sigma_h^2$ and $\mathrm{var}(e_i^{(v)}) = \sigma_v^2$. The residuals within both models are uncorrelated: $\mathrm{cov}(e_i^{(h)}, e_{i'}^{(h)}) = 0$ and $\mathrm{cov}(e_i^{(v)}, e_{i'}^{(v)}) = 0$ for $i \neq i'$. These assumptions would also be made if the two models were estimated separately. In addition, we assume $\mathrm{cov}(e_i^{(v)}, e_i^{(h)}) = \varphi$, where φ is the *cross-model covariance* of the residual errors to be estimated. For example, it might be realistic to assume that if height is greater than was expected by the systematic part of the height model, then also the volume is greater than expected by the systematic part of the volume model. In an SUR model, we allow correlation among responses. This fixed, constant correlation is assumed to be common to all observations.

To formulate the model in a matrix form, let us consider the following M individual linear models:

$$\begin{aligned}
\boldsymbol{y}_1 &= \boldsymbol{X}_1\boldsymbol{\beta}_1 + \boldsymbol{e}_1 \\
\boldsymbol{y}_2 &= \boldsymbol{X}_2\boldsymbol{\beta}_2 + \boldsymbol{e}_2 \\
&\vdots \\
\boldsymbol{y}_M &= \boldsymbol{X}_M\boldsymbol{\beta}_M + \boldsymbol{e}_M\,,
\end{aligned}$$

where $\mathrm{E}(\boldsymbol{e}_m) = \boldsymbol{0}$, $\mathrm{var}(\boldsymbol{e}_m) = \sigma_m^2\boldsymbol{I}$ for $m = 1, \ldots, M$, and $\mathrm{cov}(\boldsymbol{e}_m, \boldsymbol{e}'_{m'}) = \varphi_{mm'}\boldsymbol{I}$ for every pair $m \neq m'$.

By defining

$$\boldsymbol{y} = \begin{bmatrix} \boldsymbol{y}_1 \\ \boldsymbol{y}_2 \\ \vdots \\ \boldsymbol{y}_M \end{bmatrix}, \boldsymbol{X} = \begin{bmatrix} \boldsymbol{X}_1 & \boldsymbol{0} & \cdots & \boldsymbol{0} \\ \boldsymbol{0} & \boldsymbol{X}_2 & \cdots & \boldsymbol{0} \\ \vdots & \vdots & \ddots & \vdots \\ \boldsymbol{0} & \boldsymbol{0} & \cdots & \boldsymbol{X}_M \end{bmatrix}, \boldsymbol{\beta} = \begin{bmatrix} \boldsymbol{\beta}_1 \\ \boldsymbol{\beta}_2 \\ \vdots \\ \boldsymbol{\beta}_M \end{bmatrix}, \boldsymbol{e} = \begin{bmatrix} \boldsymbol{e}_1 \\ \boldsymbol{e}_2 \\ \vdots \\ \boldsymbol{e}_M \end{bmatrix}$$

we can write the model system in the form of univariate linear model (4.6) (p. 71) as

$$\boldsymbol{y} = \boldsymbol{X}\boldsymbol{\beta} + \boldsymbol{e}\,, \tag{9.3}$$

where the variance-covariance matrix of residual errors is

$$\mathrm{var}(\boldsymbol{e}) = \begin{bmatrix} \sigma_1^2\boldsymbol{I} & \varphi_{12}\boldsymbol{I} & \cdots & \varphi_{1M}\boldsymbol{I} \\ \varphi_{12}\boldsymbol{I} & \sigma_2^2\boldsymbol{I} & \cdots & \varphi_{2M}\boldsymbol{I} \\ \vdots & \vdots & \ddots & \vdots \\ \varphi_{1M}\boldsymbol{I} & \varphi_{2M}\boldsymbol{I} & \cdots & \sigma_M^2\boldsymbol{I} \end{bmatrix} = \begin{bmatrix} \sigma_1^2 & \varphi_{12} & \cdots & \varphi_{1M} \\ \varphi_{12} & \sigma_2^2 & \cdots & \varphi_{2M} \\ \vdots & \vdots & \ddots & \vdots \\ \varphi_{1M} & \varphi_{2M} & \cdots & \sigma_M^2 \end{bmatrix} \otimes \boldsymbol{I} = \boldsymbol{\Sigma} \otimes \boldsymbol{I}\,. \tag{9.4}$$

Here terms $\sigma_1^2, \ldots, \sigma_M^2$ are the error variances of each component and $\varphi_{mm'}$ is the cross-model covariance between residual errors (and the response) of models m and m'.

It is straightforward to extend the model formulation to a situation where the diagonal elements of the response-specific variance-covariance matrices are used to model the heteroscedastic errors (recall Section 4.3). In that case, it is common to allow the cross-model covariances to vary in such a way that the cross-model correlation for each pair of models is constant.

9.2.2 Estimation

As the above specification showed, the SUR model is a special case of the linear model, and so the methods presented in Chapter 4 for estimation and inference also apply. Thus, the estimation involves the initial estimation of the parameters specifying matrix $\boldsymbol{W} = \text{var}(\boldsymbol{e})$ using REML, and then $\boldsymbol{\beta}$ using GLS. Notice that even though matrix $\boldsymbol{W}$ may be large, its inverse is easy to find as $\boldsymbol{\Sigma}^{-1} \otimes \boldsymbol{I}$. It is also possible to use the iterative generalized least squares:

1. Estimate $\boldsymbol{\beta}$ by fitting each model separately using OLS.
2. Estimate $\boldsymbol{W}$ as the empirical variance-covariance matrix of the observed residuals of the current model.
3. Re-estimate the model parameters using GLS.
4. Repeat steps (2) and (3) until there is no change in the estimates.

Sometimes the estimation is carried out without step 4, as suggested in the classical paper by Zellner (1962).

If the same predictor variables are used in all equations of the system, or if the cross-model correlations are zero, then the parameter estimates obtained by the SUR fit are the same as one would get from separate fits, and their standard errors are also the same. If different predictor variables are used and a cross-model correlation exists, then the estimates differ and the estimates from the SUR fit have a smaller variance than the estimates from separate fits. However, if the models are well formulated and the most important predictors are included in all models, the differences in the estimates of regression coefficients between jointly and separately fitted models should not be very large. If the differences are large, it means that some of the equations include a predictor that can explain the variation in some of those responses where that predictor is not used. In such cases, the parameter estimates obtained using SUR are biased and the formulation of the fixed part should be updated.

Because the improvement in the estimates of $\boldsymbol{\beta}$ from fitting models jointly is not large, an acceptable strategy for model fitting is to estimate $\boldsymbol{\beta}$ from separate fits and, thereafter, estimate the cross-model correlation. Such an approach allows more flexibility in formulation of the variance-covariance structure. An example of such situation can ge found in Lappi (2006) (see page 288) in the context of taper curves, where the means, variances and covariances of the 13-equation system were functions of tree diameter. The estimation started by fitting models for the mean of the response variable using smoothing splines. Thereafter, the residuals of these models were extracted. The estimation of variance function was based on modeling the squared residuals (because $\text{var}(X) = \text{E}(X^2)$, if $\text{E}(X) = 0$) and the covariance was based on modeling the cross-products of the residuals of all possible pairs of observations (because $\text{cov}(X, Y) = \text{E}(XY)$ if $\text{E}(X) = \text{E}(Y) = 0$)). The cost of more flexible modeling of the means, variances, and covariances, in this study, was that the properties of the estimates were not that well known. However, this was considered a minor problem compared to the benefit gained from a realistic model for the variance-covariance structure, which allowed realistic predictions of taper curves through cross-model calibration.

The correlation can be estimated as the empirical correlation of the model residuals, as was done in step 2 of the iterative procedure above. It can also be estimated by utilizing a model for the sum of the two variables. Consider a model with two response variables, y_1 and y_2. Start by fitting the univariate models for both responses and denote the residual standard errors from these models as $\widehat{\sigma}_1$ and $\widehat{\sigma}_2$. Then also fit a model for the sum $y_1 + y_2$, resulting in estimated residual standard error $\widehat{\sigma}_{12}$. The cross-model covariance can now be estimated using $\text{var}(X+Y) = \text{var}(X) + \text{var}(Y) + 2\,\text{cov}(X,Y)$ as

$$\text{cov}(X,Y) = \frac{\widehat{\sigma}_{12}^2 - \widehat{\sigma}_1^2 - \widehat{\sigma}_2^2}{2} .$$

This idea will be even more beneficial later when we discuss the multivariate models with random intercept (see Section 9.4.2).

In R, estimation of seemingly unrelated regression models using the iterative procedure presented above can be carried out by using function `systemfit::systemfit`; by default the function only carries out steps 1–3, but iteration can be conducted by using argument `maxit`. Function `nlme::gls` allows specification of the variance-covariance structure (9.4) through arguments `cor` and `weights`, and parameter estimation using REML+GLS (and ML).

Example 9.1. Consider data set `stumplift`. The processing time t_i of a stump was previously explained by machine and stump diameter (see examples in Chapter 4). The data also includes variable "productivity", denoted by p_i. It measures the total volume of the stump per effective processing hour of the machine. The total volume is based on a model that includes tree diameter and species as predictors. The effective processing time also includes the time taken by the machine to move between successive stumps, in addition to the processing time. We model these two variables as seemingly unrelated variables, ignoring the fact that the responses are related through $p_i = \frac{v_i}{t_i+m_i}$ where v_i is the predicted stump volume and m_i is the time needed to move to the location of stump i from the location of stump $i-1$.

Let us consider the following models for these two variables:

$$t_i = \beta_1^{(t)} + \beta_2^{(t)} d_i + \beta_3^{(t)} M2_i + \beta_4^{(t)} M2_i d_i + \beta_5^{(t)} M3_i + \beta_6^{(t)} M3_i d_i + e_i^{(t)}$$
$$p_i = \beta_1^{(p)} + \beta_2^{(p)} d_i + \beta_3^{(p)} M2_i + \beta_4^{(p)} M2_i d_i + \beta_5^{(p)} M3_i + \beta_6^{(p)} M3_i d_i + e_i^{(p)} .$$

These models are initially fitted separately using OLS. The empirical correlation among the residuals is negative and rather large.

```
> modT<-gls(Time~Machine*Diameter,data=stumplift)
> modP<-gls(Productivity~Machine*Diameter,data=stumplift)
> cor(resid(modP),resid(modT))
[1] -0.7345888
```

The same estimate of cross-model correlation is obtained by utilizing the model for the sum of processing time and productivity:

```
> modTplusP<-gls(I(Productivity+Time)~Machine*Diameter,data=stumplift)
> covTP<-(modTplusP$sigma^2-modT$sigma^2-modP$sigma^2)/2
> covTP/(modT$sigma*modP$sigma)
[1] -0.7345888
```

The joint model can be fitted using IGLS, which is implemented in function `systemfit::systemfit`. Here, we use 10 iterations.

```
> library(systemfit)
> eqT<-Time~Machine*Diameter
> eqP<-Productivity~Machine*Diameter
> system<-list(Time=eqT,Productivity=eqP)
```

```
> modSUR<-systemfit(system,"SUR",data=stumplift,maxit=10)
> summary(modSUR)
The correlations of the residuals
                  Time Productivity
Time          1.000000    -0.734589
Productivity -0.734589     1.000000

SUR estimates for 'Time' (equation 1)
                    Estimate Std. Error  t value   Pr(>|t|)
(Intercept)        -9.509293   7.723040 -1.23129 0.21881907
Machine2          -44.931988  10.771147 -4.17151 3.5930e-05 ***
Machine3          -43.777689  13.110970 -3.33901 0.00090634 ***
Diameter            1.621759   0.218281  7.42968 5.0360e-13 ***
Machine2:Diameter   2.334425   0.300030  7.78064 4.4631e-14 ***
Machine3:Diameter   1.907349   0.367602  5.18863 3.1340e-07 ***

Residual standard error: 26.203231 on 479 degrees of freedom
Number of observations: 485 Degrees of Freedom: 479
SSR: 328885.858283 MSE: 686.609307 Root MSE: 26.203231
Multiple R-Squared: 0.61 Adjusted R-Squared: 0.605929

SUR estimates for 'Productivity' (equation 2)
                    Estimate Std. Error  t value   Pr(>|t|)
(Intercept)       -0.7488020  0.9340094 -0.80171  0.4231199
Machine2           2.2378383  1.3026415  1.71792  0.0864568 .
Machine3           2.6487354  1.5856150  1.67048  0.0954780 .
Diameter           0.3360912  0.0263985 12.73145 < 2.22e-16 ***
Machine2:Diameter -0.1941110  0.0362850 -5.34961 1.3683e-07 ***
Machine3:Diameter -0.1707520  0.0444570 -3.84083  0.0001391 ***

Residual standard error: 3.168968 on 479 degrees of freedom
Number of observations: 485 Degrees of Freedom: 479
SSR: 4810.288085 MSE: 10.042355 Root MSE: 3.168968
Multiple R-Squared: 0.428163 Adjusted R-Squared: 0.422194
```

In addition, **systemfit** provided the same estimate of cross-model correlation to that computed earlier using the residuals of separate OLS fits. Both models have the same predictors, so the parameter estimates are the same as in the separate OLS fits, as can be seen comparing the coefficients for processing time below to those from the fit using `systemfit`.

```
> summary(modT)
                      Value Std.Error   t-value p-value
(Intercept)        -9.50929  7.723040 -1.231289  0.2188
Machine2          -44.93199 10.771147 -4.171514  0.0000
Machine3          -43.77769 13.110970 -3.339012  0.0009
Diameter            1.62176  0.218281  7.429681  0.0000
Machine2:Diameter   2.33442  0.300030  7.780640  0.0000
Machine3:Diameter   1.90735  0.367602  5.188633  0.0000
```

We could also fit the model using REML. For that purpose, we define a data set that corresponds to the model formulation of Equation (9.3). The diagonal elements of matrix (9.4) are defined using variance function **varIdent**, which assumes that the variance for level k of the categorical argument is

$$\mathrm{var}(e_i) = \sigma^2 \delta_k^2 .$$

The off-diagonal elements are parameterized using the argument `cor=corSymm(form=~1|obs)`, which specifies a separate correlation coefficient for each pair of observations within the groups specified by the grouping factor `obs`. If there were more than two responses, the order of the observations within each class of the grouping variable would determine which pairs of observations correspond to the same pair. Therefore, it is important that the responses within each observation are in the same order and that all responses are observed for all sampling units. For implementation in the case of missing data, see Web Example 9.6.

```
> # Observation id (obs) is needed to tell R which rows are associated with the same sampling unit
> stumplift$obs<-1:dim(stumplift)[1]
> stumplift$M1<-as.numeric(stumplift$Machine==2)

> # Generate data sets for both responses and pool them
> dat1<-data.frame(response=1,obs=stumplift$obs,y=stumplift$Time,
+                  model.matrix(Time~Machine*Diameter,data=stumplift),0,0,0,0,0,0)
```

```
> dat2<-data.frame(response=2,obs=stumplift$obs,y=stumplift$Productivity,0,0,0,0,0,0,
+                  model.matrix(Productivity~Machine*Diameter,data=stumplift))
> names(dat1)[-(1:3)]<-names(dat2)[-(1:3)]<-c("IT","M2T","M3T","DT","M2DT","M3DT",
+                                              "IP","M2P","M3P","DP","M2DP","M3DP")
> mvdat<-rbind(dat1,dat2)

> # Show the lines for first five observations and the last for both responses
> mvdat[c(1:5,485:490,970),]
     response obs     y IT M2T M3T   DT M2DT M3DT IP M2P M3P   DP M2DP M3DP
1           1   1 43.80  1   0   0 37.0    0  0.0  0   0   0  0.0    0  0.0
2           1   2 71.40  1   0   0 50.0    0  0.0  0   0   0  0.0    0  0.0
3           1   3 51.60  1   0   0 28.5    0  0.0  0   0   0  0.0    0  0.0
4           1   4 16.20  1   0   0 19.0    0  0.0  0   0   0  0.0    0  0.0
5           1   5 26.40  1   0   0 20.5    0  0.0  0   0   0  0.0    0  0.0

485         1 485 27.00  1   0   1 19.5    0 19.5  0   0   0  0.0    0  0.0
1100        2   1 10.13  0   0   0  0.0    0  0.0  1   0   0 37.0    0  0.0
2100        2   2 16.17  0   0   0  0.0    0  0.0  1   0   0 50.0    0  0.0
3100        2   3  5.24  0   0   0  0.0    0  0.0  1   0   0 28.5    0  0.0
486         2   4  6.75  0   0   0  0.0    0  0.0  1   0   0 19.0    0  0.0
510         2   5  5.24  0   0   0  0.0    0  0.0  1   0   0 20.5    0  0.0

4851        2 485  3.47  0   0   0  0.0    0  0.0  1   0   1 19.5    0 19.5

> library(nlme)
> mvmod1<-gls(y~-1+IT+M2T+M3T+DT+M2DT+M3DT+IP+M2P+M3P+DP+M2DP+M3DP,
+             weights=varIdent(form=~1|response),
+             corr=corSymm(form=~1|obs),
+             data=mvdat)
> summary(mvmod1)
Correlation Structure: General
 Formula: ~1 | obs
 Parameter estimate(s):
 Correlation:
  1
2 -0.735
Variance function:
 Structure: Different standard deviations per stratum
 Formula: ~1 | response
 Parameter estimates:
       1        2
1.000000 0.120938

Coefficients:
          Value Std.Error   t-value p-value
IT     -9.50929  7.723041 -1.231289  0.2185
M2T   -44.93199 10.771147 -4.171514  0.0000
M3T   -43.77769 13.110970 -3.339012  0.0009
DT      1.62176  0.218281  7.429681  0.0000
M2DT    2.33442  0.300030  7.780640  0.0000
M3DT    1.90735  0.367602  5.188633  0.0000
IP     -0.74880  0.934009 -0.801707  0.4229
M2P     2.23784  1.302641  1.717923  0.0861
M3P     2.64874  1.585615  1.670478  0.0952
DP      0.33609  0.026398 12.731450  0.0000
M2DP   -0.19411  0.036285 -5.349615  0.0000
M3DP   -0.17075  0.044457 -3.840835  0.0001

Residual standard error: 26.20323
Degrees of freedom: 970 total; 958 residual
```

The model output is practically identical to that of `modSUR`. The residual variance for the processing time is $\widehat{\sigma}_t^2 = (1 \times 26.20)^2 = 26.20^2$. For productivity, it is $\widehat{\sigma}_p^2 = (26.20 \times 0.1209)^2 = 3.169^2$.

In our example, the fixed parts of both models included the same terms. Therefore, the parameter estimates from a SUR fit are the same as those from the separate OLS fits. If, for example, the model for processing time would also include second-order terms, then the parameter estimates from SUR would differ from those based on separate fits.

We already know from Example 4.13 that the variance of processing time increases as a function of fitted value. This is also the case with productivity. Therefore, it might be realistic to assume that

$$\mathrm{var}(e_i^{(r)}) = \sigma_r^2 |\widetilde{y}_i|^{2\delta_r}$$

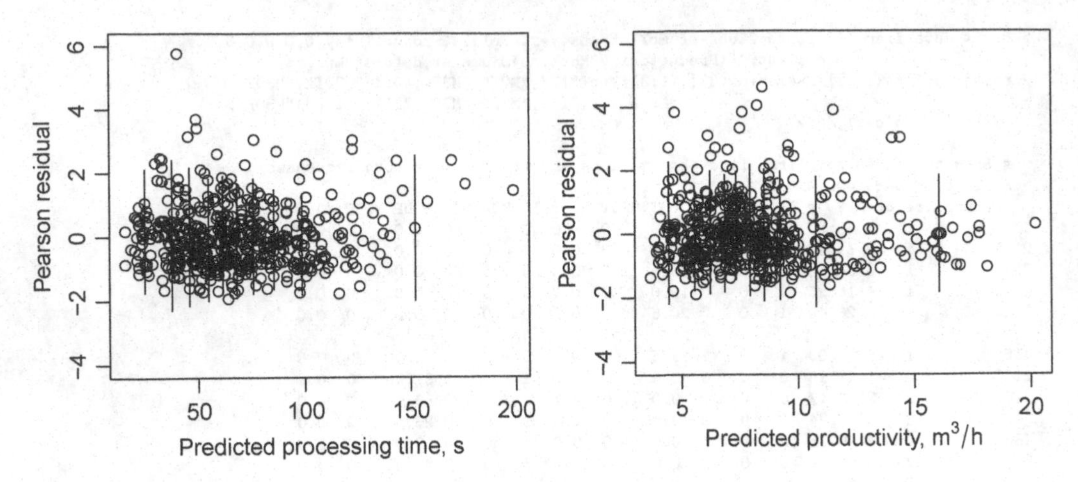

FIGURE 9.1 Pearson residuals of the multivariate model mvmod2 plotted against fitted values, separately for both response variables in Example 9.1.

where δ_r is the power parameter specified separately for the two responses ($r \in \{t, p\}$) and σ_r^2 is, correspondingly, the scaling factor of the variance function. To fit a model with this assumption, we first fit separate models for both responses to estimate the power parameters. The estimates were $\widehat{\delta}_t = 0.933$ and $\widehat{\delta}_p = 1.096$. We use these estimates and predicted values from model **mvmod1** to compute the value of $|\widetilde{y}_i|^{\widehat{\delta}_r}$, which is proportional to the standard error, to variable **z** of data set **mvdat**. Thereafter, we specify the variance function as a product of variance functions **varIdent** and **varFixed** using **varComb** to re-estimate the multivariate model, see details of these functions in Pinheiro and Bates (2000, Section 5.2). After this procedure, predictor z of the variance function utilizes the regression coefficients of the separately fitted models, which is a drawback for computations that need the estimated variance. However, this drawback can be overcome by an iterative procedure, where z is recalculated using the new model, and thereafter, the model is estimated again iteratively until the parameter estimates stabilize. In the script below, the estimate of σ^2 appears to stabilize after ten iterations.

Figure 9.1 shows that the variance is well modeled using this approach. The parameter estimates of the fitted model now differ from the previous estimates because the models for residual variance are different. For example, the estimated cross-model correlation is now even stronger than in the previous model.

```
> # Fit univariate models using varPower
> modt<-gls(Time~Machine*Diameter,
+           weights=varPower(),
+           data=stumplift)
> modp<-gls(Productivity~Machine*Diameter,
+           weights=varPower(),
+           data=stumplift)

> # Save the estimated response-specific variance function power parameters to mvdat
> mvdat$delta<-attributes(modt$apVar)$Pars[1]
> mvdat$delta[mvdat$response==2]<-attributes(modp$apVar)$Pars[1]
> mvdat$z<-fitted(mvmod1)^(2*mvdat$delta)
>
> mvmod2<-update(mvmod1,weights=varComb(varIdent(form=~1|response),varFixed(~z)))
> print(mvmod2$sigma)
[1] 0.4787793
> for (i in 1:10) {
+     mvdat$z<-fitted(mvmod2)^(2*mvdat$delta)
+     mvmod2<-update(mvmod2)
+     print(mvmod2$sigma)
+     }
[1] 0.4713546
[1] 0.4713246
.
```

```
.
[1] 0.4712704

> # left subplot of the residual plot
> plot(fitted(mvmod2)[mvdat$response==1],resid(mvmod2,type="p")[mvdat$response==1],
+       xlab="Predicted processing time, s",ylab="Pearson residual")
> mywhiskers(fitted(mvmod2)[mvdat$response==1],
+            resid(mvmod2,type="p")[mvdat$response==1],add=TRUE,se=FALSE)

> summary(mvmod2)
Correlation Structure: General
 Formula: ~1 | obs
 Parameter estimate(s):
 Correlation:
  1
2 -0.811
Combination of variance functions:
 Structure: Different standard deviations per stratum
 Formula: ~1 | response
 Parameter estimates:
      1       2
1.00000 0.63551
Variance function:
 Structure: fixed weights
 Formula: ~z

Coefficients:
           Value Std.Error   t-value p-value
IT    -15.224308  3.987833 -3.817690  0.0001
M2T   -15.345953  8.166921 -1.879038  0.0605
M3T   -21.964472  8.519718 -2.578075  0.0101
DT      1.799110  0.140665 12.790042  0.0000
M2DT    1.456560  0.274049  5.314958  0.0000
M3DT    1.262178  0.278896  4.525614  0.0000
IP     -0.878149  1.005404 -0.873429  0.3826
M2P     1.210653  1.177425  1.028220  0.3041
M3P     1.260169  1.415685  0.890148  0.3736
DP      0.340247  0.034776  9.783995  0.0000
M2DP   -0.164334  0.039824 -4.126545  0.0000
M3DP   -0.130786  0.046133 -2.834951  0.0047

Residual standard error: 0.4787793
```

9.2.3 Inference

Inference on a seemingly unrelated system of equations is based on the same procedures undertaken for the univariate model. However, the model system also allows inference on parameters of several models simultaneously. For example, this allows for hypothesis tests on whether a certain predictor has a significant effect on some of the response variables. If that hypothesis shows a significant effect, then one could continue to test the effect separately for each response variable.

It is important to note that the number of observations in model (9.3) (p. 289) is nM. Therefore, the denominator degrees of freedom in the Wald F-test (4.36) (p. 114) are much higher than in the univariate model, and the test results for an individual coefficient of the system are different compared to the results about the same coefficient based on the univariate model, even though the parameter estimates and their standard errors are the same. The difference may be substantial if M is large and n is small but vanishes as n increases. This difference does not exist if the Wald χ^2 test (4.39) (p. 115) is used (where n is implicitly assumed to be infinite). However, this is not a justified argument to favor Wald χ^2 over the Wald F test. We suggest that if inference is made about the coefficients of one equation only, it is better to use the univariate model for hypothesis testing. Otherwise, the Wald F-test for model (9.3) should be used. For more discussion about hypothesis testing of model systems, see Greene (2018).

Example 9.2. To test whether there are differences between machines in either processing time or productivity (see Example 9.1), we could consider the hypothesis:

$$H_0 : \beta_3^{(t)} = \beta_4^{(t)} = \beta_5^{(t)} = \beta_6^{(t)} = \beta_3^{(p)} = \beta_4^{(p)} = \beta_5^{(p)} = \beta_6^{(p)} = 0 .$$

Such a hypothesis is tested below using the conditional F-test for model `mvmod2`.

```
> anova(mvmod2,Terms=c("M2T","M3T","M2DT","M3DT","M2P","M3P","M2DP","M3DP"))
Denom. DF: 958
 F-test for: M2T, M3T, M2DT, M3DT, M2P, M3P, M2DP, M3DP
  numDF  F-value p-value
1     8 26.75275  <.0001
```

The small p-value indicates that there are differences between machines in either processing time or productivity. We could now progress further to test separately whether `Machine` has an effect on processing time or productivity by using separately fitted models for these responses. If the p-value from the test was so large that we could not reject the null hypothesis, then we would not undertake any further tests related to the effect of `Machine`. In this way, we could dramatically reduce the number of tests needed, thus protecting ourselves against the multiple testing problem. This approach has been used in Thitz *et al.* (2020) to test the effect of different treatments on 20 chemical compounds in birch (*Betula* spp.) leaves.

9.2.4 Prediction

Prediction from a SUR model that utilizes only the systematic part of the model is very similar to the prediction from any other linear model. We just input the values of the predictors to the model and make a prediction. The prediction intervals are calculated in a similar manner to the univariate models (recall Section 4.7.1).

However, the prediction can be improved if the value of one or more responses of the model system has been observed from a sampling unit, and other responses are being predicted for that unit. In that case, the estimated cross-model correlation can be used to carry information from one model to another to predict the residual error for the unobserved response. The prediction is a straightforward application of the linear predictor described in Section 3.5.2, and applies the methods presented in Section 4.7.2 to model (9.3).

Assume that the response variables of models $m, \ldots, M$ have been observed from the sampling unit of interest, and the other responses are to be predicted for that sampling unit. Let $\boldsymbol{y} = (y_1, \ldots, y_M)'$ be the response vector for the sampling unit in question, and $\boldsymbol{x}_1, \ldots, \boldsymbol{x}_M$ be the vectors of predictors for models $1, \ldots, M$. We define

$$\boldsymbol{h}_1 = \begin{bmatrix} y_1 \\ \vdots \\ y_{m-1} \end{bmatrix} \quad \text{and} \quad \boldsymbol{h}_2 = \begin{bmatrix} y_m \\ \vdots \\ y_M \end{bmatrix}$$

so that

$$\boldsymbol{\mu}_1 = \begin{bmatrix} \boldsymbol{x}_1'\boldsymbol{\beta}_1 \\ \boldsymbol{x}_2'\boldsymbol{\beta}_2 \\ \vdots \\ \boldsymbol{x}_{m-1}'\boldsymbol{\beta}_{m-1} \end{bmatrix} \quad \text{and} \quad \boldsymbol{\mu}_2 = \begin{bmatrix} \boldsymbol{x}_m'\boldsymbol{\beta}_m \\ \boldsymbol{x}_{m+1}'\boldsymbol{\beta}_{m+1} \\ \vdots \\ \boldsymbol{x}_M'\boldsymbol{\beta}_M \end{bmatrix} .$$

The required variance-covariance matrices are blocks of matrix $\mathbf{\Sigma}$ in Equation (9.4) (p. 289):

$$\boldsymbol{V}_1 = \begin{pmatrix} \sigma_1^2 & \varphi_{1,2} & \cdots & \varphi_{1,m-1} \\ \varphi_{1,2} & \sigma_2^2 & \cdots & \varphi_{2,m-1} \\ \vdots & \vdots & \ddots & \vdots \\ \varphi_{1,m-1} & \varphi_{2,m-1} & \cdots & \sigma_{m-1}^2 \end{pmatrix},$$

$$\boldsymbol{V}_2 = \begin{pmatrix} \sigma_m^2 & \varphi_{m,m+1} & \cdots & \varphi_{m,M} \\ \varphi_{m,m+1} & \sigma_{m+1}^2 & \cdots & \varphi_{m+1,M} \\ \vdots & \vdots & \ddots & \vdots \\ \varphi_{m,M} & \varphi_{m+1,M} & \cdots & \sigma_M^2 \end{pmatrix},$$

$$\boldsymbol{V}_{12} = \begin{pmatrix} \varphi_{1,m} & \varphi_{1,m+1} & \cdots & \varphi_{1,M} \\ \varphi_{2,m} & \varphi_{2,m+1} & \cdots & \varphi_{2,M} \\ \vdots & \vdots & \ddots & \vdots \\ \varphi_{m-1,m} & \varphi_{m-1,m+1} & \cdots & \varphi_{m-1,M} \end{pmatrix}.$$

By replacing the parameters in the above equations with their estimates, the EBLUP of $\boldsymbol{h}_1$ can be computed using Equation (3.19) (p. 60). The prediction variance that ignores the estimation errors of parameters can be estimated using Equation (3.20). The estimation errors of fixed effects can be taken into account through the formulas presented in Section 4.7.2.

Example 9.3. We illustrate the prediction using the system `mvmod2` of two models, which was previously fitted in Example 9.1. Assume that a prediction of processing time is made for a 40-cm stump processed using machine 1. The conventional prediction only utilizes information of the fixed effects. The predicted processing time is 57 seconds, and the 95% prediction interval, based on normality, is 16 – 98 seconds. The required computations are shown below.

```
> beta<-coef(mvmod2)
> x1<-c(1,0,0,40,0,0,rep(0,6))
> mu1<-x1%*%beta
> y1pred0<-mu1                              # 56.74
> V1<-sigmat^2*mu1^(2*deltat)               # 429.40
> vary1pred0<-t(x1)%*%mvmod2$varBeta%*%x1+V1 # 433.88
> # Prediction interval
> as.numeric(y1pred0)+qnorm(c(0.025,0.975))*as.numeric(sqrt(vary1pred0))
[1] 15.91423 97.56594
```

Let us then consider a hypothetical case where the productivity is known to be 17.27 m^3/h, but the processing time is unknown. This might be interesting, e.g. in some imputation situation. However, this example should be considered more as a technical illustration of the procedure, which was applied, for example, by Mehtätalo (2005), Lappi (2006), and Siipilehto (2011) under more realistic situations.

The observed productivity is greater than the expected productivity based on the systematic part of the model (12.79 m^3/h). As processing time is negatively correlated with productivity, the processing time should, therefore, be shorter than expected by the fixed part only (56.35 seconds). The script below defines the required components for prediction; we use general matrix definitions for wider applicability of the script, even though objects $\boldsymbol{\mu}_1$, $\boldsymbol{\mu}_2$, $\boldsymbol{V}_1$, $\boldsymbol{V}_2$, and $\boldsymbol{V}_{12}$ are all scalars for this two-model system. The predicted value of the processing time is 41.07 seconds, and the 95% prediction interval is much shorter than previously: 17 – 65 seconds. Our data includes this example tree (tree id 51) for which the true processing time was 41.4 seconds. This value falls well within our prediction interval.

```
> # Prediction of processing time for a stump of diameter 40 cm processed using machine 1
> # Extract the variance-covariance parameters from the fitted models
> phi<-2*plogis(attributes(mvmod2$apVar)$Pars[1])-1
> sigmat<-mvmod2$sigma
```

```
> sigmap<-exp(attributes(mvmod2$apVar)$Pars)[2]*sigmat
> deltat<-attributes(modt$apVar)$Pars[1]
> deltap<-attributes(modp$apVar)$Pars[1]
>
> # Prediction using known productivity of 17.27
> h2<-17.27
> x2<-c(rep(0,6),1,0,0,40,0,0)
> mu2<-x2%*%beta                              # 12.73
> V2<-sigmap^2*mu2^(2*deltap)                 # 23.68
> V12<-phi*sqrt(V1*V2)                        # -81.75
> y1pred1<-mu1+V12%*%solve(V2)*(h2-mu2)       # 41.07
> vary1pred1<-t(x1)%*%mvmod2$varBeta%*%x1+V1-V12%*%solve(V2)%*%t(V12) # 151.63
> as.numeric(y1pred1)+qnorm(c(0.025,0.975))*as.numeric(sqrt(vary1pred1))
[1] 16.93624 65.21108
```

9.3 Simultaneous equations

Multivariate models where the equations are directly related are called *simultaneous equation models* or *structural equation models*. They are useful in econometrics or in social sciences for studying causal paths among several dependent variables. For example, the so called Klein's small macroeconomic model includes equations for consumption, demand, and investment (Greene 2018). In forestry, model systems are sometimes used, for example, in developing a forest growth model (e.g. Hasenauer *et al.* 1998).

In simultaneous equations, the dependent variables are called *endogenous variables* which are determined within the model. *Exogenous variables* correspond to the regressor or explanatory variables in ordinary regression models. In addition, the model contains *disturbance variables* or *equation errors.*

A simultaneous equation model for observation i can be presented in a matrix equation:

$$\boldsymbol{B}\boldsymbol{y}_i + \boldsymbol{\Gamma}\boldsymbol{x}_i = \boldsymbol{e}_i \,, \tag{9.5}$$

where $\boldsymbol{B}$ is a matrix of fixed coefficients, $\boldsymbol{y}_i$ is the vector of endogenous variables, $\boldsymbol{\Gamma}$ is a matrix of fixed coefficients, $\boldsymbol{x}_i$ is a vector of exogenous variables, and $\boldsymbol{e}_i$ is the vector of random equation errors (disturbances) with mean zero and with fixed covariance matrix (see Example 9.4, p. 298). Ones appear on the diagonal of $\boldsymbol{B}$, which means that the jth row describes how y_{ij} depends on the other y-variables and x-variables. There are no intercepts in the model, i.e. both x and y-variables are centered. Generally, the disturbances of the different equations can be correlated. The disturbances are assumed to be uncorrelated with the x-variables. If the parameters of the jth row are estimated with OLS, the estimates are not consistent because the error term is correlated with the other y-variables, which contain an error that is correlated with the error term of jth row. The estimation can be carried out using such x-variables that do not appear on the jth row as *instrumental variables* (see Greene 2018). Instrumental variables are such variables that are correlated with the intended predictor variables but are uncorrelated with the error term. Instrumental variables are used as substitutes for the intended predictor variables. The identification of the model requires that there are sufficient zeroes in matrix $\boldsymbol{B}$ and in the covariance matrix of $\boldsymbol{e}$.

Example 9.4. Consider the following simultaneous equation model

$$\log v_i = \beta_0^{(1)} + \beta_1^{(1)} \log d_i + \beta_2^{(1)} \log h_i + e_i^{(1)} \tag{9.6}$$

$$\log h_i = \beta_0^{(2)} + \beta_1^{(2)} \log d_i + e_i^{(2)} \,. \tag{9.7}$$

Now, $\log h_i$ appears on the right-hand side of Equation (9.6) and on the left-hand side of Equation (9.7). Let us define $v_i^* = \log v_i - \overline{\log v}$, $h_i^* = \log h_i - \overline{\log h}$ and $d_i^* = \log d_i - \overline{\log d}$. The model can be equivalently written as

$$\begin{aligned} v_i^* &= \beta_1^{(1)} d_i^* + \beta_2^{(1)} h_i^* + e_i^{(1)} \\ h_i^* &= \beta_1^{(2)} d_i^* + e_i^{(2)}, \end{aligned}$$

which can further be reorganized to

$$\begin{aligned} v_i^* - \beta_2^{(1)} h_i^* - \beta_1^{(1)} d_i^* &= e_i^{(1)} \\ h_i^* - \beta_1^{(2)} d_i^* &= e_i^{(2)}. \end{aligned}$$

The model can be written in the form of (9.5) by defining

$$\boldsymbol{y}_i = \begin{pmatrix} v_i^* \\ h_i^* \end{pmatrix}, \boldsymbol{B} = \begin{pmatrix} 1 & -\beta_2^{(1)} \\ 0 & 1 \end{pmatrix}, x_i = d_i, \boldsymbol{\Gamma} = \begin{pmatrix} -\beta_1^{(1)} \\ -\beta_1^{(2)} \end{pmatrix} \text{ and } \boldsymbol{e}_i = \begin{pmatrix} e_i^{(1)} \\ e_i^{(2)} \end{pmatrix}.$$

We are in this book interested in models which can be used for prediction. If y-variables are not known in an application, all the predictive power in a simultaneous equation model comes from the x-variables. It is then much more straightforward to do predictions using ordinary (seemingly unrelated, see Section 9.2) regression models. Also even if some y-variables are known in a multivariate model, the prediction can utilize these values, as presented in Sections 9.2.4 and 9.4.3. That is why we do not present simultaneous equation models more closely.

However, the simultaneous equation idea can be applied when specifying an ordinary regression equation. Assume that we know that

$$y = f(x, z) + e$$

is a reasonable model for predicting y when x and z are known. If z is not known, assume that it can be predicted using equation

$$z = g(x) + u,$$

where u is a random residual error. After estimating parameters of f and g, then y can be predicted as

$$\widetilde{y} = \widehat{f}(x, \widehat{g}(x)).$$

However, it is better to estimate directly y as a function of x using model

$$y = f(x, g(x)) + e.$$

No advantage can be obtained using the two-stage procedure. The simultaneous equation idea can just be used to specify a reasonable model in terms of the exogenous variables.

Example 9.5. Let H be tree height, V be tree volume, and D be tree diameter. Assume that one wants to model the standing tree volume.

If the model is to be used for prediction in a situation where both height and diameter of the target trees are known, then a univariate model of form

$$V = f(D, H) + e$$

for the volume is sufficient, where the true height and diameter are used as predictors.

If the model is to be used for prediction in a situation where only the diameter of the target tree is known, then one might be tempted to fit a model system of form

$$V = f_1(D, H) + e_1 \tag{9.8}$$
$$H = f_2(D) + e_2. \tag{9.9}$$

In the prediction situation, the unknown H in (9.8) would be replaced by prediction $\widehat{H}$ based on the fitted model (9.9). However, $\widehat{H}$ is just a transformation of D, an estimate of $\mathrm{E}(H|D)$. In particular, it is a different random variable to H, with the same expected value (if estimation errors are ignored) but a smaller variance. That is because the total variance of H can be written as $\mathrm{var}(H) = \mathrm{var}\left[\mathrm{E}(H|D)\right] + \mathrm{E}\left[\mathrm{var}(H|D)\right]$ (Equation (2.46), p. 25) but $\mathrm{var}(\widehat{H})$ is the estimate of the first component $\mathrm{var}(\mathrm{E}(H|D))$ only. Therefore, the prediction of V using $\widehat{H}$ has less variability than the prediction based on H, leading to larger prediction variance than the estimated residual standard error of the model (9.8) would suggest. Furthermore, the prediction would also be biased even if the systematic part of the model (9.8) would give the correct expectation of $\mathrm{E}(V|D,H)$, unless the volume model (9.8) is, unrealistically, a linear function of H.

A better multivariate solution is to replace model (9.8) with

$$V = f_1(D, \widehat{H}) + e_1 \tag{9.10}$$

where predicted height from model (9.9) is now used as a predictor in (9.10), in addition to D. If model (9.10) is formulated and estimated so that the systematic part gives the correct expectation of $\mathrm{E}(V|D,\widehat{H})$, the model would be unbiased and give lower prediction variance than model (9.8).

However, the role of the height model is only to specify such a transformation of the diameter that realistically describes the height-diameter relationship. Therefore, the model system could be replaced with a univariate model

$$V = f_3(D) + e_3$$

where the function $f_3(D)$ is formulated so that it fits well with the volume-diameter relationship. Knowledge of the well-fitting height-diameter function could be used in formulation of the model, but no parameters of that model would need be fixed based on a fitted height-diameter model. This approach would lead to better fitting models than the system based on Equations (9.10) and (9.9).

9.4 Multivariate mixed-effects models

9.4.1 Model formulation

Seemingly unrelated mixed-effects models are an extension of the SUR model to grouped data, where group effects are modeled as random effects. Assume that we have M individual mixed-effects models for group i:

$$\begin{aligned}
\boldsymbol{y}_i^{(1)} &= \boldsymbol{X}_i^{(1)}\boldsymbol{\beta}_1 + \boldsymbol{Z}_i^{(1)}\boldsymbol{b}_i^{(1)} + \boldsymbol{\epsilon}_i^{(1)} \\
\boldsymbol{y}_i^{(2)} &= \boldsymbol{X}_i^{(2)}\boldsymbol{\beta}_2 + \boldsymbol{Z}_i^{(2)}\boldsymbol{b}_i^{(2)} + \boldsymbol{\epsilon}_i^{(2)} \\
&\vdots \\
\boldsymbol{y}_i^{(M)} &= \boldsymbol{X}_i^{(M)}\boldsymbol{\beta}_M + \boldsymbol{Z}_i^{(M)}\boldsymbol{b}_i^{(M)} + \boldsymbol{\epsilon}_i^{(M)} ,
\end{aligned}$$

where $\mathrm{var}(\boldsymbol{\epsilon}_i^{(m)}) = \sigma_m^2 \boldsymbol{I}_{n_i \times n_i}$ and $\mathrm{var}(\boldsymbol{b}_i^{(m)}) = \boldsymbol{D}_*^{(m)}$ for $m = 1, \ldots, M$ (recall Section 5.2.1). The random effects may be correlated across the responses, i.e. $\mathrm{cov}(\boldsymbol{b}_i^{(m)}, \boldsymbol{b}_i^{(m')}{}') = \boldsymbol{C}_i^{(mm')}$ may be non-zero. We assume that the cross-model correlation is the same for all groups and, therefore, denote $\boldsymbol{C}_i^{(mm')} = \boldsymbol{C}_*^{(mm')}$. The cross-model covariance of the random effects models the joint behavior of the group-specific curves. For example, if the models include random intercepts, so that $\boldsymbol{Z}_i^{(m)}\boldsymbol{b}_i^{(m)} = \boldsymbol{1}b_i^{(m)}$ and $\boldsymbol{C}_*^{(mm')}$ is scalar, a positive cross-model covariance between models m and m' means that, for groups where the group-specific curve is above the curve of a typical group for response m, the group-specific curve for response m' is usually above the curve of a typical group as well.

In addition to the cross-model correlation of random effects, the residual errors may also be correlated across models, similar to model (9.3) on p. 289. That is, we may assume that $\text{cov}(\epsilon_{ij}^{(m)}, \epsilon_{ij}^{(m')}) = \varphi_{mm'}$ is non-zero. For example, a positive correlation between the residual errors of models m and m' means that for an individual sampling unit j within group i, a positive residual error of model m is usually associated with a positive residual error of model m'. If the same individual sampling units are not observed for all responses, the cross-model correlation for the residual errors cannot be estimated.

For estimation, the model system is written as a univariate model. To do that for a single group, we define the following matrices and vectors:

$$\boldsymbol{y}_i = \begin{bmatrix} \boldsymbol{y}_i^{(1)} \\ \boldsymbol{y}_i^{(2)} \\ \vdots \\ \boldsymbol{y}_i^{(M)} \end{bmatrix}, \boldsymbol{X}_i = \begin{bmatrix} \boldsymbol{X}_i^{(1)} & \boldsymbol{0} & \cdots & \boldsymbol{0} \\ \boldsymbol{0} & \boldsymbol{X}_i^{(2)} & \cdots & \boldsymbol{0} \\ \vdots & \vdots & \ddots & \vdots \\ \boldsymbol{0} & \boldsymbol{0} & \cdots & \boldsymbol{X}_i^{(M)} \end{bmatrix}, \boldsymbol{\beta} = \begin{bmatrix} \boldsymbol{\beta}_1 \\ \boldsymbol{\beta}_2 \\ \vdots \\ \boldsymbol{\beta}_M \end{bmatrix},$$

$$\boldsymbol{Z}_i = \begin{bmatrix} \boldsymbol{Z}_i^{(1)} & \boldsymbol{0} & \cdots & \boldsymbol{0} \\ \boldsymbol{0} & \boldsymbol{Z}_i^{(2)} & \cdots & \boldsymbol{0} \\ \vdots & \vdots & \ddots & \vdots \\ \boldsymbol{0} & \boldsymbol{0} & \cdots & \boldsymbol{Z}_i^{(M)} \end{bmatrix}, \boldsymbol{b}_i = \begin{bmatrix} \boldsymbol{b}_i^{(1)} \\ \boldsymbol{b}_i^{(2)} \\ \vdots \\ \boldsymbol{b}_i^{(M)} \end{bmatrix}, \boldsymbol{\epsilon}_i = \begin{bmatrix} \boldsymbol{\epsilon}_i^{(1)} \\ \boldsymbol{\epsilon}_i^{(2)} \\ \vdots \\ \boldsymbol{\epsilon}_i^{(M)} \end{bmatrix},$$

$$\boldsymbol{R}_i = \boldsymbol{\Sigma}_* \otimes \boldsymbol{I},\ \boldsymbol{D}_* = \begin{bmatrix} \boldsymbol{D}_*^{(1)} & \boldsymbol{C}_*^{(12)} & \cdots & \boldsymbol{C}_*^{(1M)} \\ \boldsymbol{C}_*^{(12)\prime} & \boldsymbol{D}_*^{(2)} & \cdots & \boldsymbol{C}_*^{(2M)} \\ \vdots & \vdots & \ddots & \vdots \\ \boldsymbol{C}_*^{(1M)\prime} & \boldsymbol{C}_*^{(2M)\prime} & \cdots & \boldsymbol{D}_*^{(M)} \end{bmatrix},$$

where (recall (9.4), p. 289)

$$\boldsymbol{\Sigma}_* = \begin{bmatrix} \sigma_1 & \varphi_{12} & \cdots & \varphi_{1M} \\ \varphi_{12} & \sigma_2 & \cdots & \varphi_{2M} \\ \vdots & \vdots & \ddots & \vdots \\ \varphi_{1M} & \varphi_{2M} & \cdots & \sigma_M \end{bmatrix}$$

if observations of all response variables are taken from the same sampling units. If the observations are taken from different sampling units of the group, then the off-diagonal elements of $\boldsymbol{\Sigma}_*$ are zeroes.

We can now state the model in the familiar form (see Section 5.2.1)

$$\boldsymbol{y}_i = \boldsymbol{X}_i\boldsymbol{\beta} + \boldsymbol{Z}_i\boldsymbol{b}_i + \boldsymbol{\epsilon}_i\,, \tag{9.11}$$

where $\text{var}(\boldsymbol{b}_i) = \boldsymbol{D}_*$ and $\text{var}(\boldsymbol{\epsilon}_i) = \boldsymbol{R}_i$. Once the model for a single group is defined, the model for all data is defined in the same way as for the univariate model (see Section 5.2.3). It should be noted that the number of parameters increases rapidly when the number of responses increases, especially if random slopes are included in addition to random intercepts.

An alternative to the above-specified formulation for a single group is to define $\boldsymbol{y}_i$ so that the different responses of each sampling unit are first stacked on each other, and the resulting vectors are further stacked on each other to formulate vector $\boldsymbol{y}_i$. That would change the order of observations, so that matrix $\boldsymbol{R}_i$ is block-diagonal. Parameter estimation of the model formulated in this way is computationally less intensive than in the formulation presented above.

Model definition for multiple levels of grouping is possible by combining the ideas presented above with the ideas from Sections 6.1.1 and 6.2. For discussion of such models, see e.g. Goldstein (2003) and Hox *et al.* (2018).

9.4.2 Estimation and inference

Model (9.11) is a special case of the univariate linear mixed-effects model. Therefore, the parameter estimation and inference can be based on REML and GLS, as described in Section 5.3. Using this estimation procedure in `R` requires organizing the data in such a way that the univariate fitting procedures can be applied, see Example 9.6 below for illustration.

If the number of responses is large, convergence problems frequently occur due to the large number of parameters in matrices $\boldsymbol{D}_*$ and $\boldsymbol{\Sigma}_*$. As a solution, Fieuws and Verbeke (2006) proposed a pairwise fitting approach. Instead of fitting the large system of M mixed-effects models jointly, one could formulate all the $M(M-1)/2$ pairs of mixed-effects models and fit each of them jointly using REML. Each of these bivariate models would provide estimates of the cross-model covariances of random effects and residuals for the specific pair in question. In addition, one would obtain $M-1$ different unbiased estimates of $\boldsymbol{\beta}$, $\boldsymbol{D}_*^{(m)}$, and σ_m^2 for $m = 1, \ldots, M$, which are averaged over the fitted model pairs to obtain the final estimates. They also provided a way to estimate the variance of the averaged estimates. A forestry application of their procedure can be found in Mehtätalo *et al.* (2008).

Often, a joint estimation of the model system does not provide improved accuracy of the fixed effect estimates. Moreover, the main motivation may be in the estimation of the cross-model covariances or correlations of the random effects and residual errors to be used in prediction or simulation. One might be tempted to estimate the individual models separately, and, thereafter, estimate the cross-model covariances using the predicted random effects and residuals. However, because of the shrinkage of the predicted random effects (recall Section 5.3), the variances and covariances estimated in this way are downward biased, and this method is not acceptable. An acceptable strategy is to estimate the model so that the group effects are fixed, and, thereafter, estimate the variances based on these fixed group effects (see Section 5.3.1).

A simple alternative for the estimation of models with a random intercept is to use the method proposed by Searle and Rounsaville (1974). They suggested that the univariate mixed-effects model is estimated for all response variables separately. In addition, a similar model is formulated for all possible sums of two response variables. For example, consider a system of two variance component models with random intercept:

$$
\begin{aligned}
y_{ij}^{(1)} &= \mu_1 + b_i^{(1)} + \epsilon_{ij}^{(1)} \\
y_{ij}^{(2)} &= \mu_2 + b_i^{(2)} + \epsilon_{ij}^{(2)} .
\end{aligned}
$$

Summing the left- and right-hand sides separately and reorganizing, we can see that the model for the sum $z_{ij} = y_{ij}^{(1)} + y_{ij}^{(2)}$ can be stated as

$$
z_{ij} = \mu_z + b_i^{(z)} + \epsilon_{ij}^{(z)}
$$

where $\mu_z = \mu_1 + \mu_2$, $b_i^{(z)} = b_i^{(1)} + b_i^{(2)}$, and $\epsilon_i^{(z)} = \epsilon_i^{(1)} + \epsilon_i^{(2)}$. Fitting these three models separately gives estimates of $\mathrm{var}(b_i^{(1)})$, $\mathrm{var}(b_i^{(2)})$, and $\mathrm{var}(b_i^{(1)} + b_i^{(2)}) = \mathrm{var}(b_i^{(z)})$. Based on the equation

$$\mathrm{var}(X+Y) = \mathrm{var}(X) + \mathrm{var}(Y) + 2\,\mathrm{cov}(X,Y)\,,$$

the estimated variances can be used to estimate the cross-model covariance of random effects as

$$\widehat{\mathrm{cov}}(b_i^{(1)}, b_i^{(2)}) = \frac{\widehat{\mathrm{var}}(b_i^{(z)}) - \widehat{\mathrm{var}}(b_i^{(1)}) - \widehat{\mathrm{var}}(b_i^{(2)})}{2}\,.$$

Correspondingly, the cross-model covariance of residual errors can be estimated as

$$\widehat{\mathrm{cov}}(\epsilon_{ij}^{(1)}, \epsilon_{ij}^{(2)}) = \frac{\widehat{\mathrm{var}}(\epsilon_{ij}^{(z)}) - \widehat{\mathrm{var}}(\epsilon_{ij}^{(1)}) - \widehat{\mathrm{var}}(\epsilon_{ij}^{(2)})}{2}\,.$$

This idea also extends to multiple levels of grouping. If the fixed predictors of the two models differ, then the sum model should include all predictors that are included in either of the models. If both models fit well, the regression coefficients of the sum model should be approximately equal to the sum of the estimates from the two component models. If that is not the case, it may indicate problems in the systematic part similar to the case of SUR model. An application of this approach in a forestry context can be found in Lappi (1986). An extension to the model with random intercept and slope is, however, not possible without additional restrictions to $\boldsymbol{D}_*$.

Example 9.6. Maltamo *et al.* (2012) developed a system of seemingly unrelated mixed-effects models for five tree specific characteristics: Tree diameter at breast height (D, cm), tree height (H, m), tree stem volume (V, m^3), height of dead branch (HDB, m), and crown base height (HCB, m). The predictors were tree-specific quantities based on airborne laser scanning. The data included a total of 1510 individual Scots Pine trees from 56 sample plots in a study area in Eastern Finland. The ultimate aim is to use the models to cross-calibrate the models (see Example 9.8). The component models for tree j on plot i were as follows

$$\begin{aligned}
D_{ij} &= \beta_1^{(1)} + \beta_2^{(1)} h100_{ij} + \beta_3^{(1)} \bar{h}^*_{ij} + \beta_4^{(1)} \sqrt{h30_{ij}} + b_i^{(11)} + b_i^{(12)} h100_{ij} + \epsilon_{ij}^{(1)} \\
H_{ij} &= \beta_1^{(2)} + \beta_2^{(2)} h100_{ij} + \beta_3^{(2)} h80_{ij} + b_i^{(21)} + b_i^{(22)} h100_{ij} + \epsilon_{ij}^{(2)} \\
\ln V_{ij} &= \beta_1^{(3)} + \beta_2^{(3)} h100_{ij} + \beta_3^{(3)} \ln h30^*{}_{ij} + \beta_4^{(3)} \ln h70_{ij} + b_i^{(31)} + b_i^{(32)} h100_{ij} + \epsilon_{ij}^{(3)} \\
\sqrt{HDB_{ij}} &= \beta_1^{(4)} + \beta_2^{(4)} h100_{ij} + \beta_3^{(4)} h70^*_{ij} + b_i^{(41)} + b_i^{(42)} h100_{ij} + \epsilon_{ij}^{(4)} \\
HCB_{ij} &= \beta_1^{(5)} + \beta_2^{(5)} h20_{ij} + \beta_3^{(5)} \bar{h}^*_{ij} + \beta_4^{(5)} veg^*_{ij} + b_i^{(51)} + b_i^{(52)} h20_{ij} + \epsilon_{ij}^{(5)}\,,
\end{aligned}$$

where hxx refers to the xx^{th} quantile and $\bar{h}$ to the mean of the ALS return height; veg is the proportion of returns from vegetation. An asterisk is used for predictors that are based on a 250 m^2 neighborhood around the tree, the other predictors are based on the returns from the crown of the tree in question. All models include both random intercept and slope, therefore the variance-covariance matrix of random effects is a 10 by 10 matrix with 10 variances and 45 covariances. The matrix $\boldsymbol{\Sigma}_*$ is an unstructured, positive definite 5 by 5 matrix.

There are few missing values of HDB and HCB in the data, therefore the residual variance-covariance matrix for the data where missing values are removed is obtained from $\boldsymbol{\Sigma}_* \otimes \boldsymbol{I}$ by removing the rows and columns corresponding to the missing values.

The modeling data set is available in data set `alsTree`. To fit the model, we reorganize the data so that the five responses are all in variable y and variable `index` specifies the type of response for each row. In addition, the predictors of each of the responses are in separate blocks so that columns $5, \ldots, 22$ include the model matrix of the model system.

```
> data(alsTree)
> dat<-rbind(
+     cbind(plot=alsTree$plot,tree=alsTree$tree,index=1,y=alsTree$DBH,
+           conD=1,hmaxD=alsTree$hmax,a_hmeanD=alsTree$a_hmean,sqh30D=sqrt(alsTree$h30),
+           conH=0,hmaxH=0,h80H=0,
+           conV=0,hmaxV=0,lna_h30V=0,lnh70V=0,
```

```
+           conHDB=0,hmaxHDB=0,a_h70HDB=0,
+           conHCB=0,h20HCB=0,a_hmeanHCB=0,lna_vegHCB=0),
+      cbind(plot=alsTree$plot,tree=alsTree$tree,index=2,y=alsTree$H,
+           0,0,0,0,1,alsTree$hmax,alsTree$h80,0,0,0,0,0,0,0,0,0,0,0),
+      cbind(plot=alsTree$plot,tree=alsTree$tree,index=3,y=log(alsTree$V),
+           0,0,0,0,0,0,0,1,alsTree$hmax,log(alsTree$a_h30),log(alsTree$h70),0,0,0,0,0,0,0),
+      cbind(alsTree$plot,alsTree$tree,index=4,y=sqrt(alsTree$HDB),
+           0,0,0,0,0,0,0,0,0,0,0,1,alsTree$hmax,alsTree$a_h70,0,0,0,0),
+      cbind(alsTree$plot,alsTree$tree,index=5,y=alsTree$HCB,
+           0,0,0,0,0,0,0,0,0,0,0,0,0,0,1,alsTree$h20,alsTree$a_hmean,log(alsTree$a_veg)))
> dat<-as.data.frame(dat)
> dat<-dat[!is.na(dat$y),]
> head(dat)
  plot tree index     y conD hmaxD a_hmeanD   sqh30D conH hmaxH h80H conV hmaxV
1    1   35     1 24.70    1 22.80 14.62099 4.034848    0     0    0    0     0
2    1   30     1 24.60    1 20.87 14.43929 3.865230    0     0    0    0     0
  lna_h30V lnh70V conHDB hmaxHDB a_h70HDB conHCB h20HCB a_hmeanHCB lna_vegHCB
1        0      0      0       0        0      0      0          0          0
2        0      0      0       0        0      0      0          0          0
```

The model is fitted below with function `nlme::lme` using REML. The common random intercept is dropped and the response-specific random intercepts are specified through the random slopes for the `con`-variables. The response-specific random slopes are defined by the other terms in the argument `random`. The `weight` argument specifies different variances for each response, i.e. the structure for the diagonal of matrix $\boldsymbol{\Sigma}_*$. The off-diagonal structure is determined by the argument `corr`. Variable `index` is used as the covariate to specify which variables are present for each tree for estimation of the cross-model covariances of residual errors (compare to Example 9.1). Because of the large number of parameters and observations, the estimation took about 30 minutes using a modern PC at the time of writing. To monitor the fitting progress, we have requested function `lme` to produce some intermediate results by using argument `control`. On each row, the first value is the negative log-likelihood at the iteration and the other values are current estimates of the model parameters.

```
> syssur2 <- lme(y ~ -1+conD+hmaxD+a_hmeanD+sqh30D+conH+hmaxH+h80H+conV+hmaxV+lna_h30V+
+                lnh70V+conHDB+hmaxHDB+a_h70HDB+conHCB+h20HCB+a_hmeanHCB+lna_vegHCB,
+
+                random = ~ -1+conD+hmaxD+conH+hmaxH+conV+hmaxV+conHDB+hmaxHDB+conHCB+h20HCB|plot,
+                weights=varIdent(form = ~1|index),corr=corSymm(form = ~index|plot/tree),data = dat,
+                control=list(msVerbose=TRUE,maxIter=5000,msMaxIter=5000,msMaxEval=5000,niterEM=50))
  0:     34302.509:  1.27227  2.55406  2.04874  3.96844  2.55755  4.39270 ...
.
.
528:     28771.050:  2.55430  4.18152  3.56751  4.37642  3.39000  4.90482 ...
```

The estimated model is printed below. The output shows first the estimates of $\boldsymbol{\beta}$. Thereafter there are the standard deviations and correlations based on the estimate of $\boldsymbol{D}_*$. Term `Residual` specifies the scaling factor σ of the residual variance-covariance matrix. Thereafter the output includes the estimated correlation matrix corresponding to the matrix $\boldsymbol{\Sigma}_*$. Finally, there are the parameters of the variance function based on the diagonal elements of $\boldsymbol{\Sigma}_*$. The response-specific residual standard errors are obtained by multiplying these by $\widehat{\sigma} = 2.785$.

```
> syssur
Linear mixed-effects model fit by REML
  Data: dat
  Log-restricted-likelihood: -7192.327
  Fixed: y ~ -1 + conD + hmaxD + a_hmeanD + sqh30D + conH + hmaxH + h80H + ...
       conD       hmaxD    a_hmeanD      sqh30D        conH       hmaxH        h80H
-2.54985573  2.00148362 -0.86760449  0.14440159  1.30989628  0.68447222  0.36215703
       conV       hmaxV    lna_h30V      lnh70V      conHDB     hmaxHDB    a_h70HDB
-4.76956602  0.19724320 -1.35296653  1.31101779 -0.76934968  0.02642351  0.09144978
     conHCB      h20HCB  a_hmeanHCB  lna_vegHCB
 1.16220250  0.59044392  0.21335231  1.30428268

Random effects:
 Formula: ~-1 + conD + hmaxD + conH + hmaxH + conV + hmaxV + conHDB + ... | plot
 Structure: General positive-definite, Log-Cholesky parametrization
        StdDev     Corr
conD    4.58022765 conD   hmaxD  conH   hmaxH  conV   hmaxV  conHDB hmxHDB conHCB
```

```
hmaxD    0.28042428 -0.933
conH     0.72119756 -0.543  0.474
hmaxH    0.04981984  0.430 -0.328 -0.890
conV     0.40077687 -0.167  0.227  0.081  0.002
hmaxV    0.02637301  0.241 -0.182  0.027 -0.040 -0.896
conHDB   1.02818763  0.929 -0.914 -0.556  0.403 -0.430  0.440
hmaxHDB  0.06523172 -0.743  0.797  0.301 -0.201  0.432 -0.436 -0.899
conHCB   0.60566879 -0.749  0.718  0.224 -0.308 -0.124 -0.057 -0.681  0.724
h20HCB   0.05427928 -0.346  0.210 -0.055  0.186 -0.097  0.043 -0.209  0.083 -0.025
Residual 2.78508950

Correlation Structure: General
 Formula: ~index | plot/tree
 Parameter estimate(s):
 Correlation:
  1      2      3      4
2  0.113
3  0.916  0.226
4 -0.147  0.047 -0.150
5 -0.294  0.171 -0.259  0.076
Variance function:
 Structure: Different standard deviations per stratum
 Formula: ~1 | index
 Parameter estimates:
        1         2         3         4         5
1.0000000 0.2475122 0.1143484 0.1241762 0.3824491
Number of Observations: 7167
Number of Groups: 56
```

If the estimation had failed to converge, we could have fitted the model using the pairwise fitting approach. With 5 responses, we would get a total of 10 pairs of models, each of which would be fitted separately. Thereafter the results of the 10 models would have been pooled to a single model through averaging whenever several alternative estimates for the same parameter are available. Implementing the pairwise fitting approach is left as an exercise.

Example 9.7. To illustrate the approach of Searle and Rounsaville (1974) in the estimation of cross-model covariances of random effects and residuals, consider the model system for tree diameter and height in data **alsTree** (see Example 9.6). The univariate models for D, H, and $D + H$ are

$$\begin{aligned} D_{ij} &= \boldsymbol{x}_{ij}^{(1)\prime}\boldsymbol{\beta}_1 + b_i^{(1)} + \epsilon_{ij}^{(1)} \\ H_{ij} &= \boldsymbol{x}_{ij}^{(2)\prime}\boldsymbol{\beta}_2 + b_i^{(2)} + \epsilon_{ij}^{(2)} \\ D_{ij} + H_{ij} &= \boldsymbol{x}_{ij}^{(3)\prime}\boldsymbol{\beta}_3 + b_i^{(3)} + \epsilon_{ij}^{(3)} , \end{aligned}$$

We use random intercept models because the method cannot be applied to the random slope models. The estimates of fixed effects for the sum model are approximately equal to the sum of fixed effects for the two component models. The cross-model correlations estimated using Searle and Rounsaville are $\widehat{\text{cor}}_{SR}(b_i^{(1)}, b_i^{(2)}) = 0.122$ and $\widehat{\text{cor}}_{SR}(\epsilon_{ij}^{(1)}, \epsilon_{ij}^{(2)}) = 0.080$. Estimation using REML gives $\widehat{\text{cor}}_R(b_i^{(1)}, b_i^{(2)}) = 0.164$ and $\widehat{\text{cor}}_R(\epsilon_{ij}^{(1)}, \epsilon_{ij}^{(2)}) = 0.079$.

```
> modD<-lme(DBH~hmax,random=~1|plot,data=alsTree)
> modH<-lme(H~hmax,random=~1|plot,data=alsTree)
> modDplusH<-lme(I(DBH+H)~hmax,random=~1|plot,data=alsTree)

> fixef(modDplusH)
(Intercept)        hmax
  -7.401627    2.810454
> fixef(modD)+fixef(modH)
(Intercept)        hmax
  -7.446093    2.813016

> covb<-(getVarCov(modDplusH)[1]-getVarCov(modD)[1]-getVarCov(modH)[1])/2
> corb<-covb/sqrt(getVarCov(modD)[1]*getVarCov(modH)[1])
> cove<-(modDplusH$sigma^2-modD$sigma^2-modH$sigma^2)/2
> core<-cove/(modD$sigma*modH$sigma)
> corb
[1] 0.122
> core
```

```
[1] 0.0803

> datDH<-rbind(cbind(plot=alsTree$plot,tree=alsTree$tree,index=1,y=alsTree$DBH,
+                    conD=1,hmaxD=alsTree$hmax,a_hmeanD=alsTree$a_hmean,
+                    sqh30D=sqrt(alsTree$h30),conH=0,hmaxH=0,h80H=0),
+              cbind(plot=alsTree$plot,tree=alsTree$tree,index=2,y=alsTree$H,
+                    0,0,0,0,conH=1,hmaxH=alsTree$hmax,h80H=alsTree$h80))
> datDH<-as.data.frame(datDH)

> sysDH<-lme(y ~ -1 + conD + hmaxD + conH + hmaxH, random = ~ -1 + conD + conH|plot,
+            weights=varIdent(form = ~1|index), corr=corSymm(form = ~1|plot/tree), data = datDH)
> sysDH
Random effects:
 Formula: ~-1 + conD + conH | plot
 Structure: General positive-definite, Log-Cholesky parametrization
         StdDev    Corr
conD     2.7796740 conD
conH     0.3661644 0.164
Residual 2.8612900

Correlation Structure: General
 Formula: ~1 | plot/tree
 Parameter estimate(s):
 Correlation:
  1
2 0.079
```

9.4.3 Prediction

Prediction based on the fixed part of the model and group-specific prediction using individual components of the system follows the principles that we have presented earlier in Sections 5.4.2 and 6.4.

The model system also allows for the utilization of the cross-model correlations of random effects to predict the group effects for all response variables using measurements of some of the responses of the target group. In addition, the model system allows for the utilization of the cross-model correlation of residual errors to predict the residual errors for all response variables using measurements of some responses of the target observation. Because these predictions are based on the cross-model correlations, they are beneficial only if these correlations are not close to zero. Furthermore, if the residual errors of individual models are (e.g. spatially or temporally) correlated, the correlations can be utilized in a similar way to our approach in Sections 4.7.2 and 6.5. The prediction is based on the EBLUP (see Section 3.5.2), and the required expectations, variances and covariances are based on the fitted model as described below.

For simplicity, consider first the situation where only the random effects are being predicted and all response variables have been observed for the sampling units $1, \ldots, n$ that are used for prediction. Start by defining the unobserved ($\boldsymbol{h}_1$) and observed ($\boldsymbol{h}_2$) parts of vector $\boldsymbol{h}$ of (3.19) (p. 60). We consider a single group i but drop the group index i from the following presentation in order to simplify notations. The unobserved part includes the random effects $\boldsymbol{h}_1 = \boldsymbol{b}$ of all models. The observed part includes y for all responses for sampling units $1, \ldots, n$ of group i, i.e. $\boldsymbol{h}_2 = \boldsymbol{y} = (\boldsymbol{y}_i^{(1)\prime}, \boldsymbol{y}_i^{(2)\prime}, \ldots, \boldsymbol{y}_i^{(M)\prime})'$. Let $\boldsymbol{X}$, $\boldsymbol{Z}$, $\boldsymbol{D}$, and $\boldsymbol{R}$ be the corresponding model matrices for the fixed and random parts and the variance-covariance matrices of random effects and residuals, with similar structure that was defined before Equation (9.11) (p. 301). Now, $\mathrm{var}(\boldsymbol{y}) = \boldsymbol{ZDZ}' + \boldsymbol{R}$ and $\mathrm{cov}(\boldsymbol{b}, \boldsymbol{y}') = \boldsymbol{DZ}'$, and the prediction of random effect can be done using the procedures presented earlier in Section 5.4.1.

To consider the case where all responses have not been observed, we adjust the procedure as follows. Denote the vector of observations by $\boldsymbol{y}_o$. It is constructed from $\boldsymbol{y}$ by removing the elements that have not been observed. For example, if responses 1 and 3 have not been observed, then $\boldsymbol{y}_o = (\boldsymbol{y}_i^{(2)\prime}, \boldsymbol{y}_i^{(4)\prime}, \ldots, \boldsymbol{y}_i^{(M)\prime})'$. We define the corresponding matrices and vectors $\boldsymbol{X}_o$, $\boldsymbol{Z}_o$, $\boldsymbol{D}_o$, $\boldsymbol{R}_o$, and $\boldsymbol{\beta}_o$, by removing the rows and columns that correspond to the unobserved responses from $\boldsymbol{X}$, $\boldsymbol{Z}$, $\boldsymbol{D}$, $\boldsymbol{R}$, and $\boldsymbol{\beta}$. Furthermore, we define matrix $\boldsymbol{C}$ by removing the columns that correspond to the unobserved response variables from $\boldsymbol{D}$ but retaining all rows. Now,

$$\begin{pmatrix} \boldsymbol{b} \\ \boldsymbol{y}_o \end{pmatrix} \sim \left[\begin{pmatrix} \boldsymbol{0} \\ \boldsymbol{X}_o\boldsymbol{\beta}_o \end{pmatrix}, \begin{pmatrix} \boldsymbol{D} & \boldsymbol{C}\boldsymbol{Z}_o' \\ \boldsymbol{Z}_o\boldsymbol{C}' & \boldsymbol{Z}_o\boldsymbol{D}_o\boldsymbol{Z}_o' + \boldsymbol{R}_o \end{pmatrix} \right].$$

The BLP of random effect vector $\boldsymbol{b}$ and its prediction variance become

$$\begin{aligned} \mathrm{BLP}(\boldsymbol{b}) &= \boldsymbol{C}\boldsymbol{Z}_o' \left(\boldsymbol{Z}_o\boldsymbol{D}_o\boldsymbol{Z}_o' + \boldsymbol{R}_o\right)^{-1} (\boldsymbol{y}_o - \boldsymbol{X}_o\boldsymbol{\beta}_o) \\ \mathrm{var}(\mathrm{BLP}(\boldsymbol{b}) - \boldsymbol{b}) &= \boldsymbol{D} - \boldsymbol{C}\boldsymbol{Z}_o' \left(\boldsymbol{Z}_o\boldsymbol{D}_o\boldsymbol{Z}_o' + \boldsymbol{R}_o\right)^{-1} \boldsymbol{Z}_o\boldsymbol{C}' . \end{aligned} \tag{9.12}$$

In practice, $\boldsymbol{\beta}$ and variance-covariance matrices are replaced by their estimates, which leads to EBLUP. The variance formula above is underestimate because it ignores the related estimation errors.

In some cases, it may be interesting to predict the residuals of the observed sampling units for all models, in addition to the random effects. This was the approach used, for example, in the taper curve model system of Lappi (1986), which includes random stand effects and tree-level residual errors. If measurements of tree diameters are available for some trees of a stand, they will include information of (a) the mean stem form in the stand, and (b), the form of that particular tree stem. The former can be estimated by using the predicted stand effects of the model, and the latter can further use the predicted residual errors for all component models. To do so, we can construct the BLUP using

$$\begin{pmatrix} \boldsymbol{\epsilon} \\ \boldsymbol{y}_o \end{pmatrix} \sim \left[\begin{pmatrix} \boldsymbol{0} \\ \boldsymbol{X}_o\boldsymbol{\beta}_o \end{pmatrix}, \begin{pmatrix} \boldsymbol{R} & \boldsymbol{B} \\ \boldsymbol{B}' & \boldsymbol{Z}_o\boldsymbol{D}_o\boldsymbol{Z}_o' + \boldsymbol{R}_o \end{pmatrix} \right], \tag{9.13}$$

where $\boldsymbol{B}$ is constructed from $\boldsymbol{R}$ by removing those columns for which we do not have measurements, but we retain all rows. Defining $\boldsymbol{h}_1 = (\boldsymbol{\epsilon}', \boldsymbol{b}')'$ and $\boldsymbol{h}_2 = \boldsymbol{y}$, we can easily see the structure of $\boldsymbol{V}_1$, $\boldsymbol{V}_2$, and $\boldsymbol{V}_{12}$, which are needed for the application of Equation 3.19. Replacing unknown parameters with their estimates leads once again, to EBLUP.

Simultaneous prediction of $\boldsymbol{b}$ and $\boldsymbol{\epsilon}$ can be based on

$$\begin{pmatrix} \boldsymbol{\epsilon} \\ \boldsymbol{b} \\ \boldsymbol{y}_o \end{pmatrix} \sim \left[\begin{pmatrix} \boldsymbol{0} \\ \boldsymbol{0} \\ \boldsymbol{X}_o\boldsymbol{\beta}_o \end{pmatrix}, \begin{pmatrix} \boldsymbol{R} & \boldsymbol{0} & \boldsymbol{B} \\ \boldsymbol{0} & \boldsymbol{D} & \boldsymbol{C}\boldsymbol{Z}_o' \\ \boldsymbol{B}' & \boldsymbol{Z}_o\boldsymbol{C}' & \boldsymbol{Z}_o\boldsymbol{D}_o\boldsymbol{Z}_o' + \boldsymbol{R}_o \end{pmatrix} \right],$$

The predictions based on this formula are identical to those based on separate prediction of $\boldsymbol{b}$ using (9.12) and $\boldsymbol{\epsilon}$ using (9.13), although this formulation also allows an estimation of $\mathrm{cov}(\widetilde{\boldsymbol{b}} - \boldsymbol{b}, (\widetilde{\boldsymbol{\epsilon}} - \boldsymbol{\epsilon})')$.

Web Example 9.8. Consider the model system that was estimated in Example 9.6. Assume that three trees have been measured for D, HDB, and HCB, with the following measurements and values of the predictors.

```
> obs
  plot tree   DBH  HDB  HCB  hmax   h20   h30   h70   h80 a_hmean  a_veg a_h30 a_h70
2    1   30 24.60 0.75 12.3 20.87 13.74 14.94 18.42 18.84   14.44 0.8952 14.51 17.53
5    1   43 28.25 0.40 13.0 20.00 13.33 14.77 17.77 18.47   14.75 0.8533 14.89 18.52
7    1   18 25.40 2.30 14.1 21.50 13.43 15.58 18.63 19.50   14.51 0.9268 14.25 17.84
```

The aim is to predict the random effects of all five models of the system using these data. Detailed description and R-scripts and outputs are shown in the extended example in the book homepage.

To illustrate the benefit from using the predicted random effects, Table 9.1 shows the observed values of all five response variables for five new trees of the same target stand. Predictions were

Fixed effects only					Fixed part + random effects				
D	H	V	HDB	HCB	D	H	V	HDB	HCB
31.2	24.4	1.0	4.8	13.4	30.8	24.4	0.9	3.3	14.2
27.3	22.8	0.6	4.5	12.5	26.7	22.8	0.6	3.1	13.2
25.3	21.7	0.5	4.1	12.3	24.5	21.7	0.5	2.9	13.0
30.7	24.9	1.0	4.7	14.7	30.3	24.8	1.0	3.2	15.6
30.3	23.8	0.8	4.7	12.4	29.8	23.8	0.8	3.3	13.1

Field measurements				
D	H	V	HDB	HCB
26.9	24.1	0.63	0.4	15.9
28.3	22.4	0.65	1.7	13.2
21.2	21.2	0.34	0.8	12.4
29.8	24.2	0.77	4.0	14.0
32.4	23.9	0.90	0.4	13.1

TABLE 9.1 Predictions using the fixed part of the model only, using the fixed part of the model and the predicted random effects, and field-measured values of the five response variables for five selected trees from a target stand. The random effects were predicted by using the observations of D, HDB, and HCB for the three sample trees in Example 9.8.

made using (a) fixed part of the model only and (b) the fixed and random parts of the model. In most cases, the predictions using fixed and random effects are closer to the observed values than the predictions based on the fixed part only. Prediction of the residual errors of the unobserved trees and adjusting the predictions of V and HDB for the back-transformation bias (see Section 10.2 and Example 5.19) are left as an exercise.

9.5 Multivariate nonlinear and generalized linear (mixed) models

The multivariate nonlinear and GLMM models can be analyzed by specifying the model as a univariate model for the data in a similar way as the linear model was specified earlier. There are different ways to stack multivariate observations into a single data matrix. Stacking can be carried out either by initially stacking the responses of each observation and stacking observations thereafter, or by stacking initially the observations of each response and thereafter the responses. In our presentation for multivariate linear mixed-effects models, the stacking was a mixture of both approaches. In each group, the stacking was first carried out by stacking observations and then the responses. Finally, the groups were stacked on each other. How the data should be organized depends on the software employed.

Formulation of the multivariate nonlinear model is a rather straightforward generalization of what we did with the linear models and can be easily implemented using function `nlme`. In the context of multivariate GLMM, it is useful if the software allows the different responses to have different distributions and link functions. The SAS procedure GLIMMIX and MLwiN have this property. Forestry application of multivariate GLMM can be found in Miina and Saksa (2006) and Lappi and Luoranen (2016) who used MLwiN and SAS GLIMMIX, respectively. In `R`, one possibility is to formulate GLMM as a nonlinear mixed-effect model, using the variance function supported by each response (recall Example 8.12). Such a model would lead to Penalized Quasi Likelihood fitting of the multivariate GLMM. Also package `MCMCglmm` supports multivariate modeling.

10

Additional Topics on Regression

10.1 Random regressors

In Chapters 4-9 we considered regression models where the x-variables are treated formally as fixed variables. However, in the examples that we have presented thus far, we have treated models where it is more natural to think that all or most x-variables are random variables. In this section, we consider more closely the implications of assuming that x-variables are random variables. The main conclusion is that practically all that we have said about fixed regressors also applies to random regressors, although some differences exist.

10.1.1 Connection between OLS linear regression and BLP

Thus far we have presented two approaches to make predictions, which are linear with respect to independent or regressor x-variables. First we presented the Best Linear Predictor (BLP) in Section 3.5.2 which is linear in regard to the independent variables. It was also indicated that we can obtain the same predictor assuming that the dependent variable and the independent variables have multivariate normal distribution. In that case, the predictor is Best Predictor (BP), not only BLP. The BLP is BP, provided the conditional expectation of y given the x-variables is linear with respect to the x-variables. This condition does not make any assumptions about the distribution of the x-variables. Thus, BLP can also be BP without multinormality.

Second, in Chapter 4, we presented the linear model, which provides us with a regression equation that can be used for prediction. In this section, we show that these two ways to make linear predictions are essentially equivalent as shown by Christensen (2011, p. 137), for example.

Let us start from a random vector

$$\begin{pmatrix} y \\ \boldsymbol{x} \end{pmatrix} = \begin{pmatrix} y \\ x^{(1)} \\ \vdots \\ x^{(p)} \end{pmatrix} .$$

If we predict y from $\boldsymbol{x}$ using BLP Equation (3.19) we obtain :

$$\begin{aligned} \widetilde{y} &= \mathrm{E}(y) + \mathrm{cov}(y, \boldsymbol{x}')\,\mathrm{var}(\boldsymbol{x})^{-1}(\boldsymbol{x} - \mathrm{E}(\boldsymbol{x})) \\ &= \mathrm{E}(y) - \mathrm{cov}(y, \boldsymbol{x}')\,\mathrm{var}(\boldsymbol{x})^{-1}\mathrm{E}(\boldsymbol{x}) + \mathrm{cov}(y, \boldsymbol{x}')\,\mathrm{var}(\boldsymbol{x})^{-1}\boldsymbol{x} . \end{aligned} \tag{10.1}$$

Thus, the predictor can be presented as:

$$\widetilde{y} = \alpha_0 + \boldsymbol{\alpha}'\boldsymbol{x} = \alpha_0 + \boldsymbol{x}'\boldsymbol{\alpha} , \tag{10.2}$$

where:

$$\boldsymbol{\alpha} = \left(\text{cov}\left(y, \boldsymbol{x}'\right) \text{var}\left(\boldsymbol{x}\right)^{-1}\right)' = \text{var}\left(\boldsymbol{x}\right)^{-1} \text{cov}\left(\boldsymbol{x}, y\right) \tag{10.3}$$

$$\alpha_0 = \text{E}\left(y\right) - \boldsymbol{\alpha}' \text{E}\left(\boldsymbol{x}\right) . \tag{10.4}$$

When the means in Equation (10.4) are estimated with Best Linear Unbiased Estimator (BLUE), the obtained predictor(10.2) is the Best Linear Unbiased Predictor (BLUP). If the variances and covariances are also estimated, the obtained predictor is Estimated BLUP (EBLUP). We will now demonstrate that the EBLUP predictor is exactly the same as that obtained from OLS regression.

If we have n observed values of y and $\boldsymbol{x}$ available, we obtain the following unbiased estimates for the means:

$$\widehat{\text{E}}\left(y\right) = \bar{y} \tag{10.5}$$

$$\widehat{\text{E}}\left(\boldsymbol{x}\right) = \bar{\boldsymbol{x}} = \begin{pmatrix} \bar{x}^{(1)} \\ \vdots \\ \bar{x}^{(p)} \end{pmatrix} . \tag{10.6}$$

With standard assumptions (see Section 3.4.2) these estimates are BLUE. The covariances can be estimated in an unbiased manner as follows. Denote first

$$\boldsymbol{y_c} = \begin{pmatrix} y_1 - \bar{y} \\ \vdots \\ y_n - \bar{y} \end{pmatrix} \tag{10.7}$$

and

$$\boldsymbol{X}_c = \begin{pmatrix} x_1^{(1)} - \bar{x}^{(1)} & \cdots & x_1^{(p)} - \bar{x}^{(p)} \\ \vdots & x_i^{(k)} - \bar{x}^{(k)} & \vdots \\ x_n^{(1)} - \bar{x}^{(1)} & \cdots & x_n^{(p)} - \bar{x}^{(p)} \end{pmatrix} . \tag{10.8}$$

Then

$$\widehat{\text{cov}}\left(y, \boldsymbol{x}\right) = \tfrac{1}{n-1} \boldsymbol{X_c}' \boldsymbol{y_c} \tag{10.9}$$

$$\widehat{\text{var}}\left(\boldsymbol{x}\right) = \tfrac{1}{n-1} \boldsymbol{X_c}' \boldsymbol{X_c} . \tag{10.10}$$

Using Equations (10.9) and (10.10) in Equation (10.3), and Equations (10.6) and (10.5) in Equation (10.4) we obtain:

$$\widehat{\boldsymbol{\alpha}} = \left(\tfrac{1}{n-1}\left(\boldsymbol{X_c}'\boldsymbol{X_c}\right)\right)^{-1} \left(\tfrac{1}{n-1}\boldsymbol{X_c}'\boldsymbol{y_c}\right) = \left(\boldsymbol{X_c}'\boldsymbol{X_c}\right)^{-1} \boldsymbol{X_c}'\boldsymbol{y_c} \tag{10.11}$$

$$\widehat{\alpha}_0 = \bar{y} - \bar{\boldsymbol{x}}'\widehat{\boldsymbol{\alpha}} \tag{10.12}$$

As the $n-1$ term cancels out in Equation (10.11), we will obtain the same predictor if the maximum likelihood estimates with n in the denominator are used for variances and covariances instead of $n-1$, which produces unbiased estimates.

The estimated prediction variance is

$$\widehat{\text{var}}\left(y - \widetilde{y}\right) = \widehat{\text{var}}(y) - \widehat{\text{cov}}\left(y, \boldsymbol{x}'\right) \widehat{\text{var}}\left(\boldsymbol{x}\right)^{-1} \widehat{\text{cov}}\left(y, \boldsymbol{x}\right) , \tag{10.13}$$

where $\widehat{\text{cov}}\left(y, \boldsymbol{x}\right)$ is given in Equation (10.9), $\widehat{\text{var}}\left(\boldsymbol{x}\right)$ in Equation (10.10), and $\widehat{\text{var}}\left(y\right) = \boldsymbol{y}_c'\boldsymbol{y}_c/(n-1)$. In estimating the prediction variance, we get a different estimator, provided the variances and covariances are estimated with ML, i.e. using n in the denominator.

Box 10.1 Regression through the origin

A linear regression model has an explicit intercept if the column of ones is in the model matrix and an implicit intercept if the column of ones can be computed as a linear combination of other columns. If a model does not include an intercept, then we have *regression through the origin* (Eisenhauer 2003). A regression model should contain an intercept unless there are good theoretical reasons for the regression through the origin. Such reasons could be, for example, that y only has positive values and that when x-variables go to zero, then the y-variable also must go to zero. The intercept should not be dropped merely because it does not differ significantly from zero. For simplicity, assume now that there is only one x-variable.

The first consequence of regression through the origin is that the mean of residuals is not zero in the OLS regression. This implies that the model is biased only in cases where the intercept would be significant if it were included.

A question when using regression through the origin is: how the coefficient of determination should be computed? In Section 4.5.3, the coefficient of determination was defined as the relative decrease of the residual sum of squares after fitting the intercept-only model. Some people do not like this definition in regression through the origin as it can produce negative values for poorly fitting models. So, the R-square has also been defined for regression through origin (RTO) as:

$$R^2_{RTO} = 1 - \frac{\sum_{i=1}^{n}(y_i - \widetilde{y}_i)^2}{\sum_{i=1}^{n} y_i^2}$$

This definition can produce misleadingly high values for poorly fitting models. We prefer the adjusted R-square of Equation (4.34). We think that it is nice that it produces negative values for models, which are worse than the intercept-only model. Different statistical software produce different R-square values for regression through the origin, e.g. in `R` function `lm` uses the misleading definition defined above.

If regression through the origin is a reasonable model for the mean of y, and y can obtain only positive values, then the OLS assumption of constant residual error variance is never valid, if x has small values. If the mean of a positive y-variable goes to zero, then the residual variance must also go to zero. Thus, regression through the origin would also require a nonconstant variance function, even if this is often ignored.

Nonlinear regression functions often go through the origin. It may be a good strategy to test whether an additive intercept term could be included. A significant intercept implies that the assumed model is not valid near the origin.

In linear regression, we assume that the observations can be expressed as:

$$y_i = \beta_0 + \beta_1 {x_i}^{(1)} ... \beta_p {x_i}^{(p)} + e_i = \beta_0 + \boldsymbol{x}_i' \boldsymbol{\beta} + e_i$$

Note that the notation is slightly different from Equation (4.1) as the intercept is now β_0 and not β_1. It can be shown that when the OLS estimates of the betas are computed from the centered data we will get Equation (10.11) for $\boldsymbol{\beta}$ and Equation (10.12) for β_0. Thus, the equivalence of EBLUP and OLS can be shown by changing the orders of the variance and covariance matrices in Equation (10.3) using the transpose and centered data in OLS estimation. Notice that the formal equivalence also works when the x-variables are considered to be fixed variables. It is essential for the derivations that the intercept is included in the OLS model (for regression through the origin, see Box 10.1).

The estimated prediction variances differ between the BLUP and OLS approaches. We note first that the prediction variance of the BLUP predictions does not depend on the values of the x-variables. In the OLS approach for the estimation, the error variance of the estimated parameters is taken into account in the prediction variance $\text{var}(\widetilde{y}_0 - y_0) = \boldsymbol{x}' \text{var}(\widehat{\boldsymbol{\beta}})\boldsymbol{x} + \sigma^2$ (see Equation (4.45), p. 126), where the variance is dependent on the values of the x-variables. However, the estimated residual variance in the OLS regression, $\widehat{\sigma^2}$, is almost identical to the estimated prediction variance of BLUP. In the OLS approach, the residual variance σ^2 is estimated by dividing the sum of squared residuals by $n - p$, where p is the number of regression coefficients. In the BLUP estimator, the denominator in the variance and covariance estimators is $n - 1$. This implies that the ratio between $\widehat{\sigma^2}$ in OLS and the BLUP prediction variance, Equation (10.13), is $(n-1)/(n-p-1)$ (recall that we now have $p+1$ regression coefficients).

It is important to recognize the connection between EBLUP and OLS. However, there are several reasons why we need both methods in our statistical toolbox. First, the regression approach is intuitively more appealing. It provides us with regression coefficients that are easy to interpret and has a simple statistical theory. OLS regression leads directly to GLS regression. When a regression equation estimated with GLS is used for prediction then there is no similar simple connection to BLP as in OLS regression.

On the other hand, BLP can be applied when there is no direct regression equation behind the data. It can be applied when we have a statistical model that includes any kind of correlation, e.g. spatial or temporal (Example 3.11, p. 60). It can also be applied when the variable to be predicted is a latent (non-measurable) variable, as is the case with random effects in mixed models.

Even if the prediction with BLP and OLS regression leads to the same prediction equations, they are based on different assumptions. In the OLS regression, we assume that the regression equation with the true parameters gives unbiased predictions for all values of the x-variables. In contrast, BLP is guaranteed to give only marginally unbiased predictions, i.e. over the distribution of the x-variables. Thus, exactly the same prediction equation can be valid if interpreted as BLP but non-valid if interpreted as a regression equation.

Web Example 10.1. Illustration that estimation of the parameters of model 4.7 using the formula for BLUP leads to the same estimates of $\boldsymbol{\beta}$ and a slightly different estimate of the error variance compared to OLS.

10.1.2 Random regressors in a linear model

Let us then consider more closely what difference it makes when regressor variables are considered to be random variables. First, as we have discussed already, it is always correct to treat regressors variables as fixed even if we think that they are eventually random. If we treat the variables as fixed, we are making reasoning conditional on the observed values.

The predictor variables are only really fixed in experiments where the experimenter decides the values of the x-variables. In observational studies, the values of the x-variables are randomly picked from the population and, thus, they can be assumed to be random. Fortunately, if we can make some natural assumptions of the distributional properties of the x-variables, the theory of fixed-x regression directly applies. We can start with the same model as earlier: $\boldsymbol{y} = \boldsymbol{X}\boldsymbol{\beta} + \boldsymbol{e}$. Now, the fixed-$x$ regression assumption $\text{E}(\boldsymbol{e})$ is replaced with the assumption that $\text{E}(\boldsymbol{e}\,|\boldsymbol{X}) = \boldsymbol{0}$ for all possible realizations of the matrix $\boldsymbol{X}$. The assumption $\text{var}(\boldsymbol{e}) = \sigma^2\boldsymbol{I}$ is replaced with the assumption that $\text{var}(\boldsymbol{e}\,|\boldsymbol{X}) = \sigma^2\boldsymbol{I}$ where σ^2 does not depend on $\boldsymbol{X}$. Then, for each observed $\boldsymbol{X}$ the estimation is exactly as for a fixed $\boldsymbol{X}$, so the estimation of parameters, the variances of the parameters and the inference (assuming multivariate normality of $\boldsymbol{e}$) can be carried out exactly as with fixed x. As the parameter estimates are unbiased for each observed $\boldsymbol{X}$, they are also marginally unbiased, i.e. taking the distribution of $\boldsymbol{X}$ into account. Notice that no assumptions of the distribution of $\boldsymbol{X}$ are needed.

The obtained estimates for the standard errors of regression parameters are conditional estimates for the obtained $\boldsymbol{X}$. However, because $\boldsymbol{e}$ is independent of $\boldsymbol{X}$ and p-values are based only on the distribution of $\boldsymbol{e}$, the p-values are unconditionally valid (Fox 2008).

The marginal distribution of $\widehat{\boldsymbol{\beta}}$ is not generally known, and specifically it is not multivariate normal. The testing theory becomes more complicated and only approximate results can be obtained, if testing is considered in terms of the marginal distribution of $\widehat{\boldsymbol{\beta}}$. However, the conditional argument above is sufficient for practical research.

The above assumption $\mathrm{E}\left(\boldsymbol{e}\,|\boldsymbol{X}\right)=\boldsymbol{0}$ can be also relaxed in some cases. If the error term is contemporaneously uncorrelated with x, i.e. $\mathrm{cov}\left(x_i^{(k)}, e_i\right)=0$ for all k and i, then under some mild regularity conditions for $\boldsymbol{X}$, the OLS estimate for $\widehat{\boldsymbol{\beta}}$ is biased but consistent.

An example where a predictor of observation i is correlated with the error term of observation $i-1$ is the case when a temporal variable y_t is explained by y_{t-1}, e.g. the growth of the current year is explained by the growth of the previous year. This means that an autoregressive AR(1) process is part of the regression function (see Section 2.12). In that case, $\mathrm{cov}(y_{t-1}, e_{t-1}) = \mathrm{var}(e_{t-1})$. Thus, the predictor y_{t-1} of observation t is correlated with the error term of the observation $t-1$, and, therefore, condition $\mathrm{E}\left(\boldsymbol{e}\,|\boldsymbol{X}\right)=\boldsymbol{0}$ does not hold. But as the correlation is not for the error term and predictor of the same observation, OLS regression can be used. In econometric theory, there are more complicated models with lagged variables that cannot be treated with OLS regression (see e.g. Johnston and Dinardo (1997)).

If condition $\mathrm{cov}\left(x_i^{(k)}, e_i\right)=0$ is not satisfied for some k and i then the OLS estimate of $\boldsymbol{\beta}$ is both biased and inconsistent. If it is possible to find such variable z, which is highly correlated with $x^{(k)}$ but uncorrelated with the error term e, then it is possible to use instrumental variables (IV) regression or generalized instrumental variables regression (e.g. Greene (2018)); see Walters *et al.* (1989) for a forest application.

There are distinctions between fixed-x and random-x regression when we consider the variables that should be included in the model. When x are fixed, we assume that there is a fixed true underlying model that we should estimate. When we consider if we should have a specific regressor in the model or not, we must balance two kinds of errors. If we omit a fixed regressor that should be in the model, we get a biased model that does not include all the regressors in the valid model. If we include a regressor in the model that should not be in the model, we also obtain a theoretically invalid model, even though the estimates for the regressors, which are correctly in the model, are unbiased. Furthermore, the estimation variances of coefficients that are correctly in the model will increase.

In forestry and environmental sciences, we do not usually consider that there is a true underlying model with regard to random regressors. The purpose in many cases is merely to obtain an equation that is practical for prediction. Thus, even if we apply the OLS or GLS methodology based on fixed regressors, our motivation comes from the idea to obtain the conditional mean of the y-variable for the observed values of x-variables. Thus, we can consider that there are simultaneously several different 'true' models with different regressors. When deciding what model is acceptable for practical use, we can take criteria, such as the measurement cost and simplicity of the model, into account.

When considering which variables are unnecessary and should be dropped from a model, the reasoning can generally be similar to the fixed regressor case. There is, however, a subtle difference. With fixed regressors, the unnecessary regressors always increase the variances of parameters, which are correctly in the model. With random regressors, this is not always the case (Binkley and Abbot 1987).

Fixed regressors only appear in experiments where the experimenter has fixed the values. But when modeling experimental data, we usually need random regressors in the model as well. These regressors are typically called covariates and can serve two different purposes. First, they can reduce the residual error variance of the model and ensure that the inference of the treatment effects is more efficient. But if the model contains treatment × covariate interactions, then the treatment-specific coefficients of a covariate are an essential part of the treatment effects.

10.2 Modeling nonlinear responses using transformations in y

In linear regression (as discussed in Section 4.2), the regression is linear in two respects. First, the regression equation is linear in parameters, i.e. the regression equation is in the form $\beta_1 f_1(\boldsymbol{x}) + ... + \beta_p f_p(\boldsymbol{x})$, as we have already discussed in Section 7.1. Often, $f_k(\boldsymbol{x}) = x_k$, but not necessarily. With a nonlinear f we can express relationships that are nonlinear in x. Second, the standard OLS and GLS estimators of regression coefficients in linear regression are linear in the sense that the estimator of β_k can be expressed in the form $\boldsymbol{w}'\boldsymbol{y}$, where $\boldsymbol{w}$ does not depend on $\boldsymbol{y}$.

In Section 10.4, we will present methods where relationships, which are nonlinear in x, can be modeled using linear regression. In this section we consider making a nonlinear transformation to the dependent variable and applying linear regression to the transformed variable. A drawback of such an approach is that the predictions will be unbiased in the transformed scale, and biased in the original scale because $\mathrm{E}(g(Y)) \neq g(\mathrm{E}(Y))$ for a nonlinear transformation g. Generally, we prefer using that variable as the dependent variable that we plan to predict using the regression equation, even if that would lead to nonlinear regression with non-homogeneous variances. Nowadays, the application of generalized linear (mixed) may also provide a better approach than linear regression with a transformed dependent variable. But in some cases, transformation of the dependent variable may be reasonable. Moreover, it should be kept in mind that transformation bias also occurs with nonlinear and generalized linear mixed-effect models, because the random effects are written inside a nonlinear function (recall Example 8.10).

Some authors have selected a transformation for the dependent variable so that a maximum R^2, i.e. the coefficient of determination, is obtained. This is a completely invalid approach. R^2 values based on different transformations of the dependent variable are not comparable, as we have discussed in Section 4.5.3.

10.2.1 Logarithmic regression

Logarithmic regression may often be reasonable and is based on the assumption that y can be described with function:

$$y \approx A \exp\left(\sum_{k=1}^{p} \alpha_k x_k\right)$$

or with function:

$$y \approx B z_1^{\beta_1} ... z_r^{\beta_r} .$$

Putting both together we obtain:

$$y \approx C \exp\left(\sum_{k=1}^{p} \alpha_k x_k\right) z_1^{\beta_1} ... z_r^{\beta_r} . \tag{10.14}$$

If we assume a multiplicative error term, we obtain the following model:

$$\log(y) = c + \sum_{k=1}^{p} \alpha_k x_k + \sum_{k=1}^{r} \beta_k \log(z_k) + e\,, \tag{10.15}$$

where $c = \log(C)$. From the residual error, we assume that $\mathrm{E}(e_i) = 0$ and $\mathrm{var}(e_i) = \sigma^2$ for all observations i. The assumption that the error variance is constant in the logarithmic scale is approximately equivalent to the assumption that the relative error term of Equation (10.14) would have a constant variance. Equation (10.15) is an ordinary linear regression model and, thus, the OLS estimates are the BLUE for the regression coefficients.

If the estimated regression model is used directly to predict y in the original scale, the predictions (i.e. $\exp\left(\widehat{\log(y)}\right)$) are biased (on average too small). If the logarithmic residual is normally distributed, the obtained predictions are, however, median unbiased, that is, for each combination of predictor variables, 50% of the predictions are, on average, too small and 50% are too large. An almost unbiased predictor is obtained by utilizing the following property of the normal distribution: if e is normally distributed with mean μ and variance σ^2, then $\exp(e)$ is log-normally distributed with mean $\exp\left(\mu + \frac{1}{2}\sigma^2\right)$.

The value of the estimated regression function for given x variables is an unbiased estimate for μ and the logarithmic regression provides an unbiased estimate for σ^2. Thus, an almost unbiased predictor for y in the original scale is obtained by:

$$\widetilde{y} = \exp\left(\widehat{\mu} + \tfrac{1}{2}\widehat{\sigma}^2\right)\,, \tag{10.16}$$

where

$$\widehat{\mu} = \widehat{c} + \sum_{k=1}^{p} \widehat{\alpha}_k x_k + \sum_{k=1}^{r} \widehat{\beta}_k \log(z_k)\,.$$

Even if unbiased estimates are used for μ and σ^2, the obtained predictor is not unbiased for the corresponding theoretical quantity $\exp\left(\mu + \frac{1}{2}\sigma^2\right)$ because unbiasedness is not transmitted in the nonlinear exp-transformation. That is why the resulting prediction is only almost unbiased.

We need the same bias correction in the generalized linear mixed models with a log link (see Example 8.14, p. 285). There, we consider models with such linear predictors in the log-scale that contain normally distributed random effects. There, the unbiased prediction in the original scale can be analyzed and corrected in the same way as here in the logarithmic regression. The random effect corresponds to the residual error present here.

An alternative bias correction in the logarithmic regression can be carried out as follows:

$$\widetilde{y} = \lambda e^{\widehat{\mu}}\,,$$

where:

$$\lambda = \frac{\sum y_i}{\sum \exp(\widehat{\mu}_i)}\,.$$

This bias correction is not as sensitive to the deviations from the distributional assumptions, as the bias correction above (Snowdon 1991). Two other bias corrections, based on the Taylor series and the two-point distribution, will be presented in the next subsection. These methods can be applied for any nonlinear transformation of the dependent variable (the Taylor series correction is only used if the function is twice differentiable with respect to the error term). Notice also that the bias correction Equation (10.16) can be presented in the form $\lambda e^{\widehat{\mu}}$ where $\lambda = \exp(\frac{1}{2}\widehat{\sigma}^2)$.

10.2.2 Correcting back transformation bias using the Taylor series

If a regression model is developed for a nonlinear transformation $h(y)$, then the back-transformed predictor $h^{-1}\left(\widehat{E}\left(h\left(y\right)\right)\right)$ is biased for y, as we have already discussed in this chapter. Let us assume that

$$h\left(y\right) = f\left(\boldsymbol{x}\right) + e\,,$$

where $f\left(\boldsymbol{x}\right)$ is the regression function and e is the random residual error with mean zero and variance σ^2. Let us denote that $z = h\left(y\right)$, $\mu = f\left(\boldsymbol{x}\right)$ and g is h^{-1}, i.e. the inverse function. Using the second-order Taylor series approximation around μ (see the Box 7.1, p. 211), we obtain:

$$\begin{aligned} \mathrm{E}\left(y\right) = \mathrm{E}\left(g\left(z\right)\right) = \mathrm{E}\left(g\left(\mu + e\right)\right) &\approx \mathrm{E}\left(g\left(\mu\right) + g'\left(\mu\right)e + \tfrac{1}{2}g''\left(\mu\right)e^2\right) \\ &= g\left(\mu\right) + \tfrac{1}{2}g''\left(\mu\right)\sigma^2 \end{aligned} \tag{10.17}$$

because $\mathrm{E}\left(e\right) = 0$ and $\mathrm{E}\left(e^2\right) = \mathrm{var}\left(e\right) = \sigma^2$.

Logarithmic regression

Let us consider again logarithmic regression. If

$$z = \log\left(y\right) = f\left(\boldsymbol{x}\right) + e$$

and we wish to obtain the expected value of y, then $g\left(z\right) = \exp\left(z\right)$ and $g''\left(z\right) = \exp\left(z\right)$. From Equation (10.17) we obtain:

$$\begin{aligned} \mathrm{E}\left(y\right) &= \mathrm{E}\left(\exp\left(\log\left(y\right)\right)\right) \\ &\approx \exp\left(f\left(\boldsymbol{x}\right)\right) + \tfrac{1}{2}\exp\left(f\left(\boldsymbol{x}\right)\right)\sigma^2 \\ &= \exp\left(f\left(\boldsymbol{x}\right)\right)\left(1 + \tfrac{1}{2}\sigma^2\right). \end{aligned} \tag{10.18}$$

Thus, we obtain a different bias correction from the Taylor series than assuming that the residual error in the logarithmic regression is normally distributed. To compare the two bias corrections, the bias correction in Equation (10.16) is written in the form:

$$\exp\left(\mu + \tfrac{1}{2}\sigma^2\right) = \exp\left(\mu\right)\exp\left(\tfrac{1}{2}\sigma^2\right).$$

The bias correction based on the Taylor series, $\left(1 + \frac{1}{2}\sigma^2\right)$, is compared in Table 10.1 to the bias correction $\exp\left(\frac{1}{2}\sigma^2\right)$ based on the lognormal distribution.

Logit transformation

If the dependent variable is the relative frequency (grouped binary variable), then logit transformation is sometimes used (see Box 8.1, p. 246):

$$\mathrm{logit}(y) = \log\left(\frac{y}{1-y}\right) = f\left(\boldsymbol{x}\right) + e\,. \tag{10.19}$$

Then the inverse function is the logistic function

$$g(z) = \frac{\exp(z)}{1+\exp(z)} = \frac{1}{1+\exp(-z)}\,.$$

The second derivative of g is:

$$g''(z) = g(z)g(-z)\left(g(-z) - g(z)\right).$$

100σ ($\sim$ %)	$\exp\left(\frac{1}{2}\sigma^2\right)$	$1+\frac{1}{2}\sigma^2$
1	1.00005	1.00005
2	1.00020	1.00020
5	1.00125	1.00125
10	1.00501	1.00500
20	1.0202	1.0200
40	1.0833	1.0800
70	1.2776	1.2450

TABLE 10.1 Comparison of bias corrections in Equations (10.16) and (10.18). We see that the difference is small. This follows from the fact that $e^x \approx 1 + x$ when x is small. This is utilized in the interpretation of logistic and Poisson generalized linear models in Sections 8.2.2 and 8.2.3 (see also Figure 8.2, p. 252). The correction is also small when the relative error or the standard deviation of the logarithmic error is under 10%.

Thus, we get an approximation to $\mathrm{E}(y)$

$$\mathrm{E}(y) \approx \frac{1}{1+\exp(-\mu)} + \tag{10.20}$$
$$\frac{\sigma^2}{2}\left(\frac{1}{1+\exp(-\mu)}\right)\left(\frac{1}{1+\exp(\mu)}\right)\left(\frac{1}{1+\exp(\mu)} - \frac{1}{1+\exp(-\mu)}\right).$$

Similarly, as the bias correction in log-transformation is of interest in GLMM, the bias correction of the logit transformation is also of interest in GLMM. In GLMM a random effect in the logit scale corresponds to the residual error in Equation (10.19). Example 10.2 (p. 318) compares the Taylor series correction to the two-point distribution method (presented in Section 10.2.3).

Square root transformation

If the dependent variable is a count, then the square root transformation may stabilize the residual error variance. If

$$\sqrt{y} = f(\boldsymbol{x}) + e$$

then the bias correction is obtained directly:

$$\mathrm{E}(y) = \mathrm{E}\left(\sqrt{y}^2\right) = \mathrm{E}\left((f(\boldsymbol{x}) + e)^2\right) = \mathrm{E}\left(f(\boldsymbol{x})^2 + e^2 + 2f(\boldsymbol{x})\,e\right) = f(\boldsymbol{x})^2 + \sigma^2. \tag{10.21}$$

The same bias correction is obtained using the Taylor series: $g(z) = z^2$ and $g''(z) = 2$. The bias correction is exact and this coincides with the fact that square function can be exactly presented with the second-order Taylor series.

Reciprocal transformation

Let us assume that:

$$1/y = f(\boldsymbol{x}) + e.$$

Then, $g(z) = 1/z$ and $g''(z) = 2/z^3$. Thus:

$$\mathrm{E}(y) \approx \frac{1}{f(\boldsymbol{x})} + \frac{\sigma^2}{f(\boldsymbol{x})^3}.$$

Näslund's height equation

Näslund (1937) presented a height-diameter model:

$$h = \frac{d^m}{(\beta_0 + \beta_1 d)^m} + 1.3\,,$$

where h is the height of a tree, d is the breast height diameter (DBH), and $m = 2$, but others have also used $m = 3$. However, there is no reason to only limit consideration to integer values. The value 1.3 (meters) is the *breast height*. The model parameters can be estimated using the following equation:

$$y = \frac{d}{(h-1.3)^{1/m}} = \beta_0 + \beta_1 d + e\,.$$

The bias correction is (see Siipilehto and Kangas (2015) for $m = 2$ and $m = 3$, see also Example 5.19, p. 163):

$$\widetilde{h} = \frac{d^m}{(\beta_0 + \beta_1 d)^m} + \frac{\frac{1}{2}m\,(m+1)\,d^m}{(\beta_0 + \beta_1 d)^{m+2}}\sigma^2 + 1.3\,. \tag{10.22}$$

10.2.3 Correcting back transformation bias using the two-point distribution

In the previous section, we estimated $\mathrm{E}\,(g\,(\mu + e))$ by approximating function g by the second-order Taylor series. Thereafter, the expected value of this approximation could be computed exactly for all distributions. An alternative method is to approximate the distribution and then compute the expected value of the correct function with respect to the approximating distribution. The simplest approximating distribution with correct variance is the two-point distribution of Rosenblueth (1981), which sets probability mass $\frac{1}{2}$ to $\mu - \sigma$ and probability mass $\frac{1}{2}$ to $\mu + \sigma$. The two-point distribution has the same mean and variance as the original distribution. Using this method, the following approximation is obtained for $\mathrm{E}\,(y)$:

$$\mathrm{E}(y) \approx \frac{1}{2}\left(g\,(\mu - \sigma) + g\,(\mu + \sigma)\right) \tag{10.23}$$

The two-point distribution is easy to apply and is similar for all nonlinear transformations (see also Example 6.6).

In the following example, we compare the two-point distribution and the Taylor series approximation for the logit transformation; the 'exact' solution is obtained with numerical integration.

Example 10.2. With modern computers, it is easy to carry out the bias correction by also using the available numerical integration routines or through Monte Carlo integration, i.e. by generating random normal errors into the function in the transformed scale before applying the inverse function (mean function). Consider the logistic model $y_i = Bernoulli(p_i)$ where $\mathrm{logit}(p_i) = \mu + e_i$ and $e_i \sim N(0, \sigma^2)$. The effect of the bias correction using either the Taylor approximation, numerical integration or the two-point method for the logit transformation is shown in Figure 10.1, produced with the R code below.

```
> mu<-seq(-6,6,0.01)
> # prediction without bias-correction
> y0<-plogis(mu)
> for (sigma in c(1,2)) {
+     # Unbiased relationshhip under normality
+     # - Using integrate
+     y1true<-sapply(mu,function(x)
```

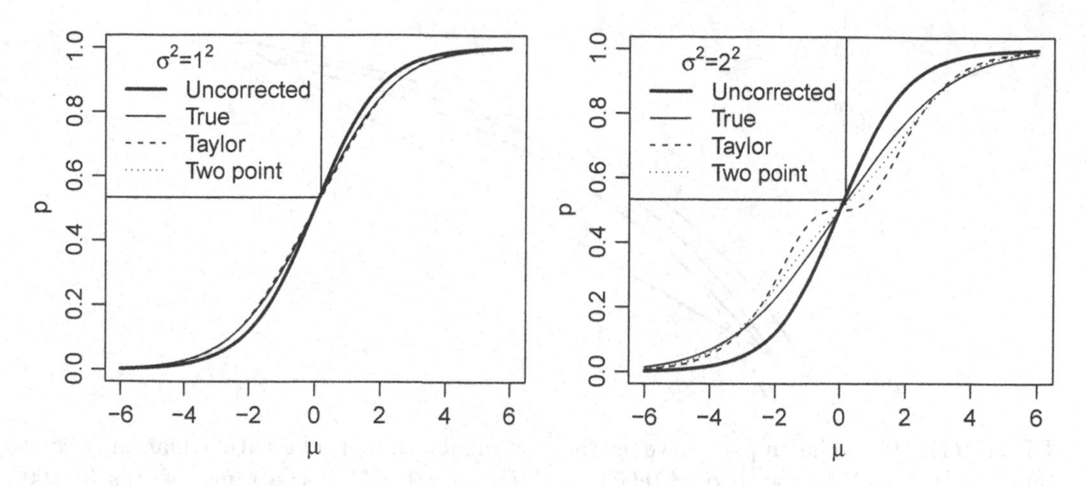

FIGURE 10.1 Bias correction for the logit transformation. The y-axes are the predicted probability. The x-axes are the estimated regression function $\mu = f(\boldsymbol{x})$. The solid thick line shows the prediction without the bias correction. The line also applies when $\sigma^2 = 0$. The thin solid lines are computed by averaging the back-transformed prediction over the normal distribution of errors. These present the correct bias-corrected curves under normality of disturbances. The dashed and dotted lines show the bias-corrected curves based on the Taylor and two-point approaches. The left-hand graph is based on $\sigma = 1$ and the right-hand graph on $\sigma = 2$. When $\sigma = 1$, the correct bias-corrected curve and the Taylor and two-point approximations almost coincide. The Taylor approximation and the two-point method work very well for small σ. For larger σ, the two-point method is better. It should be kept in mind that numerical integration may also provide poor results if the true distribution is not normal.

```
+                   integrate(function(u) dnorm(u,sd=sigma)*plogis(x+u),-Inf,Inf)$value)
+       # - Using simulation (= Monte Carlo integration)
+       #y1true<-sapply(mu,function(x) mean(plogis(x+rnorm(10000,sd=sigma))))
+       # Taylor and Two-point corrections
+       y1Taylor<-plogis(mu)+sigma^2/2*plogis(mu)*plogis(-mu)*(plogis(-mu)-plogis(mu))
+       y1Twopoint<-(plogis(mu+sigma)+plogis(mu-sigma))/2
+       # graph
+       plot(x,y0,type="l",xlab=expression(mu),ylab="p",lwd=2)
+       lines(x,y1true)
+       lines(x,y1Taylor,lty="dashed")
+       lines(x,y1Twopoint,lty="dotted")
+       legend("topleft",title="",legend=c("Uncorrected","True","Taylor","Two point"),
+              lwd=c(2,1,1,1),lty=c("solid","solid","dashed","dotted"))
+       text(-4,0.97,bquote(sigma^2*"="*.(sigma)^2))
+       }
```

10.3 Predicting the mean of a nonlinear function over the distribution of x

In the previous section, we considered bias problems when estimating a model for nonlinear transformation and using the estimated model in the original scale to predict the mean of y for fixed values of the x-variables. In this section, we consider a closely related problem that has, however, a different origin. It is assumed that we want to predict the mean of y over the distribution of x-variables, not over the distribution of random residual errors. More

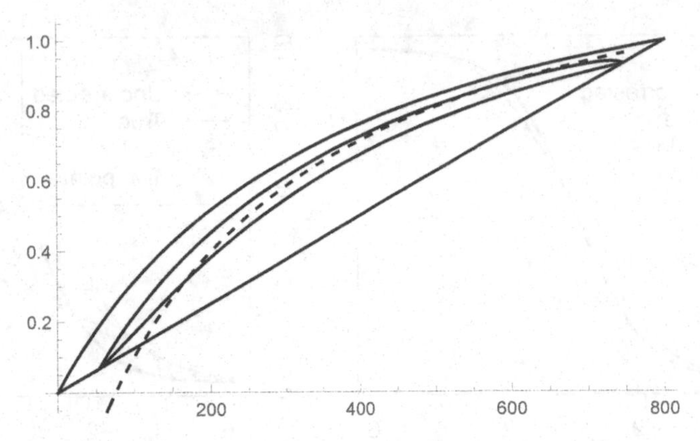

FIGURE 10.2 The upper curve is the rectangular hyperbola with equation $f(x) = \alpha x F_{max}/(\alpha x + F_{max})$ with $\alpha = 0.0044$ and $F_{max} = 1.455$. If the range of x is $(0, 800)$, the mean of f is between the curve $f(x)$ and the line connecting the extreme points. If it is known that $\sigma = 200$, the mean of $f(x)$ is within the narrow belt in the middle. The dashed line is the Taylor approximation (Equation (10.17)), which is not always within the feasible area and not always positive. If σ of x is 200, $\bar{x}$ must be within the range $(54, 746)$ (why?).

specifically, we assume that we are interested in the average of y in a given joint realization (population) of y and x-variables.

For simplicity assume that we have a nonlinear regression equation with only one random predictor x and with known parameter values. That is

$$y_t = f(x_t) + e_t\,,$$

where $t = 1, \ldots, N$ when the population consists of a fixed number of individuals, or $t \in [T_0, T_1]$ for a continuous process. We assume that the population is so large and the autocorrelations are so weak that the random errors cancel each other. Thus we are interested in:

$$\bar{y} = \frac{\sum_{t=1}^{N} f(x_t)}{N}\,,$$

or

$$\bar{y} = \int_{x_{min}}^{x_{max}} f(x)\, dG\,,$$

where G is the cumulative distribution of x and x_{min}, and x_{max} are the lower and upper limits of x, respectively.

Let us assume that we know the mean $\bar{x}$ of x in the population of interest (for instance the mean diameter in a stand, the mean temperature or light intensity during a period). We do not assume any distribution for x, i.e. our knowledge for x is based on measurements of x. Furthermore, let us assume that function f is concave (i.e. if two points on the response curve are connected by a line, the line never goes above the curve between those two points). We can make the following remarks:

1. $\bar{y} \leq f(\bar{x})$.

2. Equality $\bar{y} = f(\bar{x})$ occurs only if there is no variation in x, or the function is linear in the range of x.

3. If x can have only values within the range (x_{min}, x_{max}), the smallest possible value of $\bar{y}$ occurs when x is only assigned values x_{min} and x_{max}. This smallest value of $\bar{y}$ is the value of the linear function that passes through points $(x_{min}, f(x_{min}))$ and $(x_{max}, f(x_{max}))$ at point $\bar{x}$.

4. If x can only have values within the range (x_{min}, x_{max}), and $\bar{x}$ and the variance of x, σ^2 are known, the feasible range of $\bar{y}$ can be solved with mathematical programming (see Lappi and Smolander (1984)).

If f is a convex function, then $-f$ is a concave function. Thus, the properties of convex functions follow directly from the properties of concave functions.

Let us then consider the estimation of $\bar{y}$. If only $\bar{x}$ is known, then we cannot hope for an accurate estimate. A minimax estimator that minimizes the maximum error is obtained by taking the average of $f(\bar{x})$ and the point on the line connecting the extreme points. If σ is known, in addition to $\bar{x}$, then the minimax estimator is again one possible estimator. It can be computed by taking the average of the feasible range of $\bar{y}$. Another estimator can be obtained from using the second-order Taylor approximation, i.e. using Equation (10.17). A third estimator is based on the two-point distribution (see Eq. (10.23)). Now, σ is the sd of x, not the sd of the residual error.

Smolander and Lappi (1985) estimated the average photosynthesis rate from the mean and variance of radiation using those three methods. They found that the best method was the two-point method.

The utility of the Taylor approximation depends on the magnitude of the third derivative. For regular looking functions, the Taylor approximation may provide predictions that are outside the feasible range provided by the mathematical programming (see Figure 10.2). Both the Taylor approximation and the two-point method can be extended to functions of several variables. These extensions are of interest when estimating the mean of a nonlinear function of several x-variables, but they are not needed when correcting the back-transformation bias. They can also be extended to take skewness into account. In engineering, the two-point method has had many applications and extensions in reliability calculations.

10.4 Modeling nonlinear responses using transformations in x

General principles regarding modeling of nonlinear responses using transformations in the x-variables were discussed in Section 4.2.2. In this section, we extend that discussion by taking a closer look at splines and response surfaces.

10.4.1 Regression splines

With regression splines, we can estimate regression functions with very few assumptions on the shape of the function. Let us first consider univariate splines, i.e. the case where we have only one explanatory variable. To start with, let us consider the simplest possible regression spline. In Figure 10.3, the relationship between y and x can be clearly described with two linear segments; the linear relationship changes at approximately $x = 30$. This can be described by the equation

$$y = \beta_1 + \beta_2 x + \beta_3 \max(x - 30, 0) + e \tag{10.24}$$

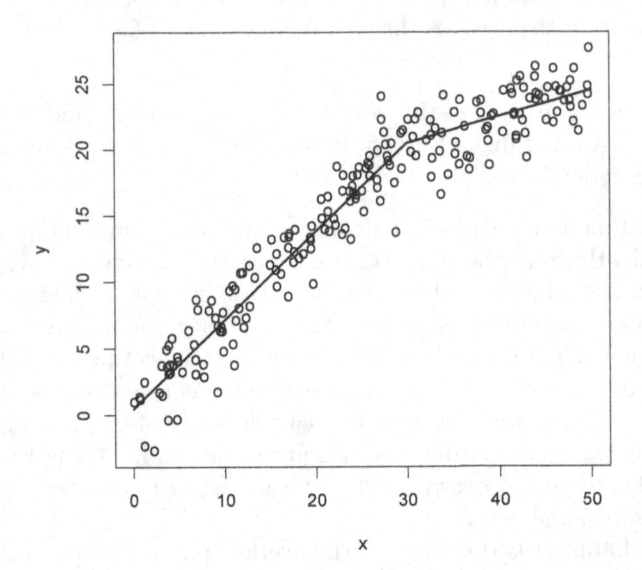

FIGURE 10.3 A regression equation consisting of two linear segments.

The resulting function is a linear (first degree) spline with one knot at $x = 30$. The function is continuous, and it can be estimated with OLS regression if standard assumptions for the error terms hold. If they do not hold, then GLS can be used. The knot point is an additional nonlinear parameter whose optimal value can be searched iteratively. Software for estimating nonlinear models may fail in estimation of the knot point because the function cannot be differentiated with respect to the knot point. In R, Equation (10.24) can be easily fitted by defining the model formula as `y~x+pmax(x-30,0)`.

Let us then move to a more general presentation. Let us assume that the regression function can be presented as:

$$f(x) = \sum_{j=1}^{J} \beta_j h_j(x), \tag{10.25}$$

where functions $h_j(x)$ are called *basis functions*. Because Equation (10.25) is linear in parameters, the parameters can be estimated with OLS regression. Note that the same function can also be used in linear and generalized linear mixed-effects models. Equation 10.24 used $h_1(x) = 1$, $h_2(x) = x$ and $h_3(x) = \max(x - 30, 0)$ as basis functions.

One possibility is to use the powers $0, \ldots, J - 1$ of x as basis functions, which leads to a polynomial regression model. Powers of x are an example of a global basis function. Using higher-order powers (e.g. greater than 3) of x would cause two problems. First, the resulting regression functions can behave wildly outside the range of x values in the data. Furthermore, each observation influences the regression function over the entire range of x values. The Taylor series with many terms is not a useful approximation of a function over a wide range of x values.

Splines avoid both problems inherent in higher-order polynomials. Splines are formed by joining low degree D polynomials (usually $D = 3$ leading to *cubic splines*) so that the resulting function is continuous and has continuous derivatives up to degree $D - 1$. The *order of the spline* is $D + 1$. Usually, splines are described in terms of order, but we regard the description in terms of degree as easier to understand. The points where the spline switches from one polynomial to the next are called *knots*. Knots are assumed to be known and are usually equidistant, or correspond to quantiles of the x-values in the data.

There are two kinds of basis functions in common use (in principle there are many alternatives). The *truncated power basis* is easy to understand, but it is not suitable for stable numeric computations. The *B-spline basis* is computationally better but its recursive definition is more difficult to understand. Let us discuss first the truncated power basis. It can be presented as:

$$h_j(x) = x^{j-1}, \qquad j = 1, ..., D+1 \tag{10.26}$$

$$h_{k+D+1}(x) = (x - \xi_k)_+^D, \qquad k = 1, ..., K \tag{10.27}$$

where ξ's are the knots, K is the number of knots and $(x - \xi_k)_+ = \max(x - \xi_k, 0)$. The degrees of freedom of the regression function is $D + 1 + K$.

If $D \geq 3$ the splines estimated using all basis functions tend to be wild outside the x-values in the data. *Natural splines* are formed so that the function is linear outside the knot range. With cubic splines, this means that there are four constraints for the parameters, and the degrees of freedom for the spline are equal to the number of knots (K). The basis functions for natural cubic splines obtained from the truncated power basis functions (Hastie *et al.* 2009, pp. 145-146) are as follows: $h_1(x) = 1$, $h_2(x) = x$, and

$$h_{k+2}(x) = d_k(x) - d_{K-1}(x), \quad k = 1, ..., K-2, \tag{10.28}$$

where:

$$d_k(x) = \left((x - \xi_k)_+^3 - (x - \xi_K)_+^3\right) / (\xi_K - \xi_k)$$

The reader may verify that $f''(x) = 0$ if $x \leq \xi_1$ or $x \geq \xi_K$. Harrell (2015) presents an alternative but equivalent presentation of the basis functions. He calls natural splines *restricted* splines. The basis functions used by Harrell (2015) are obtained by multiplying the basis functions described in Hastie *et al.* (2009) by a constant.

There are two problems in numeric computations with the truncated power basis. First, quite large numbers are raised to power three. Then, two large numbers cannot be properly separated in computations as only the most significant decimals are kept in the cubic terms. Second, the regressor variables are highly correlated and the resulting moment matrix $\boldsymbol{X}'\boldsymbol{X}$ is close to singular. This leads to low accuracy of its inverse.

The *B-spline* basis functions are numerically stable, because they have *small support*, i.e. the basis functions $h_j(x)$ are zero outside the range (ξ_i, ξ_{i+D+1}). With B-splines, the knots need to be augmented with additional knots. B-spline basis functions are defined recursively using divided differences. They are more difficult to understand, and we refer readers to de Boor (1978) or Hastie *et al.* (2009).

The B-spline basis has an additional advantage over truncated power basis. With B-splines, the continuity requirements for the derivatives can be decreased using duplicated knots, i.e. knots with the same value for consecutive knots. This may be useful when the regression function clearly makes sharp curves. Regression splines are generally more stable than *interpolating splines* (see Box 10.2, p. 324).

Regression splines have the advantage that their use is just a special case of regular OLS regression and they can also be directly used in mixed models or in generalized linear (mixed) models. Perhaps the biggest problem with regression splines is that we do not know the knots in practice, which are, however, assumed to be known. Having knots that are too close to each other causes extra variance to the estimated regression function. Knots that are too far away from each other cause bias in the estimated function. See also Examples 4.14 (p. 108) and 6.6 (p. 198).

Box 10.2 Interpolating splines

Interpolating splines are smooth curves that pass through given points $(x_i, y_i), i = 1, ...n$. The points are called knots. For interpolating splines, y needs to be specified for each knot. Let us now consider only cubic splines, i.e. piecewise third degree polynomials that pass through the knot points and have continuous second derivative everywhere. With n knots there are $n-1$ polynomials between the knots. They have $(n-1)4 = 4n-4$ coefficients. The requirement that the function passes through the knots and is continuous gives $n+n-2 = 2n-2$ constraints for the coefficients.The requirement that first and second derivatives are continuous at the inner knots gives $(n-2)2 = 2n-4$ constraints for the coefficients. Thus, we need two additional constraints in order to calculate all the coefficients. In *natural cubic splines* the additional two constraints state that the second derivative is zero at the first and last knot.

Interpolating cubic splines can produce wild curves, and their behavior always needs to be checked visually. *Taut splines* have an extra parameter by which one can get extra regularity up to linear interpolation. Statistical *smoothing splines* have a similar smoothness parameter. Interpolating splines have been used in forestry to interpolate stem curves (e.g. Lahtinen and Laasasenaho (1979)).

10.4.2 Smoothing splines

The knot selection problem can be avoided using *smoothing splines*, which can use all distinct x-values as knots. Let us assume that all x values are distinct. The knot selection problem is replaced with the problem of selecting the proper degree of *regularization.* A smoothing spline minimizes the penalized residual sum of squares (RSS):

$$\mathrm{RSS}(\lambda) = \sum_{i=1}^{n} (y_i - f(x_i))^2 + \lambda \int_{-\infty}^{\infty} f''(t)^2 dt, \qquad (10.29)$$

where λ is a fixed *smoothing parameter.* The first term penalizes the residual sum of squares and the second term penalizes the curvature and, thus, controls the smoothness of the function. If $\lambda = 0$, then the solution is any function that passes through the data points, and the resulting RSS is zero. If $\lambda = \infty$, then a simple regression line is obtained. If λ is between zero and infinity, then the minimum value of RSS is provided by a natural cubic spline where all x-values as knots.

Using the truncated power basis, the spline function can be presented as shown in Equations (10.25) and (10.28). Let $\boldsymbol{X}$ denote the obtained model matrix (i.e. element ij is the value of the jth basis function for the ith x-value). Note that the model matrix is a $n \times n$ matrix. The optimal solution of the parameter vector for a given λ is

$$\widehat{\boldsymbol{\beta}} = (\boldsymbol{X}'\boldsymbol{X} + \lambda\boldsymbol{D})^{-1}\boldsymbol{X}'\boldsymbol{y}, \qquad (10.30)$$

where:

$$D_{jk} = \int_{x_1}^{x_n} h_j''(t)\, h_k''(t)\, dt\,.$$

When $\lambda = 0$, then Equation (10.30) provides the interpolating natural spline. When we compare Equation (10.30) to Equation (4.23) (p. 91), we note that the smoothing spline is computed as a generalized ridge regression. If we compare Equation (10.30) to Equation (5.36) (p. 162), we note that the smoothing spline is computed similarly to the way the random effects are predicted in the mixed linear model. The connection between the smoothing splines and mixed models has been analyzed, for example, by Wang (1998).

The estimated parameters in Equation (10.30) are linear combinations of the y_i. Similarly the fitted values are linear in $\boldsymbol{y}$:

$$\widehat{\boldsymbol{f}} = \boldsymbol{X}(\boldsymbol{X}'\boldsymbol{X} + \lambda \boldsymbol{D})^{-1}\boldsymbol{X}'\boldsymbol{y} = \boldsymbol{S}_\lambda \boldsymbol{y}, \tag{10.31}$$

The matrix $\boldsymbol{S}_\lambda$ is called a *smoother matrix.* The *effective degrees of freedom*, which corresponds to the number of parameters in the ordinary regression function, is defined as $\mathrm{df}_\lambda = \mathrm{trace}\,(\boldsymbol{S}_\lambda)$, where trace is the sum of diagonal elements.

If the effective degrees of freedom are given, then λ can be solved numerically. This is a good and intuitively appealing method for determining the value of λ. Another method for determining λ is cross validation. In the forestry application of Lappi (2006), the cross validation did not produce consistent curves for closely related functions, while the effective degrees of freedom produced logical curves.

When using B-spline basis with the additional knots, the formulas change slightly (see Hastie *et al.* (2009)). When the number of observations is large (e.g. over 50), it may be reasonable, for computational reasons, to merge adjacent points. Smoothing splines can be generalized to the case where observations have unequal variances or are correlated (Wang 1998). Even if the residual error variance is initially equal for the original observations, it will vary when different numbers of observations are put together when making merged observations.

For smoothing splines, the basis for natural cubic splines based on the truncated power functions (Equation (10.28)) can be transformed to an orthogonal *Demmler-Reinsch basis* (Hastie *et al.* 2009). This basis gives new insight to the properties of the smoothing splines. For instance, it shows that high frequency basis functions, which change many times (from increasing to decreasing), are shrunken more than functions that change less frequently.

10.4.3 Multivariate splines

Thus far, we have only considered one-dimensional regression and smoothing splines. Now, let us briefly introduce multivariate regression and smoothing splines. First, let us consider regression splines with two regressor variables, say x_1 and x_2. Let M_1 and M_2 be the numbers of basis functions for $x_1$and x_2, respectively. Then, the *tensor product basis* is defined by:

$$h_{jk}\left(x_1, x_2\right) = h_{1j}\left(x_1\right) h_{2k}\left(x_2\right), j = 1, ..., M_1, k = 1, ..., M_2 .$$

This leads to a two-dimensional spline:

$$f\left(x_1, x_2\right) = \sum_{j=1}^{M_1} \sum_{k=1}^{M_2} \beta_{jk} h_{jk}\left(x_1, x_2\right).$$

The model is once again an ordinary linear regression model, which can be solved with OLS, and the estimation is directly generalizable to GLS, LMM, and GLMM. This can be generalized to higher dimensions, but by increasing the dimensionality, we now have a problem in that the number of basis functions increases exponentially. The multivariate splines can also be used in smoothing splines. When the number of dimensions increases then a feasible alternative is to start using generalized additive models (to be presented in Section 10.5).

10.4.4 Second-order response surface

If the relationship between a response variable y and an independent variable x is described with a regression line, it is assumed that a one-unit change in x always causes a change of the same size in y. In most cases, the assumption of linearity holds only for some range of the x-variable. Perhaps the simplest model for a nonlinear relationship is the quadratic function:

$$y = a\,x^2 + b\,x + c + e\,,$$

where e is the random error term (recall Example 4.4 and other examples in the `stumplift` data set). The model is linear with respect to its parameters, so it can be estimated with linear regression. The response function has a *stationary point*, when the derivative is zero, i.e. when: $2\,a\,x + b = 0$, or

$$x = -b/2a\,.$$

The stationary point of a second-order polynomial is an extremum, i.e. either a minimum or maximum. This property can be used, for example, to find the optimal temperature for growth in models of net photosynthesis (Laine *et al.* 2019). In that case, all other terms, except for temperature and squared temperature, are included in term c of the polynomial.

In a linear regression model with two regressors, it is assumed that the regression surface is a plane. If there are more than two regressors, the surface is a *hyper plane.* The simplest regression function, which is nonlinear on regressors, is perhaps the *second-order response surface*, which is a generalization of the quadratic polynomial. The second-order surface is easy to interpret, even when there are several regressors. The model is especially useful when optimal process parameters are searched using *response surface methodology* (RSM) (see Myers *et al.* (2016)).

The second-order surface with respect to $x_1 \ldots x_p$ is the function:

$$f\left(x_1, \ldots, x_p\right) = b_0 + \sum_{j=1}^{p} b_j x_j + \sum_{j=1}^{p}\sum_{k=1}^{j-1} b_{jk} x_j x_k + \sum_{j=1}^{p} b_j x_j^2\,,$$

which can be stated in matrix form as:

$$f\left(\boldsymbol{x}\right) = b_0 + \boldsymbol{x}'\boldsymbol{b} + \boldsymbol{x}'\boldsymbol{B}\boldsymbol{x}\,,$$

where:

$$\boldsymbol{x} = \begin{pmatrix} x_1 \\ x_2 \\ \vdots \\ x_p \end{pmatrix}, \boldsymbol{b} = \begin{pmatrix} b_1 \\ b_2 \\ \vdots \\ b_p \end{pmatrix}, \text{ and } \boldsymbol{B} = \begin{pmatrix} b_{11} & & & (sym) \\ \frac{1}{2}b_{21} & b_{22} & & \\ \vdots & \vdots & \ddots & \\ \frac{1}{2}b_{p1} & \frac{1}{2}b_{p2} & \cdots & b_{pp} \end{pmatrix}.$$

The stationary point of the function can again be obtained by stating the derivative as equal to zero:

$$\frac{\partial f}{\partial \boldsymbol{x}} = \boldsymbol{b} + 2\boldsymbol{B}\boldsymbol{x} = \boldsymbol{0}\,.$$

If $\boldsymbol{B}$ has the inverse matrix, the stationary point $\boldsymbol{x}_0$ is obtained as

$$\boldsymbol{x}_0 = -\tfrac{1}{2}\boldsymbol{B}^{-1}\boldsymbol{b}\,. \tag{10.32}$$

In practice, the estimated $\boldsymbol{B}$ always has the inverse matrix. A case where $\boldsymbol{B}$ is almost singular is discussed later. The value of the function at point $\boldsymbol{x}_0$ is (show this!):

$$f_0 = b_0 + \tfrac{1}{2}\boldsymbol{x}_0{}'\boldsymbol{b}\,. \tag{10.33}$$

The stationary point can be minimum, maximum or saddle point. The analysis of the stationary point and the behavior of the function is carried out using the *canonical analysis* where the function is presented in a new *canonical* coordinate system. The origin of the new coordinates is the point $\boldsymbol{x}_0$. Stating that $\boldsymbol{z} = \boldsymbol{x} - \boldsymbol{x}_0$, the value of the function is as follows:

$$f = f_0 + \boldsymbol{z}'\boldsymbol{B}\boldsymbol{z}.$$

After shifting the origin of the coordinate system, the coordinates are rotated so that the function can be presented without first-order and cross product terms. This kind of rotation is obtained by linear transformation:

$$\boldsymbol{w} = \boldsymbol{M}'\boldsymbol{z}, \tag{10.34}$$

where $\boldsymbol{M}$ is the matrix where columns are the eigenvectors of matrix $\boldsymbol{B}$ (see Section 7.3.6). It can be derived that the value of the function is

$$f = f_0 + \sum_{j=1}^{p} \lambda_j w_j^2, \tag{10.35}$$

where w_j is according to Equation (10.34)

$$w_j = \boldsymbol{m}_j{}'\boldsymbol{z} = \boldsymbol{m}_j{}'(\boldsymbol{x} - \boldsymbol{x}_0) \tag{10.36}$$

and $\boldsymbol{m}_j$ is column j of *eigenmatrix* $\boldsymbol{M}$, and λ_j is the jth *eigenvalue.*

The new coordinate system provides a good description of the behavior of function f. If λ_j is positive, then moving from the stationary point $\boldsymbol{x}_0$ in direction $\boldsymbol{m}_j{}'(\boldsymbol{x} - \boldsymbol{x}_0)$ the function increases. Similarly, a negative λ_j value indicates a decreasing function.

If all eigenvalues $\lambda_1, \ldots, \lambda_p$ are positive, the stationary point is the minimum. If all eigenvalues are negative, the stationary point is the maximum. If some eigenvalues are positive and some are negative, the stationary point is the *saddle point.* In practice, a saddle point may indicate that the true response function has multiple peaks. A function with multiple peaks cannot be described with a second-order surface.

If matrix $\boldsymbol{B}$ is singular, then at least one eigenvalue is zero. As described earlier, when $\boldsymbol{B}$ is estimated with regression analysis, it is never exactly singular. When interpreting the response surface, one needs to be prepared for the case where $\boldsymbol{B}$ is almost singular, i.e. at least one eigenvalue is close to zero.

If the stationary point $\boldsymbol{x}_0$ is within the observation region or in the region of interest, and some eigenvalue λ_j is close to zero, then function f is almost constant when moving from point $\boldsymbol{x}_0$ to point $\boldsymbol{x}_0 + \boldsymbol{m}_j{}'(\boldsymbol{x} - \boldsymbol{x}_0)$. Then, the response surface has a *stationary ridge* and there is no clear minimum or maximum. For instance, when studying the growth response of plants to different nutrients, a stationary ridge indicates that the ratios of different nutrients are more important than the absolute amounts.

If the stationary point $\boldsymbol{x}_0$ is far from the observation region and some eigenvalue λ_j is close to zero and other eigenvalues are negative, then the response surface is a *rising ridge.* If other eigenvalues are positive, then the response surface is a *falling ridge*

Example 10.3. Let us assume that the following second-order response function was obtained with regression analysis:

$$f = 12 + 8x - 7x^2 + 124z + 8x\,z - 13z^2. \tag{10.37}$$

Then

$$\boldsymbol{b} = \begin{pmatrix} 8 \\ 124 \end{pmatrix} \text{ and } \boldsymbol{B} = \begin{pmatrix} -7 & 4 \\ 4 & -13 \end{pmatrix}.$$

The stationary point is according to Equation (10.32)

$$\boldsymbol{x}_0 = \begin{pmatrix} x_0 \\ z_0 \end{pmatrix} = \begin{pmatrix} 4 \\ 6 \end{pmatrix}.$$

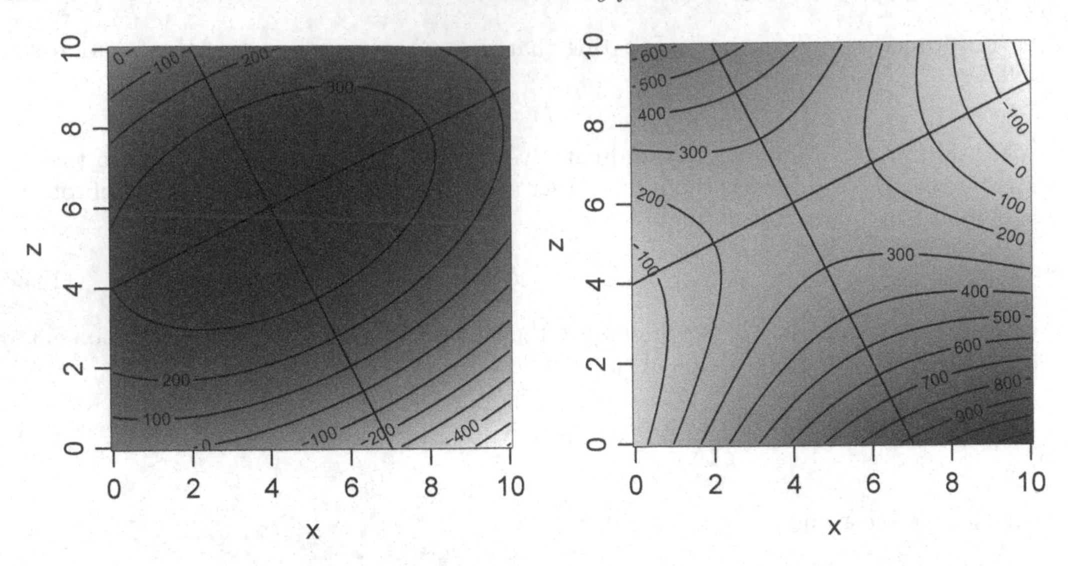

FIGURE 10.4 Response surfaces of Equation (10.37) (left) and Equation (10.37) (right), and the axes of the canonical coordinates. In the left-hand graph, point (4, 6) is the maximum. The behavior of the surface can be seen from Equation (10.38). In the right-hand graph, point (4, 6) is the saddle point.

The value of the response function at the stationary point is obtained either from Equation (10.37) or from Equation (10.33) as $f_0 = 400$. The eigenvalues of $\boldsymbol{B}$ (obtained e.g. with the R-function `eigen`) are -5 and -15, and the corresponding eigenvectors are:

$$\boldsymbol{m}_1 = \begin{pmatrix} -0.894 \\ -0.447 \end{pmatrix}, \boldsymbol{m}_2 = \begin{pmatrix} -0.447 \\ 0.894 \end{pmatrix} .$$

Eigenvectors can be multiplied by -1 to get equivalent eigenvectors.

According to Equations (10.35) and (10.36), the function can then be stated as:

$$f = 400 - 5(-0.894\,(x-4) - 0.447\,(z-6))^2 - 15(-0.447\,(x-4) + 0.894\,(z-6))^2 . \quad (10.38)$$

Because both eigenvalues are negative, the stationary point is maximum. When moving from the stationary point (4, 6) in the direction $(-0.894, -0.447)$ or in the opposite direction, i.e. when moving along the line with slope $m_{1,2}/m_{1,1} = -0.447/-0.894 = 0.5$, the function decreases as slowly as possible: the function is decreased by $5d^2$ units when moving d units. When moving from the stationary point in the direction $(-0.447, 0.894)$ or in the opposite direction, i.e. when moving along the line with slope $m_{2,2}/m_{2,1} = 0.894/-0.447 = -2$, the function decreases as quickly as possible: the function decreases $15d^2$ units when moving d units (Figure 10.4, left).

If the equation of the response surface is:

$$f = 50 + 160x - 5x^2 - 40z - 20x\,z + 10z^2 , \quad (10.39)$$

then the stationary point and the new canonical variables are the same as in the previous case. The eigenvalues of $\boldsymbol{B}$ are 15 and -10, which means that the stationary point is a saddle point. When moving from the stationary point (4, 6) along a line with slope 0.5, the value of the function decreases. When moving on the line with slope -2, the function increases (Figure 10.4, right).

The R-scripts of this example are available on the website of the book.

10.5 Generalized additive models

In Section 10.4.2, we considered smoothing splines to present flexible regression functions for one predictor variable. In Chapter 8, we presented generalized linear models (GLM) where a linear model is formed for $g(\mu)$, where g is the link function and μ is the expected value of y. *Generalized Additive Models* (GAM) are a generalization of smoothing splines to several predictor variables in the GLM framework. In GAM, the linear predictor of GLM is presented as a sum of univariate functions. It would also be possible to use component functions other than smoothing splines but we assume here that the smoothing splines are used.

Let us first consider an additive model

$$y_i = \alpha + \sum_{j=1}^{p} f_j\left(x_i^{(j)}\right) + e_i\,, \tag{10.40}$$

where the error term has mean zero. Let us assume that the component functions are smoothing cubic splines. The penalized sum of squares to be minimized is a generalization of Equation (10.29):

$$\text{PSS} = \sum_{i=1}^{n}\left(y_i - \alpha - \sum_{j=1}^{p} f_j\left(x_i^{(j)}\right)\right)^2 + \sum_{j=1}^{p}\lambda_j \int_{-\infty}^{\infty} f_j''(t_j)^2 dt_j\,.$$

There is no unique solution with respect to α: one can add an arbitrary constant to any f_j and subtract the same constant from α and obtain the same PSS. A natural solution to this identifiability problem is to define $\widehat{\alpha} = \bar{y}$ and require for each j that $\sum_{i=1}^{n} \widehat{f_i^{(j)}} = 0$.

Equation (10.40) can then be estimated by the following *backfitting algorithm*:

1. Initialize $\widehat{\alpha} = \bar{y}$, $\widehat{f^{(j)}} = 0$ for $j = 1, \ldots, p$.
2. Iterate through $j = 1, \ldots, p$ until $\widehat{f^{(j)}}$ changes marginally.
   (a) $\widehat{f_i^{(j)}}$ is the smoothing spline where y values are equal to $y_i - \widehat{\alpha} - \sum_{k \neq j} \widehat{f}_k\left(x_i^{(k)}\right)$
   (b) Subtract from $\widehat{f}_j$ the value $\frac{1}{n}\sum_{i=1}^{n} \widehat{f}_j\left(x_i^{(j)}\right)$.

When the residual variance is not constant, a weighted smoothing spline is used.

When moving to a GLM context, the model of Equation (8.3) (p. 259) is:

$$g(\mu_i)) = \alpha + \sum_{j=1}^{p} f_j\left(x_i^{(j)}\right).$$

The back-fitting algorithm is then modified into a *local scoring* algorithm. The algorithm uses a working target variable instead of the residual in step 2a. Then, the weighted back-fitting algorithm is used. See Hastie and Tibshirani (1986), Hastie and Tibshirani (1990) or Hastie *et al.* (2009) for details. A forestry application of GAM models can be found in Robinson *et al.* (2011).

10.6 Modeling continuous proportions

Binomial distribution can usually be applied when modeling relative frequencies, i.e. discrete proportions, as we have discussed in Section 8.2.2. But for continuous proportions, such as proportion of winter wood in a wood sample or vegetation cover data, the binomial distribution cannot be applied. When modeling the number of ones (y) in N observations with given values of predictor variables, there is only parameter p, which determines both the mean of the relative frequency, $\mathrm{E}(\frac{y}{N}) = p$, and the variance, $\mathrm{var}(\frac{y}{N}) = \frac{p(1-p)}{N}$. When y is a continuous proportion, we may assume that we have a similar model for $\mathrm{E}(y)$ as we have for p in the binomial data. If $\mathrm{E}(y)$ is zero or one, then there cannot be any variation in y, i.e. $\mathrm{var}(y) = 0$. Thus, we may assume that $\mathrm{var}(y)$ is proportional to $\mathrm{E}(y)(1 - \mathrm{E}(y))$. But for continuous proportions, there is no *a priori* way to scale the variance in a similar way as N scales the variance in binomial data.

One possibility to model the continuous proportion y, is to use iteratively re-weighted nonlinear least squares regression. Thus, we may assume that observation i is described as:

$$y_i = f(\boldsymbol{x}_i) + e_i \,, \tag{10.41}$$

where f is a function with values between 0 and 1, e.g. the logistic function $f(x) = 1/\left(1 + e^{a(b-x)}\right)$, which corresponds to the use of logit link for discrete proportions. Iteratively re-weighted nonlinear least squares proceed as follows. First, the parameters of f are estimated with nonlinear OLS. Then, both sides of Equation (10.41) are divided by $w = \sqrt{\widehat{f}\left(1 - \widehat{f}\right)}$, and the resulting equation is re-estimated with nonlinear OLS. The iteration continues until the parameters converge. Testing of parameters can be carried out as presented in Chapter 8. If $\widehat{f}$ is assigned values very close to 0 or 1, then a tiny constant should be added to w to prevent division with almost zero. The above procedure was applied, for example, by Luoranen *et al.* (2004). In R, this can be implemented using function `nme::gnls` (see model `ipsmodNLS` of Example 8.4 (p. 260) for a corresponding model formulation with count data).

Nowadays, it may be more appropriate, however, to also analyze the continuous proportions using generalized linear (mixed) models. The standard distribution in such an analysis is the beta distribution. The pdf of the beta distribution can be stated as

$$f(y; \alpha, \beta) = \frac{1}{B(\alpha, \beta)} y^{\alpha-1} (1-y)^{\beta-1} \,, \tag{10.42}$$

where $0 < y < 1$, $\alpha, \beta > 0$, and $B(\alpha, \beta) = \int_0^1 t^{\alpha-1} (1-t)^{\beta-1} dt$ is the *beta function.* The mean of y is $\mathrm{E}(y) = \alpha/(\alpha+\beta)$ and $\mathrm{var}\,(y) = \alpha\beta/(\alpha+\beta+1)(\alpha+\beta)^2$. It is more convenient to parameterize the beta distribution in terms of $\mu = \alpha/(\alpha+\beta)$ and a scale parameter $\phi = \alpha+\beta$. The variance of y in terms of the new parameters is $\mathrm{var}\,(y) = \mu(1-\mu)/(1+\phi)$. As the variance of a relative frequency is $p(1-p)/N$ the parameter ϕ corresponds to $N-1$, where N is the total number of observations used to define a single relative frequency in the data. For continuous proportions, the logit link is the standard link function. GLMM with beta distribution and logit link is often also called *beta regression.*

Frequently, there is one additional complication when applying the beta regression to continuous proportions. According to any continuous distribution, the probability of any specific value is zero. Non-zero probabilities are obtained only for some intervals. Moreover, the pdf of beta distribution is not even defined for 0 or 1. However, for continuous proportions it often happens that there are several zeroes and/or ones in the data. To model such data, we can use zero-one inflated beta regression. It works similarly to zero inflated Poisson or negative binomial model discussed in Section 8.2.9. The difference is that all zeroes are extra zeroes.

In the zero-one inflated beta regression, three simultaneous stochastic processes are assumed. The first process determines whether y is assigned the value 0. The second process determines whether y is assigned the value 1, if it has not obtained value 0. The third process determines the distribution of y, if it has not obtained value 0 or 1. Thus, the overall distribution is a mixture of the three distributions. Each of the three distributions can be modeled with a separate GLMM model. Of course, if zeroes or ones are not possible, then only two of the three models are needed. The zero-one inflated beta regression is described, for example, by Ospina and Ferrari (2012). R package `Zoib` can be used for the estimation (Liu and Kong 2015).

Data consisting of several continuous proportions are called *compositional data*. The main difficulty in the analysis of compositional data comes from the fact that the variance matrix of proportions or transformations of proportions is singular. The methods needed in the analysis of compositional data are not simple extensions of the methods covered in this book, so we refer readers to van den Boogaart and Tolosana-Delgado (2013). The R package `Compositions` (van den Boogaart *et al.* 2009) can be used in the analysis of compositional data.

10.7 Effect size

After estimating a statistical model, the standard procedure is to test the statistical significance of the estimated parameters. In testing, the assumption that the true parameter values deviate from some reference values or that parameter values deviate from each other is evaluated. It is a common misconception that if an effect is statistically significant, then the effect is also strong. The statistical significance is dependent both on the amount of data and the size of the effect. In biological sciences, almost all scientifically interesting factors have some effect. So, with a sufficiently large data set, practically all factors will become statistically significant. In this section, we briefly discuss the evaluation of the importance of an effect. In the literature, there are tens of different measures of the effect size. For a more complete discussion readers are referred to Grissom and Kim (2012).

A starting point when considering effect size is that, in addition to the statistical significance of an estimated parameter, one should pay attention to the estimated value of the parameter. Thus, the estimated parameters should also be published and properly discussed. Many standard statistics are in effect size literature classified as effect size measures, e.g. correlation coefficient and coefficient of determination. When comparing two population means, a popular effect size index is Cohen's d defined as:

$$d = \frac{\bar{x}_2 - \bar{x}_1}{s},$$

where $\bar{x}_1$ and $\bar{x}_2$ are population averages, and s is a pooled standard deviation. The motivation premise in this index is that the differences of population (group) means should be compared to the natural variation in the populations. Assuming normal distribution, one can calculate from Cohen's d the probability that an individual randomly picked from one population has a larger value than an individual randomly picked from the other population. The principle behind Cohen's d can be extended to more complicated situations.

When considering the regression models discussed in this book, the Cohen's d can be interpreted to mean the estimated difference between a treatment effect and a control or reference treatment effect divided by the standard error of the regression. Of course, any two treatment levels can be compared. In a linear mixed-effects model with a variance component error structure, i.e. $y_{ij} = \boldsymbol{x}'\boldsymbol{\beta} + b_i + \epsilon_{ij}$, the estimated difference can be compared either to the estimated standard deviation of the total variation, i.e. square root of $\text{var}(b_i) + \text{var}(\epsilon_{ij})$ or to the estimated standard deviation of the residual error, square root of $\text{var}(\epsilon_{ij})$. Of course, a treatment effect can also be compared to any component of the random term. As the estimated variance of a random effect is quite often zero or close to zero, division by the estimated standard deviation of the random effect may not be an appropriate way to do the comparison. If there are also random slope parameters in the model, the comparison can, without a more complicated analysis, be carried out only to the standard deviation of the residual error.

We have indicated above that in order to compare fixed treatment effects to random effects, the standard deviations of the random effects should be used instead of variances. However, variances in a variance component model have the enviable property that they are additive, while standard deviations are not. A variance component divided by the sum of variance components gives the proportion of variance associated with a given factor. Let us consider a way to compare fixed class effects to variance components so that the additivity of variances is retained. Let us assume that we want to compare the difference between a treatment effect μ_1 and the control μ_0 to the variance components. Then $((\mu_1 - \mu_0)/2\,)^2$ is the variance of such a random variable, which is assigned the value μ_1 with a probability 0.5 and μ_0 with a probability 0.5. We suggest that this variance can be compared to the variance components. A forestry application is described in Petäistö and Lappi (1996).

If there are several treatment effects, one can compute the 'treatment variance' for several treatment pairs. An option would be to include treatment effects into the model as if they were random effects. This should be done only after estimating the treatment effects first as fixed effects. The estimation of the treatment variance, assuming treatment effects as random, would help in comparing the differences between fixed effects and variance components in a variance component model.

When considering effect sizes, one should pay attention to the confidence intervals of the parameters, in addition to the estimated values of the parameters. In practice it is often sufficient to compute the standard error of the estimate and compute the confidence interval as the estimate $\pm$ twice the standard error. In significance testing the interest lies in whether the confidence interval includes zero or not. From the effect-size point of view, the interest lies on all values in the confidence interval compared to the variation of the error term.

Different effect-size calculations and indexes are popular in medical or psychological statistics. In forestry analyses, we suggest that they may be less useful than the effect-size concepts discussed above, or the properties of statistical models discussed elsewhere in this book. The effect-size concept reminds us that significance tests and p-values are not the entire result of a statistical analysis. The practical importance of estimated effects cannot be determined solely on statistical grounds. It is also dependent on subject matter theory and practical decision-making problems. Effect-size measures are widely used in meta-analysis (see e.g. Schwarzer *et al.* (2015) or Pigott (2012)). Unfortunately, we are not aware of satisfactory meta-analysis studies in forestry.

11

Modeling Tree Size

Tree size can be characterized by many variables. Tree stem diameter at the breast height (1.3 m, DBH) is probably the most commonly used variable because it is easy to determine in the field and is strongly correlated with tree volume and biomass. Other widely used variables are tree diameter at some other reference height, tree volume, and biomass. Nowadays, forests are commonly measured using aerial data, such as the airborne laser scanning (ALS) (Maltamo *et al.* 2014). In such cases, the size variable that is most easy to measure is some dimension of the crown, such as tree height or the maximum crown diameter.

Tree allometry describes the relationship between tree size variables. Commonly used allometric models are, for example, the height-diameter relationship (modeled in the examples in Chapter 4), the relationship between volume or biomass and stem diameter and height, or the relationship between maximum crown diameter area and tree diameter.

This chapter discusses the modeling of tree size in a fixed area of interest, usually a sample plot or forest stand, through *size distribution.* The most common example is diameter distribution, which is needed in many stand-level calculations, such as estimation of standing volume or biomass, providing initial trees for a stand-level growth simulator based on tree-level growth models, or as a quantitative measure of stand structure and biodiversity. We mainly discuss stand-level diameter distributions here, but the majority of the ideas also apply to other size measures and to other levels of aggregation. The allometric relationships determine how the different size variables (and their distributions) are related.

We use metric units for the dimensions in this chapter as follows. Tree diameter is expressed in cm, height in meters and tree-level volume usually in dm^3. The plot and stand level basal area is expressed in m^2 and volume in m^3.

The possible methods for describing the diameter distribution largely depend on the application and on the available data. Notably, a measured sample of diameters from the target stand may be available, or alternatively only some stand characteristics of the target stand may be known. If a sample of tree diameter is available, it is a natural basis for estimating the diameter distribution. If the sample is large enough and does not include substantial sampling error, then it can be used as such in the computations (e.g. Pienaar and Harrison 1988). If the sample is small, it can be smoothed using e.g. *kernel smoothing* (e.g. Droessler and Burk (1989), see Box 11.1, p. 334) or a theoretical distribution function can be fitted to it (e.g. Bailey and Dell 1973) (see Section 11.2).

The measurement of a diameter sample is time-consuming for many inventories. In these cases, the diameter distribution can be estimated based on some easily measurable stand variables, such as (quadratic) mean diameter, stand age, basal area, stand density, or dominant height. These estimation methods can be classified to parameter prediction methods (PPM), parameter recovery methods (PRM) (Hyink and Moser 1983) and imputation methods. In the PPM, the parameters of the assumed distribution function are predicted with some measured stand variables using estimated regression models (e.g. Schreuder *et al.* 1979). These models need to be available from a previous study. In PRM, the parameters are recovered from some stand variables using mathematical relationships between the stand variables and distribution parameters (e.g. Ek *et al.* 1975). The stand variables used

Box 11.1 Kernel smoothing

Smoothing can be used to produce nice smooth plots of observed data without assuming any specific functional form for the distribution. The most common method for smoothing is the kernel method (e.g. Härdle 1990). The idea of kernel smoothing is to place each observed value onto the x-axis, replace these points with a density function called *kernel* (e.g. Gaussian density), sum up these densities and rescale so that the area under the curve is unity. This results in an estimate of the density that is smoother than the histogram of the sample but is much more flexible than any commonly used parametric density function. The *cdf* corresponding to the kernel density is similarly found by replacing each observation with the *cdf* of the applied kernel, summing these up, and rescaling to have the limit of one in the right-hand tail.

Therefore, the *pdf* and *cdf* of a kernel-smoothed empirical distribution based on sample of size n can be expressed as:

$$f_h(x) = \frac{1}{n}\sum_{i=1}^{n} k((x - x_i)/h)$$

$$F_h(x) = \frac{1}{nh}\sum_{i=1}^{n} K((x - x_i)/h)\,,$$

where k_h and K_h are, respectively, the density and distribution function which have zero mean and unit-variance called *kernel function*, h is the smoothing parameter called *bandwidth*, and x_i, $i = 1, \ldots, n$, are the sample values of the variable of interest. The utilized kernel function is usually a symmetric function. The degree of smoothing is determined by the bandwidth, which is set by the user. Often, Silverman's rule of thumb, which has been implemented in `R`-function `stats::bwnrd0`, is used to determine the bandwidth (Silverman 1986, p. 48). The rule is

$$h_s = 0.9 \min\left(\widehat{\sigma}, \frac{\widehat{R}}{1.34}\right) n^{1/5}\,,$$

where $\widehat{\sigma}$ is the estimated standard deviation and $\widehat{R}$ is the interquartile range of the sample.

in PRM may be, for example, percentiles or moments of the diameter distribution, and they can be either directly measured or predicted by regression models in a similar way as in PPM. Recovery is possible only for as many parameters as there are measured stand variables that can be mathematically derived from the diameter distribution. See Section 11.5.2 for more discussion on PPM and PRM. In imputation methods, the diameter distribution of the target stand is estimated as the mean of the measured diameter distributions from similar stands using the k nearest neighbor or most similar neighbor approaches (e.g. Haara *et al.* 1997). Imputation methods are not covered in this book.

The approaches described above can also be combined. For example the neighbors in imputation may be smoothed diameter distributions instead of true distributions. Furthermore, sample information can be used to improve a predicted distribution, as we will discuss in Section 11.5.3.

Example 11.1. The diameters of 57 trees were measured in a sample plot 22 of data set `spati`. The distribution was illustrated using a histogram and kernel smoothing based on the Gaussian kernel (see Box 11.1). Three alternative bandwidths h were used. Silverman's rule of thumb gives us a value $h = 2.63$. The two other values were obtained by multiplying this value by 2 and 0.5, resulting in $h = 5.26$, and $h = 1.31$, respectively. Figure 11.1 shows the original data and histogram (left-hand graph) and empirical *cdf* (right-hand graph) together with the smoothed *pdf* and *cdf*.

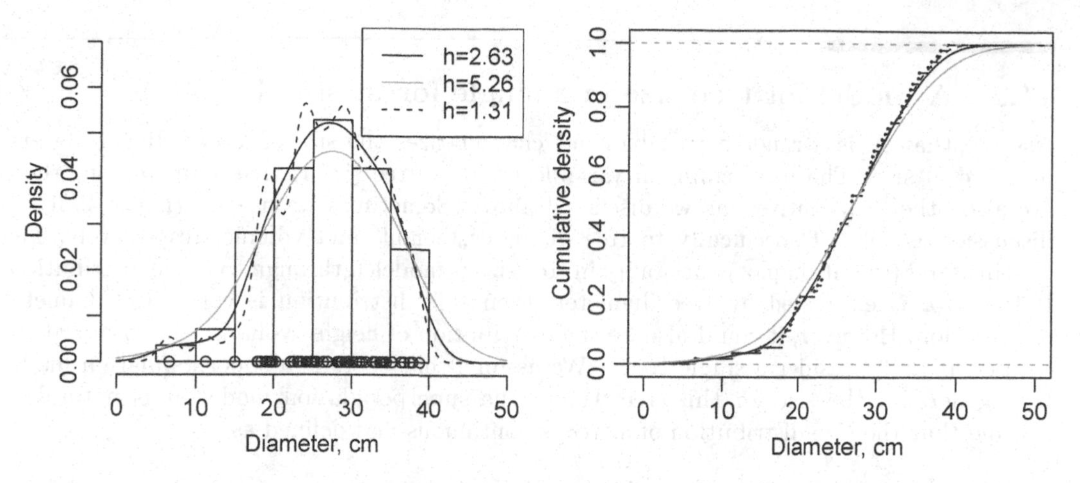

FIGURE 11.1 Illustration of the diameter distribution in Example 11.1. The dots in the left-hand graph show the observed tree diameters (cm). The density is illustrated using a histogram and kernel-smoothed density with three different values of the smoothing parameter h. The right-hand graph shows the empirical *cdf* using a step function and the smoothed *cdf*s using the three smooth lines.

The function `pgkernel` defined below was used to compute the kernel *cdf*. The *pdf* results from replacement of `pnorm` by `dnorm` within the function. Function `density` will give the same density.

```
> pgkernel<-function(x,d,bw=bw.nrd0(d)) {
+         n<-length(d)
+         sapply(x,function(a) 1/n*sum(pnorm((a-d)/bw)))
+ }
> d22<-spati$d[spati$plot==22]
> plot(ecdf(d22),cex=0.2,ylim=c(0,1),main="",
+      xlab="Diameter, cm",ylab="Cumulative density",xlim=c(0,50))
> lines(x,pgkernel(x,d22,bw=bw.nrd0(d22)))
```

For illustration, the script below computes the estimated proportion of trees in the six 5 cm diameter classes within $(0, 50)$ using kernel density of $h = 2.63$. The observed proportion of trees in those diameter classes is also computed.

```
> F<-pgkernel(seq(0,50,5),d22)
> data.frame(d=seq(2.5,47.5,5),
+           kernel=round(F[-1]-F[-length(F)],3),
+           observed=round(c(0,hist(d22)$counts/length(d22),0,0),3))
+

      d kernel observed
1   2.5  0.005    0.000
2   7.5  0.017    0.018
3  12.5  0.032    0.035
4  17.5  0.118    0.140
5  22.5  0.219    0.211
6  27.5  0.256    0.263
7  32.5  0.217    0.211
8  37.5  0.118    0.123
9  42.5  0.018    0.000
10 47.5  0.000    0.000
```

11.1 A model for tree size in a single forest stand

Assume that X is a random variable that characterizes the size of a tree. It can be any measure of size. The most common variable used for tree size is tree diameter but there are also other alternatives, as we discussed above. Sometimes we use D, H and V if the discussion is related specifically to tree diameter, height, and volume, respectively. The within stand (or within plot) variability in tree size is modeled through tree size distribution. If tree size is expressed by tree diameter, then this distribution is called the diameter distribution. However, it could also be the distribution of height, volume, or crown area.

Let us first consider a single tree i. We assume that there is a superpopulation model that generated the size for this tree. Under the superpopulation model, it is natural to assume that the size distribution of a tree is continuous and defined as

$$F(x) = P(X_i \leq x) .$$

The corresponding *pdf* is

$$f(x) = F'(x) .$$

The size distribution gives the probability for the size of tree i to have a value less than x. If all trees within the area of interest (stand, plot) are identically distributed according to the above-specified model, then the (marginal) diameter distribution over the area of interest is equivalent to the above-specified distribution of tree i, and index i is not needed. Then, the diameter distribution can also be interpreted as the expected proportion of trees with a size less than x.

It might be further assumed that the diameters are mutually independent, i.e. the size of a tree does not depend on the sizes of the other trees (e.g. the neighbors). This is not true if tree size has spatial autocorrelation within the area. The spatial autocorrelation may be negative (because of inter-tree competition) or positive (e.g. because of small-scale variability in growth conditions). If the sample plot/stand is large compared to the range of the spatial autocorrelation, the observed tree sample can be used to represent the marginal distribution of the diameters and the autocorrelation may be ignored. However, autocorrelation should be taken into account, for example, when estimating the variances of the parameter estimates.

The next sections describe the approaches that are used in fitting an assumed model to observed tree diameter data, and the models that are commonly used for that purpose. Thereafter, we will discuss the transformations and weighted distributions. These two issues, in some instances, appear mixed. The distinction can be demonstrated as follows.

Consider a histogram, where trees of a given area are classified in diameter classes of width 2 cm, and the heights of the bars show the proportion of the total number of stems in each diameter class, such as in the histogram in the left-hand graph in Figure 11.1. However, one could classify the trees according to size-related characteristics other than diameter. For example, the histogram could show the proportion of total number of stems in classes of basal area, which specifies the distribution of the basal area. This change in the x-axis is related to *transformation.*

In contrast, *weighting* changes the y-axis of the histogram. For example, instead of specifying the proportion of the total number of trees that is included in a given diameter class, we may specify the proportion of the total basal area that is included in the class, which leads to a basal area weighted diameter distribution. We could use tree height or volume as weight as well. As an example, and as an exercise, one could consider the meaning and potential use of the following distributions:

- basal-area weighted diameter distribution
- diameter-weighted basal area distribution
- diameter-weighted diameter distribution
- height-weighted diameter distribution
- diameter-weighted height distribution
- height-weighted distribution of crown diameters
- crown area-weighted height distribution
- volume-weighted distribution of crown area

Most of these may not be useful in practice, but some of them might.

In this book, and also in other statistical literature, the density of a distribution is not weighted if that is not explicitly mentioned. The frequency of a size class, therefore, refers to the number of stems by default.

11.2 Fitting an assumed model to tree diameter data

Quite often, one needs to estimate the true diameter distribution using observed tree diameter data. To do so, we first need to select the distribution function that we will assume for the tree diameter data and, thereafter, estimate the parameters using some justified method. The ML method is a well justified choice in such a case. However, there are also other methods that are commonly used in this context.

11.2.1 Maximum likelihood

The ML method was described in detail in Section 3.4.4. If the diameter sample is an independent, identically distributed sample from an assumed distribution with *pdf* $f(x|\boldsymbol{\theta})$, the log-likelihood is the sum of densities over the observed values of x:

$$l(\boldsymbol{\theta}|\boldsymbol{x}) = \sum_{i=1}^{n} \log f(x_i|\boldsymbol{\theta}) \,.$$

The large-sample variance of the ML estimates can be found using the second derivatives of the likelihood, as described in Section 3.4.4. However, it may be unrealistic to assume that the observations are independent, and asymptotic inference based on the likelihood under independence may be biased. Depending on the degree and type of the dependence and size of the sample, the standard errors of the ML estimates may be either overestimates or underestimates.

Example 11.2. ML fit of Weibull and logit-logistic distributions to fixed area plot data. Let us consider sample plot 22 of data set `spati`. Let us assume that tree diameter X follows the two-parameter Weibull distribution, which has the *pdf*

$$f(x|\alpha,\beta) = \frac{\alpha}{\beta^\alpha} x^{\alpha-1} e^{-(x/\beta)^\alpha} \qquad \alpha,\beta > 0, x > 0\,,$$

parameters α and β are commonly called the shape and scale parameters, respectively. The script below defines the negative log-likelihood and does the fitting using function `stats4::mle`.

```
> nll<-function(shape,scale) -sum(log(dweibull(d22,shape,scale)))
> estMLWE<-mle(nll,start=list(shape=3,scale=20))
Warning messages:
1: In dweibull(d22, shape, scale) : NaNs produced
2: In dweibull(d22, shape, scale) : NaNs produced
> summary(estMLWE)

Coefficients:
       Estimate Std. Error
shape  4.627609  0.4924432
scale 29.365069  0.8818998

-2 log L: 378.6704
```

The warning messages indicate that during the fitting procedure, negative values of α or β were attempted, which led to an undefined log-likelihood value. However, this is not a problem at the end, because the algorithm could continue the iteration and produce a solution at the end. Such warnings could be avoided by reparameterizing the model in terms of $\log(\alpha)$ and $\log(\beta)$, as we did in Example 3.8 (p. 55).

The two-parameter Weibull function may not be sufficiently flexible for the tree diameter data. A more flexible alternative is the logit-logistic distribution (Wang and Rennolls 2005), which has the *pdf*

$$f(x|\xi,\lambda,\mu,\sigma) = \frac{\lambda}{\sigma}\frac{1}{(x-\xi)(\xi+\lambda-x)}\frac{1}{e^{-\frac{\mu}{\sigma}}\left(\frac{x-\xi}{\xi+\lambda-x}\right)^{\frac{1}{\sigma}} + e^{\frac{\mu}{\sigma}}\left(\frac{x-\xi}{\xi+\lambda-x}\right)^{-\frac{1}{\sigma}} + 2}, \tag{11.1}$$

for $\xi < x < \xi + \lambda$ and zero otherwise; $\lambda, \xi > 0$. The parameter ξ is the minimum, λ is the range, $-\infty < \mu < \infty$ controls skewness, and $0 < \sigma < 1$ controls kurtosis; values $\mu = 0$ and $\sigma = 0.5$ provide a symmetric distribution that might provide good starting values of the parameters for estimation. We implement the above-specified bounds of parameters by reparameterizing the distribution using $\log(\xi)$, $\log(\lambda)$, and $\text{logit}(\sigma)$. The script below defines the necessary negative log-likelihood function, and estimates the parameters, utilizing the logit-logistic *pdf* implemented in `lmfor::dll`.

```
> # negative log likelihood
> nllLL<-function(lxi,llambda,lsigma,mu) {
+         sigma<-exp(lsigma)/(1+exp(lsigma))
+         xi<-exp(lxi)
+         lambda<-exp(llambda)
+         value<--sum(log(dll(d22,mu,sigma,xi,lambda)))
+ #       cat(lxi,llambda,lsigma,mu,value,"\n")
+         value
+ }
>
> estMLLL<-mle(nllLL,start=list(lxi=log(min(d22)-1),
+               llambda=log(max(d22)-min(d22)+2),lsigma=0.5,mu=0))
> summary(estMLLL)

Coefficients:
         Estimate Std. Error
lxi     1.5270856  0.7120727
llambda 3.5606048  0.1191349
lsigma  0.2952300  0.5186791
mu      0.6302712  0.2037138

-2 log L: 375.8692
```

The output shows that the estimate of the minimum diameter is $\widehat{\xi} = e^{1.527} = 4.6$ cm, the range $\widehat{\lambda} = e^{3.561} = 35.2$ cm, $\widehat{\sigma} = \frac{e^{0.295}}{1+e^{0.295}} = 0.573$, and $\widehat{\mu} = 0.630$. Figure 11.2 shows the tree diameter data, overlaid by the Weibull and logit-logistic *pdf*s using the ML estimates of the parameters.

The ML principle is the best justified approach for finding point estimates for unknown parameters. Especially, it is efficient (at least asymptotically) and provides general tools to assess the accuracy of the obtained estimates. Furthermore, it is easy to extend to a situation where the parameters are functions of some predictors (recall Examples 2.2 and 2.3, pp. 10-11). Such an extension will be illustrated in Section 11.5.2.

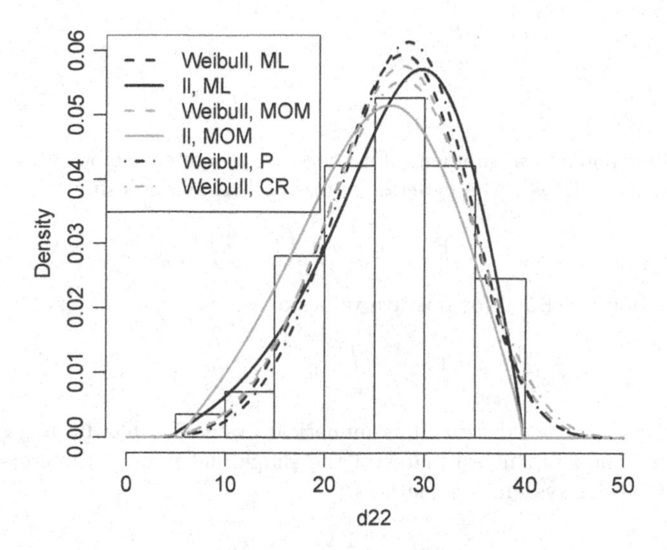

FIGURE 11.2 The histogram of tree diameters on plot 22 of data set `spati` and the fitted logit-logistic and Weibull functions from Examples 11.2 – 11.7.

11.2.2 The method of moments

The method of moments is maybe the oldest and simplest method for fitting an assumed distribution function to observed data. It was first used in the late 1800s by Carl Pearson (Casella and Berger 2001). The earliest application to tree diameter data dates back to 1914 (Cajanus 1914). Burk and Newberry (1984) presented the method of moments for the three-parameter Weibull distribution.

Let $X_1, X_2, \ldots, X_n$ be a sample from a population with density $f(x|\theta_1, \theta_2, \ldots, \theta_p)$. For example, it may be a sample of tree diameters, and the assumed distribution may be a Weibull density function with $\theta_1 = \alpha$ and $\theta_2 = \beta$. The method of moments finds parameter estimates by setting the first p moments of the assumed distribution to equal the corresponding sample moments. This leads to a system of p equations, which is solved for $\theta_1, \ldots, \theta_p$. With a two-parameter model, such as the two-parameter Weibull distribution or Normal distribution, only the first two moments are needed, and the number of equations is two. With four-parameter distributions, such as the logit-logistic distribution, the number of parameters and equations would be four.

For a formal definition, we define the sample moments m_k and corresponding population moments μ'_k as follows (Casella and Berger 2001)

$$m_k = \frac{1}{n}\sum_{i=1}^{n} X_i^k \qquad \mu'_k = \mathrm{E}(X^k) \qquad k = 1, 2, \ldots, p$$

The population moments $\mathrm{E}(X^k)$ can be computed using Equation (2.10) (p. 15). These expectations are functions of parameters $\theta_1, \ldots, \theta_p$. The moment estimators of these parameters result from solving the system of p equations of form

$$m_k = \mu'_k(\theta_1, \ldots, \theta_p)$$

for $\theta_1, \ldots, \theta_p$. The method does not provide tools to evaluate the accuracy of parameter estimates.

Example 11.3. Moment-based estimation of Weibull parameters. The sample moments of the tree diameter data from sample plot 22 of data set `spati` are computed below:

```
> mean(d22)
[1] 26.83333
> mean(d22^2)
[1] 766.0942
```

The corresponding population moments for the two-parameter Weibull distribution cannot be written in closed form. However, a general expression is available using the gamma function as

$$E(X^n) = \beta^n \Gamma\left(1 + \frac{n}{\alpha}\right).$$

The gamma function is defined for positive values of z as

$$\Gamma(z) = \int_0^\infty u^{z-1} e^{-u} du.$$

It cannot be written in closed form, but numerical evaluation has been implemented, e.g. in R-function `gamma`. The moment estimates of the shape and scale parameters are found as the solution of the following system of equations

$$\begin{aligned} \beta\Gamma\left(1 + \frac{1}{\alpha}\right) &= 26.83 \\ \beta^2\Gamma\left(1 + \frac{2}{\alpha}\right) &= 766.09. \end{aligned}$$

The system was solved numerically using the Newton-Raphson method as implemented in function `lmfor::NRnum`. The function requires the functions $f_i(\boldsymbol{x})$ as a list of R-functions, and finds solutions to equations $f_i(\boldsymbol{x}) = 0$. Initial values for the solution need to be given by the user. The required derivatives are computed numerically within function `NRnum`. In our case, the system includes the differences between the theoretical and sample moments. The parameter estimates are $\widehat{\alpha}_{MOM} = 4.48$ and $\widehat{\beta}_{MOM} = 29.4$. The *pdf* based on these parameter estimates is shown in Figure 11.2 with a gray dashed line.

```
> wmom<-function(n,alpha,beta) beta^n*gamma(1+n/alpha)
>
> fnlist<-list(function(theta) wmom(1,theta[1],theta[2])-mean(d22),
+              function(theta) wmom(2,theta[1],theta[2])-mean(d22^2))
>
> print(estMOMWE<-NRnum(c(3,20),fnlist))
$par
[1]  4.484167 29.410049

$value
[1] -8.395808e-10  7.979895e-08
```

Web Example 11.4. Moment-based estimation of logit-logistic function parameters.

Notice that the method of moments matches $E(X)$ to the sample mean of x and $E(X^2)$ to the sample mean of x^2. Therefore, the variance of the resulting distribution will be $\frac{1}{n}\sum(x-\bar{x})^2$ and not $\frac{1}{n-1}\sum(x-\bar{x})^2$. The method can, however, be adjusted so that the moments of the fitted distribution are matched to the unbiased estimators of sample moments, as illustrated in the following example.

Example 11.5. Matching sample mean and variance. The following script matches the sample mean and sample variance to the population mean and population variance in the two-parameter Weibull distribution. The estimates $\widehat{\alpha} = 4.44$ and $\widehat{\beta} = 29.43$ differ from the moment-based estimates and the resulting distribution has a slightly larger variance.

```
> fnlist<-list(function(theta) wmom(1,theta[1],theta[2])-mean(d22),
+  function(theta) wmom(2,theta[1],theta[2])-
+  wmom(1,theta[1],theta[2])^2-var(d22))
> print(estMOMWE2<-NRnum(c(3,20),fnlist))
$par
[1]  4.440702 29.426675
```

11.2.3 Other methods

The method of percentiles is analogous to the method of moments, but instead of moments, it matches the sample percentiles with the quantiles of the assumed distribution function. For example, Bailey and Dell (1973) used (in addition to ML) the percentile-based method because of its simplicity. They also presented closed-form solutions for the percentile equations.

Example 11.6. Dubey (1967) showed that the 24th and 93rd percentiles are asymptotically best for estimating the shape parameter α of the Weibull distribution with no prior knowledge on the scale parameter. However, the efficiency is only 44% compared to the ML, and the percentile method was suggested only for finding starting values for ML estimation. In this example, we use these two percentiles to estimate both the shape and scale parameters. Technically, the estimation procedure is very similar to that carried out in Example 11.3. The estimates are $\widehat{\alpha}_P = 4.85$ and $\widehat{\beta}_P = 29.8$. The fitted density is shown in Figure 11.2 with the black dot-dashed line.

```
> quants<-quantile(d22,probs=c(0.24,0.93))
>
> fnlist<-list(function(theta) qweibull(probs[1],theta[1],theta[2])-quants[1],
+              function(theta) qweibull(probs[2],theta[1],theta[2])-quants[2])
>
> print(estPWE<-NRnum(c(3,20),fnlist))
$par
[1]  4.854709 29.784369
```

In addition to moments and percentiles, other quantities can also be used for estimation. In principle, any system of p equations can be used to find estimates of the p parameters. Some of the equations may be based on percentiles, others on moments. Also other characteristics, such as the quadratic mean diameter $DQM = 2\sqrt{G/(\pi N)}$, where G is the basal area (m^2 per hectare) and N the stand density (trees per hectare), could be set to equal those derived from the assumed distribution. These approaches are discussed in more detail in Section 11.4.5.

Some studies have used the method of *cdf* regression to fit an assumed distribution function to tree diameter data. In that approach, the diameter observations are ordered, and the ith smallest observation of a sample of size n is interpreted as $i/(n+1)^{\text{th}}$ quantile of the distribution. The estimated values are obtained by fitting the *cdf* of the assumed distribution function to the data of diameters and the corresponding values of the empirical *cdf* using ordinary nonlinear least squares (see Chapter 7). One can also transform the quantiles and percentage values so that the model is linear and then use linear OLS. A problem with both these methods is that the sample quantiles are dependent, and that the variance is heteroscedastic. Some authors have proposed methods where the dependence and heteroscedasticity are taken into account in a theoretically justified way (see Engeman and Keefe 1982).

Example 11.7. The following code implements the *cdf* regression for the data set described in the previous examples, ignoring dependency and heteroscedasticity. The estimates are $\widehat{\alpha}_{CR} = 4.33$ and $\widehat{\beta}_{CR} = 29.6$. The fitted density is shown in Figure 11.2 with a gray dot-dashed line.

```
> dord<-sort(d22)
> quants<-(1:length(dord))/(length(dord)+1)
> estCRWE<-nls(quants~pweibull(dord,shape,scale),start=list(shape=3,scale=20))
> estCRWE
Nonlinear regression model
  model: quants ~ pweibull(dord, shape, scale)
   data: parent.frame()
shape scale
 4.33 29.61
 residual sum-of-squares: 0.01619
```

Distribution	Parameters and functions
2-parameter Weibull	$\alpha, \beta > 0;\ x > 0$ $F(x\|\alpha, \beta) = 1 - \exp\left\{-\left(\frac{x}{\beta}\right)^{\alpha}\right\}$ $f(x\|\alpha, \beta) = \frac{\alpha}{\beta^{\alpha}} x^{\alpha-1} \exp\left\{-\left(\frac{x}{\beta}\right)^{\alpha}\right\}$
3-parameter Weibull	$-\infty < \xi < \infty$ (min), $\alpha, \beta > 0;\ x > 0$ $F(x\|\alpha, \beta) = 1 - \exp\left\{-\left(\frac{x-\xi}{\beta}\right)^{\alpha}\right\}$ $f(x\|\alpha, \beta) = \frac{\alpha}{\beta^{\alpha}} (x-\xi)^{\alpha-1} \exp\left\{-\left(\frac{x-\xi}{\beta}\right)^{\alpha}\right\}$
Non-standard beta	$-\infty < \xi < \infty$ (min), $\lambda > 0$ (max-min), $\alpha, \beta > 0$ $f(x\|\xi, \lambda, \alpha, \beta) = \frac{1}{B(\alpha,\beta)} \frac{(x-\xi)^{\alpha-1}(\xi+\lambda-x)^{\beta-1}}{\lambda^{\alpha+\beta-1}}$
Johnson's SB	$-\infty < \xi < \infty$ (min), $\lambda > 0$ (max-min), $-\infty < \mu < \infty$, $\sigma > 0$ $f(x\|\xi, \lambda, \mu, \sigma) = \frac{1}{\sqrt{2\pi}} \frac{\lambda}{\sigma(\xi+\lambda-x)(x-\xi)} \exp\left\{-\frac{1}{2}\left[\frac{\log\left(\frac{x-\xi}{\xi+\lambda-x}\right)-\mu}{\sigma}\right]^2\right\}$
Logit-logistic	$-\infty < \xi < \infty$ (min), $\lambda > 0$ (max-min), $-\infty < \mu < \infty$, $0 < \sigma < 1$ $F(x\|\xi, \lambda, \mu, \sigma) = \frac{1}{1+\exp\left(\frac{\mu}{\sigma}\right)\left(\frac{x-\xi}{\xi+\lambda-x}\right)^{-\frac{1}{\sigma}}}$ $f(x\|\xi, \lambda, \mu, \sigma) = \frac{\lambda}{\sigma} \frac{1}{(x-\xi)(\xi+\lambda-x)} \frac{1}{e^{-\frac{\mu}{\sigma}}\left(\frac{x-\xi}{\xi+\lambda-x}\right)^{\frac{1}{\sigma}} + e^{\frac{\mu}{\sigma}}\left(\frac{x-\xi}{\xi+\lambda-x}\right)^{-\frac{1}{\sigma}} + 2}$

Note: See `FAdist::dweibull3` (Aucoin 2015), `extraDistr::dnsbeta` (Wolodzko 2018), `suppDists::dJohnson` (Wheeler 2016), and `lmfor::dll` (Mehtätalo 2020) for R-implementation of 3-parameter Weibull, nonstandard beta, Johnson's SB, and logit-logistic distributions, respectively. The parameterizations in these packages may differ from those presented above.

TABLE 11.1 Some distributions that have been used in modeling the diameter distributions. The *cdf* of non-standard Beta and Johnson's SB distributions cannot be written in closed form.

11.3 Model forms

11.3.1 Distribution models for tree size

Examples 11.2 – 11.7 used the two-parameter Weibull and logit-logistic models for tree diameter data. These are not the only alternatives, and a number of distribution functions have been used to model tree diameter data. Based on Wang and Rennolls (2005), we summarize in Table 11.1 some of the unimodal functions that have been used for tree diameter data. Of these functions, the Johnson's SB distribution can actually have a bimodal shape, but the shape of the bimodal form is so restricted that the function is not suggested for serious modeling of bimodal diameter data.

A good tool to evaluate the flexibility of a distribution is to analyze which combinations of skewness (asymmetry) and kurtosis can be described by a given distribution function (Hafley and Schreuder 1977, Johnson *et al.* 2005). Skewness and kurtosis are defined as

$$\pm\sqrt{\beta_1} = \frac{\mu_3}{\mu_2^{3/2}}$$

$$\beta_2 = \frac{\mu_1}{\mu_2^2},$$

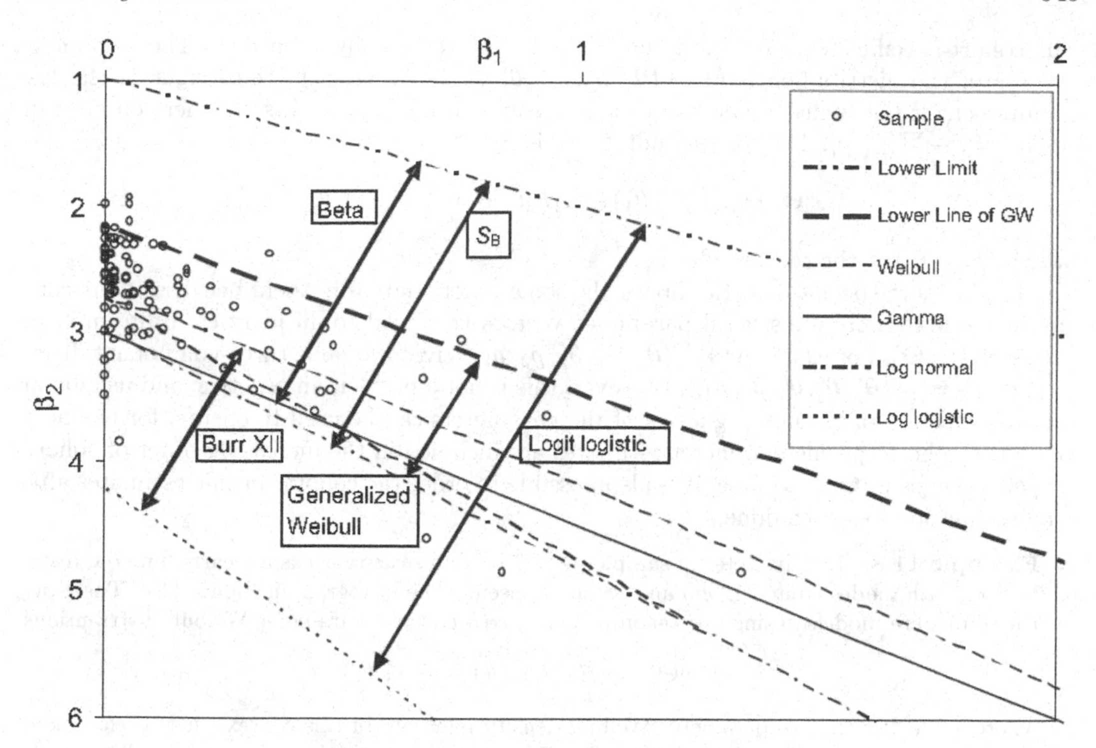

FIGURE 11.3 Range of possible skewness-kurtosis combinations for different diameter distribution models and observed values of skewness and kurtosis in an example data set. Adapted from Wang and Rennolls (2005).

where

$$\mu_k = \mathrm{E}(X - \mathrm{E}(X))^k$$

is the kth central moment of X (recall Section 2.2.6). The greater the value of $\sqrt{\beta_1}$, the more asymmetric is the distribution. Solving the equation of skewness for β_1 gives

$$\beta_1 = \frac{\mu_3^2}{\mu_2^3}$$

where a positive value indicates a long tail to the right and a negative value indicates a long tail to the left. Kurtosis measures how peaked the distribution is, high values indicating a peaked shape and low values a flat shape. Figure 11.3 illustrates the skewness-kurtosis combinations that can be described by different distribution models.

The unimodal *pdf*'s can be combined to model bimodal or multimodal data. In that case, we can think that the marginal distribution of tree size is a sum of p cohorts, each of which is distributed according to a unimodal model, such as the Weibull model. The cohorts can originate, for example, from variation in tree age or tree species. If information of the cohort membership (e.g. tree species) is available, each cohort can be modeled separately. If such information is not available, the marginal density can then be written as a *finite mixture* of cohort-specific *pdf*'s,

$$f(x|\boldsymbol{\Theta}) = \rho_1 f_1(x|\boldsymbol{\theta}_1) + \rho_2 f_2(x|\boldsymbol{\theta}_2) + \ldots + \rho_p f_p(x|\boldsymbol{\theta}_p)\,,$$

where $f_i(x|\boldsymbol{\theta}_i)$ is the *pdf* for cohort i, described by parameter $\boldsymbol{\theta}_i$ and $\rho_1, \rho_2, \ldots, \rho_p$ are positive weights that sum up to 1. It is common to assume the same functional form for all cohort-specific densities, even though it is not technically necessary. The parameter vector of the distribution is $\boldsymbol{\Theta} = (\boldsymbol{\theta}'_1, \boldsymbol{\theta}'_2, \ldots, \boldsymbol{\theta}'_p, \rho_1, \rho_2, \ldots, \rho_{p-1})'$; the weight of the last component is not included because it can be solved using the weights of other components as $\rho_p = 1 - \sum_{i=1}^{p-1} \rho_i$. The corresponding *cdf* is

$$F(x|\boldsymbol{\Theta}) = \rho_1 F_1(x|\boldsymbol{\theta}_1) + \rho_2 F_2(x|\boldsymbol{\theta}_2) + \ldots + \rho_p F_p(x|\boldsymbol{\theta}_p),$$

where $F_i(x|\boldsymbol{\theta}_i)$ is the cdf for cohort i.

If all cohort-specific densities are of the same functional form, then, because of the commutative law, there are several parameter vectors that lead to an identical finite mixture density $f(x|\boldsymbol{\Theta})$. For example $\boldsymbol{\Theta} = (\boldsymbol{\theta}'_2, \boldsymbol{\theta}'_1, \boldsymbol{\theta}'_3, \rho_2, \rho_1)'$ gives the same three-component distribution as $\boldsymbol{\Theta} = (\boldsymbol{\theta}'_1, \boldsymbol{\theta}'_2, \boldsymbol{\theta}'_3, \rho_1, \rho_2)'$. However, this is not a problem in practice, and usually an astute selection of the initial guesses of the parameters can be used to ensure, for example, that the cohort-specific parameter estimates are included in the increasing order of cohort-specific means in $\boldsymbol{\Theta}$. Of course, it is also possible to order the cohort-specific estimates after the estimation has been done.

Example 11.8. Tree diameter in sample plot 27 of data set `spati` has a clearly bimodal distribution, with modes around 12 cm and 38 cm, as seen in the histogram in Figure 11.4. Therefore, the data were modeled using a two-component mixture of two-parameter Weibull distributions

$$f(x|\boldsymbol{\Theta}) = \rho f(x|\boldsymbol{\theta}_1) + (1-\rho) f(x|\boldsymbol{\theta}_2),$$

where $f(x|\boldsymbol{\theta}_i)$ is the two-parameter Weibull density function. In this context, it is preferable to use such parameterization of the Weibull distribution where one of the parameters is, directly, the expected value μ. Such parameterization is available in function `gamlss.dist:dWEI3` (Stasinopoulos and Rigby 2018). The other parameter `sigma` is the original shape parameter α, so that we have $\boldsymbol{\theta}_i = (\mu_i, \alpha_i)$. Using this function, we construct the density of the mixture to function `dfmw2` and find the ML estimates using `mle`. We use values 11 and 28 as the initial guesses of cohort-specific means. The initial estimate for the share of the first cohort is 0.7. The shape parameters for both cohorts are initially set to 3. The ML estimates are $\widehat{\mu}_1 = 11.94$, $\widehat{\alpha}_1 = 3.446$, $\widehat{\mu}_2 = 37.76$, $\widehat{\alpha}_2 = 8.977$ and $\widehat{\rho}_1 = 0.7638$. The smooth dashed gray line in Figure 11.4 illustrates the *pdf* based on these parameter estimates; the black lines illustrate the cohort-specific Weibull densities.

```
> d27<-spati$d[spati$plot==27]
> library(gamlss.dist)
> dfmw2<-function(x,mu1,sigma1,mu2,sigma2,rho) {
+         rho*dWEI3(x,mu1,sigma1)+(1-rho)*dWEI3(x,mu2,sigma2)
+ }

> nll<-function(lmu1,lsigma1,lmu2,lsigma2,logitr) {
+      mu1<-exp(lmu1); sigma1<-exp(lsigma1)
+      mu2<-exp(lmu2); sigma2<-exp(lsigma2)
+      rho<-exp(logitr)/(1+exp(logitr))
+      val<--sum(log(dfmw2(d27,mu1,sigma1,mu2,sigma2,rho)))
+      cat(mu1,sigma1,mu2,sigma2,rho,val,"\n") #print current estimates to monitor estimation
+      val
+ }

> estfm<-mle(nll,start=list(lmu1=log(12),lsigma1=log(3),
+                          lmu2=log(38),lsigma2=log(3),logitr=log(0.7/(1-0.7))))
> summary(estMLWE)

Coefficients:
        Estimate Std. Error
lmu1    2.479690  0.0388597
lsigma1 1.237288  0.1029757
lmu2    3.631311  0.0328582
lsigma2 2.194716  0.2091023
logitr  1.173589  0.2397425

-2 log L: 702.5583
```

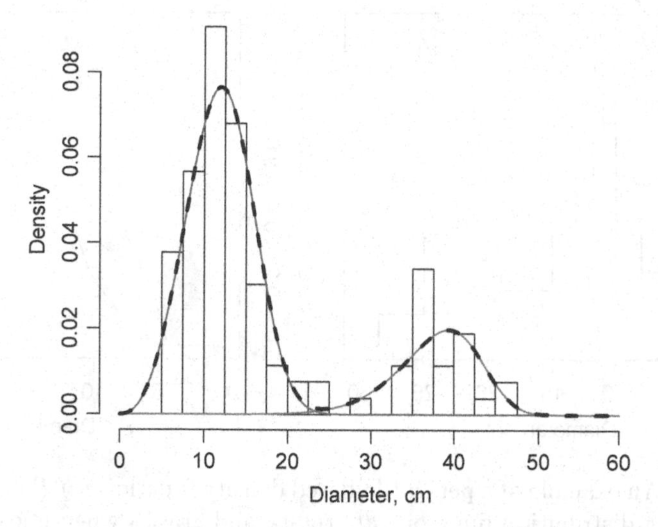

FIGURE 11.4 Histogram of tree diameters on sample plot 27 of data set `spati` and the two-component Weibull mixture (dashed black line) fitted in Example 11.8. The thin gray lines illustrate the two components of the mixture.

A flexible distribution family for modeling tree diameter distributions is provided by the percentile-based method (e.g. Borders *et al.* 1987, Maltamo *et al.* 2000) where the distribution is characterized by a fixed number of percentiles $\boldsymbol{\xi} = (\xi_1, \xi_2, \ldots, \xi_k)'$ that correspond to predefined values $\boldsymbol{p} = (p_1, p_2, \ldots, p_k)'$ of the *cdf*, where $p_1 = 0$ and $p_k = 1$ (Figure 11.5). The continuous *cdf* is obtained by interpolating between the percentiles $\boldsymbol{\xi}$. The flexibility of the model is improved by increasing k. For example, Borders *et al.* (1987) used $k = 12$; such a model can represent a large variability in shapes, including multimodal shapes. If interpolation is carried out using linear interpolation, the resulting *cdf* becomes (Mehtätalo 2005)

$$F(x|\boldsymbol{\xi}) = \begin{cases} 0 & x < \xi_1 \\ a_i + b_i x & \xi_i \le x < \xi_{i+1} \text{ for } i = 1, \ldots, k-1 \\ 1 & x > \xi_k \end{cases} \tag{11.2}$$

where $b_i = \frac{p_{i+1} - p_i}{\xi_{i+1} - \xi_i}$ and $a_i = p_i - b_i \xi_i$. The *pdf* results from the differentiation of the *cdf* with respect to x, giving

$$f(x|\boldsymbol{\xi}) = \begin{cases} 0 & x < \xi_1 \\ b_i & \xi_i \le x < \xi_{i+1} \text{ for } i = 1, \ldots, k-1 \\ 0 & x > \xi_k \end{cases} \tag{11.3}$$

It is easy to show that

$$\mathrm{E}(X) = \frac{1}{2} \sum_{i=1}^{k-1} b_i \left(\xi_{i+1}^2 - \xi_i^2 \right)$$

and

$$\mathrm{E}(X^2) = \frac{1}{3} \sum_{i=1}^{k-1} b_i \left(\xi_{i+1}^3 - \xi_i^3 \right),$$

the variance can then be computed as $\mathrm{var}(X) = \mathrm{E}(X^2) - \mathrm{E}(X)^2$. Mehtätalo (2004b) also presented formulas for other characteristics of interest from the percentile-based model. The

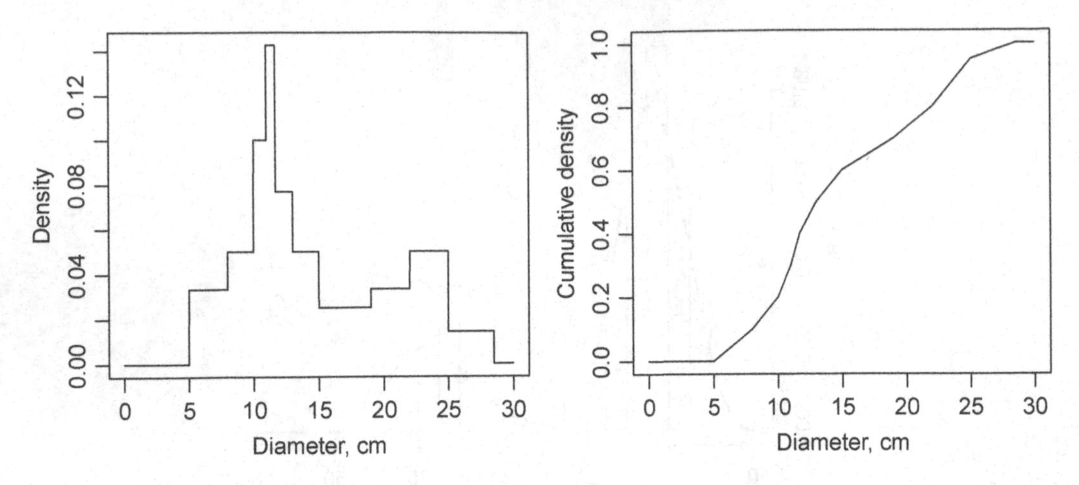

FIGURE 11.5 An example of a percentile-based density function *pdf* (left-hand graph) and the corresponding distribution function *cdf* (right-hand graph) when linear interpolation is used.

percentile-based model with linear interpolation can also be seen as a segmented uniform distribution or a finite mixture of $k-1$ uniform distributions. Kangas and Maltamo (2000b) used a monotonicity-preserving spline for interpolation between the percentiles (see Box 10.2, p. 324). Kangas *et al.* (2007) and Mehtätalo *et al.* (2008) used second- and third-order polynomials in the tails and linear interpolation for the intermediate percentiles. Functions for modeling percentile-based distribution are available in `lmfor::dPercbas`. See Web Examples 2.23 and 2.23 (p. 32) and 11.33 (p. 376).

11.3.2 Model comparison criteria

For comparison of alternative models, several test statistics and performance criteria have been used. If ML is used in model fitting, the criteria can be based on the evaluated likelihood. The simplest alternative is the value of the log-likelihood at the solution, $l(\widehat{\theta})$ (or the negative log-likelihood), which is the higher (the lower) the better the fit. However, this criterion does not penalize model complexity. The AIC and BIC (see Section 3.6.4) could be used when models with a different number of parameters are compared.

Reynolds *et al.* (1988) suggested the so-called error index for comparing two diameter distributions. It is the sum of absolute differences of fitted and observed size class frequencies over all diameter classes:

$$ei = \sum_{k=1}^{K} \left| f_k - \hat{f}_k \right| ,$$

where f_k is the true proportion of trees in class k and $\hat{f}_k$ is the predicted proportion. The value depends on the utilized diameter class width and sample size. Because the class proportions sum up to one, the value of the index falls between 0 (complete match) and 2 (complete mismatch). Another option is to use the number of stems per diameter class instead of proportions, so that the index would give the sum of differences in class-specific numbers of trees. Alternatively, the number of stems could be replaced with basal area or volume to give more weight to the right end of the distribution. If two continuous densities are compared, then no diameter classes are needed, and the sum can be replaced by the integral. If $\left|f_k - \hat{f}_k\right|$ is replaced by $\left(f_k - \hat{f}_k\right)^2$, the statistic becomes the χ^2 goodness-of-fit statistic.

In some cases, it makes sense to partition the differences between the observed sample diameter data and the estimated distribution to differences in the mean, in the variance, and in the shape of the distribution. Such an approach has been implemented in function `lmfor::ddcomp`. The shape is compared using the error index, which is computed using both the original predicted density, and using such predicted density that has been switched and rescaled to have the same mean and variance as the diameter data.

The Kolmogorov-Smirnov (KS) test has also been widely used to test whether a sample follows a specified continuous distribution function. For the test, the empirical distribution function F_n for n independent, identically distributed observations X_i is defined as

$$F_n(x) = \frac{1}{n}\sum_{i=1}^{n} I_{X_i \leq x}$$

where the indicator function $I_{X_i \leq x}$ is equal to 1 if $X_i \leq x$ and 0 otherwise. The null hypothesis of the test is that the data were sampled from a continuous distribution with *cdf* $F(x)$. The test statistic is

$$D_n = \sup_x (F_n(x) - F(x))$$

where $\sup(S)$ is the supremum ($\approx$ maximum) of set S. It is implemented in `stats4::ks.test`. For formal testing, the statistic is compared to the Kolmogorov's distribution (see e.g. Zar 2010).

Other criteria that have been used for comparison include the basal area or number of stems in certain diameter ranges, such as for sawtimber-sized trees and pulpwood-sized trees. In addition, one could explore the RMSE of volume, if a volume model based on tree diameter is available. One could also use the RMSE of second and third powers of diameters, which are related to basal area and volume, respectively.

All these criteria are data-specific. Thus, they can be used for comparisons only when different distributions are fitted to the same data set using the same method. When comparing different fitting methods, caution is needed. The criteria should not be based on the function that is used as the objective function in fitting. For example, criteria based on likelihood would naturally favor ML, and criteria that compare the matching of moments would favor the method of moments. We recommend that the known statistical properties of estimators are used in selecting the estimation methods instead of empirical results from fitting to a specific data set.

Example 11.9. We compared the Weibull and logit-logistic distributions that were fitted using ML in Example 11.2. The AIC and KS-statistic both indicate that the Weibull distribution better fits to the data.

```
> AIC(estMLWE,estMLLL)
        df      AIC
estMLWE  2 382.6704
estMLLL  4 383.8692

> ks.test(d22,"pweibull",shape=coef(estMLWE)[1],scale=coef(estMLWE)[2])$statistic
         D
0.05708233

> ks.test(d22,"pll",xi=exp(coef(estMLLL)[1]),lambda=exp(coef(estMLLL)[2]),
+         sigma=exp(coef(estMLLL)[3])/(1+exp(coef(estMLLL)[3])),
+         mu=coef(estMLLL)[4])$statistic
         D
0.06985039
```

The results of ddcomp show that the mean of the logit-logistic distribution is rather different from the sample mean. The variance/standard deviation differ more in the Weibull distribution than in the logit-logistic distribution. The error index using 5 cm classes is lower for logit-logistic distribution. However, if the distributions are rescaled to have the same mean and variance as the diameter data, then the error indices (ei2 in the output) are very similar for both models.

```
> unlist(ddcomp(d22,"dweibull",shape=coef(estMLWE)[1],scale=coef(estMLWE)[2],
>               limits=seq(0,100,5)))
       mudif       vardif        sddif          ei1          ei2 
-0.007999693  3.360823068  0.249965455  0.180113603  0.161057972 
>
> unlist(ddcomp(d22,"dll",xi=exp(coef(estMLLL)[1]),lambda=exp(coef(estMLLL)[2]),
+        sigma=exp(coef(estMLLL)[3])/(1+exp(coef(estMLLL)[3])),
+        mu=coef(estMLLL)[4],limits=seq(0,100,5)))
     mudif     vardif      sddif        ei1        ei2 
 0.1387851 -1.4851625 -0.1075993  0.1593353  0.1634730 
```

In Example 11.5, we fitted the Weibull distribution using mean and variance matching. In that case, the sample means and variances are equal to the expected value and variance of the distribution, and the two versions of the error index are, therefore, also the same. The zero differences in sample mean and variance, however, do not indicate a better performance of the method of moments compared to the ML, which had much larger differences in mean and variance.

```
> unlist(ddcomp(d22,"dweibull",shape=estMOMWE2$par[1],scale=estMOMWE2$par[2],
>               limits=seq(0,100,5)))
        mudif        vardif         sddif           ei1           ei2 
 2.220482e-09 -3.318142e-07 -2.422865e-08  1.613067e-01  1.613067e-01 
```

11.4 Mathematics of size distributions

11.4.1 Transformation of tree size

In some cases, it is useful to change the applied measure of size. For example, one might be willing to specify such distribution of tree heights or basal area that corresponds to a certain distribution of tree diameters. To do so, information about the relationships between these two size characteristics is needed. It is given by an allometric model, such as the height-diameter curve, or by the transformation of diameter to basal area $g = \frac{\pi}{4}d^2$. If the allometric relationship is deterministic, then it can be treated as a transformation, which we have discussed in Section 2.2.3.

Example 11.10. Let D be tree diameter and H tree height. Assume that the diameter follows the two-parameter Weibull distribution, which has the *cdf*

$$F_D(d) = 1 - \exp\left[-\left(\frac{d}{\beta}\right)^{\alpha}\right],$$

and the height-diameter curve is

$$h(d) = 1.3 + a\exp\left(\frac{b}{d}\right), \tag{11.4}$$

where parameters have known values $\alpha = 4$, $\beta = 15$, $a = 25$, and $b = -5$.

Solving Equation (11.4) for d gives

$$g^{-1}(h) = \frac{b}{\log\left(\frac{h-1.3}{a}\right)}.$$

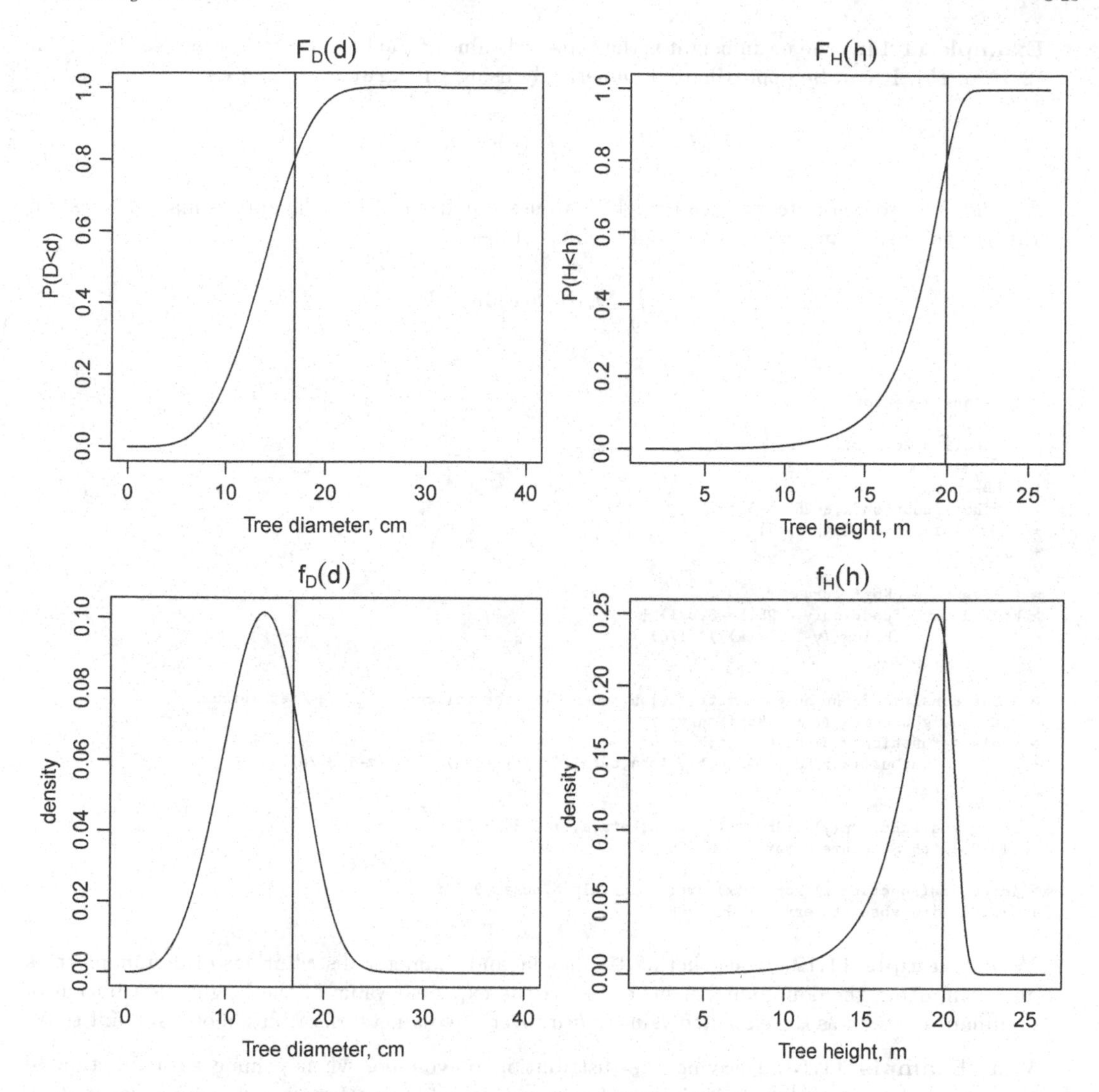

FIGURE 11.6 The diameter distribution and density (left) and the height distribution and density (right) in Example 11.10. The vertical line shows the limit of dominant trees (see Example 11.12).

Using it in the diameter distribution gives the *cdf* of tree height as

$$F_H(h) = 1 - \exp\left[-\left(\frac{b}{\beta \log\left(\frac{h-1.3}{a}\right)}\right)^{\alpha}\right],$$

which is no longer a Weibull distribution (c.f. Example 2.8). The *pdf* of height is obtained by differentiating $F_H(h)$ with respect to h,

$$f_H(h) = \frac{\alpha}{\beta}\left(\frac{b}{\beta \log\left(\frac{h-1.3}{a}\right)}\right)^{\alpha-1} \exp\left[-\left(\frac{b}{\beta \log\left(\frac{h-1.3}{a}\right)}\right)^{\alpha}\right] \frac{-b}{(h-1.3)\left[\log\left(\frac{h-1.3}{a}\right)\right]^2}.$$

These functions are illustrated in the right-hand graphs in Figure 11.6.

Example 11.11. The mean height is the expected value of the height distribution (see Equation (2.9), p. 15). It can be approximated numerically using **integrate**, which gives

$$\bar{H} = \int_0^{H_{max}} u f_H(u) \mathrm{d}u = 18.121 \,.$$

Another way to compute the mean height is to use Equation (2.10); the approximation based on **integrate** is the same at least up to 4 decimal places:

$$\bar{H} = \int_0^{\infty} f_D(u) h(u) \mathrm{d}u = 18.121 \,.$$

```
> # x- and y-vectors
> x<-seq(0,40,0.1)
> y<-seq(1.3,26.2,0.1)

> # HD-curve
> korfhd<-function(x,a=25,b=-5,c=1) {
+         1.3+a*exp(b*x^(-c))
+         }

> # Inverse of Korf curve
> korfhd.inv<-function(y,a=25,b=-5,c=1) {
+             (b/log((y-1.3)/(a)))^(1/c)
+             }

> # The density of the height distribution, when diameter follows a 2-parameter Weibull
> # and Height comes from a Korf curve
> dhdist<-function(x,alpha,beta,a,b) {
+         dweibull(korfhd.inv(x,a,b),alpha,beta)*(-b)/(x-1.3)/(log((x-1.3)/a))^2
+         }

> integrate(function(y) y*dhdist(y,alpha,beta,a,b),1.3,26.3)
18.12103 with absolute error < 1.9e-05

> integrate(function(x) korfhd(x)*dweibull(x,alpha,beta),0,Inf)
18.12104 with absolute error < 9.4e-06
```

Web Example 11.12. Constructing the height and diameter distributions of dominant trees and computing the dominant height, either as the expected value of the height distribution of dominant trees, or as the expected value of $h(d)$ over the diameter distribution of dominant trees.

Web Example 11.13. Deriving the distribution of volume, when volume as a function of diameter is known, although its inverse function (diameter as a function of volume) cannot be expressed in a closed form.

Examples 11.10 – 11.13 assumed that the variable of interest is obtained through transformation of the diameter. The concept can be easily generalized to any other allometric relationships among trees to find the distribution of basal area, volume, crown area, crown height, or root biomass using the distribution of any size-related characteristics. Treating the allometric relationship as a transformation is well justified for basal area, which can be expressed exactly as a function of diameter as $\pi d^2/4$, if the diameter is defined as in Chapter 12. If the allometric model includes a considerable amount of random noise, treating the models as transformations leads to excessively narrow distributions. However, the expected value of that distribution is equal to the true expected value for the variable of interest.

In applications where the marginal variance or the entire distribution of the tree size is of interest, one should also take the randomness in the allometric relationship into account. To do so, the bivariate distribution of the variables of interest is needed. For example, the height distribution results from marginalization of the joint distribution of tree diameter and height over the diameter using Equation (2.17), as illustrated in the following example.

Example 11.14. Let us allow variability in the height-diameter relationship in Example 11.10. More specifically, let us assume

$$\begin{aligned} D &\sim Weibull(4, 15) \\ (H|D = d) &\sim N(\mu(d), \sigma(d)) \\ \mu(d) &= 1.3 + 25 \exp\left(\frac{-5}{d}\right) \\ \sigma^2(d) &= (0.3d^{0.68})^2 . \end{aligned}$$

The applied variance function parameters are taken from the model of Mehtätalo *et al.* (2015) for the teak data set. Notice that because the normally distributed residuals have a rather large variance compared to the expected height, there is a non-zero probability for negative heights for large-diameter trees, which is a drawback of this model formulation. A better justified model might be, for example, a generalized nonlinear model that assumes gamma or lognormal distribution for heights. Ignoring this small problem, we assume that tree diameters and the residuals of the height model are independent, with the distributions specified above.

We have now defined the complete joint distribution of tree heights and diameters. The only difference to the previous examples is in the last row; the previous example assumed that $\sigma(d) = 0$. Denoting the marginal *pdf* of D by $f_D(d)$ and the conditional *pdf* and *cdf* of height by $f_{H|D}(h, d)$ and $F_{H|D}(h, d)$, the marginal *pdf* and *cdf* of tree height are

$$\begin{aligned} f_H(h) &= \int_0^\infty f_D(u) f_{H|D}(h, u) du \\ F_H(h) &= \int_0^\infty f_D(u) F_{H|D}(h, u) du . \end{aligned}$$

The R-functions implemented below approximate these *pdf*'s and *cdf*'s using `integrate`. The expected value and variance of height were computed using these functions, resulting in $\mathrm{E}(H) = 18.12$ m and $\mathrm{var}(H) = 8.41$ m^2. The mean height is the same as in Example 11.11, but the variance is larger. The marginal distribution, which also takes the variability of the height-diameter curve into account, is illustrated by the wider *pdf* in Figure 11.7. The narrow *pdf* illustrates the *pdf* from Example 11.10.

```
> fh<-function(hvec,alpha,beta,a,b,sigma,delta) {
+     f1<-function(d,h) dweibull(d,alpha,beta)*dnorm(h,mean=korfhd(d,a,b),
+         sd=sigma*d^delta)
+     fhscalar<-function(h) integrate(function(u) f1(u,h),0,
+               qweibull(1-1e-10,alpha,beta))$value
+     sapply(hvec,fhscalar)
+     }

> Fh<-function(hvec,alpha,beta,a,b,sigma,delta) {
+     f1<-function(d,h) dweibull(d,alpha,beta)*pnorm(h,mean=korfhd(d,a,b),
+         sd=sigma*d^delta)
+     fhscalar<-function(h) integrate(function(u) f1(u,h),0,
+               qweibull(1-1e-10,alpha,beta))$value
+     sapply(hvec,fhscalar)
+     }

> Hmin<--10; Hmax<-40
> EH<-integrate(function(h) fh(h,alpha,beta,a,b,sigma,delta)*h,Hmin,Hmax)$value
> EH2<-integrate(function(h) fh(h,alpha,beta,a,b,sigma,delta)*h^2,Hmin,Hmax)$value
> varH<-EH2-EH^2
> EH
[1] 18.12102
> varH
[1] 8.412796
```

The variables `Hmin` and `Hmax` above specify the integration limits. You could use `Inf` as maximum, but such a choice often leads to poorer approximations than astutely selected finite values. Notice that `Hmin` is negative because there is some probability mass on the negative side due to the normal conditional distribution of tree height.

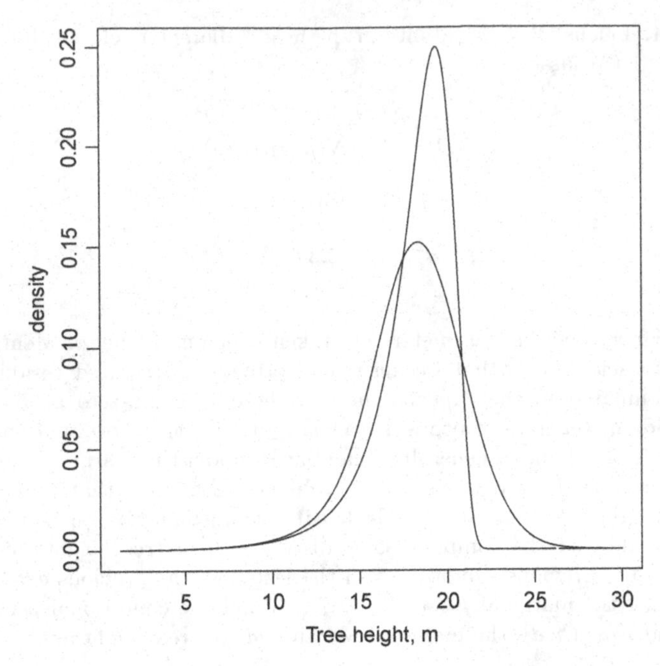

FIGURE 11.7 Distribution of tree heights (m)) in Examples 11.10 (narrow *pdf*) and 11.14 (wide *pdf*).

The variance could also be computed using the law of total variance (2.46) (p. 25) as

$$\text{var}(H) = \text{E}(\text{var}(H|D)) + \text{var}(\text{E}(H|D)) .$$

The computation in R is shown below:

```
> Dmin<-0; Dmax<-qweibull(1-1e-10,alpha,beta) # diameter limits of integration
> EHD<-integrate(function(u) dweibull(u,alpha,beta)*korfhd(u),Dmin,Dmax)$value
> EHD2<-integrate(function(u) dweibull(u,alpha,beta)*korfhd(u)^2,Dmin,Dmax)$value
> varEHD<-EHD2-EHD^2 # var(E(H|D))
> EvarHD<-integrate(function(u) dweibull(u,alpha,beta)*sigma^2*u^(2*delta),
+          Dmin,Dmax)$value

> varEHD
[1] 5.219077
> EvarHD              # E(var(H|D))
[1] 3.192981
> varEHD+EvarHD       # var(E(H|D))+E(var(H|D))
[1] 8.412059
```

Notice that the first term, $\text{var}(\text{E}(H|D)) = 5.219$, is the variance of the height distribution of Example 11.10 (the narrow *pdf* of Figure 11.7).

We could also compute the dominant height using the joint distribution of height and diameter. In Web Example 11.12, we obtained the same result when dominant trees were defined as the 100 thickest or as the 100 tallest trees per hectare, because the order of trees according to height was the same as the order according to diameter. The situation changes if we take the random noise in heights into account.

If we define dominant height as the expected height of the 100 thickest trees per hectare, we get the same value as in Example 11.12 ($H_{dom} = 20.45$), because $\text{E}(H|D)$ are the same in both examples. However, the mean height of the 100 tallest trees per hectare is much greater, as illustrated in the script below. First, the script finds the limit of dominant trees (the 80th percentile of the height distribution) which is 20.45 meters; it is approximately equal to the mean height of the 100 thickest trees by coincidence. We use the step-halving algorithm implemented in function `lmfor::updown` (see Web Example 11.13 for an explanation).

Second, after finding the height limit of the dominant trees, we compute the dominant height as the expected value of the height-distribution that is left-truncated at this limit. We obtain $H_{dom} = 21.9$m, which is almost 1.5 m greater than when dominance was defined by diameter.

```
> Domlim2<-updown(5,30,function(h) Fh(h,alpha,beta,a,b,sigma,delta)-0.8)
> Domlim2
[1] 20.45362
> integrate(function(h) 5*fh(h,alpha,beta,a,b,sigma,delta)*h,Domlim2,40)
21.91598 with absolute error < 4.2e-05
```

11.4.2 Basal-area weighted distribution

The previous section demonstrated the use of transformations to change the variable on the x-axis of a distribution graph (recall the discussion in Section 11.1). In this section, we discuss weighting, which changes the variable of the y-axis of the density function or histogram of the size distribution. With tree size distributions, the most commonly used weighting variable is the basal area or equivalently, the second power of the diameter.

Let us denote the *pdf* of the (unweighted) size distribution by $f_D(d)$. The *pdf* of the basal-area weighted diameter distribution is based on Equation (2.52), (p. 34), where $w(d) = \frac{\pi}{40000}d^2$. It gives

$$\begin{aligned} f_D^g(d) &= \frac{\frac{\pi}{40000}d^2 f_D(d)}{\int_0^\infty \frac{\pi}{40000}u^2 f_D(u)du} \\ &= \frac{d^2 f_D(d)}{\int_0^\infty u^2 f_D(u)du} \\ &= \frac{d^2 f_D(d)}{\mathrm{E}(D^2)} . \end{aligned} \tag{11.5}$$

Either the second or third form is useful in computations. Note the additional superscript g in the *pdf* to express the weighting by basal area.

As an example, consider the Weibull distribution, which has the density

$$f_D(d|\alpha,\beta) = \frac{\alpha}{\beta^\alpha}d^{\alpha-1}e^{-(d/\beta)^\alpha} .$$

Gove and Patil (1998) showed that the corresponding basal-area weighted diameter distribution is

$$f_D^g(d|\alpha,\beta) = \frac{d^2\left(\frac{\alpha}{\beta^\alpha}\right)d^{\alpha-1}e^{-(d/\beta)^\alpha}}{\beta^2\Gamma(2/\alpha+1)} \tag{11.6}$$

$$= \frac{1}{\Gamma(k)}y^{k-1}e^{-y} , \tag{11.7}$$

where $y = (d/\beta)^\alpha$ and $k = 2/\alpha + 1$. We notice that this is the *pdf* of the standard *Gamma* distribution. That is, our assumption that the diameter distribution is of the form of the two-parameter Weibull function, implies that the basal-area weighted distribution of transformation $y = (d/\beta)^\alpha$ is of the form of the standard Gamma distribution, and the basal-area weighted distribution of diameter is of the generalized gamma distribution.

A generalized framework for distributions of form

$$f_\alpha^*(x) = \frac{x^\alpha f(x)}{E(x^\alpha)} \tag{11.8}$$

is provided by the size-biased distribution theory (Gove and Patil 1998, Gove 2003). The basal-area weighted diameter distribution, which is obtained by specifying $\alpha = 2$, is the most common example of this theory in forestry. This is called the size-biased distribution of order 2.

Size-biased distributions of order 2 are directly linked with angle-count sampling (relascope sampling, horizontal point sampling, Bitterlich sampling) (Bitterlich 1949, Grosenbaugh 1952). In angle-count sampling, large trees have a greater sampling probability than small trees. More specifically, the inclusion probability is proportional to tree basal area; see Gregoire and Valentine (2008) for a description of this sampling method.

In angle-count sampling, the sampling distribution of the trees is the basal-area weighted version of the true diameter distribution. The mean diameter of the sampled trees is greater than the mean diameter in the population and the shape is, of course, also different. If a specific distribution is fitted to the angle-count sampling data, the weighting of the sample needs to be taken into account in estimation (Van Deusen 1986, Gove and Patil 1998, Gove 2000, 2003), as illustrated in the examples below. One can either assume that the (unweighted) diameter distribution is of the assumed form and take the basal-area weighting into account in estimation, or assume that the basal-area weighted distribution is of the assumed functional form and determine the corresponding unweighted distribution.

Example 11.15. Distribution obtained by angle-count sampling. Consider a simulated, 1-hectare forest stand with 1000 trees that uniformly and independently distributed in the forest area of interest (recall Example 8.6 p, 266) with independent, identically distributed diameters (marks) from the *Weibull*$(3, 20)$ distribution. Furthermore, assume that a sample of trees is taken using a relascope with BAF=1.

The following R-code presents a simulation for such a stand.

```
> set.seed(1234567)          # ensure exactly same sample every time
> dpop<-rweibull(1000,3,20) # diameters, cm
> x<-runif(1000,-50,50)     # x-coordinates, meters
> y<-runif(1000,-50,50)     # y-coordinates
>
> # Simulate relascope sample at location (0,0) using BAF=1.
> dist<-sqrt(x^2+y^2)
> dsample<-dpop[2*dist<dpop]
>
> h1<-hist(dpop)
> h2<-hist(dsample)
> plot(h1$mids-0.5,h1$density,type="h",lwd=10,col=gray(0.6),
+      ylim=c(0,max(h2$density)),xlab="Diameter, cm", ylab="Density")
> points(h2$mids+0.5,h2$density,type="h",lwd=10)

> x<-seq(0,50,0.1)
> lines(x,dweibull(x,3,20),lwd=2,col=gray(0.6))
> lines(x,x^2*dweibull(x,3,20)/(20^2*gamma(2/3+1)),lwd=2)
>
> mean(d)
[1] 17.6769
> mean(dsample)
[1] 22.41906
```

The histogram of the diameters of all 1000 trees is shown in gray in Figure 11.8, and the histogram of the angle-count sample of size 24 trees is shown in black. The mean diameter in the stand is 17.7 cm, and the mean of the sample is 22.41 cm. The lines show the true Weibull-distribution of the population and the basal-area weighted (generalized gamma) sampling distribution. Both distributions are quite symmetric, but the sampling distribution is slightly more peaked than the distribution of tree diameters.

Example 11.16. Fitting Weibull distribution to angle count sampling data. Assume that tree diameter follows the Weibull distribution, and the diameter sample of Example 11.15 is available from the stand. The diameters have been tabulated below.

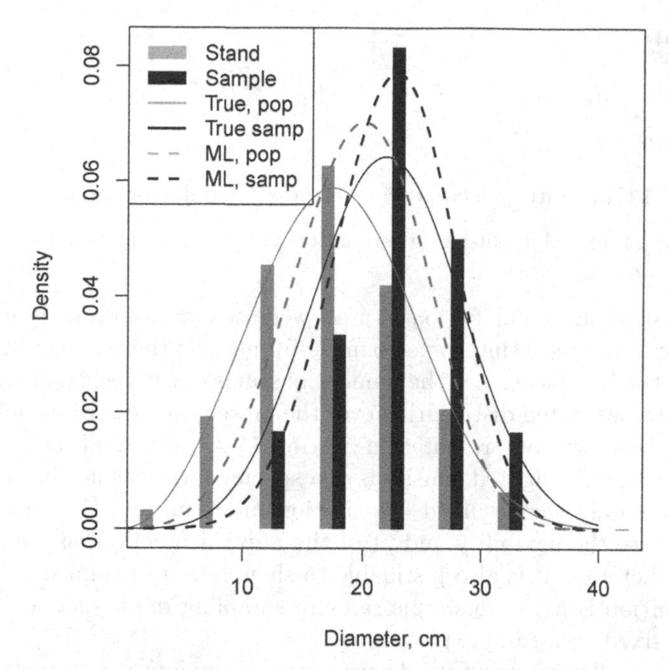

FIGURE 11.8 Illustration of Examples 11.15 and 11.16. The histograms show the diameter distribution over the simulated 1-hectare stand and the angle-count sample. The solid lines show the true diameter and sampling distributions. The dashed lines show the estimated population and sampling distributions.

18.8	23.7	23.4	20.1	26.4	26.2	25.0	30.4
22.6	25.1	23.9	10.8	21.7	31.0	22.3	24.9
22.4	15.0	15.1	22.2	29.8	17.3	12.6	27.3

If the unweighted distribution is of the Weibull form, then the sampling distribution is of the form (11.6). Fitting that distribution to the observed angle-count sample and using the resulting parameter estimates in the Weibull function provides an estimate of the unweighted distribution of the plot. The log-likelihood of the sample is:

$$l(\alpha, \beta | d_1, \ldots, d_n) = \sum_{i=1}^{n} \log(f_D^g(d_i|\alpha, \beta)),$$

where n is the number of trees in the angle count sample. The ML fit using function `mle` resulted in point estimates $\widehat{\alpha}_{ML} = 3.997$ and $\widehat{\beta}_{ML} = 21.67$. The estimated variance of the point estimates was

$$\text{var}\begin{pmatrix} \widehat{\alpha}_{ML} \\ \widehat{\beta}_{ML} \end{pmatrix} = \begin{pmatrix} 0.8088^2 & 0.777 \times 0.8088 \times 1.437 \\ 0.777 \times 0.8088 \times 1.437 & 1.437^2 \end{pmatrix}.$$

The population and sampling *pdf*s based on these values are shown in Figure 11.8.

```
> dsampling<-function(x,shape,scale) {
+           x^2*dweibull(x,shape,scale)/(scale^2*gamma(2/shape+1))
+           }
>
> ll2<-function(shape,scale) {
+       -sum(log(dsampling(dsample,shape,scale)))
+       }
>
> library(stats4)
> print(est.ml<-mle(ll2,start=list(shape=3,scale=20)))

Coefficients:
```

```
    shape     scale
 3.996933 21.666591
> vcov(est.ml)
          shape     scale
shape 0.6542356 0.9034208
scale 0.9034208 2.0635600
```

Web Example 11.17. Fitting a basal area weighted Weibull function to an angle count sample.

Web Example 11.18. Moment-based estimation of the Weibull distribution parameters using an angle-count sample.

Assuming a specific model for basal-area weighted distribution is convenient, if the modeling data are collected using angle-count sampling and the scaling of the distribution is performed using the basal area, not the number of stems. However, no other justifications to favor the basal-area weighted distribution over the unweighted distribution exists; a solution to this technical issue will be presented in Section 11.4.4. On the other hand, justification for assuming that the unweighted density is of a specific form has not been presented either. Weighting changes only the assumed distribution function, and the only justified criteria for model choice are the flexibility and fit of the applied function, as we have discussed in Section 11.3.1. Therefore, it is also justifiable to show how an assumed basal-area weighted diameter distribution is fitted to simple random sampling data, such as the tree diameter data based on a fixed-area sample plot.

If the data are collected using fixed area-plots and one wants to make assumptions on the basal-area weighted distribution, then we need to implement the weighting in the opposite direction. The unweighted density is obtained from the basal area weighted diameter distribution, f_D^g, through

$$f_D(x) = \frac{d^{-2} f_D^g(d)}{\int_0^\infty u^{-2} f_D^g(u) \mathrm{d}u} . \tag{11.9}$$

The equation is very similar to Equation (11.5), except that exponent 2 has now been replaced by -2. Both are special cases of the size-biased density (11.8). With the two-parameter Weibull distribution, this distribution becomes (c.f. Equation (11.7))

$$f_D(d|\alpha, \beta) = \frac{d^{-2} \left(\frac{\alpha}{\beta^\alpha}\right) d^{\alpha-1} e^{-(d/\beta)^\alpha}}{\beta^{-2}\Gamma(1 - 2/\alpha)} \tag{11.10}$$

The gamma function is only defined with a positive value of its argument. Thus, in contrast to the Weibull density, the generalized gamma distribution of Equation (11.10) is only defined when $\alpha > 2$ and $\beta > 0$. Fitting a basal-area weighted Weibull distribution to a fixed-area plot sample is left as an exercise.

The discussion in this section can also be extended to tree growth. Lappi and Bailey (1987) discuss the estimation of tree increment using angle-count samples. Because large trees have, on average, grown faster than smaller trees in the same even-aged stand, the resulting correlation of sampling probability and past growth needs to be taken into account in computations.

11.4.3 Weighted distributions

The basal-area weighted and size-biased distributions are both special cases of a weighted distribution. The density of a weighted distribution (recall Equation (2.52), p. 34) is:

$$f_X^w(x) = \frac{w(x) f_X(x)}{\mathrm{E}(w(X))} . \tag{11.11}$$

The size-biased distribution is such a special case, where $w(x) = x^{\alpha}$. The corresponding *cdf* is found using Equation (2.2) (p. 10).

Weighted distributions can be used for fitting assumed distributions to a sample obtained through weighted sampling. The basal-area weighting, which was considered in the previous section, is one example, but there are many other examples of sampling where the inclusion probability is proportional to the tree size according to a known function. For example, the sample plot radius may vary according to tree diameter. For example, in an aerial inventory the small trees may have a smaller probability of being observed compared to the larger trees. In addition, the data may be truncated, for example, because the trees may need to have a fixed minimum diameter for inclusion in the sample, which means that the trees that have diameter less than the limit have zero inclusion probability. This leads to a *left-truncated sampling distribution* of tree diameters.

If a fixed minimum t is used for tree size measurements, then the weighting is

$$w(x) = \begin{cases} 0 & x \leq t \\ 1 & x > t \end{cases}$$

and the *pdf* of the sampling distribution becomes

$$f^w(x|\boldsymbol{\theta}) = \begin{cases} 0 & x \leq t \\ \frac{1}{\int_t^\infty f(u|\boldsymbol{\theta})du} f(d|\boldsymbol{\theta}) & d > t \end{cases},$$

which is called a *left-truncated pdf.* Note that the denominator can be written as $\int_t^\infty f(u|\boldsymbol{\theta})du = 1 - F(t|\boldsymbol{\theta})$. If the number of trees below the measurement limit were known, the data would be *censored.* In that case, one could additionally require that $1 - F(t|\boldsymbol{\theta})$ is equal to the observed proportion of trees with $x > t$.

In some cases, we need to find an unweighted distribution that corresponds to a certain weighted distribution. Such a distribution is found using

$$f_X(x) = \frac{\frac{1}{w(x)} f_X^w(x)}{\int_0^\infty \frac{1}{w(u)} f_X^w(u)du}. \tag{11.12}$$

A special case of this formula was (11.9). For future reference, we denote the numerator $\frac{1}{w(x)} f_X^w(x)$ by $h(x)$.

Example 11.19. Fitting an assumed distribution to truncated data. In sample plot 27 of data set `spati`, a minimum diameter of 5 cm was used in callipering. Therefore, there may have been trees with a diameter less than 5 cm in the plot, even though such trees are not present in the data. To adjust our estimation procedure for the missing (small) trees (recall Example 11.8), we fit a truncated version of the two-component mixture distribution to the data. An R-function for the sampling density is defined below in function `dfmw2tr`, where the argument `limit` specifies the diameter limit of truncation. For implementation, we also need the *cdf* of the two-component mixture, which is defined in function `pfmw2` using `gamlss.dist::pWEI3`. Function `dfmw2` was defined already in Example 11.8 (p. 344).

```
> pfmw2<-function(x,mu1,sigma1,mu2,sigma2,rho) {
+        rho*pWEI3(x,mu1,sigma1)+(1-rho)*pWEI3(x,mu2,sigma2)
+ }

> dfmw2tr<-function(x,mu1,sigma1,mu2,sigma2,rho,limit=0) {
+          f<-dfmw2(x,mu1,sigma1,mu2,sigma2,rho)/(1-pfmw2(limit,mu1,sigma1,mu2,sigma2,rho))
+          f[x<limit]<-0
+          f
+ }

> nll<-function(lmu1,lsigma1,lmu2,lsigma2,logitr) {
+      mu1<-exp(lmu1); sigma1<-exp(lsigma1)
```

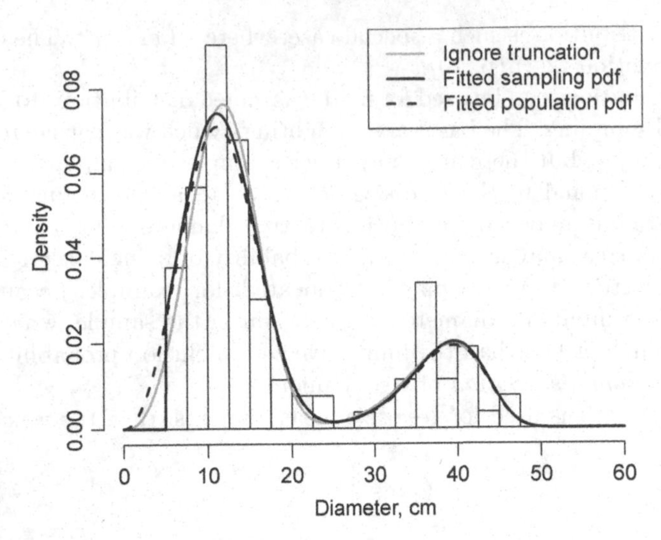

FIGURE 11.9 Observed diameter distribution of sample plot 27 of data set `spati`. The lines show the fitted densities from Example 11.8, and the fitted sampling and the population *pdf*'s in Example 11.19.

```
+        mu2<-exp(lmu2); sigma2<-exp(lsigma2)
+        rho<-exp(logitr)/(1+exp(logitr))
+        -sum(log(dfmw2tr(d27,mu1,sigma1,mu2,sigma2,rho,limit=5)))
+ }

> estfmtr<-mle(nll,start=list(lmu1=log(12),lsigma1=log(3),
+              lmu2=log(38),lsigma2=log(3),logitr=log(0.7/(1-0.7))))

> summary(estfmtr)
Maximum likelihood estimation

Coefficients:
         Estimate Std. Error
lmu1     2.446125 0.04602866
lsigma1  1.102232 0.11505666
lmu2     3.638358 0.02835839
lsigma2  2.247005 0.17863012
logitr   1.257399 0.23711414
```

The parameter estimates differ from the previous estimates, especially for the first component of the mixture, for which the estimated mean decreased from 11.93 cm to 11.54 cm. Figure 11.9 illustrates the fitted population *pdf* that is adjusted for the truncation and the sampling *pdf*. The fitted sampling/population *pdf* without adjusting for the truncation is also shown.

Variable-radius sample plots are concentric plots where the plot radius varies according to tree size. They are commonly used in forest inventories. For example, the Spanish National Forest Inventory (NFI) uses concentric circular plots, where the plot radius varies between 5 and 25 m according to the tree diameter. Trees with a diameter less than $t = 7.5$ cm are not measured at all. Thus, the sampling weights are

$$w(d) = \begin{cases} 0 & \text{if} \quad d < 7.5 \\ 5^2/25^2 & \text{if} \quad 7.5 \leq d < 12.5 \\ 10^2/25^2 & \text{if} \quad 12.5 \leq d < 22.5 \\ 15^2/25^2 & \text{if} \quad 22.5 \leq d < 42.5 \\ 1 & \text{if} \quad d \geq 42.5 \end{cases},$$

Denoting the *cdf* and *pdf* of the population of trees by $F(d|\boldsymbol{\theta})$ and $f(d|\boldsymbol{\theta})$, the sampling *pdf* can be defined by parts as (Mehtätalo *et al.* 2011)

$$f^w(d|\boldsymbol{\theta}) = \begin{cases} 0 & \text{if} \quad d < 7.5 \\ \frac{5^2}{25^2 I} f(d|\boldsymbol{\theta}) & \text{if} \quad 7.5 \leq d < 12.5 \\ \frac{10^2}{25^2 I} f(d|\boldsymbol{\theta}) & \text{if} \quad 12.5 \leq d < 22.5 \\ \frac{15^2}{25^2 I} f(d|\boldsymbol{\theta}) & \text{if} \quad 22.5 \leq d < 42.5 \\ \frac{1}{I} f(d|\boldsymbol{\theta}) & \text{if} \quad d \geq 42.5 \end{cases}, \tag{11.13}$$

where $I = \int_0^\infty w(u) f(u|\boldsymbol{\theta}) du = 1 - F(42.5|\boldsymbol{\theta}) + \frac{15^2}{25^2}[F(42.5|\boldsymbol{\theta}) - F(22.5|\boldsymbol{\theta})] + \frac{10^2}{25^2}[F(22.5|\boldsymbol{\theta}) - F(12.5|\boldsymbol{\theta})] + \frac{5^2}{25^2}[F(12.5|\boldsymbol{\theta}) - F(7.5|\boldsymbol{\theta})]$.

Example 11.20. Fitting Weibull function to variable radius plot data. A sample plot according to the Spanish NFI design was established into the forest stand simulated in Example 11.15. The black histogram of Figure 11.10 shows the diameter distribution of the resulting 35 sampled trees. The shape of the histogram is very different from the shape of the underlying diameter distribution, which is shown by the gray histogram. ML-fit was performed by maximizing the likelihood of the sampling distribution (11.13) in the observed sample. The resulting estimates were $\widehat{\alpha} = 3.93$ and $\widehat{\beta} = 21.11$. The approximate 95% confidence intervals of parameters, $[2.77, 5.58]$ and $[18.99, 23.47]$, include the true values 3 and 20. The dashed lines of Figure 11.10 show the fitted sampling distribution, and the corresponding population distribution. The solid lines show the distributions by applying the true parameter values of 3 and 20.

```
> dsample2<-c(dpop[dist<25&dpop>=42.5],dpop[dist<15&dpop<42.5&dpop>=22.5],
+             dpop[dist<10&dpop<22.5&dpop>=12.5],dpop[dist<5&dpop<12.5&dpop>=7.5])
>
> pwtr<-function(x,shape,scale,limit=0) {
+      F<-(pweibull(x,shape,scale)-pweibull(limit,shape,scale))/(1-pweibull(limit,shape,scale))
+      F[x<limit]<-0
+      F
+      }

> dwtrPPS<-function(x,shape,scale,limit=0) {
+         f<-dweibull(x,shape,scale)/(1-pweibull(limit,shape,scale))
+         f[x<limit]<-0
+         w<-rep(1,length(x))
+         w[x<42.5]<-(15/25)^2
+         w[x<22.5]<-(10/25)^2
+         w[x<12.5]<-(5/25)^2
+         Iwf<-(1-pwtr(42.5,shape,scale,limit=limit))+
+              (15/25)^2*(pwtr(42.5,shape,scale,limit=limit)-
+                          pwtr(22.5,shape,scale,limit=limit))+
+              (10/25)^2*(pwtr(22.5,shape,scale,limit=limit)-
+                          pwtr(12.5,shape,scale,limit=limit))+
+              (5/25)^2 *(pwtr(12.5,shape,scale,limit=limit)-
+                          pwtr(7.5,shape,scale,limit=limit))
+         w*f/Iwf
+         }
>
> # Negative log-likelihood, parameters in log scale
> nLLwtrPPS<-function(lshape,lscale) {
+            shape<-exp(lshape)
+            scale<-exp(lscale)
+            nLL<--sum(log(dwtrPPS(dsample2,shape=shape,scale=scale,limit=7.5)))
+            nLL
+ }

> est<-mle(nLLwtrPPS,start=list(lshape=log(4),lscale=log(20)))
> exp(coef(est))
   lshape    lscale
 3.930691 21.113744
> sd<-sqrt(diag(vcov(est)))
> exp(cbind(coef(est)-1.96*sd,coef(est)+1.96*sd))
            [,1]      [,2]
lshape  2.768318  5.581126
lscale 18.990337 23.474580
```

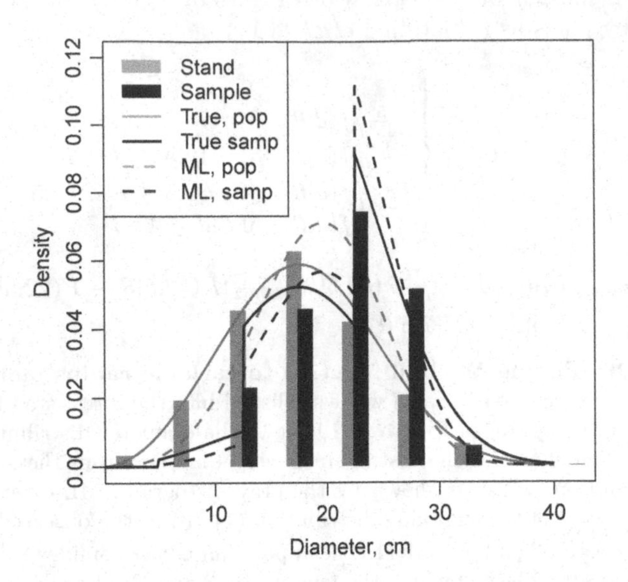

FIGURE 11.10 The distribution of the stand and the variable-radius plot in Example 11.20. The solid lines show the true population and sampling distributions, and the dashed lines show the corresponding estimates obtained using ML fit to the sampled trees.

Web Example 11.21. In aerial forest inventories, small trees may not be detectable from above because they are covered by larger tree crowns. Therefore, the distribution of observed crown diameters is based on a weighted sampling from the population of tree crowns. This web example illustrates the solution of Mehtätalo (2006) to the problem. The approach has also been extended by Kansanen *et al.* (2016), see also function `lmfor::HTest`.

11.4.4 Scaling the size distribution

The (unweighted) size distribution gives the proportion of trees in a given size-class as the difference between the value of the *cdf* at the lower l and upper u endpoints of the diameter class as $F_D(u) - F_D(l)$. The total number of stems (per hectare) in that class is obtained by multiplying the proportion by the total number of stems per hectare as

$$N_{[l,u]} = N(F_D(u) - F_D(l)) . \tag{11.14}$$

For the basal-area weighted distribution, the diameter class probabilities give the proportion of the total basal area, and the basal area of that class (per hectare) is obtained by multiplying the class proportion by the total basal area G per hectare:

$$G_{[l,u]} = G(F_D^g(u) - F_D^g(l)) . \tag{11.15}$$

Consider a general measure x for tree size and weight w in the distribution with the *cdf* $F_X^w(x)$. If the weighting variable w in the size distribution is the same as the variable used for measuring the total growing stock T (i.e. $T = N\int_{-\infty}^{\infty} w(u)f(u)du$) then the growing stock in size class $l < x \leq u$ is

$$T_{[l,u]} = T(F_X^w(u) - F_X^w(l)) . \tag{11.16}$$

Example 11.22. In Example 11.2 the estimates of the Weibull diameter distribution parameters for sample plot 22 of data set `spati` were $\widehat{\alpha} = 4.62$ and $\widehat{\beta} = 29.37$. There is a total of 57 trees in a 50m×50m; the estimated stand density is therefore $\widehat{N} = 190$ trees per hectare. The estimated number of stems within diameter class $(10, 20]$ is $N_{(10,20]} = 28.26$ trees per hectare.

```
> shape<-coef(estMLWE)[1]
> scale<-coef(estMLWE)[2]
> 190*diff(pweibull(c(10,20),coef(estMLWE)[1],coef(estMLWE)[2]))
[1] 28.26208
```

If the unweighted diameter distribution is of the Weibull form, the basal-area weighted distribution (11.5) (p. 353) takes the form (11.7). The basal area within diameter class $(10, 20]$ is computed below both by numerically integrating the general formula (11.5) and by using the generalized *Gamma* distribution (11.6). The basal area on this plot is $G = 11.43$ m^2/ha, and the basal area of the class $(10, 20]$ is $G_{(10,20]} = 0.63$ m^2/ha. Approximation by assuming that all trees of the diameter class are of diameter 15 cm $G_{(10,20]} \approx 0.50$ m^2/ha is a severe underestimate.

```
> print(G<-sum(pi*d22^2/4)/(50*60))
[1] 11.43209
> G*integrate(function(x) x^2*dweibull(x,shape,scale)/(scale^2*gamma(2/shape+1)),
+             10,20)$value
[1] 0.6332198
> G*diff(pgamma((c(10,20)/scale)^shape,2/shape+1))
[1] 0.6332198
> 190*diff(pweibull(c(10,20),coef(estMLWE)[1],coef(estMLWE)[2]))*pi*(0.15/2)^2
[1] 0.4994321
```

If angle-count sampling is used in a forest inventory, then it is convenient to measure the total growing stock by basal area, an estimate of which is obtained by multiplying the observed number of trees on the plot by the basal-area factor. As we have already discussed, it is then convenient to model the basal-area weighted distribution instead of the unweighted distribution, because the scaling can then be carried directly with the basal area, and fitting a model to the angle count sample is straightforward (see Web Example 11.17). Similar justification can be given to the use of unweighted distribution and number of stems as the measure of the total growing stock when fixed-area sample plots are used. However, it is quite common that the distribution needs to be scaled with a total variable that is not the sum of the weights.

Consider the transformation $t(d)$ of tree diameter, such as the basal area, volume, or biomass. Denote the total (per hectare) by $T = N \int_{-\infty}^{\infty} t(u)f(u)du$. The mean of $t(d)$ per tree is the expected value $\mathrm{E}(t(D)) = \int_{-\infty}^{\infty} t(u)f(u)du$. It can also be written as

$$\mathrm{E}(t(D)) = \frac{T}{N},$$

which gives

$$T = N\,\mathrm{E}(t(D))$$

and

$$N = \frac{T}{\mathrm{E}(t(D))}. \tag{11.17}$$

To scale a diameter distribution by a known total T, we can first use Equation (11.17) to compute the number of stems that corresponds to the known T and scale the distribution using that value.

For example, the expected value of the basal area is

$$\mathrm{E}(g(D)) = \mathrm{E}\left(\frac{\pi}{4}D^2\right) = \int_{-\infty}^{\infty} \frac{\pi}{4}u^2 f_D(u)du,$$

which can also be written as G/N. Equation (11.17) gives N that corresponds to the known G as

$$N = \frac{G}{\int_{-\infty}^{\infty} \frac{\pi}{4} u^2 f_D(u) du} = \frac{G}{\frac{\pi}{4}\,\mathrm{E}(D^2)}\,.$$

The transformation $\sqrt{E(D^2)} = 2\sqrt{\mathrm{E}(g(D)/\pi)} = 2\sqrt{G/(N\pi)}$ is commonly called *the quadratic mean diameter* and denoted by DQM. We use DQM often in the form of mean tree basal area $\frac{G}{N} = \frac{\pi}{4} DQM^2$.

In aerial forest inventories, for example, we may need to scale the diameter distribution by total volume instead of the number of stems or basal area. The above-specified principles can be also used in that situation. In order to compute the average tree volume, a function to predict individual tree volume on diameter is needed. Alternatively, we may use a stand-specific height-diameter curve and a volume function based on tree diameter and height.

The total of t within specified size limits x_1 and x_2 can be computed either by using the number of stems and the unweighted density, or by using the total T and the weighted density $f^t(x)$:

$$T_{(x_1,x_2]} = N \int_{x_1}^{x_2} f(u)t(u)\mathrm{d}u = T(F^t(x_2) - F^t(x_1))\,.$$

Example 11.23. Scaling the diameter distribution by basal area. Consider the sample that was used in Example 11.22. Using the estimated Weibull *pdf*, the mean basal area per tree, `GperN`, is 0.060 m^2. Dividing the total basal area by the mean basal area per tree gives the corresponding $\widehat{N} = 190.5$ trees per hectare.

```
> gdfd<-function(d,shape,scale) {
+       pi*(d/200)^2*dweibull(d,shape,scale)
+ }

> GperN<-integrate(gdfd,0,Inf,shape,scale)
> GperN
0.06000327 with absolute error < 2.6e-05

> N<-G/GperN$value
> N
[1] 190.5245
```

This number of stems is compatible with the estimated basal area and the diameter distribution. However, the estimated number of stems differs slightly from the measured number of stems (190 trees per hectare), because the second moment of the Weibull distribution differs slightly from that of the sample. If moment-based estimates are used (Example 11.3), the sample moments match the population moments, and the measured and observed number of stems also match.

```
> shape2<-estMOMWE$par[1]
> scale2<-estMOMWE$par[2]
>
> GperN2<-integrate(gdfd,0,Inf,shape2,scale2)
> GperN2
0.0601689 with absolute error < 3.4e-05
>
> N2<-G/GperN2$value
> N2
[1] 190
```

Example 11.24. Scaling unweighted diameter distribution by total volume. A hypothetical forest stand has the following height-diameter curve (see Example 11.10)

$$h = 1.3 + 25\exp\left(\frac{-5}{d}\right),$$

and the parameters of the Weibull diameter distribution are $\alpha = 4$ and $\beta = 15$. The total volume is 200 m^3 per hectare. All these values could be predictions from previously fitted regression

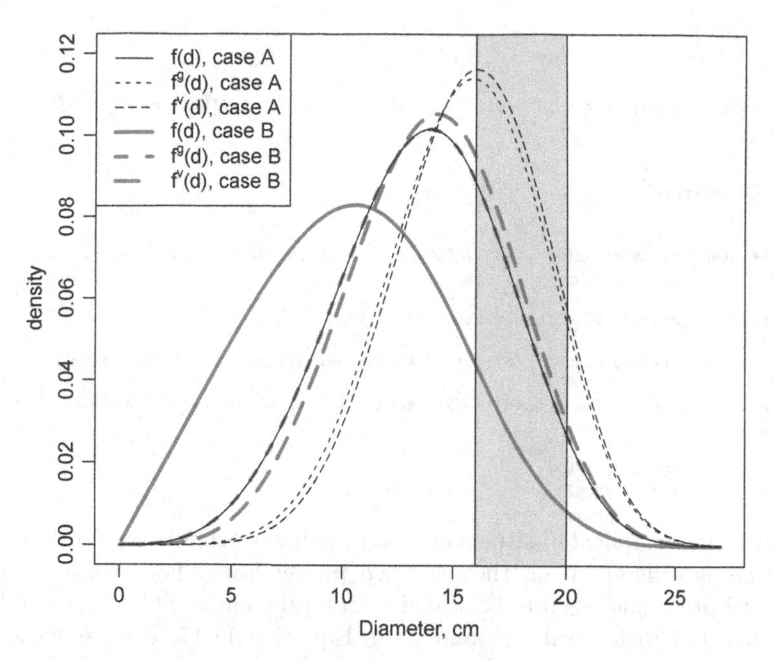

FIGURE 11.11 The unweighted, basal-area weighted, and volume-weighted densities of the diameter distribution when the unweighted distribution is *Weibull*(4,15) (case A) and when the basal-area-weighted distribution is *Weibull*(4,15) (case B). The gray shading illustrates the diameter class $(16, 20]$.

models that utilize remotely sensed data as predictors. In such cases, the estimate of volume is usually more accurate than the estimates of basal area or the number of stems. That is why it would be desirable to scale the diameter distribution by total volume instead of basal area or the number of stems.

We use the following volume function based on diameter and height (Laasasenaho 1982)

$$v(d,h) = 0.0361 d^{2.014} 0.997^d h^{2.070} (h - 1.3)^{-1.072} .$$

The mean tree volume is

$$\mathrm{E}(v) = \int_0^\infty v(u, h(u)) f_D(u) \mathrm{d}u .$$

Numerical evaluation gives $\mathrm{E}(v) = 145.9$ dm^3. Thus, the total number of stems that is compatible with the predicted volume is $1000V/\mathrm{E}(v) = 1371$ trees per hectare. The predicted diameter distribution was scaled with this value. The corresponding total basal area, based on $G = N\,\mathrm{E}(\frac{\pi}{4}D^2)$, is 21.46 m^2 per hectare.

The number of stems, basal area and total volume were computed for the diameter class $(16, 20]$. The values were $N_{(16,20]} = 317.5$ trees per hectare, $G_{(16,20]} = 7.78$ m^2 per hectare and $V_{(16,20]} = 75.6$ m^3 per hectare. These values comprise 26%, 36%, and 38% of the total N, G, and V, respectively. The black lines in Figure 11.11 (case A) demonstrate the unweighted, basal-area weighted, and volume-weighted densities.

A weakness of this analysis is that the variability in the height-diameter relationship and volume model are not taken into account. In addition, prediction errors are present in the stand characteristics and parameters based on remote sensing data. Because the applied functions are nonlinear, ignoring prediction errors leads to bias in predictions, the magnitude and direction of which would need careful analysis.

```
> vol2<-function(d) {
+        v<-predvol(species=1,d=d,h=korfhd(d),model=2)
+        v[d<=1.3]<-0 # Avoid illogical behaviour of the volume model
```

```
+         v
+         }

> print(VperN<-integrate(function(d) vol2(d)*dweibull(d,4,15),0,Inf)$value)
[1] 145.8982

> V<-200
> print(N<-1000*V/VperN)
[1] 1370.819

> print(G<-N*integrate(function(d) pi*d^2/40000*dweibull(d,4,15),0,Inf)$value)
[1] 21.46829

> print(Nclass<-N*(pweibull(20,4,15)-pweibull(16,4,15)))
[1] 317.5084
> print(Gclass<-N*integrate(function(d) pi*d^2/40000*dweibull(d,4,15),16,20)$value)
[1] 7.780359
> print(Vclass<-N*integrate(function(d) vol2(d)*dweibull(d,4,15),16,20)$value/1000)
[1] 75.60875

> c(Nclass/N, Gclass/G, Vclass/V)
[1] 0.2316196 0.3624117 0.3780437
```

A slightly more complicated situation arises when a weighted distribution is scaled using a variable that is different from the one used in weighting. For example, the basal-area weighted distribution may need to be scaled with total volume. Start by using both weights, $t_1(d)$ and $t_2(d)$ and their totals, T_1 and T_2 in Equation 11.17. Both equations specify an expression for N. Setting them as equal gives

$$\frac{T_1}{\mathrm{E}(t_1(D))} = \frac{T_2}{\mathrm{E}(t_2(D))} .$$

If a distribution weighted by $t_1(d)$ needs to be scaled using a known value of T_2, we need an expression for fraction $\tau_{12} = T_1/T_2$ based on the estimated diameter distribution:

$$\tau_{12} = \frac{T_1}{T_2} = \frac{\mathrm{E}(t_1(D))}{\mathrm{E}(t_2(D))} . \tag{11.18}$$

The total T_1 that corresponds to the observed T_2 is then

$$T_1 = \tau_{12} T_2 .$$

The expectations in Equation (11.18) need to be computed using the distribution weighted by t_1. The expected values of $t_2(D)$ and $t_1(D)$ using the weighted density $f_D^{t_1}(d)$ become

$$\begin{aligned} \mathrm{E}(t_2(D)) &= \int_{-\infty}^{\infty} t_2(u) f_D(u) du \\ &= \frac{\int_{-\infty}^{\infty} \frac{t_2(u)}{t_1(u)} f_D^{t_1}(u) du}{\int_{-\infty}^{\infty} \frac{1}{t_1(u)} f_D^{t_1}(u) du} \end{aligned} \tag{11.19}$$

$$\mathrm{E}(t_1(D)) = \frac{1}{\int_{-\infty}^{\infty} \frac{1}{t_1(u)} f_D^{t_1}(u) du} , \tag{11.20}$$

and the ratio (11.18) now simplifies to

$$\tau_{12} = \frac{1}{\int_{-\infty}^{\infty} \frac{g(u)}{t(u)} f^T(u) du}$$

Web Example 11.25. Scaling basal-area-weighted diameter distribution by total volume. The gray lines (case B) in Figure 11.11 demonstrate the unweighted, basal-area weighted, and volume-weighted densities in this case.

11.4.5 Utilizing the arithmetic relationships of stand variables

This section includes additional examples related to the utilization of diameter distribution in computing stand variables of interest, and in parameter recovery.

Example 11.26. The effect of harvest on total volume. Consider a forest stand where F_D^g is a Weibull distribution with $\alpha = 4$ and $\beta = 15$ and the basal area is $G = 20$ m^2/ha. The height-diameter curves and volume functions are as in Example 11.24. 50% of the trees will be harvested from below, and the aim is to compute the remaining volume and basal area.

The numerator of the *pdf* of the unweighted diameter distribution (11.12) (p. 357) is

$$h(d) = \frac{1}{\frac{\pi}{4}d^2} f_D^g(d)$$

and the numerator of the corresponding *cdf* is

$$H(d) = \int_0^d h(u)du \,.$$

These are defined below as R-functions. Because they are unscaled, the area under $h(d)$ is equal to the number of stems per m^2 of basal area. The total number of stems is obtained by multiplying that value by G.

```
> a<-25; b<--5
> alpha<-4; beta<-15
> G<-20

> hNw<-function(x,alpha,beta) {
+      val<-rep(0,length(x))
+      val[x!=0]<-40000*dweibull(x[x!=0],alpha,beta)/(pi*x[x!=0]^2)
+      val
+ }

> # The cumulative version
> HNw<-function(x,alpha,beta) {
+      integrate(f=function(u) hNw(u,alpha,beta),lower=0,upper=x)$value
+ }

> # The total number of stems for G=20
> print(N<-G*HNw(100,alpha,beta))
[1] 2006.007
```

The harvest diameter limit is the median of the unweighted distribution $f_D(d)$, which is found by solving

$$GH(d) = 0.50N$$

for d. The step-halving algorithm of **lmfor::updown**) gives the solution $d = 10.36$.

```
> print(Hlim<-updown(0.1,100,function(x) G*HNw(x,alpha,beta)-0.50*N))
[1] 10.35909
```

The remaining volume and basal area are found by evaluating

$$V_r = G \int_{10.36}^{\infty} h(u)v(u, h(u))du$$

and

$$G_r = G \int_{10.36}^{\infty} f^G(u)du \,,$$

which results in $G_r = 15.9$ m^2/ha and $V_r = 147.8$ m^3/ha.

```
> hVw<-function(x,alpha,beta,a,b) {
+      hNw(x,alpha,beta)*predvol(species=1,x,korfhd(x,a,b),model=2)
+ }

> # The volume function behaves illogically for dbh<1,
> # therefore we assume zero volume for them
```

```
> HVw<-function(x,alpha,beta,a,b) {
+       integrate(f=function(u) hVw(u,alpha,beta,a,b),lower=1,upper=x)$value
+ }
> # Total volume
> print(Vtot<-G*HVw(100,alpha,beta,a,b)/1000)
[1] 177.8053
> # The remaining volume
> print(Vrem<-Vtot-G*(HVw(Hlim,alpha,beta,a,b)/1000))
[1] 147.7682
> # The remaining basal area
> print(Grem<-G*(pweibull(100,alpha,beta)-pweibull(Hlim,alpha,beta)))
[1] 15.93095
```

Example 11.27. Recovery of basal-area weighted Weibull parameters for known G, N and DGM. Assuming that the basal-area weighted diameter distribution is of the Weibull form, find the parameter values that simultaneously fulfill the values $\widehat{G} = 24$, $\widehat{DGM} = 22.99$, and $\widehat{N} = 765$, where DGM is the median diameter weighted by basal area. These values are based on the angle-count sample in Example 11.16 (p. 354).

Assuming a basal-area weighted diameter distribution $f^g(d)$, the unweighted density is from (11.9) (p. 356)

$$f(d|\alpha,\beta) = \frac{d^{-2} f^g(d|\alpha,\beta)}{\int_0^\infty u^{-2} f^g(u|\alpha,\beta)du} .$$

The mean tree basal area is

$$\frac{G}{N} = \int_0^\infty g(d) f(d|\alpha,\beta)du ,$$

where $g(d) = \frac{\pi}{40000} d^2$ gives the basal area (m^2) for diameter d (cm). The number of stems that corresponds to the measured basal area $\widehat{G}$, results from dividing the measured basal area by the expected tree basal area:

$$N = \frac{\widehat{G}}{\int_0^\infty g(u) f(u|\alpha,\beta)du} .$$

It is a function of parameters α and β. Matching this with the measured number of stems, gives the first recovery equation

$$\frac{\widehat{G}}{\int_0^\infty g(d) f(d|\alpha,\beta)} - \widehat{N} = 0 .$$

We could equivalently match the theoretical DQM by the observed DQM, or the theoretical ratio G/N by the observed mean tree basal area.

The second equation is based on DGM. The basal-area median diameter is the 0.5th quantile of $f^g(d|\alpha,\beta)$, which is matched with the median diameter of the angle count sample. This gives

$$F^{g,-1}(0.5) - \widehat{DGM} = 0 ,$$

where $F^{g,-1}$ is the quantile function of the Weibull distribution. The resulting system of two equations was solved numerically using the two-dimensional Newton-Raphson algorithm, which was used before for moment-based recovery of Weibull parameters (see Example 11.18, p. 356). The resulting parameter estimates were $\widehat{\alpha} = 4.73$ and $\widehat{\beta} = 24.83$. Figure 11.12 shows the recovered distribution, together with the histogram of the angle-count sample. The script below utilizes function HNw, which was defined in Example 11.26.

```
> print(G<-length(dsample))
[1] 24
> print(DGM<-mean(dsample[order(dsample)][12:13]))
[1] 22.98554
> print(N<-sum(1/(pi*(dsample/2)^2))*10000)
[1] 764.6316
> # Recovery of BA-weighted diameter distribution
> f1<-function(theta) G*HNw(1000,theta[1],theta[2])-N
> f2<-function(theta) qweibull(0.5,theta[1],theta[2])-DGM

> fn<-list(f1,f2)
> print(theta<-NRnum(c(4,DGM),fn))
$par
[1]  4.732124 24.836578
```

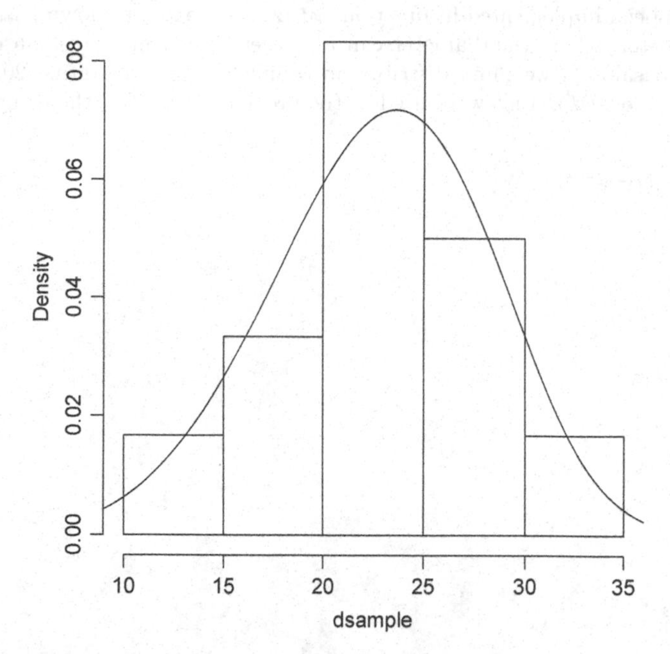

FIGURE 11.12 Histogram of the angle-count sample and recovered Weibull distribution based on $\widehat{G} = 24$ m^2/ha, $\widehat{N} = 765$ trees per ha, and $\widehat{DGM} = 23$ cm in Example 11.27.

Example 11.28. Example 11.27 continued. Take a closer look at the situation in Example 11.27. We assume that G, N, and DGM are known. When $f^G(d)$ is the Weibull density function, the system of equations in Example 11.27 simplifies to

$$\beta = \frac{DGM}{(\log 2)^{1/\alpha}}$$

$$\frac{\pi}{4\Gamma(1-2/\alpha)(\log 2)^{2/\alpha}} = \frac{G}{N \cdot DGM^2},$$

where $\Gamma()$ is the gamma function. The shape parameter can be first solved using the lower equation. Using that value in the upper equation gives the estimate for scale. The numerical values of the estimates are the same as in the earlier example. To solve the lower equation, we use function `lmfor::NR`, which solves a single nonlinear equation using the Newton-Raphson algorithm. It also requires a gradient of the equation with respect to α. This solution is faster and might be useful, for example, if the recovery is implemented in a large-scale forest simulator.

```
> func<-function(shape,G,N,DGM) {
+       pi/(4*gamma(1-2/shape)*log(2)^(2/shape))-G/(N*DGM^2)
+       }

> grad<-function(shape) {
+       pi/4*(-1)*
+       (gamma(1-2/shape)*log(2)^(2/shape))^(-2)*
+       (gamma(1-2/shape)*digamma(1-2/shape)*2*shape^(-2)*log(2)^(2/shape)+
+       log(2)^(2/shape)*log(log(2))*(-2)*shape^(-2)*gamma(1-2/shape))
+       }

> print(shape<-NR(5,fn=function(x) func(x,G=10000*G,N,DGM),
+             gr=grad,crit=10,range=c(2.1,Inf))$par)
[1] 4.732124
> print(scale<-DGM/(log(2)^(1/shape)))
[1] 24.83658
```

Example 11.29. It is more common to assume the Weibull form for the unweighted distribution than for the basal-area weighted distribution. Recovery of Weibull parameters under this assumption has been implemented in function `lmfor::recweib` for known basal area, number of stems and diameter, where the diameter can be either the mean or median diameter of either unweighted or basal-area weighted distribution (Siipilehto and Mehtätalo 2013). For the same values of G, N, and DGM that were used in the previous examples, the estimates are $\widehat{\alpha} = 3.18$ and $\widehat{\beta} = 21.10$.

```
> recweib(G,N,DGM,Dtype="D")
$shape
[1] 3.175324

$scale
[1] 21.10493

$G
[1] 24

$N
[1] 764.6316

$D
[1] 22.98554

$Dtype
[1] "D"

$val
[1] -1.705303e-13
```

Web Example 11.30. Recovering Weibull parameters using volume, mean height, and basal area.

Web Example 11.31. Simultaneous recovery of the height-diameter curve and diameter distribution using stand density, basal area median diameter, basal area mean tree height and total volume (Mehtätalo *et al.* 2007) (see Figure 11.13).

11.5 Modeling tree size in a population of stands

Often, a diameter sample is not available but some stand characteristics, such as stand age, site fertility, basal area, and mean diameter have been assessed instead. Some of these are directly related to diameter distribution (e.g. basal area and mean diameter), whereas others are related only indirectly (e.g. stand age and site fertility). In this case, PPM and PRM (see p. 333) are alternatives for the prediction of the diameter distribution. We start by defining a statistical model for a population of plots. Such a model is implicitly assumed in the literature of parameter prediction methods, even though it has not usually been expressed in the way that we do in the following paragraphs.

11.5.1 The statistical model

Section 11.1 formulated a statistical model for tree size within an area of interest. The area of interest is commonly a stand or large sample plot, but we call it stand here for simplicity. We now extend that model to a population of stands. Similar to Section 11.1, we assume that the diameter of tree j in stand i, d_{ij}, is a random variable with a continuous univariate distribution. All trees of the stand are assumed to be identically distributed with a common *pdf*:

$$d_{ij} \sim f(d_{ij}|\boldsymbol{\theta}_i) \,. \tag{11.21}$$

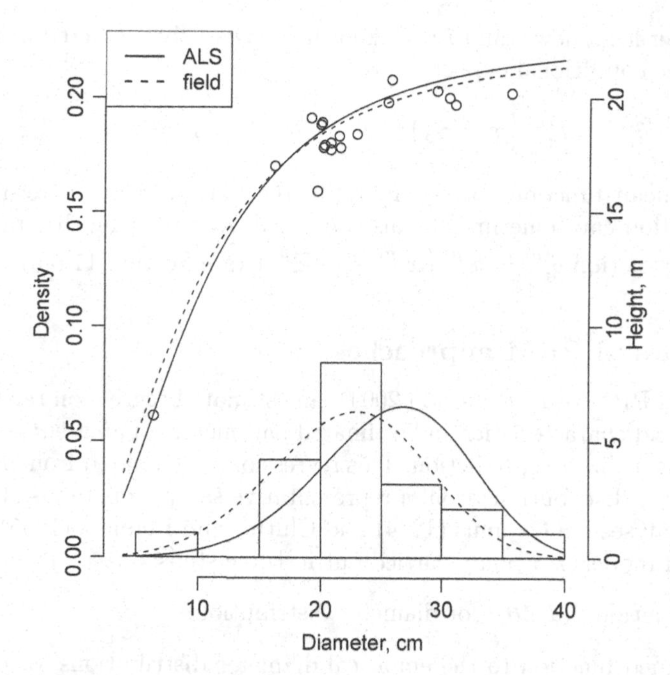

FIGURE 11.13 Histogram of field-measured diameters and observed heights, and the recovered Weibull *pdf* and H-D curve using ALS based predictions (solid) and field estimates (dashed) of $N = 473$ trees per ha, $V = 248 - 60$ m^3 per ha, $DGM = 18.69$ cm, and $HGM = 20.60$ m in Web Example 11.31.

The parameter vector $\boldsymbol{\theta}_i = (\theta_i^{(1)}, \theta_i^{(2)}, \ldots, \theta_i^{(P)})$ is assumed to be a random vector that takes a unique value for each stand. It is assumed that the value for stand i depends on stand-level predictors $\boldsymbol{x}_i^{(p)}$ according to the multivariate model

$$g_p(\theta_i^{(p)}) = \boldsymbol{x}_i^{(p)\prime}\boldsymbol{\gamma}_p + e_i^{(p)} \qquad p = 1, 2, \ldots, P \tag{11.22}$$

where g_p is a link function and $\boldsymbol{\gamma}_p$ includes the regression coefficients of the submodel for $\theta_i^{(p)}$. We use notation $\boldsymbol{\gamma}$ instead of $\boldsymbol{\beta}$ to avoid confusion with the scale parameter β of the Weibull function. Term $e_i^{(p)}$ is the stand-level random effect. For simplicity, we assume linear submodels $v_i^{(p)}(\boldsymbol{x}_i^{(p)}|\boldsymbol{\gamma}_p) = \boldsymbol{x}_i^{(p)\prime}\boldsymbol{\gamma}_p$. The stand effects allow the parameters of a stand to differ from $\boldsymbol{x}_i^{(p)\prime}\boldsymbol{\gamma}_p$. For example, the diameter distributions for two stands with identical values of $\boldsymbol{x}_i$ may be different. The stand effects are assumed to have zero mean, so that $\boldsymbol{x}_i^{(p)\prime}\boldsymbol{\gamma}_p = \mathrm{E}\left(\theta_i^{(p)}|\boldsymbol{x}_i^{(p)}\right)$. The variance-covariance matrix of the stand effects, $\mathrm{var}(\boldsymbol{e}_i)$, where $\boldsymbol{e}_i = \left(e_i^{(1)}, e_i^{(2)}, \ldots, e_i^{(P)}\right)$ specifies how much the stand effects jointly vary around the conditional expectations.

The link function g_p is used to restrict the parameters to their support. For example, if $\theta_i^{(p)}$ is strictly positive, we may use the log link and if $\theta_i^{(p)}$ is bounded from both sides, the logit link could be used. Notice that we can always reparameterize a distribution so that the new parameters have the whole real axis as the support. Therefore we can see the link function g_p as a reparameterization of the applied diameter distribution model.

The parameter for a new stand i with known $\boldsymbol{x}_i$ is usually predicted using the estimated expression for the conditional expectation

$$\widehat{\theta}_i^{(p)} = h_i^{(p)}\left(\boldsymbol{x}_i^{(p)\prime}\widehat{\boldsymbol{\gamma}}_p\right) \qquad p = 1, 2, \ldots, P,$$

where h is the mean function corresponding to the link function g (recall Section 8.1.4). Improved prediction can sometimes be also obtained by predicting the stand effects $\boldsymbol{e}_i$ and using them in prediction $\widehat{\theta}_i^{(p)} = h_i^{(p)}\left(\boldsymbol{x}_i^{(p)\prime}\widehat{\boldsymbol{\gamma}}_p + \widetilde{e}_i^{(p)}\right)$ (see Section 11.5.3).

11.5.2 PPM and PRM approaches

In the traditional PPM (e.g. Robinson 2004), an estimated regression relationship between the measured stand characteristics and estimated parameters of an assumed diameter distribution function is utilized in prediction. This regression is estimated from data that includes measured diameter distributions from a representative sample of stands. The earliest references to such analyses are Cajanus (1914) and Clutter and Bennett (1965). The estimation of the prediction models is usually carried out in three steps:

1. Select the function $f(d_{ij}|\boldsymbol{\theta}_i)$ for diameter distribution.
2. Fit the assumed function to the empirical diameter distributions for each stand of the modeling data and save the stand-specific estimates.
3. Estimate regression models that explain the variation in parameter estimates by stand characteristics.

Fitting of the assumed diameter distribution (Equation (11.21)) and building the submodels for parameters (Equation (11.22)) are two separate steps. The residual errors of the regression models of the second step are the stand effects of model (11.22). Therefore, the procedure is a straightforward combination of the diameter distribution fitting procedures, which were presented in Section 11.2, and the regression models, which have been presented in previous chapters (e.g. Chapter 4). When the models are formulated, the mathematical relationship between the applied stand characteristics and diameter distribution should be kept in mind and utilized. It might be justified to fit the models as a seemingly unrelated system of equations (Chapter 9), especially if the correlations across the equations are of interest. Often the regression models are estimated only for some of the distribution parameters and the others are recovered using the known stand characteristics, such as mean or median diameter.

The parameter recovery method utilizes the mathematical relationships between stand characteristics, percentiles, and moments, and the diameter distribution parameters to formulate a nonlinear system of equations (Hyink and Moser 1983). For the system of equations to be feasible, the number of directly related characteristics needs to be $k+1$, where k is the number of parameters to be estimated. Several examples of PRM were presented in Sections 11.2.2, 11.2.3, and 11.4.5. It is common that the moments or percentiles used in recovery are based on predictive models (e.g. Ek *et al.* 1975). From a mathematical point of view, this approach can be seen as a special case of the models specified in Equations (11.21) – (11.22) where the assumed distribution is reparameterized using the selected moments, percentiles and stand characteristics. For example, the Weibull function can be reparameterized using the first two moments or two percentiles of the user's choice, instead of the shape and scale parameters α and β.

Cao (2004) combined steps 2 and 3 of PPM procedure. His model was based on the Weibull distribution, but other distribution functions might also be used. Let $\exp\left(\boldsymbol{x}_i^{(\alpha)\prime}\boldsymbol{\gamma}_\alpha\right)$

and $\exp\left(\boldsymbol{x}_i^{(\beta)\prime}\gamma_\beta\right)$ be the functions that are used to explain the parameters α and β of the assumed two-parameter Weibull distribution, where $\boldsymbol{\gamma} = (\gamma_\alpha, \gamma_\beta)$ are the regression coefficients of these functions and $\boldsymbol{x}_i^{(\alpha)}$, $\boldsymbol{x}_i^{(\beta)}$ are the vectors of plot-specific predictors. Let $f(d|\alpha,\beta)$ be the Weibull *pdf*. Cao (2004) minimized the following weighted sum of plot-specific log-likelihoods

$$\ell(\boldsymbol{\gamma}) = \sum_{i=1}^{N} \frac{1}{n_i} \sum_{j=1}^{n_i} \log f\left(d_{ij}|\exp(\boldsymbol{x}_i^{(\alpha)\prime}\gamma_\alpha), \exp(\boldsymbol{x}_i^{(\beta)\prime}\gamma_\beta)\right) \tag{11.23}$$

where N is the total number of stands in the data and n_i is the number of trees on stand i. The function is no longer a function of the first-order parameters α and β but rather the coefficients of the PPM models, $\boldsymbol{\gamma}$, which are second-order parameters in the same meaning that we used with nonlinear models (recall Section 7.2.2). This avoids the two-step model chain of the traditional PPM.

If the weighting of stand-specific likelihoods by $\frac{1}{n}$ is removed from (11.23), the formula becomes the likelihood based on model (11.21), where $\theta_i^{(p)} = \exp\left(\boldsymbol{x}_i^{(p)\prime}\boldsymbol{\beta}_p\right)$. Therefore, this model is very similar to (11.22), but without the plot effect $e_i^{(p)}$. The formulation mimics the GLM with Weibull distribution, where log links and linear predictors are used for the shape and scale parameters. However, the parent distribution is not a member of the exponential family, and it is not parameterized using the expected value. Of course, we could parameterize the Weibull function in terms of expected value and shape parameter (recall Example 11.8).

The model does not allow variability in the shape of diameter distribution for given values of the predictors $\boldsymbol{x}_i^{(\alpha)}$ and $\boldsymbol{x}_i^{(\beta)}$. Such variability could be allowed by using the model specification of (11.22), and assuming, for example, a bivariate normal model for $\boldsymbol{e}_i$. This model specification would mimic a GLMM with Weibull parent distribution and random intercept in both parameters. The likelihood of such model would look similar to Equation (8.11) (p. 279). We have not seen such a model presented in this context.

Cao (2004) found that his method based on (11.23) performed much better than the traditional PPM and PRM approaches. Moreover, he also identified a better method whereby the fitting criteria was based on the *cdf* regression. However, the method based on likelihood is better justified from a statistical point of view, especially if the random stand effects are included.

Example 11.32. Using data set `spati`, the prediction models for the parameters of a two-parameter Weibull distribution are based on basal-area weighted mean diameter (D_i), basal area (G_i) and stand age (T_i) as candidate predictors. We drop 7 sample plots that have a clearly bimodal diameter distribution, because the Weibull distribution is not justified for those plots. The script below fits the Weibull distribution to every plot separately using ML. Examination of the data (not shown) indicates that the following models are justified for the plot-specific parameter estimates

$$\log(\alpha_i) = \gamma_1^{(\alpha)} + \gamma_2^{(\alpha)}\log(D_i) + e_i^{(\alpha)}$$
$$\log(\beta_i) = \gamma_1^{(\beta)} + \gamma_2^{(\beta)}D_i + \gamma_3^{(\beta)}T_i + e_i^{(\beta)}\,.$$

The log transformations were not necessary from a model fitting point of view, but they are justified to ensure that both parameters of the Weibull distribution are positive. The resulting transformation bias is not an issue here, because we see them only as reparameterizations of parameters α and β. The script below implements the fitting procedure. The initial guesses $\alpha = 5$ and $\beta = 20$ led to convergence of estimation using `stats4::mle` in all plots. The warning

	2-stage PPM		Cao	GLM		GLM + recovery	
	est	s.e.	est	est	s.e.	est	s.e.
$\gamma_1^{(\alpha)}$	-0.284	0.0972	-0.312	-0.260	0.0269	-	-
$\gamma_2^{(\alpha)}$	1.061	0.0349	1.077	1.057	0.00994	-	
$\gamma_1^{(\beta)}$	0.733	0.126	0.728	0.793	0.027	0.754	0.026
$\gamma_2^{(\beta)}$	0.0393	0.0133	0.0377	0.0344	0.00331	0.0412	0.00318
$\gamma_3^{(\beta)}$	-0.0058	0.0031	-0.0064	-0.0067	0.00081	-0.0083	0.00076

TABLE 11.2 Parameter estimates with estimated standard errors of the PPM models in Example 11.32 using the two-stage procedure, Cao's approach, and the GLM-type approaches.

messages arise from situations where the estimation procedure attempts to assign extremely large or small values for the logarithmic parameters during the iteration. Estimates based on OLS on the ML estimates of α and β (models `modsc` and `modsh` below) are shown in the first two columns of Table 11.2.

```
> data(spati)
> spati<-spati[!(spati$plot%in%c(8,12,15,21,27,28,44)),]
> plotdat<-unique(spati[,c("plot","Dg","Tg","G")])

> # 2 parameter Weibull -logL
> nLLweibull<-function(x, lshape,lscale) {
+     -sum(dweibull(x,shape=exp(lshape),scale=exp(lscale),log=TRUE))
+ }

> fitw2<-function(d) {
+   est<-mle(function(lshape=log(5),lscale=log(20)) nLLweibull(d,lshape,lscale))
+   if (class(est)=="try-error") rep(NA,2)
+   else exp(coef(est))
+   }
> a<-tapply(spati$d,spati$plot,fitw2)
There were 50 or more warnings (use warnings() to see the first 50)
> plotdat<-cbind(plotdat,matrix(unlist(a),ncol=2,byrow=TRUE))
> names(plotdat)[5:6]<-c("shape","scale")
> modsc<-lm(log(scale)~log(Dg),data=plotdat)
> modsh<-lm(log(shape)~Dg+Tg,data=plotdat)
```

Cao's procedure is implemented below. We use the estimates from the two-step procedure as starting values for the estimation. The fitting is performed using function `stats::mle`, even though the function evaluated by `nllCao` is not a full likelihood over the data but a weighted sum of plot-specific likelihoods. The resulting parameter estimates are shown in the third column of Table 11.2.

```
> spati$n<-round(spati$N*spati$X*spati$Y/1000000)
> # 2 parameter Weibull -logL
> nllCao<-function(x,sc1,sc2,sh1,sh2,sh3) {
+         scale<-exp(sc1+sc2*log(spati$Dg))
+         shape<-exp(sh1+sh2*spati$Dg+sh3*spati$Tg)
+         -sum(1/spati$n*dweibull(x,shape=shape,scale=scale,log=TRUE))
+         }

> start<-as.list(c(coef(modsc),coef(modsh)))
> names(start)<-c("sc1","sc2","sh1","sh2","sh3")

> estCao<-mle(function(sc1,sc2,sh1,sh2,sh3) nllCao(spati$d,sc1,sc2,sh1,sh2,sh3),
+             start=start)
```

The GLM-type approach results from removing the weighting of the likelihoods from `nllCao`. Estimation was implemented by minimizing the likelihood `nllML` below. The estimates and their standard errors are shown in columns 4 and 5 in Table 11.2. Note that the reported standard errors are underestimates because the likelihood is based on the assumption of independent and identically distributed diameter data, which is violated because there are several trees per plot. A model with plot-level random effects in the submodels of α and β would fix this problem, at least partially.

```
> # 2 parameter Weibull -logL
> nllML<-function(x,sc1,sc2,sh1,sh2,sh3) {
+       scale<-exp(sc1+sc2*log(spati$Dg))
+       shape<-exp(sh1+sh2*spati$Dg+sh3*spati$Tg)
+       -sum(dweibull(x,shape=shape,scale=scale,log=TRUE))
+       }

> estML<-mle(function(sc1,sc2,sh1,sh2,sh3) nllML(spati$d,sc1,sc2,sh1,sh2,sh3),
+          start=start)
```

The known basal-area weighted mean diameter can be computed from the diameter distribution using $D = \frac{\int_0^\infty u^3 f(u)du}{\int_0^\infty u^2 f(u)du}$. Using a two-parameter Weibull density as $f(d)$ and solving for the scale parameter gives $\beta = D\frac{\Gamma(2/\alpha+1)}{\Gamma(3/\alpha+1)}$. We can, therefore, replace the model for parameter α using this formula, which has been implemented in `lmfor::scaleDGMean`, a support function for `lmfor::recweib`. An example of the GLM-type approach is shown below. The resulting parameter estimates are shown in the last two columns of Table 11.2.

```
> nllML2<-function(x,sh1,sh2,sh3) {
+        shape<-exp(sh1+sh2*spati$Dg+sh3*spati$Tg)
+        scale<-apply(cbind(spati$Dg,shape),1,function(x) scaleDGMean(x[1],x[2]))
+        -sum(dweibull(x,shape=shape,scale=scale,log=TRUE))
+        }

> estML2<-mle(function(sh1,sh2,sh3) nllML2(spati$d,sh1,sh2,sh3),
+      start=start[3:5])
```

For an assumed diameter distribution model, parameter recovery ensures that the resulting distribution is compatible with all those stand variables that were used in the recovery. However, it is not always possible to find a solution to the system of recovery equations. In such cases, methods have been proposed to adjust the predicted diameter class frequencies so that they are compatible with all known stand variables (Nepal and Somers 1992, Cao and Baldwin 1999, Kangas and Maltamo 2000a). After this adjustment, the distribution is no longer of the originally assumed mathematical form, but it is compatible with all utilized stand variables. In the context of the percentile-based method, Mehtätalo (2004a) proposed a method that directly adjusts the predicted percentiles, taking into account the prediction errors of the percentiles and the measurement errors of the stand variables.

11.5.3 Improving the prediction using sample information

If sample information of the tree diameters is available from the sample plot in question, it can be utilized to improve the prediction based on a PPM model in a similar way that we have used sample information to predict the random effects of a mixed-effects model and the residual errors for a system of models (recall Section 9.2.4).

Mehtätalo (2005) formulated such an approach using the percentile-based diameter distribution model (see Section 11.3.1 for definition of the percentile-based distribution). The measurements that were used in calibration were sample quantiles from sample plots within the stand. Special cases are the sample minima and maxima. However, the method allows the use of the rth smallest diameter in a random sample of n trees. Several quantiles from the same sample can be utilized, and even all the trees from the sample.

The measured sample quantiles were interpreted as measured percentiles. Therefore, the response variable of the PPM model (diameter corresponding to a certain cumulative density value) has been observed for some values of the *cdf* from the sample plots. However, the quantile trees do not correspond exactly to the cumulative densities of the PPM model system. Therefore, interpolation of the predictions and the variance-covariance matrix of the prediction errors is needed, as applied by Lappi (2006) for taper curves.

Model (11.22) can be written in this case as

$$\boldsymbol{\xi} = \boldsymbol{\mu} + \boldsymbol{b}$$

where $\boldsymbol{\xi}$ is the vector of percentiles that corresponds to prespecified fixed values $\boldsymbol{p}$ of the stand-specific *cdf*, $\boldsymbol{\mu}$ are their expected values, which may be functions of stand characteristics, such as stand age, basal area and mean diameter, and $\boldsymbol{b}$ are the residuals that can be considered as stand effects. All these are specific to a single forest stand; however, to keep notations simple, we do not write the stand index explicitly. From the PPM model fitting step, we assume that $\boldsymbol{\mu}$ and $\text{var}(\boldsymbol{b}) = \boldsymbol{D}$ are known. A carefully formulated model would ensure that the percentiles are in increasing order, but that is not guaranteed in our case by model formulation.

In addition, we may have observed one or more sample quantiles, possibly from several sample plots from the stand. The quantiles are considered as measured percentiles, which follow the model

$$\boldsymbol{\xi}^* = \boldsymbol{\mu}^* + \boldsymbol{b}^* + \boldsymbol{\epsilon}$$

where $\boldsymbol{\xi}^*$ are the measured percentiles that correspond to values $\boldsymbol{p}^*$ of the *cdf*, $\boldsymbol{\mu}^*$ are their expected values, and $\boldsymbol{b}^*$ the stand effects. Term $\boldsymbol{\epsilon}$ includes the random measurement errors, which are uncorrelated for measurements from different plots but correlated for measurements of the same plot. They are also independent of the stand effects $\boldsymbol{b}^*$. We use notations $\text{var}(\boldsymbol{b}^*) = \boldsymbol{D}^*$, $\text{var}(\boldsymbol{\epsilon}^*) = \boldsymbol{R}$, and $\text{cov}(\boldsymbol{b}, \boldsymbol{b}^{*\prime}) = \boldsymbol{C}$. After finding $\boldsymbol{p}^*$, objects $\boldsymbol{\mu}^*$, $\boldsymbol{D}^*$, and $\boldsymbol{C}$ are obtained by interpolating $\boldsymbol{\mu}$ and $\boldsymbol{D}$ for the values of $\boldsymbol{p}^*$. Vector $\boldsymbol{p}_i^*$ and matrix $\boldsymbol{R}$ are found by applying the basic results from the theory of order statistics (see e.g. Casella and Berger 2001, Section 5.4) to the percentile-based parent distribution, as described below.

We assume that the true distribution of the plot is the percentile-based distribution with parameters $\boldsymbol{\xi}$, as defined in Equations (11.2) and (11.3) (p. 345); however, we replace b_i in those equations with c_i to avoid confusion with the plot effect $\boldsymbol{b}$. The *pdf* of the sample order statistic is defined by parts, in a similar way as the *pdf* and *cdf* of the percentile-based distribution. Let $Y_{r:n}$ be the rth smallest observation in a sample of size n. Within each interval $\xi_i < y \leq \xi_{i+1}$, the *pdf* of $Y_{r:n}$ is (see Figures 11.14b and 11.14c for illustration)

$$f_{r:n}^{(i)}(y) = \frac{n!}{(r-1)!\,(n-r)!} c_i \left(a_i + c_i y\right)^{r-1} \left(1 - a_i - c_i y\right)^{n-r} . \tag{11.24}$$

The corresponding expected value $\text{E}(Y_{r:n})$ can be computed analytically by summing integrals of the form $\int_{\xi_i}^{\xi_{i+1}} u f_{r:n}^{(i)}(u) du$ over all percentile intervals. This expected value is equal to the $p^* = F(\text{E}(Y_{r:n}))^{\text{th}}$ quantile of the stand-specific diameter distribution F. Therefore, the rth smallest observation in the sample of size n is regarded as the observed $100p^{*\text{th}}$ percentile. The measurement error related to this percentile has the variance $\text{var}(Y_{r:n}) = \text{E}(Y_{r:n}^2) - (\text{E}(Y_{r:n}))^2$, which can also be computed analytically. This gives $\text{var}(\epsilon)$ that corresponds to the percentile in question, and is written to the diagonal of $\boldsymbol{R}$.

The observed percentiles from the same sample plot are dependent, which needs to be taken into account if some of $\boldsymbol{\xi}^*$ are taken from the same plot. To compute the related covariance, the joint *pdf* of two percentiles from the same plot is needed. The joint *pdf* is defined by parts for the quadrangles specified by the percentile intervals. For each quadrangle $\xi_i < y_1 \leq \xi_{i+1}$ and $\xi_j < y_2 \leq \xi_{j+1}$, the joint *pdf* is (see Figure 11.14d)

$$\begin{aligned} f_{r_1:n,r_2:n}^{(i,j)}(y_1, y_2) = &\frac{n!}{(n-r_2)!\,(r_2-r_1-1)!\,(r_1-1)!} c_i c_j \left(a_i + c_j y_i\right)^{r_1-1} \\ &\times \left(a_j + c_j y_2 - a_i - c_i y_1\right)^{r_2-r_1-1} \left(1 - a_j - c_j y_2\right)^{n-r_2} . \end{aligned} \tag{11.25}$$

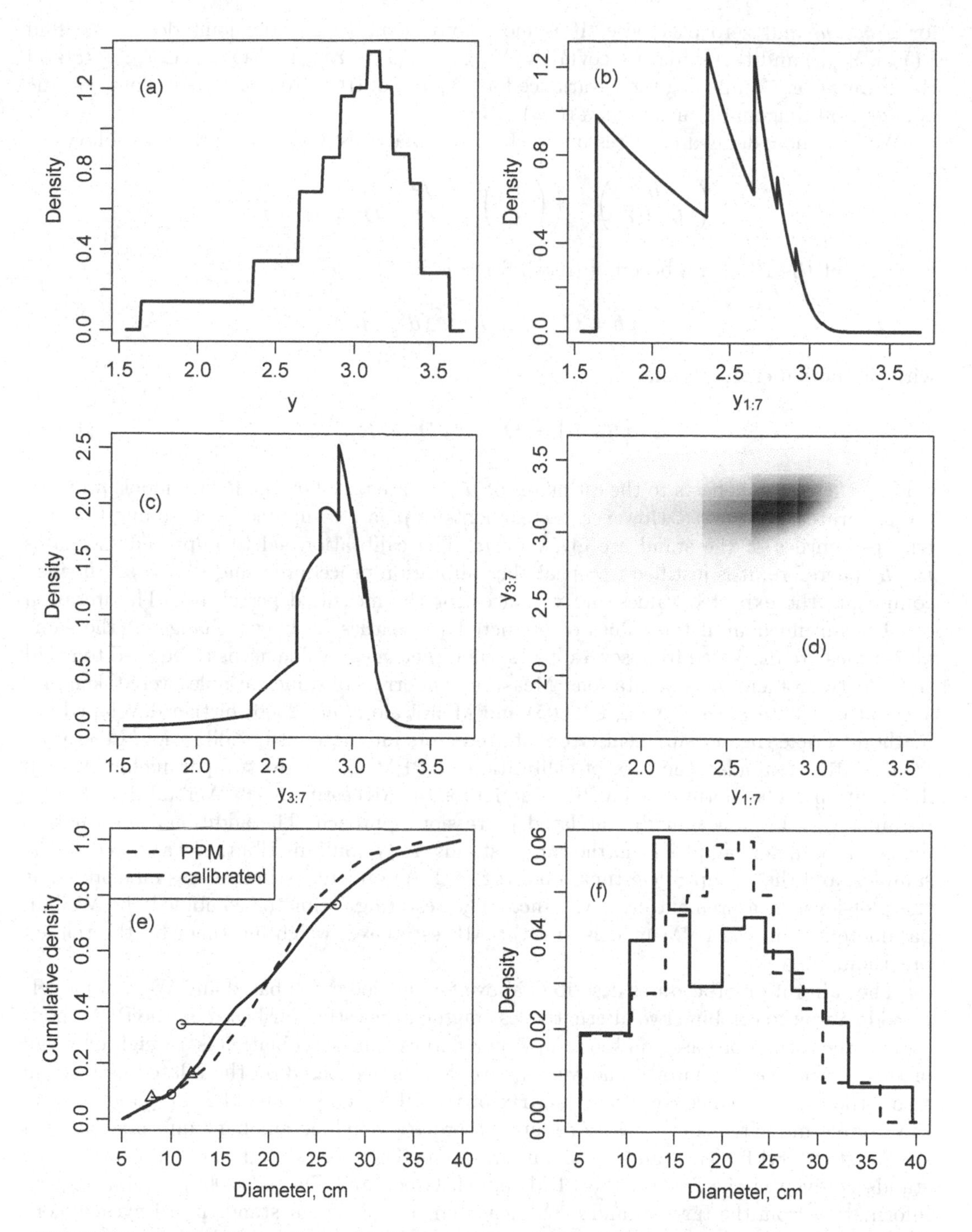

FIGURE 11.14 Percentile-based diameter distribution of logarithmic diameters (a), the corresponding marginal *pdf*'s of the minimum $y_{1:7}$ (b) and the third smallest trees $y_{3:7}$ (c) in a sample of size 7, and the joint *pdf* of $y_{1:7}$ and $y_{3:7}$ (d). The lower graphs illustrate the percentile-based diameter distribution (pdf in (e) and cdf in (f)) using the back-transformed diameters based on the PPM model (solid) and the distribution calibrated using five quantile trees from two sample plots (dashed). The observed quantile trees are illustrated by the dots in (e), with plot-specific symbols. See Web Example 11.33 for details.

for $y_j > y_i$ and zero otherwise. It is most convenient to use the joint density to find $\mathrm{E}(Y_{r_1:n}Y_{r_2:n})$ and thereafter use $\mathrm{cov}(Y_{r_1:n}, Y_{r_2:n}) = \mathrm{E}(Y_{r_1:n}Y_{r_2:n}) - \mathrm{E}(Y_{r_1:n})\,\mathrm{E}(Y_{r_2:n})$ to find the covariance. Computing the covariance for all pairs of trees from the same plot gives the non-zero off-diagonal elements of $\mathrm{var}(\boldsymbol{\epsilon}_{ij}) = \boldsymbol{R}_{ij}$.

We have now defined all necessary vectors and matrices. Organizing them as follows:

$$\begin{pmatrix} \boldsymbol{b} \\ \boldsymbol{b}^* + \boldsymbol{\epsilon} \end{pmatrix} \sim \left(\begin{pmatrix} \boldsymbol{0} \\ \boldsymbol{0} \end{pmatrix}, \begin{pmatrix} \boldsymbol{D} & \boldsymbol{C} \\ \boldsymbol{C}' & \boldsymbol{D}^* + \boldsymbol{R} \end{pmatrix} \right),$$

we see that the BLP of $\boldsymbol{e}$ becomes (recall Section 3.5.2)

$$\widetilde{\boldsymbol{b}} = \boldsymbol{C}\left(\boldsymbol{D}^* + \boldsymbol{R}\right)^{-1}\left(\boldsymbol{d}^* - \boldsymbol{\mu}^*\right)$$

with prediction error variance

$$\mathrm{var}\left(\widetilde{\boldsymbol{b}} - \boldsymbol{b}\right) = \boldsymbol{D} - \boldsymbol{C}\left(\boldsymbol{D}^* + \boldsymbol{R}\right)^{-1}\boldsymbol{C}'. \tag{11.26}$$

Adding these stand effects to the estimates of $\boldsymbol{\mu}$ that are based on the PPM models provides the calibrated percentiles. However, the elements of $\boldsymbol{p}^*$ are computed by assuming that the true percentiles of the stand are given by $\boldsymbol{\mu}$. The calibration led to improved estimates $\boldsymbol{\mu} + \widetilde{\boldsymbol{b}}$. Therefore, it is justified to repeat the calibration procedure using $\boldsymbol{\mu}^{(2)} = \boldsymbol{\mu} + \boldsymbol{b}$ when computing the expected values and variances for the measured percentiles. The iteration can be continued until the values of predicted percentiles no longer change. If the same percentiles are measured from several plots, then their mean value needs to be used to avoid singularity problems in computations. Measurement errors of stand variables were taken into account in Mehtätalo and Kangas (2005) and Mehtätalo *et al.* (2006) further developed the method to take the measurement errors of predictors into account in calibration. Mehtätalo *et al.* (2011) extended the idea of calibration of PPM models to two-parameter Weibull distributions. They formulated a PPM model for the ML estimates of Weibull distribution parameters using a seemingly unrelated regression approach. The additional information included diameter samples from the target stands. A Weibull distribution was fitted to the samples, and the asymptotic estimation error of the parameter estimates was interpreted as the plot-level measurement error variance. The resulting estimator combined the Weibull parameters from the PPM models and the ML estimates, weighting them by the related precision.

The calibration procedures described above for the percentile-based and Weibull models provide a way to combine two alternative estimators: the estimator based on the PPM models and the estimator based on the sample percentiles. The combination is a weighted mean of the alternative estimators. The weights are determined based on the related estimation error: the error variance-covariance matrix of the PPM models and the sampling error of the sample measurements. The procedures, therefore, combine auxiliary information from similar stands with local stand-specific information. The auxiliary information from similar stands is given in the form of the PPM model. Green and Clutter (2000) combined local information from the target stand with measurements of similar stands in a Bayesian context, by weighting the alternative estimators by their precision. Their basic idea was very similar to the approaches presented here, even though it has not been formulated in the context of a PPM framework.

Web Example 11.33. Illustration of the calibration procedure of Mehtätalo (2005) using the percentile-based models of Kangas and Maltamo (2000b) for Norway spruce stands in Finland.

12

Taper Curves

A *taper curve* or *stem curve* is a mathematical description for the tree stem profile. Modeling taper curves is a very popular area of forest research. Our purpose here is not to undertake a review of taper curve models (see e.g. Burkhart and Tome (2012)); instead we want to point out some statistical aspects of taper curve modeling. In spite of the large amount of taper curve literature, we feel that the taper curve modeling problem has not been satisfactorily solved. In this chapter, we will attempt to identify some requirements for a statistically sufficient solution.

12.1 Definition of tree diameter

Intuitively, a taper curve of a tree is a description of the *diameters* of the tree stem at different heights. Our first task is to define what we mean by a diameter. The cross-sectional areas of stems are not circular, so if a cross-sectional area of a tree at height h is $A(h)$, then a natural definition of the tree diameter at height h is

$$D(h) = 2\sqrt{\frac{A(h)}{\pi}}. \tag{12.1}$$

That is, the diameter is defined to be the diameter of a circle whose area is equal to the cross-sectional area either under or over bark. When defined in this way, standard methods that are used to measure diameter induce measurement errors into the data (see Chapter 13). Furthermore, all standard measurements are biased as they give overestimates for the cross-sectional area. Pulkkinen (2012) provided a thorough statistical analysis of the effects of non-circularity on the estimation of stem volumes. She recommends that diameter derived from girth measurement should be used (i.e. dividing the perimeter measurement by π) as it is the most stable and practical way to measure diameters if non-circularity is of concern. It should be noted, however, that even this measurement will always be larger than the true diameter defined in Equation (12.1). Diameter measured from the girth contains only bias, i.e. not sampling errors as with the other diameter measurement methods. Pulkkinen (2012) found that non-circularity produced an approximately 1.2% bias when a volume equation was used to predict the stem volume.

Moreover, practical and theoretical problems in measuring diameters are also caused by branches, irregular forms of trees, leaning of trees, slope of the terrain etc. For later discussion, we will assume that there is a true taper curve $D(h)$ for each tree, and we can take measurements of points on the taper curve, although the measurements can contain measurement error.

12.2 Uses of taper curves

We consider the main motivation in taper curve modeling is to make the following predictions using an estimated taper model, and measurements of stem dimensions and other stand, plot or tree variables. Let H denote the total height of the tree.

1. Predict diameter at a given height, $D(h), h \in [L, U]$ where L can be zero or some estimate of the stump height (e.g. a fixed percentage of H), and U can be the known or estimated total height, or the known or estimated merchantable timber height.
2. Predict stem volume between two heights $V_{UL} = \frac{\pi}{4}\int_L^U D^2(h)\,dh$ where L can be zero or usually the stump height, and U can be the total height, or the merchantable timber height. Furthermore L and U can be the limits for saw-logs, pulp wood and energy wood, in which case L and U are not fixed constants but random variables determined by some minimum diameter and log-length requirement or by an optimization process where the stem value is maximized.

Prediction of volumes may seem to be a trivial application of a stem curve model that is to be used to predict diameters at given heights. But there are theoretically clear differences. First, if $\widetilde{D}(h)$ is an unbiased predictor for $D(h)$ given the measured variables, then $(\widetilde{D}(h))^2$ is biased for $D^2(h)$. This follows from the fact that

$$\mathrm{E}\left(D^2(h)\right) = \left(E(D(h))\right)^2 + \mathrm{var}\left(D(h)\right).$$

Therefore, the estimation of volume is biased when it is based on integration of a taper curve fitted for $D(h)$.

The distinction between predicting the diameter and the square of the diameter is taken into account, for example, by Lappi (1986), Lappi (2006) and Kublin *et al.* (2013). Alternatively, a model can be directly developed for the cross-sectional area, as carried out by Gregoire *et al.* (2000) and Fortin *et al.* (2013).

Because $D^2(h)$ is nonlinear in h, $\int_{\widetilde{L}}^{\widetilde{U}} \widetilde{D^2}(h)dh$ is biased for $\int_L^U D^2(h)dh$, even if $\widetilde{L}$, $\widetilde{U}$ and $\widetilde{D^2}$ are unbiased predictors (we will return to this shortly).

Zakrzewski (1999) has promoted the idea of a taper curve function, which is analytically integrable.

12.3 Statistical modeling of taper curves

There are four main approaches for describing $D(h)$: 1) assuming a parametric functional form for $D(h)$, 2) using a multivariate formulation for a discrete set of dimensions and then using interpolation to obtain the entire stem curve, 3) formulating the stem curve in terms of continuous basis functions (see Section 10.4.1), and 4) using interpolating splines (see Box 10.2, p. 324, and Lahtinen and Laasasenaho (1979)).The first approach has generally dominated (see e.g. the models tested by de Miguel *et al.* (2012)) and usually leads to an application of the nonlinear (mixed-effect) models, which were discussed in Chapter 7. The multivariate approach has been applied, for example, by Kilkki *et al.* (1978), Lappi (1986) and Lappi (2006)(also see related discussion in Chapter 9). In Kilkki *et al.* (1978), the tree dimensions were diameters at relative heights and Lappi (1986) used diameters in polar coordinates. The formulation in terms of basis functions has been applied by Kublin *et al.* (2008) and Kublin *et al.* (2013).

Taper curve models use different assumptions for measured dimensions. Most models assume that *dbh* and tree height are measured. Almost all models assume that height is either measured or it is predicted with another model. In the model of Lappi (1986) or the simplified version in Varjo *et al.* (2006), any dimension can be measured and height does not need to be known. Thus, these models could be used in harvester applications. In the model of Nummi and Möttönen (2004) developed for harvesters, the lower part of the stem is used to predict the upper part.

When taper curves are estimated using statistical principles, a problem is to find an acceptable model for the residual errors, i.e. we need a model for the variance of the errors and for the correlations between errors, in order to apply GLS estimation. The residual errors of an individual tree are clearly correlated and the residual errors of trees within the same stand are correlated.

If *dbh* and height are assumed to be measured, it is evident from the studies of Lappi (2006) and Kublin *et al.* (2008) that the standard covariance models of geostatistics or for a time series are not adequate for within-tree diameter correlations. For example, the covariance structure is anisotropic, and the residuals just below and above breast height are negatively correlated, even if the distance is short. The covariance structure is determined by the relationship of points with respect to the fixed measurements, and not just as a function with distance. One option for modeling the covariance structure is to use nonparametric methods as in Lappi (2006), or semiparametric methods as in Kublin *et al.* (2008). Another alternative for describing a major part of the residual error is to use linear (e.g. Calama and Montero (2006)) or nonlinear (e.g. Garber and Maguire (2003) and Arias-Rodil *et al.* (2015)) mixed model based on some standard taper curve model. In the model of Kublin *et al.* (2013), random coefficients are used for spline basis functions.

Some random parameter models have random parameters both at the stand and tree level (e.g. Lappi (1986), Calama and Montero (2006)), and some only at the tree level (Garber and Maguire (2003), or Kublin *et al.* (2013)). Lappi (2006) predicted stem curves using only tree-level covariances but without having explicit tree-level parameters. When there are random parameters at the stand level, the mixed model formulation makes it possible to calibrate the model for a stand using additional diameter measurements for some trees. When both *dbh* and height are fixed variables in the model, there is, evidently, not so much variation between stands. With random tree parameters, the model can be calibrated only for those trees that have additional measurements.

With a mixed model, the residual error that remains after the initial residual variation has been explained with random parameters is much smaller. Thus, the correlation model for the residual errors is not so important.

The error variance of the volume prediction is the variance of a stochastic integral. The variance of an integral of a stochastic process is (see Parzen (1962), Gregoire *et al.* (2000), Lappi (2006), Kublin *et al.* (2013)):

$$\operatorname{var}\left(\int_a^b y(x)\,\mathrm{d}x\right) = \int_a^b\int_a^b \operatorname{cov}\left(y(x_1),\ y(x_2)\right)\mathrm{d}x_1\,\mathrm{d}x_2 .$$

In volume prediction, we integrate cross-sectional area. Thus, the variance of volume between fixed heights L and U is:

$$\operatorname{var}\left(\frac{\pi}{4}\int_L^U D^2(h)\,\mathrm{d}h\right) = \frac{\pi^2}{16}\int_L^U\int_L^U \operatorname{cov}\left(D^2(h_1), D^2(h_2)\right)\mathrm{d}h_1\,\mathrm{d}h_2 . \tag{12.2}$$

If the model is formulated for diameters, then the covariances of the cross-sectional areas can be obtained assuming multi-normal distribution of errors (see Lappi (2006) and Kublin *et al.* (2013) who corrected a non-significant error of Lappi (2006)). In Gregoire *et al.* (2000) and Fortin *et al.* (2013), the model was directly developed for cross-sectional areas. In Fortin *et al.* (2013) the integral in Equation (12.2) was not computed explicitly but was computed indirectly by calculating the variance of sum of small segments using a second-order Taylor series approximation of a nonlinear mixed model. In Gregoire *et al.* (2000), the estimation errors of the fixed parameters were also taken into account in the variance formulas.

For modeling the covariance structure of residuals, it is important to know what fixed measurements are in the model and how many random tree-level parameters are included. If *dbh* is assumed to be measured in an over-bark model, then the variance function should go to zero as height approaches breast height. In an under-bark model, the variance should get smaller close to breast height. If total height is assumed to be measured, then the variance should approach zero when the height approaches the total height in an over-bark model and be small in an under-bark model. For example, the variance function of the under-bark model described in Fortin *et al.* (2013) is smaller at the top of the tree, as well as near breast height.

Residual correlations have been described using a first-order autoregressive model (e.g. Garber and Maguire (2003) and Fortin *et al.* (2013)), exponential correlation structure (Gregoire *et al.* 2000), and four-banded Toeplitz (Yang *et al.* 2009). These models do not consider that the correlation of diameters, just above and below breast height, is usually negative. The semiparametric model of Kublin *et al.* (2008) and the nonparametric model of Lappi (2006) describe the covariance structure without any assumptions of the correlation function. It should be noted that similarity of trees within the same stand cannot be described using a model for autocorrelation of residuals; a mixed model with random stand parameters is required for that purpose.

If a stem curve model assumes that height is known, and in an application the height is not known but is predicted, then the standard way is to use the predicted height as if it had been measured. From a statistical point of view, this is not correct. The expected diameter at a given height h and its variance can be obtained using the law of total expectation (Equation (2.45), p. 25) and the law of total variance (Equation (2.46), p. 25).

Note that if tree height is not measured, then the expected value of diameter over height does not resemble a stem curve of any tree (the expected value has a very thin top). Kublin *et al.* (2013) showed how these formulas can be applied when predicting expected diameters or the volumes of stem segments. Lappi (2006) used a two-point distribution approach (see Section 10.2.3) to account for height when it is not measured, i.e. the distribution of height is approximated by a distribution that assigns the probability 0.5 to value μ minus the standard deviation, and probability 0.5 to point μ plus the standard deviation.

Predicted stem curves are usually used directly to compute volumes of timber assortments. But when timber assortments are defined by minimum diameter requirements, then the predicted timber volumes can be quite biased. For instance, if it is required that a stem qualifies as a saw-log if the diameter at H_0 is at least D_0, a naive approach would produce a yes or no answer, even if the correct answer would be the probability that H_0 is at least D_0 given the measurements. If a stem curve model also provides an estimate for the distribution of errors, then such probabilities can be estimated (see Lappi (1986), Lappi (2006)). Statistical modeling of taper curves still presents many challenges, e.g. in bucking optimization or laser scanning inventories.

13

Measurement Errors

In forestry and in environmental studies, as in most fields, measurements often contain errors. In standard statistical analyses, the measurement errors are usually ignored even if they can cause considerable bias in the results. In this chapter, we present several cases where measurement errors should be taken into account. Let us first consider the case where there are measurement errors in the independent variables of regression. Then, we consider how measurement errors affect distributions.

Let X denote the measured value of a variable and let x denote the true value. Let us consider first the unbiased measurements, that is:

$$X = x + u\,, \tag{13.1}$$

where u is a random measurement error with mean zero. It is a common misunderstanding that these conditions are sufficient to specify a measurement error model. Unfortunately they are not, as Equation (13.1) is equivalent to:

$$x = X + v\,,$$

where v is now a random measurement error with mean zero ($v = -u$). To correctly specify a measurement error model, it needs to be specified how measurement errors depend on the true value or on the measured value or on both. When presenting measurement errors using Equation (13.1), the implicit assumption is that u is independent of x, but often this assumption does not hold. When it holds, a bias occurs in the estimation of regression parameters.

13.1 Measurement errors in estimating regression parameters

13.1.1 Measurement errors of the regressor variables are independent of the true value

Let us first consider the case where measurement errors of the independent variables are independent of the true value, i.e. assume that Equation (13.1) and the independence assumption hold. In this case, measurement errors lead to bias in the estimated parameters. The analysis is more straightforward if we consider the regressor variables to be random (see Section 10.1.2). With fixed predictors, there is no simple solution to the bias problem caused by measurement errors. Let us first consider the case with only one predictor, i.e. assume that the true model is

$$y = \beta_1 + \beta_2 x + e\,, \tag{13.2}$$

where x and e are independent multi-normally distributed variables, and thus also y is normally distributed. The assumption of normality is not necessary in the following derivations,

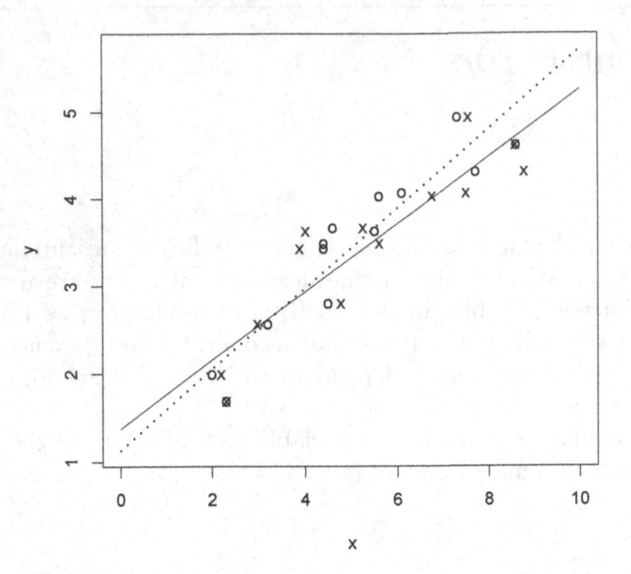

FIGURE 13.1 Circular points show observations where the x-variable is measured without error. Dashed line shows the corresponding estimated regression line. Crossed points show observations with the same y-values, but random measurement errors have been added to the x-values. The solid line shows the estimated regression line. Measurement errors decrease the estimated slope and increase the estimated intercept.

although in its absence the Best Linear Predictor (BLP) that is used is not necessarily conditionally unbiased for each value of x. According to Equation (10.2) (p. 309) and Equation (10.3) (p. 310), Equation (13.2) can be stated as:

$$y = \mu_y + \frac{\text{cov}\,(x, y)}{\text{var}\,(x)}\,(y - \mu_x)\,. \tag{13.3}$$

That is, in Equation (13.2)

$$\beta_2 = \frac{\text{cov}\,(x, y)}{\text{var}\,(x)}\,, \tag{13.4}$$

$$\beta_1 = \mu_y - \beta_2\mu_x\,. \tag{13.5}$$

However, variable x cannot be observed, only variable X, which contains the random measurement error:

$$X = x + u\,,$$

where u is assumed to be normally distributed with mean zero and is independent of x and e. This case is often called the *classical measurement error* case. Let us assume that we are estimating β_1 and β_2 from the following:

$$y = \beta_1 + \beta_2 X + e\,,$$

and β_2 is estimated with OLS, which in our case is equivalent to estimating β_2 with

$$\widehat{\beta}_2 = \frac{\text{cov}\,(y, X)}{\text{var}\,(X)}\,, \tag{13.6}$$

where $\text{cov}(y, X)$ and $\text{var}(X)$ are estimated using the sample covariance between X and y and using the sample variance of X (see Section 10.1.1). The estimation errors of variances and covariances are ignored in the following discussion. Now,

$$\text{cov}(y, X) = \text{cov}(y, x + u) = \text{cov}(y, x) + \text{cov}(\beta_1 + \beta_2 x + e, u) = \text{cov}(y, x) \, .$$

Thus, the sample covariance of y and X is unbiased for $\text{cov}(y, x)$. But

$$\text{var}(X) = \text{var}(x + u) = \text{var}(x) + \text{var}(u) \, . \tag{13.7}$$

Thus, the sample variance of X is upwardly biased as an estimator of $\text{var}(x)$. Altogether, the Equation (13.6) is a biased estimate for Equation (13.4), its absolute value is too small. As measurement errors spread the distribution of x, they lower the apparent slope (Figure 13.1).

The expected values of μ_x and μ_y in Equation (13.5) can be unbiasedly estimated from the sample, but as β_1 also depends on β_2 in Equation (13.5), the estimate of the intercept is also biased.

If we have an unbiased estimate for the measurement error variance, then according to Equation (13.7), the true variance of x can be unbiasedly estimated from:

$$\text{var}(x) = \text{var}(X) - \text{var}(u) \, .$$

If we have made several independent measurements for N subjects, then an unbiased estimate for the measurement error variance is:

$$\widehat{\text{var}}(u) = \frac{\sum_{i=1}^{N} \sum_{j=1}^{n_i} \left(X_{ij} - \bar{X}_{i.}\right)^2}{\sum_{i}^{N} n_i - N} \, , \tag{13.8}$$

where n_i is the number of measurements for subject i and $\bar{X}_{i.}$ is the average for subject i. Alternatively, $\text{var}(u)$ can be estimated using a working mixed model $X_{ij} = \mu + b_i + u_{ij}$.

Rounding errors are a special type of measurement error whose variance can be solved analytically. If d is the smallest detectable difference between the measured values and the true values vary randomly (uniformly) within each interval, the variance of the measurement errors is $d/12$. If $d = 1$, the sd is thus 0.2887. The measurement errors are independent of the true values. At first, this may sound surprising, as the true value determines the measurement error. Demonstrate this with a simulation!

Note that the measurement error bias needs to be corrected only if we are interested in the true relationship between x and y. If we are interested in obtaining predictions of y from the measured value X, and we assume that the measurement error variance is the same as in the estimation data, then we can just estimate the regression between y and X in the usual manner and use it for prediction.

The generalization of the above analysis for multiple regression, and for general BLP, is straightforward. In the case of several predictors, we are estimating the linear model (see Equation (3.19), p. 60 or Equation (10.1), p. 309):

$$y = \mu_y + \text{cov}(y, \boldsymbol{x}') \, \text{var}(\boldsymbol{x})^{-1} (\boldsymbol{x} - \boldsymbol{\mu_x}) \, ,$$

where $\text{var}(\boldsymbol{x})$ can be estimated using $\text{var}(\boldsymbol{x}) = \text{var}(\boldsymbol{X}) - \text{var}(\boldsymbol{u})$. If there are several independent measurements for N subjects, the variances in $\text{var}(\boldsymbol{u})$ can be estimated using Equation (13.8) and the covariances by:

$$\widehat{\text{cov}}(u_1, u_2) = \sum_{i=1}^{N} \sum_{j=1}^{n_i} \left(X_{1ij} - \bar{X}_{1i.}\right) \left(X_{2ij} - \bar{X}_{2i.}\right) / \left(\sum_{i}^{N} n_i - N \right) .$$

Let us then show, in a different way, why ordinary linear regression is biased when the measurement error of the regressor is independent of the true value. If $X = x + u$, then $x = X - u$ where $-u$ still is measurement error with mean zero. Equation (13.2) can be stated as:

$$y = \beta_1 + \beta_2 (X - u) + e = \beta_1 + \beta_2 X + (-\beta_2 u + e) \,,$$

where the error term $-\beta_2 u + e$ has mean zero. Thus, the equation seems to be appropriate for estimating parameters of Equation (13.2). However, the error term is correlated with the predictor X:

$$\text{cov}(-\beta_2 u + e, X) = \text{cov}(-\beta_2 u + e, x + u) = -\beta_2 \, \text{var}(u) \,.$$

Ordinary linear regression is unbiased for the regression coefficients only if the error term is uncorrelated with the regressor variables.

13.1.2 Berkson case: measurement errors independent of the measured value

The effect of measurement error is different, if the measurement error arises so that, in an experiment, the level of x is set, according to some measurement instrument, to a fixed level X, which is recorded as the measurement, and the true value x varies randomly around this measured value. Thus, let us assume that

$$\begin{aligned} y &= \beta_1 + \beta_2 x + e \,, \\ x &= X + u \,, \end{aligned}$$

where u is independent of X. Thus:

$$y = \beta_1 + \beta_2 (X + u) + e = \beta_1 + \beta_2 X + (\beta_2 u + e) \,,$$

where the error term $\beta_2 u + e$ is independent of X. Thus, β_2 can be unbiasedly estimated with ordinary linear regression using the erroneous regressor.

It is illuminating to derive the same result using Equation (13.6). Note first that $\text{var}(X) = \text{var}(x - u) = \text{var}(x) + \text{var}(u) - 2\,\text{cov}(x, u) = \text{var}(x) - \text{var}(u)$ and $\text{cov}(y, X) = \text{cov}(\beta_2 x, x - u) = \beta_2 (\text{var}(x) - \text{var}(u))$. Thus:

$$\widehat{\beta_2} = \frac{\text{cov}(y, X)}{\text{var}(X)} = \frac{\beta_2 (\text{var}(x) - \text{var}(u))}{\text{var}(x) - \text{var}(u)} = \beta_2 \,.$$

In forestry applications, there are several cases where we may assume that the measurement error behaves more like the Berkson case than the classical case. For instance, if forest variables are measured with ocular estimation, then it is probable that the measurer weights the apparent value toward the average value, which he/she thinks to be more likely. If a variable is measured using the predicted value obtained from a regression equation that is an estimator of $\text{E}(y|\boldsymbol{x})$, then the true value varies randomly around the predicted value, and the Berkson case applies. If the k nearest neighbor (k-nn) method is used to 'measure' values of variables from a satellite image, then the measured values vary less than the true values, and thus the measurement errors behave more like a Berkson case than a classical case (see Example 13.1, p. 388).

13.1.3 Measurement error correlated both with true and measured value

Let us then consider cases that fall between the classical and Berkson cases. Let us start as in the classical case:

$$y = \beta_1 + \beta_2 x + e \,,$$

$$X = x + u\,,$$

but now we do not assume that u and x are uncorrelated. Also, u and X are assumed to be correlated (so we are not in the Berkson case either). We are still interested in estimating the parameters of Equation (13.2), i.e. $\beta_1 + \beta_2 x + e$ which can be expressed as in Equation (13.3). Now:

$$\mathrm{cov}\,(X, y) = \mathrm{cov}\,(x + u, y) = \mathrm{cov}\,(x, y) + \mathrm{cov}\,(u, \beta_2 x) = \mathrm{cov}\,(x, y) + \beta_2\,\mathrm{cov}\,(u, x)\,,$$

$$\mathrm{var}\,(X) = \mathrm{var}\,(x + u) = \mathrm{var}\,(x) + \mathrm{var}\,(u) + 2\,\mathrm{cov}\,(x, u)\,.$$

From these we can derive that

$$\beta_2 = \frac{\mathrm{cov}\,(x, y)}{\mathrm{var}\,(x)} = \frac{\mathrm{cov}\,(X, y) - \beta_2\,\mathrm{cov}\,(u, x)}{\mathrm{var}\,(X) - \mathrm{var}\,(u) - 2\,\mathrm{cov}\,(x, u)}\,.$$

Solving this for β_2 we get:

$$\beta_2 = \frac{\mathrm{cov}\,(X, y)}{\mathrm{var}\,(X) - \mathrm{var}\,(u) - \mathrm{cov}\,(x, u)}\,.$$

Note that this equation gives us the classical and Berkson measurement errors as special cases (show this!).

A forestry example of a case where the measurement error is correlated both with the true and measured values was analyzed by Lappi (2005). It was assumed that tree growth is influenced by the number of trees around the tree, and the local density is measured by the density of the plot used to select sample trees or by a larger density plot. If the density of the entire stand influences the growth, and the stand density is measured with a plot density, the measured value varies randomly around the true value. This is the classical measurement error case. If the local density is influencing the growth and the density is measured from a large plot, then the true density varies randomly around the measured density. This is the Berkson case. Generally, the situation falls between the classical and Berkson cases. The analysis is rather complicated, even with simple assumptions. An interesting result in Lappi (2005) is that the plot size that produces the maximum correlation between density and the dependent variable is not equal to the size of the influence zone, even though it is close to it. The plot size that produces the maximum correlation can be far from optimal with respect to the bias in estimating the density effect. It is better to use larger density plots: the larger the better.

Kangas *et al.* (2003) studied the accuracy of visually assessed stand characteristics. They found that the measurement errors for some variables behaved more like classical measurement errors, for other variables the measurement errors were the Berkson type, and the remaining variables fell between the classical and Berkson type.

13.1.4 Measurement errors of the dependent variable

Let us then consider the case where there are measurement errors in the dependent variable. The measurement error may be correlated with the true value, with the measured value or with the value of the independent variable. Let

$$Y = y + u$$

denote the measured value of the dependent variable, and let us assume that $\mathrm{E}\,(u) = 0$. The slope obtained by ignoring the measurement error is

$$\widehat{\beta}_2 = \frac{\mathrm{cov}\,(Y, x)}{\mathrm{var}\,(x)} = \frac{\mathrm{cov}\,(y + u, x)}{\mathrm{var}\,(x)} = \frac{\mathrm{cov}\,(y, x)}{\mathrm{var}\,(x)} + \frac{\mathrm{cov}\,(u, x)}{\mathrm{var}\,(x)} = \beta_2 + \frac{\mathrm{cov}\,(u, x)}{\mathrm{var}\,(x)}\,.$$

Thus, only the correlation of u with x causes a bias, and the bias can be easily corrected if the correlation is known. Note that it is not required for unbiasedness that $\text{cov}(u, e) = 0$, even if this is often assumed when considering measurement errors of y.

If the dependent variable is not measured directly but is predicted with a regression equation, then the implied measurement error may or may not be correlated with the predictor variables.

We can calibrate mixed models for a given group by making measurements for the y-variable. In the mixed model equations, the measurement errors of the y-variable are part of the residual error term ϵ. If there are measurement errors in the calibration data, which were not present in the model estimation data, the measurement error variance (if known) can be taken into account by adding the measurement error variance to the residual error variance. Lappi (1986) and Lappi (2006) took measurement errors into account in his taper curve models. When there are measurement errors in the measured diameters or height, then the predicted taper curves do not pass through the measurements.

Let us then make a sidestep and briefly consider a bias problem that is not caused by measurement errors but from the fact that measurements of y are from a 'wrong' distribution. Let us assume that we measure diameter increments from sample trees selected with the relascope. If we regress increments on the initial diameters, we get a biased model, because with trees of the same initial diameter, the faster growing trees have a greater probability to be included in the data. The distribution of the measured increments is a weighted distribution of the original distribution (see Sections 2.8 and 6.4). Lappi and Bailey (1987) have analyzed this problem and suggest methods for correcting the bias. One solution is to weight observations proportionally to the initial basal area and inversely to the current basal area. If the data set is not very small, a practical solution is also to simulate the relascope sampling in the initial time point.

13.1.5 Measurement errors in nonlinear models

Measurement errors naturally cause bias problems in nonlinear models as well. With respect to nonlinear models, readers are referred to the monograph by Carroll *et al.* (2006). As an intuitively appealing example how measurement errors can be handled in nonlinear models, we briefly present the simulation-extrapolation (SIMEX) procedure.

The SIMEX procedure consists of two stages: simulation and extrapolation. In the simulation stage, extra errors are added to the independent variables and the parameters are estimated using the new independent variables. Then, the dependency of the parameter estimates on the scale of the measurement error is approximately estimated, for example, with a quadratic function. The value of this function is then extrapolated for the case where there is no measurement error.

Let us consider only one parameter, β, and one independent variable, x. Let $\widehat{\beta}\left(\sigma_u^2\right)$ be the estimate obtained using the original data where the measurement error variance of x is assumed to be σ_u^2. Then, extra random errors with variance v are added to x and β is estimated again. Let $V = v + \sigma_u^2$. Using different values of v, we can formulate an approximative function G so that $\widehat{\beta}(V) \approx G(V) = G\left(v + \sigma_u^2\right)$. A quadratic function is typically used. Extrapolating this function to $G(0)$ gives us an estimate of β that is corrected with respect to the measurement errors.

A forestry application of SIMEX is described in Kangas (1998). The `R` package `simex` provides an algorithm for SIMEX estimation.

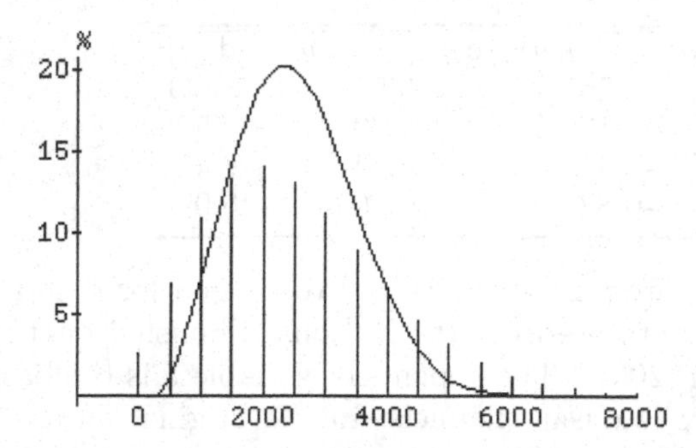

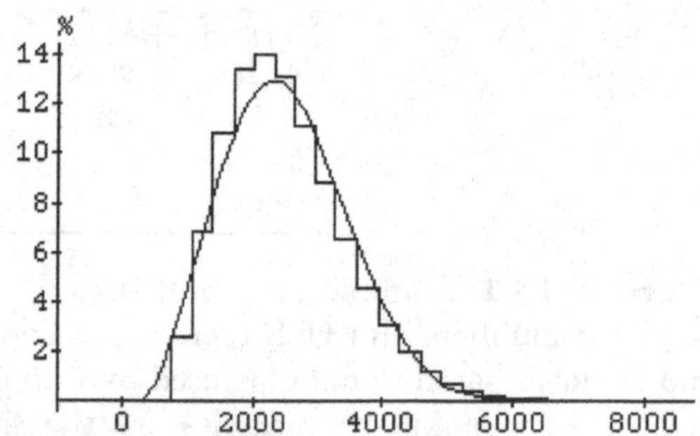

FIGURE 13.2 Continuous curves show an assumed distribution for stand densities. Tree locations are assumed to be random (Poisson process). The standard deviation of the densities is 949 trees/ha. Stand densities are measured (estimated) from 20 m^2 plots. Bars on the left graph show the estimated distribution with an sd of 1470 trees/ha. The discrete distribution on the left graph is the estimated distribution. The histogram in the right is the distribution that is estimated so that the estimated distribution has the same variance as the true distribution (Lappi 1991b).

13.2 Measurement errors in estimating distributions

In forestry, we are often interested in the properties of a distribution other than the mean of the distribution. For instance, in a forest inventory we may be interested in the proportion of stands where a stand variable is smaller or larger than a given limit. If a variable is measured with an error, then distribution of measured values may have larger or smaller variance than the distribution of the true values.

13.2.1 Measurement errors are independent of the true value

If a random variable x is measured with an error that has a zero expected value, the measured variable X has the same expected value as x. If:

$$X = x + u \, ,$$

where u is not correlated with x, then:

$$\text{var}\,(X) = \text{var}\,(x) + \text{var}\,(u) \; .$$

Thus, the distribution of X is wider than the distribution of x. In addition, the distribution of the measured values can be unimodal, even if the true distribution is bimodal. If we try to infer the properties of the distribution of x using measured values of X, the measurement errors should be considered. Sampling errors behave analogously to ordinary measurement errors, for instance, if a sample with size n_i is taken from group (e.g. stand) i and the group average is measured with sample mean $\bar{x}$. If the within group variance is σ^2, then the measurement error variance is σ^2/n_i.

Lappi (1991b) analyzed a case where the densities of plantations were studied by counting the number of seedlings from a plot with area A_i. If the true density in plantation i is λ_i, the density is measured by n_i/A_i, and the seedling locations are random, then the measurement error variance in plantation i is λ_i/A_i. Recall Example 8.6 (p. 266). The average measurement error variance is λ/A_i where λ is the average density in the entire population. The distribution of the measured densities is discrete and wider than the true

Method	RMSE	$\text{cor}(\tilde{y}-y, y)$	$\text{cor}(\tilde{y}-y, \tilde{y})$	$\text{sd}(\tilde{y})$
k=1	12.18	-0.56	0.57	10.79
k=5	9.36	-0.73	0.20	7.52
k=21	8.88	-0.80	0.04	6.45
regr	8.89	-0.83	0.00	5.09

TABLE 13.1 Example 13.1. Statistics of nearest neighbor (k-nn) regression for different neighbor numbers and OLS regression. The data are from the National Finnish Inventory and Landsat satellite data, as used by Lappi (2001). The dependent variable y is the basal area and the auxiliary variables are the six Landsat channels. In the original data set, $\text{sd}(y) = 10.24$.

distribution (Figure 13.2). Lappi (1991b) suggested that the distribution is estimated so that the variance of predicted densities is the same as the variance of true densities.

More formally, the distribution of the measured variable X is the *convolution* of the distributions of x and u (recall Section 2.10):

$$f_X(X) = \int_{-\infty}^{\infty} f_x(X-u) f_u(u)\, du = \int_{-\infty}^{\infty} f_x(x) f_u(X-x)\, dx$$

A method for obtaining nonparametric estimates for the distribution of x is called *deconvolution*. The method is used in the `R` package `decon` (Wang and Wang 2011), which can also be used for nonparametric regression when the independent variable is measured with error.

13.2.2 Measurement errors correlated with the true value

If the true values vary randomly around the measured value, i.e. the measurement errors are of the Berkson type, the measured values have smaller variance than the true values. Specifically the variance of the measured values is equal to the variance of true values minus the variance of the measurement errors. This is the case, for example, when a variable is predicted using a regression equation. This is a common practice nowadays, for example, in aerial forest inventories. The estimation errors of the regression parameters ensure that the prediction errors do not have exactly the same distribution as the true residual errors. Let us assume that the parameter estimates are so good that their estimation errors can be ignored from the current discussion. Then, the measurement error variance is approximately equal to the residual error variance when a regression equation is used to 'measure' a variable.

Example 13.1. K-nn regression is a popular method for predicting forest variables in forest inventories using satellite or radar data. One motivation for the method is that the different predicted variables have variation and covariation which is similar to the variation and covariation of the predicted variable y. However, the goals for obtaining highly realistic distributions and associated small RMSE values are in conflict: minimizing RMSE leads to a narrow distribution (Meng *et al.* 2007). Table 13.1 and Figure 13.3 show how OLS regression and k-nn regression with 21 neighbors produced the smallest RMSE values while k-nn regression with one neighbor produced similar variance for the predicted values as the variance of y.

With one neighbor, i.e. with $k = 1$, $\text{cor}(\tilde{y}-y, \tilde{y})$ is large, and it then decreases to zero when k increases, i.e. the 'measurement error' approaches the Berkson case. When $\text{cor}(\tilde{y}-y, \tilde{y})$ is non-zero, the prediction error variance can be decreased using $\tilde{y}$, obtained by k-nn method, as a predictor in OLS regression. With all values of k, $\text{cor}(\tilde{y}-y, y)$ has a large absolute value. Thus, we are very far from the classical measurement error case, where the measurement error

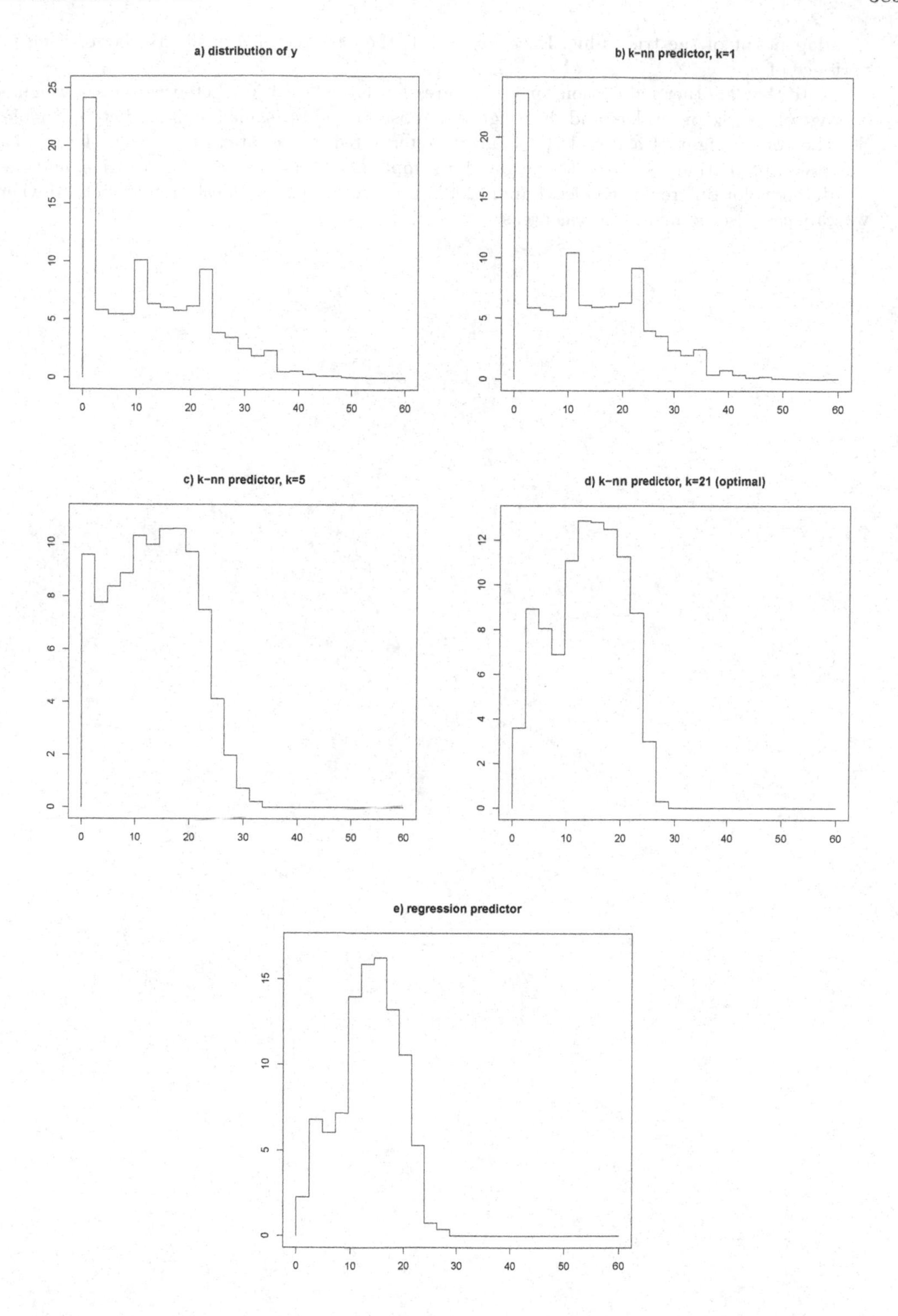

FIGURE 13.3 Example 13.1. The distribution of y (graph a), distribution of predicted values in nearest neighbor (k-nn) estimation for different values of k (graphs b-d), and the distribution of values obtained from linear regression (graph e). See Table 13.1 for explanation of variables.

is independent of the true value. Even for $k = 1$, the variance of $\tilde{y}$ is slightly larger than the variance of y.

Note that another motivation for k-nn regression is that each predicted value is a weighted average of sample plot values and the weights are positive and the same for all predicted variables. So, the total weight of a sample plot can be interpreted as the area of a similar forest. The regression prediction also provides weights for sample plots, but they can be negative, and they are different for different variables. Lappi (2001) provides a small-area estimation method where weights can also be interpreted as areas.

14

Forest and Environmental Experiments

In this book we have concentrated on modeling forestry and other environmental data. Data can be either *observational* or *experimental.* Observational data are obtained by sampling existing populations. Experimental data are obtained by first sampling *experimental units* or *treatment units* from the target population and then applying different *treatments* to different experimental units. Some response variables are then measured from the experimental units or from their subunits. Then the effects of different treatments are compared using a statistical model. We do not cover *experimental design* in detail here, but we present some comments related to the other parts of this book.

Information from experimental data is always obtained using a statistical model. Unfortunately, researchers often start to consider modeling after collection of the data. The design of an experiment should be in agreement with the intended model. Often, the final estimated model may not be exactly the same as the intended model, but such effects can only be estimated that are supported by the experimental design.

14.1 The target population

When designing an experiment, one should clearly define what is the population of interest and how the experimental findings will be extrapolated to the target population. This is often the weakest and most problematic issue in forest and environmental experiments. For practical reasons, it is not usually possible to randomly sample stands from the intended target population. As such, it requires strong assumptions to extrapolate the findings from a few experimental stands to the target population. Nonetheless, these assumptions should be explicitly made and discussed. The target population could be defined in terms of geographic area, soil type, soil fertility, elevation, tree species, age and time. In particular, time is often overlooked. For example, weather conditions vary between different years. There are long-term trends in climate. It is often difficult to separate time and age effects, especially if treatments are only established in one year and treatment units are only followed over a relatively short timeframe. Thus, it is often important to replicate over years. In addition, it is difficult to separate the effects of a climate trend and age in long-term experiments.

14.2 Design and analysis of experiments

In experiments we are interested in the effects of treatments on an output variable(s). Treatments are defined by setting different *levels* to independent variables called *factors.* Factors can be quantitative or qualitative. The *design matrix* describes how the different levels of factors are assigned to experimental units. Many people call the matrix, which contains the independent variables of a linear model used to analyze the results of the experiment, a design matrix. We call it a model matrix. A design matrix, augmented with a column of ones, is equal to the model matrix in cases where the treatment factors are quantitative, and where an ordinary linear regression model is used to analyze the data. Otherwise, the model matrix contains dummy variables used to model the effects of qualitative treatment factors and their interactions, and possible quadratic terms to model nonlinear effects, etc.

Binkley (2008) discussed, from the viewpoint of a journal editor, what are the three key points in forestry experiments. The first point is a clear definition of the population of interest, which was already discussed above. The second point was 'Based on the population of interest, be clear on what is a true replicate, and what is a subsample'. This was discussed already in Box 5.1 (p. 132). Nowadays, mixed models are natural models for the simultaneous analyses of replicates (treatment units) and subsamples (measurement units). The efforts invested in more true replicates are (almost) always more profitable than the same effort applied to obtaining more subsamples, which contribute no degrees of freedom for extrapolation to the population.

The third point was 'Analyze quantitative treatments quantitatively and not as separate categories'. We will now briefly discuss this issue. The purpose of forestry experiments is to study the regularities of reactions of forests or trees to experimental treatments. We should attempt to obtain a simple model that is consistent with the data and has as much predictive power as possible. The simplicity of a model can be measured by the number of parameters. The consistency with data can be tested formally only in special cases, but there are graphical methods, for example, that can be used to assess the consistency with the data, as we have illustrated above. A model can have predictive power both within the estimation data set and outside the estimation data set.

If a quantitative (continuous) treatment factor is analyzed as a qualitative (categorical) factor, i.e. each level of the factor is assumed to have its own parameter (recall Section 4.2.4), then there are usually more parameters than in a case where linear or quadratic relationship is assumed. Furthermore, with qualitative factors, extrapolation is not possible outside the levels used in the experiment. Interpolation between the levels can, in practice, be done, even if the assumed model in principle does not even support interpolation. Nonetheless, a model estimated with qualitative factors can provide the full model reference when testing whether the assumed quantitative model form is valid (p. 115). Too often, researchers are satisfied with significant p-values and do not aim at meaningful modeling of the data.

When designing an experiment for estimating quantitative models, we can take what we have learned about the variances of the estimated parameters into account. More specifically, let us express an OLS regression equation in the form $y_i = \beta_0 + \boldsymbol{x}_i'\boldsymbol{\beta} + e_i$ where the intercept is separated from other parameters. Then, $\widehat{\boldsymbol{\beta}} = (\boldsymbol{X}_c'\boldsymbol{X}_c)^{-1}\boldsymbol{X}_c'\boldsymbol{y}_c$, where $\boldsymbol{X}_c$ is the centered model matrix given in Equation (10.8) (p. 310) and $\boldsymbol{y}_c$ is the centered y-vector given in Equation (10.7). Then:

$$\mathrm{var}(\widehat{\boldsymbol{\beta}}) = \sigma^2 (\boldsymbol{X}_c'\boldsymbol{X}_c)^{-1} .$$

The variance of $\widehat{\boldsymbol{\beta}}$ can be made smaller by increasing the diagonal of $\boldsymbol{X}_c'\boldsymbol{X}_c$ and by making the non-diagonal elements of $\boldsymbol{X}_c'\boldsymbol{X}_c$ zero. This implies two useful principles for the design of experiments: 1) spread out the fixed treatment x-variables, and 2) make the columns of $\boldsymbol{X}_c$ orthogonal (uncorrelated). The first principle means that if error variance is constant for all y_i, it is optimal to choose $x_i = \min(x)$ for half of the data and $x_i = \max(x)$ for the other half. However, the principle needs to be applied rationally. From a mathematical point of view, it is true, provided we know that the assumed linear relation between y and x-variables holds. But in fact, we know that it holds only for some range of x-variables. A reasonable range must be evaluated by the experimenter. Nevertheless, it is a good starting principle, e.g. when calibrating a linear mixed model with a random intercept and slope.

If we know that the relation between y and an x-variable is linear, then adding design points, where x is equal to the average of x, does not provide useful information for estimating the coefficient of the variable. But if we add such center points to the design, we can test in the modeling stage, whether the assumption of linearity really holds. The testing can be carried out by adding a quadratic term into the model and then testing whether its coefficient deviates from zero. If an x-variable has only three different values in the data, then the lack of fit test of the linear model (discussed in Section 4.6.2, p. 115) is equivalent to the test of the coefficient of the quadratic term. With three x-values, the quadratic function describes the data at least as well as any other nonlinear relationship.

14.3 Nuisance factors, blocks and split-plots

The definition of treatments is often problematic owing to nuisance factors that affect the results of the experiments. These nuisance factors can be related e.g. to the following variation:

1. The same forest nursery practice will produce different kinds of seedling material in different years and in different nurseries.
2. The treatment applications cannot often be defined very precisely, so different persons apply treatments differently depending on their expertise, for example.
3. Different machines produce different results, e.g. in harvesting and soil preparation treatments.
4. The treatments affect a larger area than just the treatment plot, and the surroundings of the treatment plot also affects the result in a given experimental plot. Thus, treatment plots often need *buffer zones.*

In addition, the measurement errors (including bias) of the measured variables may vary from measurer to measurer, especially when variables are measured using ocular estimates.

Nuisance factors related to the variation among experimental units are usually controlled using *blocking.* In blocking, treatment units are grouped so that units within blocks are more homogeneous than units randomly selected from the population. For example, in laboratory experiments, growth chambers inevitably form natural blocks. The purpose in blocking is to reduce the residual error variance and, thus, to make tests more powerful.

In an analysis of blocking experiments, a basic question is whether block effects should be treated as fixed or random effects. As mixed model analysis has become more popular in recent decades, block effects are now usually treated as random effects. But, as discussed by Dixon (2016), there are often good arguments for defining block effects as fixed.

Can blocks be interpreted as a random sample from a target population? In most forest field experiments where forest stands form one level of blocking, this may not be the case. In growth chamber experiments, however, this may be a more valid assumption. An argument for favoring random block effects is that one can generalize outside the blocks present in the study. This argument has two weaknesses. First, if the blocks are not a random sample from a specific population, there is no specific population to which the results can be generalized. Second, the purpose in experiments is usually to estimate the differences between treatment means, not the treatment means in a population. The differences between treatment means can be equally well generalized using fixed block effects. Furthermore, with random block effects, plotting estimated treatment means with standard error bars may give a misleading image of treatment differences, because standard errors also contain the influence of the block effect variance. It should also be noted that the variance of the block effects can only be poorly estimated with very few blocks.

As Dixon (2016) points out, a problem with random effect models is that standard statistical software may produce erroneous ML or REML estimates even for simple variance component models owing to the multi-modality of the likelihood surface.

Since the time of Fisher, agricultural experiments have driven the theory of experimental design much more than forest experiments. Forest experiments have often special features as compared to agricultural experiments.

1. Forest experiments often have a longer time span. There are several dependent variables of interest and the entire development path of a dependent variable may be of interest.

2. Forest experiments are often more difficult to apply in a standard way, i.e. there are many nuisance factors in the experiments.

3. Forests are often more heterogeneous than agricultural fields.

Both agricultural and forest experiments may have an advantage of *split-plot* type of designs. For instance, soil preparation methods may form the main plot treatment factor, because it is not practical to use machines on a very small plot. The planting method may form a subplot treatment factor, because it is more practical to use different planting methods on smaller plots. Seedlings are then measurement units with their own level of variation. Nowadays, split-plot experiments can be easily analyzed using mixed models by including both main plot effects and subplot effect as random effects. Such an analysis automatically considers that the subplot observations are not independent replicates when analyzing the main plot treatment effects.

14.4 Too few replicates

Forestry and environmental experiments do not always allow appropriate statistical comparisons of treatments, especially in experiments that were established a long time ago. Then, it is better to speak about demonstration plots rather than experiments. Demonstration plots can, however, be important in showing what can happen with special treatments. Moreover, with one observation (i.e. without replications), we can obtain point estimates, albeit without the possibility of testing and constructing confidence intervals for the parameters. Statistical inference is not the only way that we can obtain information from nature. In the history of science, big leaps have also been made based on a single observation.

If we have a well-designed experiment in a single forest stand in a certain area, then we can make statistically valid inferences about differences in treatment effects for that stand. Generalizations to other stands are based on professional judgment. In this respect, demonstration plots and experiments are basically similar. Authors should clearly describe the approach used for leaping from experimental results to the population of interest.

Bibliography

Amrhein, V., Greenland, S., and McShane, B., 2019. Scientists rise up against statistical significance, *Nature*, 567, 305–307.

Arias-Rodil, M., Castedo-Dorado, F., Cámara-Obregón, A., and Diéguez-Aranda, U., 2015. Fitting and calibrating a multilevel mixed-effects stem taper model for Maritime pine in NW Spain, *PLOS ONE*, 10, 1–15.

Aucoin, F., 2015. *FAdist: Distributions that are Sometimes Used in Hydrology*, R package version 2.2.

Baddeley, A. and Turner, R., 2005. spatstat: An R package for analyzing spatial point patterns, *Journal of Statistical Software*, 12 (6), 1–42.

Bailey, R. and Dell, T., 1973. Quantifying diameter distributions with the Weibull function, *Forest Science*, 19, 97–104.

Bates, D., Mächler, M., Bolker, B., and Walker, S., 2015. Fitting linear mixed-effects models using lme4, *Journal of Statistical Software*, 67 (1), 1–48.

Bates, D.M. and Watts, D.G., 2007. *Nonlinear Regression Analysis and Its Applications*, New York, USA: Wiley-Interscience, 365 p.

Bell, A., Fairbrother, M., and Jones, K., 2019. Fixed and random effects models: making an informed choice, *Quality & Quantity*, 53, 1051–1074.

Berger, J.O., 2003. Could Fisher, Jeffreys and Neyman have agreed on testing?, *Statistical Science*, 18 (1), 1–32.

Berger, J.O. and Sellke, T., 1987. Testing a point null hypothesis: the irreconcilability of p values and evidence, *Journal of the American Statistical Association*, 82 (397), 112–122.

Binkley, D., 2008. Three key points in the design of forest experiments, *Forest Ecology and Management*, 255 (7), 2022–2023.

Binkley, J. and Abbot, P., 1987. The fixed X assumption in econometrics: can the textbooks be trusted?, *The American Statistician*, 41 (3), 206–214.

Bitterlich, W., 1949. Die winkelzählprobe, *Allgemeine Forst- und Holzwirtschaftliche Zeitung*, 59 (1/2), 4–5.

Bivand, R.S., Pebesma, E.J., and Gómez-Rubio, V., 2008. *Applied Spatial Data Analysis with R*, New York, USA: Springer, 405 p.

Bliss, C.I., 1934. The method of probits, *Science*, 79 (2037), 38–39.

Boos, D.D. and Stefanski, L.A., 2011. P-value precision and reproducibility, *The American Statistician*, 65 (4), 213–221.

Borders, B.E., Souter, R.A., Bailey, R.L., and Ware, K.D., 1987. Percentile-based distributions characterize forest stand tables, *Forest Science*, 36 (2), 570–576.

Box, G.E.P., Jenkins, G.M., Reinsel, G.C., and Ljung, G.M., 2015. *Time Series Analysis, Forecasting and Control*, New York: Wiley, 5th ed., 712 p.

Bretz, F., Hothorn, T., and Westfall, P., 2011. *Multiple Comparisons Using R*, Boca Raton, USA: CRC Press, 1st ed., 182 p.

Burk, T.E. and Newberry, J.D., 1984. A simple algorithm for moment-based recovery of Weibull distribution parameters, *Forest Science*, 30 (2), 329–332.

Burkhart, H.E. and Tomé, M., 2012. *Modeling Forest Trees and Stands*, Dordrecht, Netherlands: Springer, 1st ed., 457 p.

Burkhart, H.E. and Tome, M., 2012. *Modeling Forest Trees and Stands*, Springer, 472 p.

Cajanus, W., 1914. Entwicklung gleichaltiger Waldbestände. Eine statistische Studie I. a, *Acta Forestalia Fennica*, 3.

Calama, R. and Montero, G., 2006. Stand and tree-level variability on stem form and tree volume in Pinus pinea L.: A multilevel random component approach, *Invest. Agrar.: Sist. Recur. For.*, 15 (1), 24–41.

Cao, Q.V., 2004. Predicting parameters of a Weibull function for modeling diameter distribution, *Forest Science*, 50 (5), 682–685.

Cao, Q.V. and Baldwin, V., 1999. A new algorithm for stand table projection models, *Forest Science*, 45 (4), 506 – 511.

Carroll, R., Ruppert, D., Stefanski, L., and Crainiceanu, C., 2006. *Measurement Error in Nonlinear Models: a Modern Perspective, Second edition*, Boca Raton: Chapman & Hall/CRC, 447 p.

Casella, G. and Berger, R.L., 2001. *Statistical Inference*, Pacific Grove, USA: Duxbury, 2nd ed., 660 p.

Christensen, R., 2003. Significantly insignificant F tests, *The American Statistician*, 57 (1), 1–6.

Christensen, R., 2005. Testing Fisher, Neyman, Pearson, and Bayes, *The American Statistician*, 59 (2), 121–126.

Christensen, R., 2011. *Plane Answers to Complex Questions. The Theory of Linear Models*, New York: Springer, 4th ed., 494 p.

Clutter, J.L. and Bennett, F.A., 1965. Diameter distribution in old-field slash pine plantation, Report 13, Georgia Forest Research Council, 9p.

Collett, D., 1991. *Modelling Binary Data*, London, UK: Chapman & Hall, 369 p.

Collett, D., 2014. *Modelling Survival Data in Medical Research*, Boca Raton, USA: Chapman and Hall/CRC, 548 p.

Comets, E., Lavenu, A., and Lavielle, M., 2017. Parameter estimation in nonlinear mixed effect models using saemix, an R implementation of the saem algorithm, *Journal of Statistical Software, Articles*, 80 (3), 1–41.

Cressie, N.A., 1993. *Statistics for Spatial Data*, New York, USA: Wiley, 900 p.

Davidian, M. and Giltinan, D.M., 1995. *Nonlinear Models for Repeated Measurement Data*, Boca Raton, USA: Chapman & Hall/CRC Monographs on Statistics & Applied Probability, 360 p.

de Boor, C., 1978. *A Practical Guide to Splines*, New York: Chapman & Hall, 392 p.

de Miguel, S., Mehtätalo, L., Shater, Z., Kraid, B., and Pukkala, T., 2012. Evaluating marginal and conditional predictions of taper models in the absence of calibration data, *Canadian Journal of Forest Research*, 42 (7), 1383–1394.

de Souza Vismara, E., Mehtätalo, L., and Batista, J.L.F., 2016. Linear mixed-effects models and calibration applied to volume models in two rotations of eucalyptus grandis plantations, *Canadian Journal of Forest Research*, 46 (1), 132–141.

del Pino, G., 1989. The unifying role of iterative generalized least squares in statistical algorithms, *Statist. Sci.*, 4 (4), 394–403.

Demidenko, E., 2013. *Mixed Models. Theory and Applications with R.*, New York, USA: Wiley, 2nd ed., 717 p.

Diggle, P.J. and Ribeiro, P.J.J., 2007. *Model-Based Geostatistics*, New York, USA: Springer, 232 p.

Dixon, B. and Howitt, R., 1979. Continuous forest inventory using a linear filter, *Forest Science*, 25 (4), 675–689.

Dixon, P., 2016. Should blocks be fixed or random?, *Conference on Applied Statistics in Agriculture*, 39 p.

Droessler, T.D. and Burk, T.E., 1989. A test of nonparametric smoothing of diameter distributions, *Scandinavian Journal of Forest Research*, 4, 407–415.

Dubey, S.D., 1967. Some percentile estimators for Weibull parameters, *Technometrics*, 9 (1), 119–129.

Eerikäinen, K., 2009. A multivariate linear mixed-effects model for the generalization of sample tree heights and crown ratios in the Finnish national forest inventory, *Forest Science*, 55 (6), 480–493.

Eisenhauer, J.G., 2003. Regression through the origin, *Teaching Statistics*, 25 (3), 76–80.

Ek, A.R., Issos, J.N., and Bailey, R.L., 1975. Solving for Weibull diameter distribution parameters to obtain specified mean diameters, *Forest Science*, 21 (3), 290–292.

Engeman, R.M. and Keefe, T.J., 1982. On generalized least squares estimation of the Weibull distribution, *Communications in Statistics - Theory and Methods*, 11 (19), 2181–2193.

Everitt, B. and Hothorn, T., 2011. *An Introduction to Applied Multivariate Analysis with R*, Springer, 273 p.

Fahrmeir, L., Kneib, T., Lang, S., and Marx, B., 2013. *Regression Models, Methods and Applications*, Berlin Heidelberg: Springer-Verlag, 1st ed., 698 p.

Fieuws, S. and Verbeke, G., 2006. Pairwise fitting of mixed models for the joint modeling of multivariate longitudinal profiles, *Biometrics*, 62 (7), 424–431.

Finney, D.J., 1971. *Probit Analysis*, Cambridge: Cambridge University Press, 3rd ed., 256 p.

Finney, D.J., 1979. Bioassay and the practice of statistical inference, *International Statistical Review*, 47 (1), 1–12.

Fortin, M., Schneider, R., and Saucier, J.P., 2013. Volume and error variance estimation using integrated stem taper models, *Forest Science*, 59 (3), 345–358.

Fox, J., 2008. *Applied Regression Analysis and Generalized Linear Models. Second edition*, London: SAGE, 665 p.

Fox, J. and Weisberg, S., 2010. *An Appendix to an R Companion to Applied Regression*, SAGE publishers, 2nd ed., http://socserv.mcmaster.ca/jfox/Books/Companion/appendix/Appendix-Nonlinear-Regression.pdf.

Fox, J. and Weisberg, S., 2011. *An R Companion to Applied Regression*, New York, USA: SAGE Publishing, 2nd ed., 472 p.

Freedman, D., 2009. *Statistical Models : Theory and Practice*, vol. 2nd ed, Cambridge University Press.

Gałecki, A. and Burzykowski, T., 2013. *Linear Mixed-Effects Models Using R. A Step by Step Approach*, New York: Springer, 1st ed., 542 p.

Garber, S.M. and Maguire, D.A., 2003. Modeling stem taper of three central Oregon species using nonlinear mixed effects models and autoregressive error structures, *For. Ecol. Manage.*, 179, 507–522.

Gelman, A., Carlin, J., Stern, H., Dunson, D., Vehtari, A., and Rubin, D., 2013. *Bayesian Data Analysis*, Chapman and Hall/CRC, New York, 675 p.

Giesbrecht, F.G. and Burns, J.C., 1985. Two-stage analysis based on a mixed model: Large-sample asymptotic theory and small-sample simulation results, *Biometrics*, 41 (2), 477–486.

Goldberger, A.S., 1962. Best linear unbiased prediction in the generalized linear regression model, *Journal of the American Statistical Association*, 57, 369–375.

Goldstein, H., 2003. *Multilevel Statistical Models*, London, UK: Hodder Arnold, 3rd ed.

Gove, J.H., 2000. Some observations on fitting assumed diameter distributions to horizontal point sampling data, *Canadian Journal of Forest Research*, 30 (4), 521 – 533.

Gove, J.H., 2003. Moment and maximum likelihood estimators for Weibull distributions under length- and area-biased sampling, *Environmental and Ecological Statistics*, 10 (3), 455 – 467.

Gove, J.H. and Patil, G.P., 1998. Modeling basal area-size distribution of forest stands: A compatible approach, *Forest Science*, 44 (2), 285 – 297.

Green, E.J. and Clutter, M., 2000. Using auxiliary information to estimate stand tables, *Canadian Journal of Forest Research*, 30, 865–872.

Greene, W.H., 2018. *Econometric Analysis*, Pearson, 8th ed., 1176 p.

Gregoire, T., Schabenberger, O., and Kong, F., 2000. Prediction from an integrated regression equation: a forestry application, *Biometrics*, 56 (2), 414–419.

Gregoire, T.G. and Schabenberger, O., 1996. A non-linear mixed-effects model to predict cumulative bole volume of standing trees, *J. Appl. Stat.*, 23 (2/3), 257–271.

Gregoire, T.G., Schabenberger, O., and Barrett, J.P., 1995. Linear modelling of irregularly spaced, unbalanced, longitudinal data from permanent-plot measurements, *Canadian Journal of Forest Research*, 25 (1), 137–156.

Gregoire, T.G. and Valentine, H.T., 2008. *Sampling Strategies for Natural Resources and the Environment*, Boca Raton: Chapman & Hall/CRC, 1st ed., 474 p.

Grilli, L. and Rampichini, C., 2011. The role of sample cluster means in multilevel models: a view on endogeneity and measurement error issues, *European Journal of Research Methods for the Behavioral and Social Sciences*, 7 (4), 121–133.

Grissom, R.J. and Kim, J.J., 2012. *Effect Sizes for Research. Univariate and Multivariate Applications. Second edition*, New York: Routledge, 434 p.

Grosenbaugh, L.R., 1952. Plotless timber estimates–new, fast, easy, *Journal of Forestry*, 50 (1), 32–37.

Guerin, L. and Stroup, W., 2000. A simulation study to evaluate proc mixed analysis of repeated measures data, *in: Proceedings of the 12th conference on applied statistics in Agriculture*, Manhattan, KS: Kansas State University, Department of Statistics, 170–203.

Haara, A., Maltamo, M., and Tokola, T., 1997. The k-nearest-neighbour method for estimating basal-area diameter distribution, *Scandinavian Journal of Forest Research*, 12, 200 – 208.

Hadfield, J., 2010. Mcmc methods for multi-response generalized linear mixed models: The mcmcglmm R package, *Journal of Statistical Software, Articles*, 33 (2), 1–22.

Hafley, W.L. and Schreuder, H.T., 1977. Statistical distributions for fitting diameter and height data in even-aged stands, *Canadian Journal of Forest Research*, 7 (3), 481–487.

Hall, D.B. and Bailey, R.L., 2001. Modeling and prediction of forest growth variables based on multilevel nonlinear mixed models, *Forest Science*, 47 (3), 311–321.

Hall, D.B. and Clutter, M., 2004. Multivariate multilevel nonlinear mixed effects models for timber yield predictions, *Biometrics*, 60 (1), 16–24.

Hämäläinen, A., Hujo, M., Heikkala, O., Junninen, K., and Kouki, J., 2016. Retention tree characteristics have major influence on the post-harvest tree mortality and availability of coarse woody debris in clear-cut areas, *Forest Ecology and Management*, 369, 66 – 73.

Härdle, W., 1990. *Smoothing Techniques with Implementation in S*, New York, USA: Springer, 261 p.

Harrell, F.J., 2001. *Regression Modeling Strategies with Applications to Linear Models, Logistic Regression, and Survival Analysis*, New York, USA: Springer, 562 p.

Harrell, F.J., 2015. *Regression Modeling Strategies with Applications to Linear Models, Logistic Regression, and Survival Analysis (2nd edition)*, New York, USA: Springer, 582 p.

Harville, D.A., 1977. Maximum likelihood approaches to variance component estimation and to related problems, *Journal of the American Statistical Association*, 72 (358), 320–338.

Harville, D.A., 1981. Unbiased and minimum-variance unbiased estimation of estimable functions for fixed linear models with arbitrary covariance structure, *Annals of Statistics*, 9, 633–637.

Hasenauer, H., Monserud, R.A., and Gregoire, T.G., 1998. Using simultaneous regression techniques with individual-tree growth models, *Forest Science*, 44 (1), 87–95.

Hastie, T. and Tibshirani, R., 1986. Generalized additive models, *Stat. Sci.*, 1 (3), 297–318.

Hastie, T. and Tibshirani, R., 1990. *Generalized Additive Models*, London: Chapman and Hall, 352 p.

Hastie, T., Tibshirani, R., and Friedman, J., 2009. *The Elements of Statistical Learning: Data Mining, Inference, and Prediction*, New York: Springer, 2nd ed., 745 p.

Henderson, C.R., 1963. Selection index and expected genetic advance, *Statistical Genetics and Plant Breeding, National Academy of Sciences-National Research Council Publication*, 982 (982), 141–163.

Henderson, C.R., 1975. Best linear unbiased estimation and prediction under a selection model, *Biometrics*, 31, 423–447.

Hox, J.J., Moerbeek, M., and van de Schoot, R., 2018. *Multilevel Analysis, Techniques and Applications*, New York and London: Routledge, 3rd ed., 347 p.

Hsiao, C., 2014. *Analysis of Panel Data*, Econometric Society Monographs, Cambridge University Press, 3rd ed., 562 p.

Hubbard, R., Bayarri, M.J., Berk, K.N., and Carlton, M.A., 2003. Confusion over measures of evidence (p's) versus errors (α's) in classical statistical testing, *The American Statistician*, 57 (3), 171–182.

Hui, F.K.C., Müller, S., and Welsh, A.H., 2019. Testing random effects in linear mixed models: another look at the F-test (with discussion), *Australian & New Zealand Journal of Statistics*, 61 (1), 61–84.

Hyink, D.M. and Moser, J.W.J., 1983. A generalized framework for projecting forest yield and stand structure using diameter distributions, *Forest Science*, 29 (1), 85–95.

Illian, J., Penttinen, A., Stoyan, H., and Stoyan, D., 2008. *Statistical Analysis and Modelling of Spatial Point Patterns*, Chichester, UK: Wiley, 536 p.

Immer, F.R., Hayes, H.K., and Powers, L., 1934. Statistical determination of barley varietal adaptation, *Journal of the American society of agronomy*, 26, 403–419.

Johnson, N.L., Kemp, A.W., and Kotz, S., 2005. *Univariate Discrete Distributions*, New York: Wiley, 3rd ed., 654 p.

Johnston, J. and Dinardo, J., 1997. *Econometric Methods. Fourth edition.*, McGraw Hill, 531p.

Kachiashvili, K., 2019. Modern state of statistical hypotheses testing and perspectives of its development, *Biostatistics and Biometrics Open Access Journal*, 9 (2).

Kackar, R.N. and Harville, D.A., 1984. Approximations for standard errors of estimators of fixed and random effects in mixed linear models, *Journal of the American Statistical Association*, 79 (388), 853–862.

Kangas, A., 1998. Effect of errors-in-variables on coefficients of a growth model and on prediction of growth, *Forest Ecology and Management*, 102 (2-3), 203–212.

Kangas, A., Heikkinen, E., and Maltamo, M., 2003. Accuracy of partially visually assessed stand characteristics: a case study of Finnish forest inventory by compartments, *Canadian Journal of Forest Research*, 34 (4), 916–930.

Kangas, A. and Maltamo, M., 2000a. Calibrating predicted diameter distribution with additional information, *Forest Science*, 46 (3), 390–396.

Kangas, A. and Maltamo, M., 2000b. Percentile-based basal area diameter distribution models for Scots pine, Norway spruce and birch species, *Silva Fennica*, 34 (4), 371–380.

Kangas, A., Mehtätalo, L., and Maltamo, M., 2007. Modelling percentile-based basal area weighted diameter distribution, *Silva Fennica*, 41 (3), 425 – 440.

Kansanen, K., Vauhkonen, J., Lähivaara, T., and Mehtätalo, L., 2016. Stand density estimators based on individual tree detection and stochastic geometry, *Canadian Journal of Forest Research*, 46 (11), 1359–1366.

Kenward, M.G. and Roger, J.H., 1997. Small sample inference for fixed effects from restricted maximum likelihood, *Biometrics*, 53 (3), 983–997.

Kilkki, P., Saramäki, M., and Varmola, M., 1978. A simultaneous equation model to determine taper curve, *Silva Fennica*, 12 (2), 120–125.

Kilpeläinen, A., Peltola, H., Ryyppö, A., and Kellomäki, S., 2005. Scots pine responses to elevated temperature and carbon dioxide concentration: growth and wood properties, *Tree Physiology*, 25 (1), 75–83.

Kögel, T., 2017. Simpson's paradox and the ecological fallacy are not essentially the same: The example of the fertility and female employment puzzle, 12p.

Korpela, I., Mehtätalo, L., Seppänen, A., and Markelin, L., 2014. Tree species classification using directional reflectance anisotropy signatures in multiple aerial images, *Silva Fennica*, 48 (3), 1087.

Kublin, E., Augustin, N.H., and Lappi, J., 2008. A flexible regression model for diameter prediction, *Eur. J. For. Res.*, 127, 415–428.

Kublin, E., Breidenbach, J., and Kändler, G., 2013. A flexible stem taper and volume prediction method based on mixed-effects B-spline regression, *Eur. J. For. Res.*, 132, 983–997.

Kuznetsova, A., Brockhoff, P.B., and Christensen, R.H.B., 2017. lmerTest package: Tests in linear mixed effects models, *Journal of Statistical Software*, 82 (13), 1–26.

Kvålseth, T.O., 1985. Cautionary note about R^2, *The American Statistician*, 39 (4), 279–285.

Laasasenaho, J., 1982. Taper curve and volume functions for pine, spruce and birch, *Comm. Inst. For. Fenn.*, 108.

Lahtinen, A. and Laasasenaho, J., 1979. On the construction of taper curves by using spline functions, *Commun. Inst. For. Fenn.*, 95 (8), 1–63.

Laine, A.M., , Mäkiranta, P., Laiho, R., Mehtätalo, L., Penttilä, T., Korrensalo, A., Minkkinen, K., Fritze, H., and Tuittila, E.S., 2019. Warming impacts on boreal fen CO2 exchange under wet and dry conditions, *Global Change Biology*, 25 (6), 1995–2008.

Lappi, J., 1986. Mixed linear models for analyzing and predicting stem form variation of scots pine, Communicationes Instituti Forestalis Fenniae 134, FFRI, 69 p.

Lappi, J., 1991a. Calibration of height and volume equations with random parameters, *Forest Science*, 37, 781–801.

Lappi, J., 1991b. Estimating the distribution of a variable measured with error: stand densities in a forest inventory, *Canadian Journal of Forest Research*, 21 (4), 469–473.

Lappi, J., 1997. A longitudinal analysis of height/diameter curves, *Forest Science*, 43, 555–570.

Lappi, J., 2001. Forest inventory of small areas combining the calibration estimator and a spatial model, *Canadian Journal of Forest Research*, 31 (2), 1551–1560.

Lappi, J., 2005. Plot size related measurement error bias in tree growth models, *Canadian Journal of Forest Research*, 34 (5), 1031–1040.

Lappi, J., 2006. A multivariate, nonparametric stem-curve prediction method, *Canadian Journal of Forest Research*, 36, 1017–1027.

Lappi, J. and Bailey, R.L., 1987. Estimation of the diameter increment function or other tree relations using angle-count samples, *Forest Science*, 33 (3), 725–739.

Lappi, J. and Bailey, R.L., 1988. A height prediction model with random stand and tree parameters: an alternative to traditional site index methods, *Forest Science*, 34, 907–927.

Lappi, J. and Luoranen, J., 2016. Using a bivariate generalized linear mixed model to analyze the effect of feeding pressure on pine weevil damage, *Silva Fennica*, 50 (1), 8p., article id 1496.

Lappi, J. and Luoranen, J., 2018. Testing the differences of LT50, LD50, or ED50, *Canadian Journal of Forest Research*, 48 (6), 729–734.

Lappi, J. and Malinen, J., 1994. Random-parameter height/age models when stand parameters and stand age are correlated, *Forest Science*, 40 (4), 715–731.

Lappi, J. and Smolander, H., 1984. Integration of the hyperbolic radiation-response function of photosynthesis, *Photosynthetica*, 18 (3), 591–612.

Lehmann, E.L., 1993. The Fisher, Neyman-Pearson theories of testing hypotheses: one theory or two, *Journal of the American Statistical Association*, 88 (424), 1242–1249.

Leskinen, P., Miina, J., Mehtätalo, L., and Kangas, A., 2009. Model correlation in stochastic forest simulators–a case of multilevel multivariate model for seedling establishment, *Ecological Modelling*, 220 (4), 545 – 555.

Lindstrom, M.J. and Bates, D.M., 1990. Nonlinear mixed effects models for repeated measures data, *Biometrics*, 46 (3), 673–687.

Linnakoski, R., Mahilainen, S., Harrington, A., Vanhanen, H., Eriksson, M., Mehtätalo, L., Pappinen, A., and Wingfield, M.J., 2016. Seasonal succession of fungi associated with ips typographus beetles and their phoretic mites in an outbreak region of Finland, *PLOS ONE*, 11 (5), 1–14.

Liu, F. and Kong, Y., 2015. zoib: An R package for Bayesian inference for beta regression and zero/one inflated beta regression, *The R Journal*, 7 (2), 34–51.

Luoranen, J., Repo, T., and Lappi, J., 2004. Assessment of frost hardiness of shoots of silver birch seedlings with and without controlled freezing exposure, *Canadian Journal of Forest Research*, 34 (5), 1108–1118.

Maltamo, M., Kangas, A., Uuttera, J., Torniainen, T., and Saramäki, J., 2000. Comparison of percentile based prediction methods and the Weibull distribution in describing the diameter distribution of heterogeneous scots pine stands, *Forest Ecology and Management*, 133 (3), 263 – 274.

Maltamo, M., Mehtätalo, L., Vauhkonen, J., and Packalén, P., 2012. Predicting and calibrating tree attributes by means of airborne laser scanning and field measurements, *Canadian Journal of Forest Research*, 42 (11), 1896–1907.

Maltamo, M., Næsset, E., Vauhkonen, J., and (ed.), 2014. *Forestry Applications of Airborne Laser Scanning - Concepts and Case Studies*, Dordrecht: Springer, 462 p.

Mandallaz, D., 2008. *Sampling Techniques for Forest Inventories*, Boca Raton: Chapman & Hall/CRC, 1st ed., 256 p.

Martin, R.J., 1992. Leverage, influence and residuals in regression models when observations are correlated, *Communications in Statistics - Theory and Methods*, 21 (5), 1183–1212.

McCullagh, P. and Nelder, J.A., 1989. *Generalized Linear Models*, Chapman & Hall, 2nd ed., 532 p.

McCulloch, C.E., Searle, S.R., and Neuhaus, J.M., 2008. *Generalized, Linear, and Mixed Models*, New York, USA: Wiley, 2nd ed., 384 p.

Mehtätalo, L., 2004a. An algorithm for ensuring compatibility between estimated percentiles of diameter distribution and measured stand variables, *Forest Science*, 50 (1), 20–32.

Mehtätalo, L., 2004b. Predicting stand characteristics using limited measurements, FFRI Research Report 929, Finnish Forest Research Institute, 39 p. + appendices I-V. (dissertation).

Mehtätalo, L., 2005. Localizing a predicted diameter distribution using sample information, *Forest Science*, 51 (4), 292–302.

Mehtätalo, L., 2006. Eliminating the effect of overlapping crowns from aerial inventory estimates, *Canadian Journal of Forest Research*, 36 (7), 1649–1660.

Mehtätalo, L., 2020. *lmfor: Functions for Forest Biometrics*, R package version 1.5.

Mehtätalo, L., Comas, C., Pukkala, T., and Palahí, M., 2011. Combining a predicted diameter distribution with an estimate based on a small sample of diameters, *Canadian Journal of Forest Research*, 41 (4), 750–762.

Mehtätalo, L., de Miguel, S., and Gregoire, T.G., 2015. Modeling height-diameter curves for prediction, *Canadian Journal of Forest Research*, 45 (7), 826–837.

Mehtätalo, L., Gregoire, T.G., and Burkhart, H.E., 2008. Comparing strategies for modelling tree diameter percentiles from remeasured plots, *Environmetrics*, 19, 529–548.

Mehtätalo, L. and Kangas, A., 2005. An approach to optimizing data collection in an inventory by compartments, *Canadian Journal of Forest Research*, 35 (1), 100–112.

Mehtätalo, L., Maltamo, M., and Kangas, A., 2006. The use of quantile trees in predicting the diameter distribution of a stand, *Silva Fennica*, 40 (3), 501–516.

Mehtätalo, L., Maltamo, M., and Packalén, P., 2007. Recovering plot-specific diameter distribution and height-diameter curve using als-based stand characteristics, *IAPRSS*, 36/ Part 3/W52, 288–293.

Mehtätalo, L., Peltola, H., Kilpeläinen, A., and Ikonen, V.P., 2014. The response of basal area growth of scots pine to thinning: A longitudinal analysis of tree-specific series using a nonlinear mixed-effects model, *Forest Science*, 60 (4), 636 - 644.

Melin, M., Matala, J., Mehtätalo, L., Pusenius, J., and Packalen, P., 2016a. Ecological dimensions of airborne laser scanning – analyzing the role of forest structure in moose habitat use within a year, *Remote Sensing of Environment*, 173, 238 - 247.

Melin, M., Mehtätalo, L., Miettinen, J., Tossavainen, S., and Packalen, P., 2016b. Forest structure as a determinant of grouse brood occurrence - an analysis linking lidar data with presence/absence field data, *Forest Ecology and Management*, 380, 202 - 211, special section: Drought and {US} Forests: Impacts and Potential Management Responses.

Meng, Q., Cieszewski, C.J., Madden, M., and Borders, B.E., 2007. K nearest neighbor method for forest inventory using remote sensing data, *GIScience & Remote Sensing*, 44 (2), 149–165.

Meng, S.X. and Huang, S., 2009. Improved calibration of nonlinear mixed-effects models demonstrated on a height growth function, *Forest Science*, 55 (3), 238–248.

Miina, J. and Saksa, T., 2006. Predicting regeneration establishment in Norway spruce plantations using a multivariate multi-level model, *New Forests*, 32 (3), 265–283.

Montgomery, D.C., Peck, E.A., and Vining, G.G., 2012. *Introduction to Linear Regression Analysis*, New York: Wiley, 5th ed., 672 p.

Muller, K.E. and Stewart, P.W., 2006. *Linear Model Theory. Univariate, Multivariate, and Mixed Models.*, Hoboken, New Jersey, USA: Wiley, 410 p.

Murdoch, D.J., Tsai, Y., and Adcock, J., 2008. P-values are random variables, *The American Statistician*, 62, 242–245.

Myers, R., Montgomery, D., and Anderson-Cook, C., 2016. *Response Surface Methodology: Process and Product Optimization Using Designed Experiments, 4th Edition*, Wiley, 856 p.

Nakagawa, S. and Scielzeth, H., 2012. A general and simple method for obtaining R2 from generalized linear mixed-effects models, *Methods in Ecology and Evolution*, 4 (2), 133–142.

Näslund, M., 1937. Skogsförsöksanstaltens gallringsförsök i tallskog (Forest research institute's thinning experiments in Scots pine forests), Meddelanden från statens skogsförsöksanstalt Häfte 29, in Swedish.

Nelder, J.A. and Wedderburn, R.W.M., 1972. Generalized linear models, *Journal of the Royal Statistical Society. Series A (General)*, 135 (3), 370–384.

Nepal, S.K. and Somers, G.L., 1992. A generalized approach to stand table projection, *Forest Science*, 38 (1), 120 - 133.

Neuhaus, J.M. and McCulloch, C.E., 2006. Separating between- and within-cluster covariate effects by using conditional and partitioning methods, *Journal of the Royal Statistical Society, Series B*, 68 (5), 859–872.

Neuhaus, J.M. and McCulloch, C.E., 2011. Estimation of covariate effects in generalized linear mixed models with informative cluster sizes, *Biometrika*, 98 (1), 147–162.

Nummi, T. and Möttönen, J., 2004. Estimation and prediction for low degree polynomial models under measurement errors with an application to forest harvesters, *Applied Statistics*, 53 (3), 495–505.

Ospina, R. and Ferrari, S.L.P., 2012. A general class of zero-or-one inflated beta regression models, *Computational Statistics & Data Analysis*, 56 (6), 1609–1623.

Palander, T., Kärhä, K., and Mehtätalo, L., 2016. Applying polynomial regression modeling to productivity analysis of sustainable stump harvesting, *Scandinavian Journal of Forest Research*, 0 (0), 1–9.

Parzen, E., 1962. *Stochastic Processes*, San Francisco Calif.: Holden-Day, 324 p.

Patil, G., 2006. *Weighted Distributions*, American Cancer Society.

Patil, G.P. and Rao, C.R., 1978. Weighted distributions and size-biased sampling with applications to wildlife populations and human families, *Biometrics*, 34, 179–189.

Patterson, H.D. and Thompson, R., 1971. Recovery of inter-block information when block sizes are unequal, *Biometrika*, 58, 545–554.

Pebesma, E.J. and Bivand, R.S., 2005. Classes and methods for spatial data in R, *R News*, 5 (2), 9–13.

Petäistö, R.L. and Lappi, J., 1996. Capability of the European and North American race of Gremmeniella abietina to hydrolyse polygalacturonic acid in vitro, *Eur. J. For. Path*, 26, 123–132.

Pienaar, L.V. and Harrison, W.M., 1988. A stand table projection approach to yield prediction in unthinned even-aged stands, *Forest Science*, 34 (3), 804 – 808.

Piepho, H.P. and Ogutu, J.O., 2007. Simple state-space models in a mixed model framework, *The American Statistician*, 61 (3), 224–232.

Pigott, T.D., 2012. *Advances of Meta-analysis*, New York: Springer, 155 p.

Pinheiro, J., Bates, D., DebRoy, S., Sarkar, D., and R Core Team, 2016. *nlme: Linear and Nonlinear Mixed Effects Models*, R package version 3.1-128.

Pinheiro, J.C. and Bates, D.M., 2000. *Mixed-Effects Models in S and S-plus*, New York, USA: Springer-Verlag, 528 p.

Pryseley, A., Tchonlafi, C., Verbeke, G., and Molenberghs, G., 2011. Estimating negative variance components from Gaussian and non-Gaussian data: A mixed models approach, *Computational Statistics and Data Analysis*, 55, 1071–1085.

Pulkkinen, M., 2012. On non-circularity of tree stem cross-sections: effect of diameter selection on cross-section area estimation, Bitterlich sampling and stem volume estimation in Scots pine, *Silva Fennica*, 46 (5B), 747–986, article id 924.

R Core Team, 2016. *R: A Language and Environment for Statistical Computing*, R Foundation for Statistical Computing, Vienna, Austria.

Rao, C.R., 1987. Estimation in linear models with mixed effects: A unified theory, *in:* T. Pukkila and S. Puntanen, eds., *Proceedings of 2nd International Tampere Conference in Statistics*, Dept. Math. Sci., Univ. of Tampere, 73–98.

Ratkowsky, D.A., 1990. *Handbook of Nonlinear Regression*, New York, USA: Marcel Dekker inc., 241p.

Reynolds, M.R.J., Burk, T.E., and Huang, W.C., 1988. Goodness-of-fit tests and model selection procedures for diameter distribution models, *Forest Science*, 34 (2), 373–399.

Rikken, M.G.J. and Van Rijn, R.P.G., 1993. Soil pollution with heavy metals - an inquiry into spatial variation, cost of mapping and the risk evaluation of copper, cadmium, lead and zinc in the floodplains of the Meuse west of Stein, the Netherlands, Tech. rep., Utrecht University, Dept. of Physical Geography, 73 p.

Robinson, A., 2004. Preserving correlation while modelling diameter distributions, *Canadian Journal of Forest Research*, 34 (1), 221 – 232.

Robinson, A.P., Lane, S.E., and Thérien, G., 2011. Fitting forestry models using generalized additive models: a taper model example, *Canadian Journal of Forest Research*, 41, 1909–1916.

Rosenblueth, E., 1981. Two-point estimates in probabilities, *Applied Mathematical Modelling*, 5 (5), 329–335.

Särndal, C.E., Swensson, B., and Wretman, J., 1992. *Model Assisted Survey Sampling*, New York, USA: Springer, 1st ed., 695 p.

Satterthwaite, F.E., 1946. An approximate distribution of estimates of variance components, *Biometrics Bulletin*, 2 (6), 110–114.

Schenker, N. and Gentleman, J., 2001. On judging the significance of differences by examining the overlap between confidence intervals, *The American Statistician*, 55, 182–186.

Schreuder, H.T., Hafley, W.L., and Bennett, F.A., 1979. Yield prediction for unthinned natural slash pine stands, *Forest Science*, 25 (1), 25 – 30.

Schumacher, F.X. and Hall, F.S., 1933. Logarithmic expression of timber-tree volume, *Journal of Agricultural Research*, 47, 719–734.

Schwarzer, G., Carpenter, J.R., and Rücker, G., 2015. *Meta-analysis with R*, New York: Springer, 252 p.

Scott, A. and Wild, C., 1991. Transformations and R^2, *The American Statistician*, 45 (2), 127–129.

Seaman, S., Pavlou, M., and Copas, A., 2014. Review of methods for handling confounding by cluster and informative cluster size in clustered data, *Stat. Med*, 33 (30), 5371–5387.

Searle, S.R., 1982. *Matrix Algebra Useful for Statistics*, New York, USA: Wiley.

Searle, S.R., 1988. Parallel lines in residual plots, *The American Statistician*, 42 (3), 211.

Searle, S.R., Casella, G., and McCulloch, C.E., 1992. *Variance Components*, New York, USA: Wiley, 501 p.

Searle, S.R. and Rounsaville, T.R., 1974. A note on estimating covariance components, *The American Statistician*, 28 (2), 67–68.

Seber, G.A.F. and Wild, C.J., 2003. *Nonlinear Regression*, Hoboken, NJ: Wiley-Interscience, 774 p.

Shen, L., Shao, J., Park, S., and Palta, M., 2008. Between- and within-cluster covariate effects and model misspecification in the analysis of clustered data, *Statistica Sinica*, 18 (2), 731–748.

Siipilehto, J., 2011. Local prediction of stand structure using linear prediction theory in scots pine-dominated stands in Finland, *Silva Fennica*, 45 (4), 669–692.

Siipilehto, J. and Kangas, A., 2015. Näslundin pituuskäyrä ja siihen perustuvia malleja läpimitan ja pituuden välisestä riippuvuudesta suomalaisissa talousmetsissä, *Metsätieteen aikakauskirja*, (4), 215–236.

Siipilehto, J. and Mehtätalo, L., 2013. Parameter recovery vs. parameter prediction for the Weibull distribution validated for scots pine stands in Finland, *Silva Fennica*, 47 (4).

Silverman, B.W., 1986. *Density Estimation for Statistics and Data Analysis*, London, UK: Chapman & Hall, 176 p.

Sirkiä, S., Heinonen, J., Miina, J., and Eerikäinen, K., 2014. Subject-specific prediction using a nonlinear mixed model: Consequences of different approaches, *Forest Science*, 61 (2), 205–212.

Skellam, J.G., 1948. A probability distribution derived from the binomial distribution by regarding the probability of success as variable between the sets of trials, *Journal of the Royal Statistical Society: Series B (Methodological)*, 10 (2), 257–261.

Smith, D.W. and Murray, L.W., 1984. An alternative to Eisenhart's model ii and mixed model in the case of negative variance estimates, *Journal of the American Statistical Association*, 79 (385), 145–151.

Smolander, H. and Lappi, J., 1985. Integration of a nonlinear function in a changing environment: estimating photosynthesis using mean and variance of radiation, *Agricultural and Forest Meteorology*, 34, 83–91.

Snijders, T. and Bosker, R.J., 1994. Modeled variance in two-level models, *Sociological Methods Research*, 22, 342–363.

Snowdon, P., 1991. A ratio estimator for bias correction in logarithmic regression, *Canadian Journal of Forest Research*, 21 (5), 720–724.

St-Pierre, A., Shikon, V., and Schneider, D., 2018. Count data in biology—data transformation or model reformation?, *Ecology and Evolution*, 8, 3077–3085.

Stan Development Team, 2018. RStan: the R interface to Stan, R package version 2.18.2.

Stasinopoulos, M. and Rigby, R., 2018. *gamlss.dist: Distributions for Generalized Additive Models for Location Scale and Shape*, R package version 5.1-1.

Stroup, W.W., 2013. *Generalized Linear Mixed Models. Modern Concepts, Methods and Applications*, Boca Raton: CRC Press Taylor and Francis Group, 1st ed., 529 p.

Stuart, A. and Ord, K., 1991. *Kendall's Advanced Theory of Statistics*, London: Edward Arnold, 5th ed., 1323 p.

Swamy, P., 1970. Efficient inference in random coefficient regression model, *Econometrics*, 3 (2), 311–323.

Temesgen, H., Monleon, V.J., and Hann, D.W., 2008. Analysis and comparison of nonlinear tree height prediction strategies for douglas-fir forests, *Canadian Journal of Forest Research*, 38 (3), 553–565.

Thitz, P., Mehtätalo, L., Välimäki, P., Randriamanana, T., Lännenpää, M., Hagerman, A.E., Andersson, T., Julkunen-Tiitto, R., and Nyman, T., 2020. Phytochemical shift from condensed tannins to flavonoids in transgenic betula pendula decreases consumption and growth but improves growth efficiency of epirrita autumnata larvae, *Journal of Chemical Ecology*, 46 (2), 217–231.

Thompson, S.K., 2012. *Sampling*, Hoboken, New Jersey: Wiley, 3rd ed.

van den Boogaart, K.G., Tolosana, R., and Bren, M., 2009. compositions: Compositional data analysis. R package version 1.02-1.

van den Boogaart, K.G. and Tolosana-Delgado, R., 2013. *Analyzing Compositional Data with R*, Berlin: Springer, 258 p.

Van Deusen, P., 1986. Fitting assumed distributions to horizontal point sample diameters, *Forest Science*, 32 (1), 146–148.

Vanclay, J.K., 1994. *Modelling Forest Growth and Yield: Applications to Mixed Tropical Forests*, Wallingford, Berkshire, England: CAB International, 312 p.

VanDeusen, P., 1989. A model based approach to tree ring analysis, *Biometrics*, 45 (3), 763–779.

Varjo, J., Henttonen, H., Lappi, J., Heikkonen, J., and Juujärvi, J., 2006. Digital horizontal tree measurements for forest inventory, *Working Papers of the Finnish Forest Research Institute*, (40), 23p.

Venables, W.N. and Ripley, B.D., 2002. *Modern Applied Statistics with S*, New York, USA: Springer, 4th ed., 495 p.

Verbeke, G. and Molenberghs, G., 2000. *Linear Mixed Models for Longitudinal Data*, New York, USA: Springer, 1st ed., 568 p.

Vuorio, V., Tikkanen, O.P., Mehtätalo, L., and Kouki, J., 2015. The effects of forest management on terrestrial habitats of a rare and a common newt species, *European Journal of Forest Research*, 134 (2), 377–388.

Walters, D., Gregoire, T., and Burkhart, H., 1989. Consistent estimation of site index curves fitted to temporary plot data, *Biometrics*, 45 (1), 23–33.

Wang, M. and Rennolls, K., 2005. Tree diameter distribution modeling: introducing a logit-logistic distribution, *Canadian Journal of Forest Research*, 35, 1305–1313.

Wang, X.F. and Wang, B., 2011. Deconvolution estimation in measurement error models: the R package decon, *Journal of Statistical Software*, 39 (10), 1–24.

Wang, Y., 1998. Smoothing spline models with correlated random errors, *Journal of the American Statistical Association*, 93, 341–348.

Weisberg, S., 2014. *Applied Linear Regression*, New York: Wiley, 4th ed., 384 p.

Wheeler, B., 2016. *SuppDists: Supplementary Distributions*, R package version 1.1-9.4.

Willett, J.B. and Singer, J.D., 1988. Another cautionary note about R^2: Its use in weighted least-squares regression analysis, *The American Statistician*, 42 (3), 236–238.

Wolodzko, T., 2018. *extraDistr: Additional Univariate and Multivariate Distributions*, R package version 1.8.10.

Yang, Y. and Huang, S., 2011. Comparison of different methods for fitting nonlinear mixed forest models and for making predictions, *Canadian Journal of Forest Research*, 41 (8), 1671–1686.

Yang, Y., Huang, S., and Meng, S.X., 2009. Development of tree-specific stem profile model for white spruce: a nonlinear mixed model approach with a generalized covariance structure, *Forestry*, 82 (5), 541–555.

Yule, G.U. and Kendall, M.G., 1958. *An Introduction to the Theory of Statistics*, London, UK: Charles Griffin & Co, 14th ed.

Zakrzewski, W.T., 1999. A mathematically tractable stem profile model for jack pine in Ontario, *Northern Journal of Applied Forestry*, 16 (3), 138–143.

Zar, J.H., 2010. *Biostatistical Analysis*, Pearson, 5th ed., 756 p.

Zellner, A., 1962. An efficient method of estimating seemingly unrelated regressions and tests for aggregation bias, *Journal of the American Statistical Association*, 57, 348–368.

Index

ted in the United States
er & Taylor Publisher Services